# Metal Ions in Life Sciences

## Volume 27

# Lanthanide and Other Transition Metal Ion Complexes and Nanoparticles in Magnetic Resonance Imaging



*Guest Editor*
Carlos F. G. C. Geraldes
*University of Coimbra, Coimbra Portugal*

*Series Editors*
Astrid Sigel, Helmut Sigel
*Department of Chemistry*
*Inorganic Chemistry*
*University of Basel*
*St. Johanns-Ring 19*
*CH-4056 Basel*
*Switzerland*
*astrid.sigel@unibas.ch,*
*helmut.sigel@unibas.ch*

Eva Freisinger, Roland K. O. Sigel
*Department of Chemistry*
*University of Zürich*
*Winterthurerstrasse 190*
*CH-8057 Zürich*
*Switzerland*
*freisinger@chem.uzh.ch,*
*roland.sigel@chem.uzh.ch*

*Cover illustration*: Parameters influencing GBCAs properties. The figure was taken by permission of R. Tripier and M. Beyler from Chapter 1 of the book itself.

First edition published 2024
by CRC Press
2385 Executive Center Drive, Suite 320, Boca Raton, FL 33431

and by CRC Press
4 Park Square, Milton Park, Abingdon, Oxon, OX14 4RN



ISBN: 9781032438368 (hbk)
ISBN: 9781032449456 (pbk)
ISBN: 9781003374688 (ebk)

DOI: 10.1201/9781003374688

Typeset in Times
by codeMantra

# About the Editors

**Carlos F. G. C. Geraldes** is a retired Full Professor at the University of Coimbra, Portugal (Orcid number 0000-0002-0837-8329). He received his "Licenciatura" (MS equivalent) in Chemistry from the University of Coimbra in 1972 and his Doctor of Philosophy (Dr. Phil) degree in Inorganic Chemistry from the University of Oxford in 1976 with Prof. Robert J. P. Williams. In his graduate work, he studied the solution conformation of a series of mono- and dinucleotides using the so-called Lanthanide Probe Method, measuring the proton NMR-induced shifts and nuclear relaxation rates of their complexes of whole paramagnetic Ln(III) series combined with coupling constants and NOE data. Computer analysis of the combined NMR data showed their flexibility in solution for the first time and provided the corresponding population mixtures present in each case. He became Assistant Professor in the Chemistry Department of the University of Coimbra in 1976, where he became Associate Professor in 1982 and Full Professor in 1988, after completing "Agregação" (Habilitation) in 1986. His research centered from as early as in 1984 in the development of Gd(III)-based complexes as MRI contrast agents, such as Gd(DOTA), Gd(DTPA), and Gd(DTPA-bis amides), which became clinical MRI contrast media at that time. He also published on multinuclear NMR in Bio-inorganic Chemistry, e.g., $^{7}Li$ and $^{51}V$ of complexes as metallo-drugs *in vitro* and in perfused cells. He also studied Tm(DOTP) as $Na^{+}$ shift reagent (SR) in perfused cells and rat hearts and $Li^{+}$ SR in perfused human brain cells. He published on Ln(III) protein tags for structural analysis, small ligand–protein interactions by NMR techniques, Gd- and Mn-based targeted and responsive MRI contrast agents and Yb-based PARACEST agents in cells and in vivo animals. He also worked on iron oxide nanoparticles and multimodal probes for MRI-based molecular imaging. He has published more than 320 co-authored peer-reviewed articles in international SCI journals, with 9850 citations and an h factor of 52 (Scopus) or 12482 citations, an h factor of 59 (Google Scholar) (February 19, 2024), 24 peer-reviewed book chapters and one book on NMR spectroscopy. His work has been funded by the Portuguese Foundation of Science and Technology, the Robert Welsh and Fulbright Foundations (USA), the Swiss Foundation Herbette, the British EPSRC, and several EU programs, such as the European Molecular Imaging Laboratory (EMIL) Network of Excellence. He is a member of the Editorial Board of the Journal of Inorganic Biochemistry and Molecules and in the past in Metal-based Drugs and Contrast Media and Molecular Imaging. He was involved in eight EU COST Actions in Chemistry, serving in their Management Committees and was member of its Technical Committee between 1994 and 2001. He has supervised 20 PhD students and 2 ongoing. He has given over 200 oral presentations in international and national conferences, was plenary or keynote speaker at 13 international conferences and presented more than 200 poster communications. He was a member of the Scientific and/or Organizing Committee of about 20 international and national conferences. He was a member of the Board of Trustees of EUROMAR, of the Nuclear and Technological Institute Advisory Panel, Sacavém, Portugal and of the Scientific Council of the Foundation "Francisco Manuel dos Santos," Portugal. He is chairman of the NMR section of the Portuguese Chemical Society and member of the Portuguese Academy of Sciences. He was awarded the Gulbenkian Foundation Prize to Young Science Students (1966), the Excellence in Research Prize by the Portuguese Foundation for Science and Technology (2005), the Bruker "António Xavier" Prize for Excellence in NMR Research (2017) and the "Frausto da Silva" prize of the Portuguese Chemical Society for work in Bioinorganic Chemistry (2023).

**Astrid Sigel** has studied languages; she was an Editor of the *Metal Ions in Biological Systems* (*MIBS*) series (until Volume 44) and also of the *Handbook on Toxicity of Inorganic Compounds* (1988), the *Handbook on Metals in Clinical and Analytical Chemistry* (1994; both with H. G. Seiler and H.S.), and the *Handbook on Metalloproteins* (2001; with Ivano Bertini and H.S.). She is also an Editor of the *MILS* series from Volume 1 on, and she co-authored more than 50 papers on topics in Bioinorganic Chemistry.

**Helmut Sigel** is an Emeritus Professor (2003) of Inorganic Chemistry at the University of Basel, Switzerland. He is a Co-editor of the series *Metal Ions in Biological Systems* (1973–2005; 44 volumes) as well as of the Sigels' new series *Metal Ions in Life Sciences* (since 2006). He also co-edited three handbooks and published over 350 articles on metal ion complexes of nucleotides, amino acids, coenzymes, and other bio-ligands. Together with Ivano Bertini, Harry B. Gray, and Bo G. Malmström, he founded (1983) the International Conferences on Biological Inorganic Chemistry (ICBICs). He lectured worldwide and was named Protagonist in Chemistry (2002) by Inorganica Chimica Acta (issue 339). Among Endowed Lectureships, appointments as Visiting Professor (e.g., Austria, China, Japan, Kuwait, UK), and further honors, he received the P. Ray Award (Indian Chemical Society, of which he is also a Honorary Fellow), the Alfred Werner Award (Swiss Chemical Society), and a Doctor of Science *honoris causa* degree (Kalyani University, India). He is also an Honorary Member of SBIC (Society of Biological Inorganic Chemistry).

**Eva Freisinger** is Associate Professor of Bioinorganic Chemistry and Chemical Biology (2018) at the Department of Chemistry at the University of Zürich, Switzerland. She obtained her doctoral degree (2000) from the University of Dortmund, Germany, where she worked with Bernhard Lippert and spent three years as a postdoc at SUNY Stony Brook, USA, with Caroline Kisker. Since 2003, she has performed independent research at the University of Zürich where she held a Förderungsprofessur of the Swiss National Science Foundation from 2008 to 2014. In 2014, she received her *Habilitation* in Bioinorganic Chemistry. Her research is focused on the study of plant metallothioneins with an additional interest in the sequence-specific modification of nucleic acids. Together with Roland Sigel, she chaired the 12th European Biological Inorganic Chemistry Conference (2014 in Zürich, Switzerland) and the 19th International Conference on Biological Inorganic Chemistry (2019 in Interlaken, Switzerland). She also serves on a number of Advisory Boards for international conference series; since 2014, she has been the Secretary of the European Biological Inorganic Chemistry Conferences (EuroBICs), and she is currently co-Director of the Department of Chemistry. She joined the group of Editors of the *MILS* series from Volume 18 on.

**Roland K. O. Sigel** is Full Professor (2016) of Chemistry at the University of Zürich, Switzerland. In the same year, he became Vice Dean of Studies (BSc/MSc), and in 2017, he was elected Dean of the Faculty of Science. From 2003 to 2008, he was endowed with a Förderungsprofessur of the Swiss National Science Foundation, and he is the recipient of an ERC Starting Grant 2010. He received his doctoral degree *summa cum laude* (1999) from the University of Dortmund, Germany, where he worked with Bernhard Lippert. Thereafter, he spent nearly three years at Columbia University, New York, USA, with Anna Marie Pyle (now Yale University). During the six years abroad, he received several prestigious fellowships from various sources, and he was awarded the EuroBIC Medal in 2008 and the Alfred Werner Prize (SCS) in 2009. Between 2015 and 2019, he was the Secretary of the Society of Biological Inorganic Chemistry (SBIC), and since 2018, he has been the Secretary of the International Conferences on Biological Inorganic Chemistry (ICBICs). His research focuses on the structural and functional role of metal ions in ribozymes, especially group II introns, regulatory RNAs, and on related topics. He is also an editor of Volumes 43 and 44 of the *MIBS* series as well as the *MILS* series from Volume 1 on.

# Historical Development and Perspectives of the Series *Metal Ions in Life Sciences**

It is an old wisdom that metals are indispensable for life. Indeed, several of them, such as sodium, potassium, and calcium, are easily discovered in living matter. However, the role of metals and their impact on life remained largely hidden until inorganic chemistry and coordination chemistry experienced a pronounced revival in the 1950s. The experimental and theoretical tools created in this period and their application to biochemical problems led to the development of the field or discipline now known as *Bioinorganic Chemistry*, *Inorganic Biochemistry*, or more recently also often addressed as *Biological Inorganic Chemistry*. By 1970, *Bioinorganic Chemistry* was established and further promoted by the book series *Metal Ions in Biological Systems* founded in 1973 (edited by H.S., who was soon joined by A.S.) and published by Marcel Dekker, Inc., New York, for more than 30 years. After this company ceased to be a family endeavor and its acquisition by another company, we decided, after having edited 44 volumes of the *MIBS* series (the last two together with R.K.O.S.), to launch a new and broader minded series to cover today's needs in the *Life Sciences*. Therefore, the Sigels' new series is entitled.

**"Metal Ions in Life Sciences".**

After publication of 22 volumes (since 2006), we are happy to join forces from Volume 23 on in this still growing endeavor with Taylor & Francis, London, UK, a most experienced Publisher in the *Sciences*. The development of *Biological Inorganic Chemistry* during the past 40 years was and still is driven by several factors; among these are (1) attempts to reveal the interplay between metal ions and hormones or vitamins, etc.; (2) efforts regarding the understanding of accumulation, transport, metabolism, and toxicity of metal ions; (3) the development and application of metal-based drugs; (4) biomimetic syntheses with the aim to understand biological processes as well as to create efficient catalysts; (5) the determination of high-resolution structures of proteins, nucleic acids, and other biomolecules; (6) the utilization of powerful spectroscopic tools allowing studies of structures and dynamics; and (7) more recently, the widespread use

* Reproduced with some alterations by permission of John Wiley & Sons, Ltd., Chichester, UK (copyright 2006) from pages v and vi of Volume 1 of the series *Metal Ions in Life Sciences* (MILS-1).

of macromolecular engineering to create new biologically relevant structures at will. All this and more is reflected in the volumes of the series *Metal Ions in Life Sciences.*

The importance of metal ions to the vital functions of living organisms, hence, to their health and well-being, is nowadays well accepted. However, in spite of all the progress made, we are still only at the brink of understanding these processes. Therefore, the series *Metal Ions in Life Sciences* links coordination chemistry and biochemistry in their widest sense. Despite the evident expectation that a great deal of future outstanding discoveries will be made in the interdisciplinary areas of science, there are still "language" barriers between the historically separate spheres of chemistry, biology, medicine, and physics. Thus, it is one of the aims of this series to catalyze mutual "understanding." It is our hope that *Metal Ions in Life Sciences* continues to prove a stimulus for new activities in the fascinating "field" of *Biological Inorganic Chemistry.* If so, it will well serve its purpose and be a rewarding result for the efforts spent by the authors.

**Astrid Sigel and Helmut Sigel**
*Department of Chemistry, Inorganic Chemistry*
*University of Basel, CH-4056 Basel, Switzerland*

**Eva Freisinger and Roland K. O. Sigel**
*Department of Chemistry*
*University of Zürich, CH-8057 Zürich, Switzerland*

*October 2005 and March 2023*

# Preface to Volume 27

## *Lanthanide and Other Transition Metal Ion Complexes and Nanoparticles in Magnetic Resonance Imaging*

MILS-27 provides an up-to-date review of recent advances in *in vitro* and preclinical research based on metal-containing molecules and nanomaterials to develop novel diagnostic MRI contrast agents that could expand their clinical use beyond the established Gd-based agents, such as Gd(DOTA) (Dotarem®). It offers an update on the synthesis of ligands and relaxometric characterization of their Gd(III), Mn(II/III), Fe(III), and other transition metal complexes as responsive and targeted MRI contrast agents. It also provides information on the latest developments in PARASHIFT systems, including $^{19}F$ NMR effects, and paramagnetic CEST systems. Finally, it covers the most current research supporting the relevance of paramagnetic and superparamagnetic nanomaterials for MRI and bimodal MRI/PET-based theranostics. Volume 27 provides this authoritative and timely review in 12 comprehensive chapters written by 33 internationally recognized experts from 8 nations in Europe and America, containing more than 1350 references, 26 tables, and 152 illustrations, many of which are in color.

- Presents the importance of metal-containing molecules and nanomaterials for MRI-based diagnosis and the practical use of these systems.
- Discusses the structure and dynamics of Ln-based contrast agents as well as computational studies related to these agents.
- Endorses and stimulates research in the vibrant field of biological inorganic chemistry.
- Reviews the most current research supporting the relevance of metal-containing molecules and nanomaterials for diagnosis and therapy.
- The authors are preeminent bioinorganic and medicinal inorganic chemists and review the most current research in this field.

**Carlos F. G. C. Geraldes**

# Contents

# Contributors to Volume 27

**Silvio Aime**
Department of Molecular Biotechnology and Health Sciences
University of Turin
Via Nizza, 52
I-10126, Turin, Italy
and
IRCCS SDN SynLAB,
Via Gianturco 113,
Naples, Italy

**Matthew J. Allen**
Department of Chemistry
Wayne State University
5101 Cass Avenue,
Detroit, MI 48202, USA

**Maryline Beyler**
Université de Bretagne Occidentale,
UMR-CNRS 6521 CEMCA
6 avenue Victor le Gorgeu,
F-29238 Brest, France

**Célia S. Bonnet**
Centre de Biophysique Moléculaire,
CNRS UPR 4301
Université d'Orléans
Rue Charles Sadron,
F-45071 Orléans Cedex 2, France

**Fabio Carniato**
Dipartimento di Scienze e Innovazione Tecnologica
Università del Piemonte Orientale "A. Avogadro"
I-15121 Alessandria, Italy

**Rafael Torres Martin De Rosales**
School of Biomedical Engineering & Imaging Sciences
King's College London
4th Floor, Lambeth Wing, St. Thomas Hospital,
London, SE1 7EH, UK

**Kristina Djanashvili**
Department of Biotechnology
Delft University of Technology
Van der Maasweg 9,
NL-2629 HZ Delft, The Netherlands

**Zoltán Garda**
Centre de Biophysique Moléculaire,
CNRS UPR 4301
Université d'Orléans
Rue Charles Sadron,
F-45071 Orléans, France

**Carlos F. G. C. Geraldes**
Department of Life Sciences and Coimbra Chemistry Center - Institute of Molecular Sciences (CQC-IMS)
University of Coimbra
3004-535 Coimbra, Portugal
and
Department of Chemistry
University of Coimbra
3004-535 Coimbra, Portugal

**Peter Harvey**
Sir Peter Mansfield Imaging Centre – School of Medicine & School of Chemistry
University of Nottingham
Nottingham, NG9 2QL, UK

**Manon Isaac**
Centre de Biophysique Moléculaire,
CNRS UPR 4301
Université d'Orléans
Rue Charles Sadron,
F-45071 Orléans Cedex 2, France

**Md. Sydul Islam**
Department of Chemistry
Wayne State University
5101 Cass Avenue,
Detroit, MI 48202, USA

**Jan Kretschmer**
Department of Preclinical Imaging and Radiopharmacy, Werner Siemens Imaging Center
Eberhard Karls University Tübingen
Röntgenweg 13/1,
D-72076 Tübingen, Germany
and
Cluster of Excellence iFIT (EXC 2180) "Image-Guided and Functionally Instructed Tumor Therapies"
Eberhard Karls University Tübingen
Röntgenweg 13/1,
D-72076 Tübingen, Germany

**Sara Lacerda**
Centre de Biophysique Moléculaire,
CNRS UPR 4301
Université d'Orléans
Rue Charles Sadron,
F-45071 Orléans, France

**Sophie Laurent**
General, Organic and Biomedical Chemistry Unit, NMR and Molecular Imaging Laboratory
University of Mons
B-7000 Mons, Belgium
and
Center for Microscopy and Molecular Imaging
Rue Adrienne Bolland, 8,
B-6041 Gosselies, Belgium

**Luke A. Marchetti**
Centre de Biophysique Moléculaire,
CNRS UPR 4301
Université d'Orléans
Rue Charles Sadron,
F-45071 Orléans Cedex 2, France

**Andre Ferreira Martins**
Department of Preclinical Imaging and Radiopharmacy, Werner Siemens Imaging Center
Eberhard Karls University Tübingen
Röntgenweg 13/1,
D-72076 Tübingen, Germany
and
Cluster of Excellence iFIT (EXC 2180) "Image-Guided and Functionally Instructed Tumor Therapies"
Eberhard Karls University Tübingen
Röntgenweg 13/1,
D-72076 Tübingen, Germany
and
German Cancer Consortium (DKTK), partner site Tübingen
German Cancer Research Center (DKFZ)
Im Neuenheimer Feld 280,
D-69120 Heidelberg, Germany

**Janet R. Morrow**
Department of Chemistry, Natural Sciences Complex
University at Buffalo, the State University of New York
Amherst, NY 14260, USA

**Robert N. Muller**
General, Organic and Biomedical Chemistry Unit, NMR and Molecular Imaging Laboratory
University of Mons
B-7000 Mons, Belgium
and
Center for Microscopy and Molecular Imaging
Rue Adrienne Bolland, 8,
B-6041 Gosselies, Belgium

**Juan Pellico**
School of Biomedical Engineering & Imaging Sciences
King's College London
4th Floor, Lambeth Wing,
St. Thomas Hospital,
London, SE1 7EH, UK

**Joop A. Peters**
Department of Biotechnology
Delft University of Technology
Van der Maasweg 9,
NL-2629 HZ Delft, The Netherlands

**Angelina Prytula-Kurkunova**
Department of Preclinical Imaging and Radiopharmacy, Werner Siemens Imaging Center
Eberhard Karls University Tübingen
Röntgenweg 13/1,
D-72076 Tübingen, Germany
and
Cluster of Excellence iFIT (EXC 2180) "Image-Guided and Functionally Instructed Tumor Therapies"
Eberhard Karls University Tübingen
Röntgenweg 13/1,
D-72076 Tübingen, Germany

**Jaclyn J. Raymond**
Department of Chemistry, Natural Sciences Complex
University at Buffalo, the State University of New York
Amherst, NY 14260, USA

**Alexander G. Sertage**
Department of Chemistry
Wayne State University
5101 Cass Avenue,
Detroit, MI 48202, USA

**A. Dean Sherry**
Advanced Imaging Research Center
University of Texas Southwestern Medical Center
Dallas, TX 75390, USA
and
Department of Chemistry & Biochemistry
University of Texas at Dallas
Richardson, TX 75080, USA

**Dimitri Stanicki**
General, Organic and Biomedical Chemistry Unit, NMR and Molecular Imaging Laboratory
University of Mons
B-7000 Mons, Belgium

**Lorenzo Tei**
Dipartimento di Scienze e Innovazione Tecnologica
Università del Piemonte Orientale "A. Avogadro"
I-15121 Alessandria, Italy

**Enzo Terreno**
Department of Molecular Biotechnology and Health Sciences
University of Turin
Via Nizza, 52
I-10126, Turin, Italy

**Éva Tóth**
Centre de Biophysique Moléculaire, CNRS UPR 4301
Université d'Orléans
Rue Charles Sadron,
F-45071 Orléans, France

**Raphaël Tripier**
Université de Bretagne Occidentale, UMR-CRS 6521 CEMCA
6 avenue Victor le Gorgeu,
F-29238 Brest, France

**Thomas Vangijzegem**
General, Organic and Biomedical Chemistry Unit, NMR and Molecular Imaging Laboratory
University of Mons
B-7000 Mons, Belgium

**Levy Van Leuven**
General, Organic and Biomedical Chemistry Unit, NMR and Molecular Imaging Laboratory
University of Mons
B-7000 Mons, Belgium

**Mark Woods**
Department of Chemistry
Portland State University
Portland, OR 97201, USA
and
Advanced Imaging Research Center
Oregon Health & Science University
Portland, OR 97239, USA

# Handbooks and Book Series Published and (Co-)edited by the SIGELs

**"Handbook on Toxicity of Inorganic Compounds"** (ISBN: 0–8247-7727-1) Eds H. G. Seiler, H. Sigel, A. Sigel; Dekker, Inc.; New York; 1988; 1069 pp

**"Handbook on Metals in Clinical and Analytical Chemistry"** (ISBN: 0–8247– 9094-4) Eds H. G. Seiler, A. Sigel, H. Sigel; Dekker, Inc.; New York, Basel, Hong Kong; 1994; 753 pp

**"Handbook on Metalloproteins"** (ISBN: 0–8247–0520–3) Eds I. Bertini, A. Sigel, H. Sigel; Marcel Dekker, Inc.; New York, Basel; 2001; 1182 pp

**Metal Ions in Biological Systems**

Volumes 1–44

*https://www.routledge.com/Metal-Ions-in-Biological-Systems/book-series/IHCMEIOBISY*

*(see also the website given below)*

**Metal Ions in Life Sciences**

Volumes 1–27

*Details about all books (series) edited by the SIGELs, including the Guest Editors, can be found at*

*http://www.bioinorganic-chemistry.org/mils*

# 1 Synthesis and Characterization of Ligands and their Gd(III) Complexes

*Raphaël Tripier*[*] and Maryline Beyler*[*]*
Université de Bretagne Occidentale, Faculté des Sciences et Techniques, UMR-CNRS 6521 CEMCA, 6 avenue Victor le Gorgeu, F-29238 Brest, France
raphael.tripier@univ-brest.fr, maryline.beyler@univ-brest.fr

## CONTENTS



* Corresponding authors.

DOI: 10.1201/9781003374688-1

**Abstract**

This chapter explores the synthesis and characterization of Gd(III) complexes using *N*-functionalized ligands derived from tacn, cyclen, cyclam and pyclen, focusing on their performance in magnetic resonance imaging (MRI) applications. Ligands' synthesis methodologies are discussed, emphasizing key structural features for effective Gd(III) coordination. The ensuing complexes are thoroughly depicted throughout their physicochemical characterization, including thermodynamic stability and kinetic constants, relaxivity (and related parameters) and coordination geometry. A systematic evaluation of ligands sheds light on their merits and limitations in enhancing Gd(III) complex efficiency for MRI contrast. The analysis of ligand structure's influence provides valuable insights into designing optimized Gd(III) complexes for biomedical imaging. This chapter reviews 45 years of research and serves as a concise resource for advancing tailored ligands in Gd(III) coordination, contributing to the ongoing refinement of MRI techniques for enhanced medical diagnostics.

## KEYWORDS

Polyazacycloalkanes; Ligands; Organic Synthesis; Macrocyclic Chemistry; Gadolinium(III); Coordination; Thermodynamics; Kinetics; Relaxivity

## 1 INTRODUCTION

Gadolinium(III) is a chemical element that belongs to the lanthanide series of the periodic table. It is commonly used in medical imaging, specifically in magnetic resonance imaging (MRI), due to its unique magnetic properties. Indeed, gadolinium(III) possesses seven unpaired electrons ($S = 7/2$), making it highly paramagnetic. This property allows gadolinium-based contrast agents (GBCAs) to enhance the visibility of certain tissues or structures during MRI scans.

These GBCAs are administered to patients intravenously prior to medical examination. The paramagnetic nature of the cation causes it to create a local magnetic field, which enhances the relaxation rates of nearby protons of water molecules present in the body. This leads to a more intense signal in the MRI image, improving the visibility of blood vessels, tumors, inflammation and other anomalies. These lanthanide-based contrast agents then have a high relaxivity, which refers to their ability to shorten the relaxation times of nearby water protons. Higher relaxivity results in a stronger signal enhancement in MRI images, improving the diagnostic accuracy. In the last years, the safety of GBCAs has been a topic of intense discussions in many scientific research fields. It has been found that some people with decreased kidney functions or kidney illnesses may experience a rare condition called nephrogenic systemic fibrosis (NSF) when exposed over a sufficient time to certain GBCAs [1]. As a result, the use of GBCAs has been restricted in patients with severe kidney diseases. However, it is important to note that the risk of NSF is very low in patients with normal kidney function explaining the intensive use of such pharmaceutical products nowadays. More recent data also point out the fact that gadolinium(III) accumulates in various tissues of patients (bones, brain and kidney) who do not suffer from any renal dysfunction [2,3].

Free gadolinium(III) is a highly toxic heavy metal ion. To hide its toxicity, GBCAs use chelators that tightly bind to the gadolinium ion and prevent its release into the patient's body. If the mechanisms conferring toxicity to $Gd^{3+}$ and the effects of multiple GBCA exposures are still unclear, it seems admitted by the community that the *in vivo* stability of GBCAs must be one parameter to not neglect. Besides enhanced safety, chelating gadolinium for MRI offers several other advantages, especially in terms of pharmacokinetics. GBCAs are water soluble and can thereby be administered as a solution intravenously. They exhibit fast biodistribution through blood and extravascular space. Therefore, the image can be recorded a few minutes after the injection. The clearance of most of the GBCAs uses the renal route with a 90% excretion in the urine within 24 h in patients with no renal impairment [4]. Chelation of gadolinium must also be done with an increase of the circulation time of the contrast agent in the body since it helps to prevent the rapid elimination of gadolinium(III), allowing more time for the contrast agent to accumulate in the targeted tissues or organs. This prolonged circulation time enhances the effectiveness of the contrast agent in highlighting specific areas during the MRI scan. Chelated gadolinium compounds must have a relaxivity as high as possible, meaning having a strong ability to enhance the relaxation rates of nearby water protons. This leads to the expected greater signal enhancement.

To resume, chelation allows for the customization of GBCAs. By varying the chelating agent's structure, different properties and characteristics of the contrast agent can be achieved, which can be physicochemical and magnetic properties as we will see in this chapter, as well as other attributes such as different clearance times or targeted delivery to specific tissues. Designing $Gd^{3+}$ chelators for the development of GBCAs is then an important work that requires a combination of knowledge, skills and techniques in organic chemistry (protection/deprotection steps, etc.), macrocyclic chemistry, ligand design (which involves considering factors such as the structure, size, charge and coordinating groups of the ligand to ensure optimal chelation and stability) and coordination chemistry (determination of the appropriate ligand-to-metal ratio, the coordination geometry, stability of the resulting complex, etc.).

$Gd^{3+}$ chelates that are used as MRI contrast agents are gathered in two categories: the linear $Gd^{3+}$ chelates and the macrocyclic ones. It is assumed that macrocyclic chelates offer higher thermodynamic and kinetic stability/inertness than their linear analogues [5–7], and we will therefore especially focus on this second category in this book chapter. The objective is to highlight the synthetic routes settled so far to develop azamacrocyclic GBCAs based on cyclen, cyclam, tacn and pyclen platforms. If the center of the topic is to underline different organic tools reported in the literature to lead the most performing $Gd^{3+}$ chelators in terms of physicochemical properties, the thermodynamic, kinetic and relaxivity data, when available, will also be presented and discussed since the design of each ligand directly impacts such properties.

## 2 PHYSICOCHEMICAL PROPERTIES OF GBCAs: THERMODYNAMIC, INERTNESS, AND RELAXOMETRY

GBCAs are classified as "positive contrast agents". They increase both the longitudinal ($1/T_1$) and transverse ($1/T_2$) relaxation rates; however, the percentage change in tissue is higher for $1/T_1$, so $T_1$-weighted MRI is the most currently performed imaging technique resulting in a bright enhancement contrast. Unfortunately, as already discussed earlier, $Gd^{3+}$ is a toxic ion. Because of its similar size with $Ca^{2+}$, it enters in the biological calcic circle, interacts with calcium-specific enzymes and affects the voltage-gated $Ca^{2+}$ channels, leading to hostile biological effects. Then, suitable chelators must be used to "trap" the metal to form a non-toxic complex (this includes no interaction with specific enzymes of other biomolecules and a fast elimination) by preventing its presence as free metal that can lead to NSF or, as more and more highlighted, to $Gd^{3+}$ accumulation in brain, kidney and bones.

Chelates for application as $Gd^{3+}$-based contrast agents should present an important selectivity for this metal ion over endogenous cations (specially $Ca^{2+}$, $Zn^{2+}$ and $Cu^{2+}$) to avoid the release of the lanthanide *in vivo*. The thermodynamic stability of the $Gd^{3+}$ complex is of high importance and is determined by several methodologies including potentiometry and spectroscopic techniques. The higher the complexation constant (or more specifically the pGd value), the better; however, the selectivity is essential and the complexation constants with competitive cations must be as less as possible. Unfortunately, as we will see, lanthanides and especially $Gd^{3+}$ complexes are N- and/or O-based chelators which have affinities also for other metals as transition metals. Then, the kinetic inertness of the $Gd^{3+}$ complexes is probably more important and must be studied with same care as thermodynamics. If in the case of thermodynamics, comparison can be done especially by comparing the pGd values (a special attention must be paid to the experimental data including the ionic strength and the methodology/equipment used), the comparison of kinetical data is more difficult. Different experiments are indeed performed by the authors, including transmetallation using $Zn^{2+}$ as competitor, acid-assisted dissociation (usually performed with different acid concentration) or competition with another ligand. Additionally, the dissociation mechanism is different depending on the studied ligands and various investigations (first or second order, etc.), which makes any assessment difficult.

Different physicochemical factors influence the performance of GBCAs in terms of thermodynamic stability and inertness that are directly linked to the following: (1) denticity, (2) overall charge, (3) basicity and donor character of the coordinating atoms, (4) ligand fields and (5) conformational–configurational effects of the complexes.

Once thermodynamic and kinetic issues are solved, the key parameter of GBCAs is to consider their ability to increase the relaxation rates ($1/T_1$ and $1/T_2$) of water protons. This is generally translated by the relaxivity. As mentioned before, for CAs containing highly paramagnetic elements, such as GBCAs, the reduction of the longitudinal relaxation time ($T_1$) of water protons translates to brighter $T_1$-weighted (or brighter, by comparison with the term darker used for $T_2$-weighted MRI images). GBCAs indirectly provide a contrast in the analysed tissues. The global $T_1$-weighted MR signal strength is associated with the observed $T_1$ relaxation rate of the water molecule protons ($1/T_{1,obs}$), which corresponds to the paramagnetic component added to the diamagnetic contribution of water protons of the gadolinium(III) complexes. The paramagnetic contribution of those contrast agents equals the product of longitudinal relaxivity ($r_1$) and the concentration of $Gd^{3+}$. $r_1$ (in $mM^{-1}\,s^{-1}$) describes the efficiency of the GBCA at relaxing water protons and is the degree to which GBCAs decrease $T_1$ of water protons as a function of $Gd^{3+}$concentration.

The relaxivity ($r_1$) is the result of the outer-sphere relaxivity $r_1^{OS}$ and the inner-sphere relaxivity $r_1^{IS}$. The $r_1^{OS}$ relaxivity comes from water molecules that are hydrogen bonded to the Gd-based complex as well as those that are further away. $r_1^{IS}$ comes from the metal-bound water molecule(s) and the water molecules that are in exchange with the water molecule(s) directly bound to the gadolinium center. This component is predominant in the overall value of the GBCA relaxivity and could be considered as the most important. This inner-sphere relaxivity ($r_1^{IS}$) is related to three key terms (Figure 1): $q$, $k_{ex}$ ($1/\tau_M$) and $\tau_R$ ($T_{1m}$).

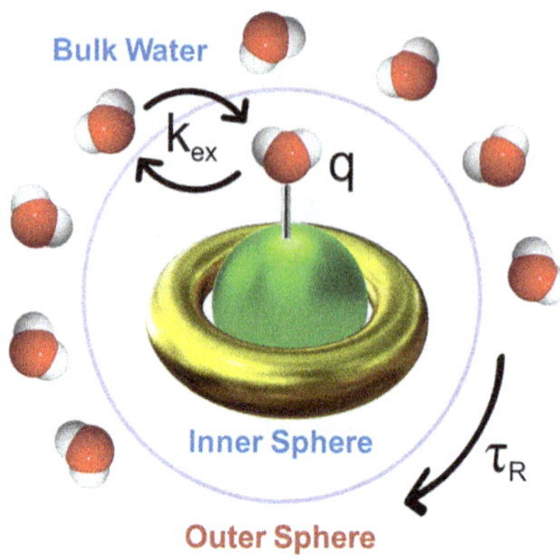


**FIGURE 1** Parameters influencing GBCAs properties.

- $q$ is the number of water molecules directly coordinated to $Gd^{3+}$ center and is usually called "hydration number". It is directly proportional to the inner-sphere relaxivity $r_1^{IS}$. In the case of clinically used GBCAs, the hydration number is $q=1$, simply because they employed octadentate chelators which prevent the release of free $Gd^{3+}$. The modulation of $q$ can be achieved simply by playing with the number of coordination sites of the ligand. Designing such complexes is quite challenging, since decreasing the number of coordination sites usually leads to a decrease in the thermodynamic stability and/or the kinetic inertness of the complexes. However, as we will see in this chapter, recent examples of bis-hydrated $Gd^{3+}$ complexes retaining good stability have been reported.
- $k_{ex}$ is the exchange rate of the inner-sphere water molecules with the bulk. This parameter is directly correlated to the residence time of water molecules ($\tau_M$) in the inner sphere ($k_{ex}=1/\tau_M$). To transmit the relaxation effect to the solvent, the coordinated water molecule should be in rapid exchange with bulk solvent. Relaxivity can be lowered if $\tau_M$ is too slow or too fast. The optimal range of $\tau_M$ for low-molecular-weight $Gd^{3+}$ complexes is around 10–30 ns [8,9]. The water molecules of the inner sphere are very close to the gadolinium center that creates a strong dipole–dipole interaction between both entities. The importance of the effect of the inner sphere is directly linked to the possibility of the water molecules to approach the $Gd^{3+}$ center and to how many potential binding sites are available (direct dependence on $q$). The design and form (in terms of size and space shape) of the metal complex also affect the residence time of each water molecule in the coordination site. The average residence time of most commercial contrast agents is about 1 µs meaning a water exchange rate of about 1 million times per second. As said before, the inner-sphere water molecules must be differentiated from those of the outer sphere. The interaction is also dipolar but not as much as inner sphere effects because these water molecules are very

far from the gadolinium center. This explains that their influence is less important in the overall relaxivity of the GBCAs. These water molecules are not directly interacting with the $Gd^{3+}$ center, but since they are also exposed to the magnetic field and interacting with the one present in the inner sphere and linked to the metal, they also affect the relaxivity. These outer-sphere water molecules are also exchanging magnetization with others in the bulk water pool more distally. The relative contributions of both sphere effects (inner and outer) to total relaxation depend on field strength, the size and type of the contrast agent molecule and the diffusion rate of water in the surroundings.

- $\tau_R$ is the rotational correlation time of the complex and is directly linked to the size and rigidity of the system (so, that means in terms of geometrical and molecular sizes). The metal-bound $H_2O$ rapidly exchanges with the water molecules from the bulk that diffuses the strong paramagnetic relaxation enhancement of $Gd^{3+}$ to neighboring water molecule. The rotational correlation time associated with the increased molecular size is about 0.1 ns for small-molecule GBCAs as DOTA-like $Gd^{3+}$ complexes. Then, large size complexes and polytopic compounds have been largely studied to influence the rotational correlation time. The larger the molecule, the slower the $\tau_R$, showing that the slow rotation allows a stronger interaction with the water molecules of the environment to involve them in the process. In this chapter, we will only focus on low-molecular weight $Gd^{3+}$-complexes meaning that different structures discussed will not fundamentally impact the $\tau_R$ parameter, which will stay close to those of the commercially available GBCAs.

## 3 "DOTA-LIKE" DERIVATIVES AND RELATED DOTA-COMPOUNDS BASED ON THE CYCLEN PLATFORM

In 2013, Long and Stasiuk titled a review "The ubiquitous DOTA and its derivatives: the impact of 1,4,7,10-tetraazacyclododecane-1,4,7,10-tetraacetic acid on biomedical imaging" [10], where they clearly demonstrated, based on an extensive bibliographic study, that this famous chelator and its derivatives are unavoidable to coordinate cations in several imaging modalities such as fluorescence, positron emission tomography (PET), single-photon emission computed tomography (SPECT) and, of course, magnetic resonance imaging (MRI).

**DOTA**$^{4-}$ is a cyclen platform bearing four acetate pendant arms (Figure 2). Its high affinity for divalent but especially trivalent cations, both in terms of thermodynamic stability and kinetic inertness, easily explains its use in several applications, and even for others than imaging, depending on the nature of the cation coordinated in the macrocyclic cavity. Let's talk about history to have an overview of the importance of $H_4$**DOTA** in coordination chemistry and to understand why there are so many "DOTA-like" derivatives presented in the literature in recent decades. By the way, what is the definition of "DOTA-like compounds"? Actually, to the best of our knowledge, there is not a real definition since several authors

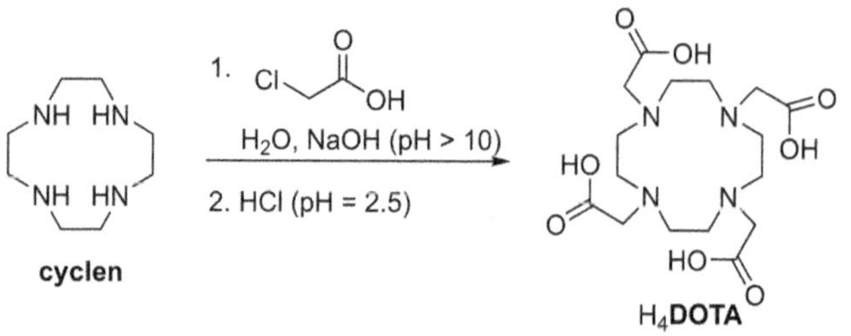

**FIGURE 2** Synthesis of $H_4$DOTA from cyclen as first described by Desreux et al. [15].

have given theirs, depending on what they wanted to describe and according to their expertise. Biologists and doctors, as well as coordination chemists, often misuse language and speak about "DOTA-like" compounds as soon as they describe an azamacrocycle with acetate pendants, regardless of the macrocyclic core (tacn, cyclen, cyclam, pyclen, etc.). Some others are using this standard terminology only when the chelator is composed of a cyclen core (or one modified *via* at least one carbon atom of the ring) bearing not less than three acetate pendants, meaning a DO3A-core, and an additional coordinating arm in the last of the four nitrogen atoms. This definition is more convenient since the compound usually preserves the coordination properties of its DOTA-parent, and it was often used to describe a chelator that resembles $H_4$**DOTA** because it was conjugated/grafted on an additional moiety that presents its own properties; in this case, the conjugation is frequently done *via* an amine function leading to an amide coordinating arm. In this case, the chelation has the same properties as **DOTA** for cations with a number of coordination less than 8, such as copper(II), and slightly lower for cations that need at least 8 coordinating atoms, such as lanthanides(III). But let's go back to the beginning of the history of this "ubiquitous DOTA".

Desreux and co-workers were at the origin of the first [Ln(**DOTA**)]$^-$ crystal structure [11], three years after the publication of the first lanthanide–DOTA complex by Bryden and collaborators in 1981 [12]. They demonstrated, thanks to this opening solid-state study, the perfect fit between the chelator and the central lanthanide cation that leads to an ideal combination of HSAB properties and ligand structural design. At the center of this study, the X-ray structure showed the presence of a water molecule on the apical position of the complex. This paved the way for the use of $H_4$**DOTA** for the complexation of gadolinium(III) in MRI applications symbolized by the first use of [Gd(**DOTA**)]$^-$ by Magerstädt et al. [13] and Caille et al. [14] thanks to its paramagnetic properties in 1986.

In terms of synthesis (Figure 2), $H_4$**DOTA** is easily obtained by the direct tetra-functionalization of **cyclen** (1,4,7,10-tetraazacyclododecane) with 2-chloroacetic acid in aqueous media (pH > 10). After acidification (pH = 2.5), filtration and ion-exchange resin, the synthesized $H_4$**DOTA** ligand is typically purified by recrystallization [15]. Some other alternatives have been developed in the literature such as the equivalent synthesis *via* its tetra-*tert*butyl ester protecting derivative prior to the final hydrolysis [16,17].

It is important to note that, as we will see later, the commercial availability of $H_4$**DOTA** and its derivatives simplifies the synthesis process, as ready-made **DOTA** ligands can be purchased from chemical suppliers, thereby bypassing the need for complete synthesis from starting materials. Furthermore, the specific synthetic procedures and conditions can vary depending on the desired modifications or derivatizations of **DOTA**. Different protecting groups or variations in the reaction conditions may be employed to achieve specific functionalization or improved yields.

Even if, in this chapter, attention will be given to alternatives to $H_4$**DOTA**, the latter is the golden standard in terms of GBCAs and it is primordial to emphasize its main properties. Some years ago, Desreux and collaborators reported different published thermodynamic stability constants for [Gd(**DOTA**)]$^-$. Most of the studies presented stability constants of about 25 log units, going from log $K_{GdL} = 25.3$ determined by spectrophotometric titration [18] to 22.1 as obtained by computing formation and dissociation rates [19]. Table 1 gives various constants reported for [Gd(**DOTA**)]$^-$.

As already mentioned earlier, to ensure the safe use of GBCAs *in vivo*, their dissociation process leading to $Gd^{3+}$ release should be much slower than their excretion from the body. Many kinetic dissociation experiments were conducted to measure the dissociation rate in serum, *in vivo*, in the presence of endogenous cations, and comparison between different $Gd^{3+}$ chelates is often rendered impossible due to the lack of similarity between experiments. However, Tweedle et al. stipulated in 1992 that the measure of the acid-catalysed dissociation rates constitutes a good probe to evaluate the *in vivo* loss of gadolinium(III) [20]. In acidic conditions, [Gd(**DOTA**)]$^-$ distinguishes by a remarkably high kinetic inertness in part due to its rigid structure and difficulty for protons to access into the cavity of the macrocycle [19]. Some half-lives of dissociation in acidic media are given in Table 1. However, when necessary in this chapter, additional kinetic data can be discussed. Generally, the relaxivity is measured by NMR relaxometry. The plot of the relaxation rate ($1/\Delta T_1$) as a function of the concentration of CA gives direct access to $r_1$ (slope of the line). However, the relaxation rate depends on many factors (temperature, magnetic field and solvent/medium). That is why, when comparing different GBCAs, attention must be given to compare $r_1$ values measured under identical conditions. As for the physicochemical properties, Table 1 gathers some relaxivities $r_1$ found in the literature for the mono-aqua [Gd(**DOTA**)($H_2O$)]$^-$ chelate. If [Gd(**DOTA**)($H_2O$)]$^-$ appears as a reference in terms of thermodynamic and dissociation kinetic, its properties to increase relaxation rate of protons are limited; this is probably due to a limited residence time of the coordinating water molecule ($\tau_M \approx 200$ ns [21]), which is out of the range of ideal $\tau_M$ values. However, an interesting fact is that [Gd(**DOTA**)($H_2O$)]$^-$ exists in the form of different isomers, and each isomer displays its proper magnetic-related structure characteristics.

$Gd^{3+}$ is coordinated by the four nitrogen atoms of the macrocycle and the four oxygen atoms of the carboxylate groups. The nitrogen and oxygen atoms form a planar and nearly parallel arrangement, resulting in N4 and O4 bases. The O4

**TABLE 1**
**Physicochemical and Magnetic Properties of [Gd(DOTA)($H_2O$)]$^-$**

| Parameter | Value | Technic | Conditions | |
|---|---|---|---|---|
| log $K_{GdL}$ | 25.3 | Spectrophotometric titration | Calculated from conditional $K_f$ at pH 4 & 7.4 | [18] |
| | 24.0 | Potentiometric titration | 0.1 M KCl, 25°C | [30] |
| | 22.1 | Computational value from $k_f$ & $k_d$ | | [19] |
| | 24.7 | Spectrophotometric titration | 0.1 M KCl, 25°C | [31] |
| | 23.6 | Spectrophotometric titration | 1 M NaCl, 37°C | [32] |
| $t_{1/2}$ | 85 d | | pH 2 | [19] |
| | > 200 d | | pH 5 | |
| | 60.2 h | | [$H^+$] = 0.1 M | [33] |
| $r_1$ | 4.2 | | 20 MHz, 25°C, pH 7.4 | [34] |
| ($mM^{-1}s^{-1}$) | 3.56 | | 20 MHz, 39°C, pH 7.3 | [35] |
| | 3.5 | | 20 MHz, 25°C | [36] |

plane is capped with a water molecule. The ethylene groups of the macrocycle adopt a gauche conformation, forming a five-membered coordination metallacycle with either a δ or a λ configuration. This leads to two possible square [3,3,3,3] conformations of the macrocycle: δδδδ and λλλλ [22,23]. The pendant acetate arms can occupy two orientations: Δ or Λ. In solution, the interconversion of the ring between the δ and λ configurations and the rotation of the acetate groups between Δ and Λ orientations result in four stereoisomers (two racemic diastereoisomers: Δδδδδ/Λλλλλ and Δλλλλ/Λδδδδ, respectively). Furthermore, the two diastereoisomers differ in the angle ω, which represents the mutual rotation of the O4 and N4 planes. In the Δλλλλ/Λδδδδ isomers, a rotation of approximately 40° leads to the square-antiprismatic (SAP) isomer (with an ideal angle of 45°). Traditionally, this diastereoisomer is referred to as the "major" (M) isomer due to its higher abundance in solutions of the [Gd(**DOTA**)($H_2O$)]$^-$ complex [24]. In the pair of enantiomers Λλλλλ/Δδδδδ, a rotation of approximately −24° corresponds to the twisted-square-antiprismatic (TSAP) isomer (with an ideal angle of −22.5°), also known as the "minor" (m) isomer.

These two diastereoisomers play a crucial role in MRI applications. It has been observed that lanthanide(III) complexes of DOTA-like ligands (with one water molecule in the first coordination sphere) exhibit different residence times ($\tau_M$) for coordinated water in SAP and TSAP isomers [9,25–29]. The water molecule exchanges 10–100 times faster in the TSAP isomer compared to the SAP isomer. This difference can be explained by the steric crowding around the coordinated water molecule in the TSAP isomer. The water exchange rate in the TSAP isomers of gadolinium(III) complexes with carboxylate ligands approaches the optimal range of $\tau_M$ (10–30 ns) [9,25,26].

Curiously, [Gd(**DOTA**)($H_2O$)]$^-$, the first azamacrocyclic complex studied for such application, was year after year confirmed as the best one for MRI uses, presenting a very high thermodynamic stability and kinetic inertness, especially for *in vivo* studies where later the dissociation process was measured to be much longer than the renal excretion. This explains the use of Dotarem® and its recent generic concurrency developed by GUERBET for many years without ever losing their leadership. That said, during three decades, researchers have given efforts to slightly change the original design of **DOTA** hoping to improve the complex properties.

## 3.1 Ways to Improve "DOTA"

As we will see in the next paragraphs, starting from the DOTA structure, several approaches have been followed to impact different $q$, $\tau_R$ and $k_{ex}$ (or $\tau_M$) parameters.

### 3.1.1 Number of Water Molecules in the First Coordination Sphere of the Metal Ion

As said before, the chelators based on a DOTA platform could not offer more than one water molecule ($q = 1$) to their $Gd^{3+}$ complexes. The sacrifice of one of the four acetate pendants releases one coordination site that can be taken by a second water molecule, but it is detrimental to the stability and inertness of the associated $Gd^{3+}$ complex. The gadolinium complex formed with $H_3$**DO3A** ($q = 2$) then presents an inertness of $t_{1/2} = 301$ s in $[H^+] = 0.1$ M *vs* 44.5 days for [Gd(**DOTA**)$(H_2O)]^-$ (meaning $k_1 = 2.3 \times 10^{-2}$ $M^{-1}s^{-1}$ for [Gd(**DO3A**)$(H_2O)_2$] and $k_1 = 1.8 \times 10^{-7}$ for [Gd(**DOTA**)$(H_2O)]^-$) [19,37,38]. The complexation constant of [Gd(**DO3A**)] is also quite far from its **DOTA** analogue with pGd = 15.9 (log $K_{GdL} = 21.56/21.1$ in 0.1 M KCl) [38]. In terms of relativity, water exchange rates ($k_{ex}$) at 298 K of $6.25 \times 10^6 s^{-1}$ [39] to $11 \times 10^6 s^{-1}$ [40] were reported, associated with $r_1 = 4.8$ $mM^{-1} s^{-1}$ (20 MHz, 40°C, pH ≈ 7) [39]. The bis-hydrated [Gd(**DO3A**)$(H_2O)_2$] chelate presents quite interesting performances that however do not allow any application as the stability is too low.

### 3.1.2 Improvement of the Rotational Correlation Time of DOTA Derivatives

At low field, low-weight monohydrated GBCAs have a $r_1$ relaxivity in the range of 3–5 $mM^{-1}s^{-1}$, which is far from the theoretical maximum of ~100 $mM^{-1}s^{-1}$. As said before, the rotational correlation time ($\tau_R$) of a $Gd^{3+}$ chelate is inversely proportional to its size. By increasing the molecular weight of the chelate, the rotational motion can be slowed down, increasing relaxivity.

The DOTA-like structure can be modified to enhance the rigidity of the complex, thereby reducing the number of possible rotational orientations. This can be achieved by introducing rigid groups or incorporating rigid scaffolds into the chelate structure. Such modifications restrict the molecular motion and enhance the relaxivity of the chelate.

Tethering GBCAs to (bio)macromolecules is probably one of the most efficient approaches to improve the rotational correlation; the encapsulation of the complex within macromolecular carriers, such as dendrimers or nanoparticles, has been successfully applied. These carriers restrict the motion of the chelates, leading to a slower rotational correlation time and increased relaxivity. In the same order of idea, Gd-DOTA-based complexes associated with proteins are a class of contrast agents used in MRI to enhance image quality and provide valuable information about physiological processes in living organisms. These adducts are designed to target specific tissues or biomolecules, making them particularly useful in molecular imaging and diagnostics. When Gd-DOTA is associated for instance with HSA, it forms a stable adduct of high molecular weight. HSA is known for its long circulation half-life, which allows the Gd-DOTA-HSA adduct to remain in the bloodstream for an extended period. This extended circulation time enhances the availability of the contrast agent for imaging, making it a valuable blood pool contrast agent. This property allows for better visualization of blood vessels and vascular structures during MRI scans.

Finally, increasing the viscosity of the surrounding medium can also improve the rotational correlation time of $Gd^{3+}$ chelates; this was achieved with DOTA derivatives by incorporating high-viscosity moieties on the structure or using additives that increase the viscosity of the imaging medium [37,41]. All these points will not be further discussed in this chapter, which will focus on molecular objects.

### 3.1.3 Residence Time in the First Coordination Sphere

Improving the water exchange rate in DOTA-like $Gd^{3+}$ chelates is an important aspect for enhancing the efficiency of contrast agents used in MRI while maintaining a stable Gd-chelating moiety. This can be done by slightly changing the ligand design. Substituting the acetate pendant arms of the DOTA-like ligand with more labile coordinating groups, such as amide or hydroxyl groups, can increase the rate of water exchange. These labile groups can facilitate the dissociation of coordinated water molecules, leading to faster exchange kinetics. In the same order of idea, introducing electron-withdrawing or electron-donating groups on the ligand scaffold can influence the electronic environment around the $Gd^{3+}$ ion. This alteration can impact the water exchange rate by affecting the coordination strength and electronic properties of the chelate. Proper selection of substituents can enhance the water exchange rate while maintaining the stability of the complex.

Finally, exploring different structural isomers of DOTA-like ligands can provide insights into their water exchange kinetics. Stereoisomeric modifications can influence the spatial arrangement of the ligand and alter the coordination environment around the $Gd^{3+}$ ion, potentially affecting the water exchange rate. These stereoisomers can interconvert through ligand exchange processes, especially in the presence of other coordinating species or under specific conditions such as pH or temperature changes. The stability and dynamics of the isomers play a crucial role in the design and optimization of DOTA-$Gd^{3+}$ complexes as MRI contrast agents. Several strategies have been followed to change the steric compression on the first coordination sphere of the [Gd(**DOTA**)($H_2O$)]$^-$ complex without straying too far from the original structure of $H_4$**DOTA** to preserve the kinetic and thermodynamic properties. Several approaches have aimed to stiffen the structure by insertion of alkyl groups/chains on the chelating ring or pendants.

## 3.2 DO3A AND DO3A-BASED DERIVATIVES

The only way to increase the number of water molecules in the inner sphere of the DOTA-like derivatives (as defined above) with the aim to enhance the relaxivity of the complex is to remove one of the four coordinating arms leading to the $H_3$**DO3A** chelator. But as discussed earlier, this is detrimental to the thermodynamics and kinetics of the associated Gd(III) chelate. That said, $H_3$**DO3A** is an important part of several DO3A-functionalized compounds that have been intensively studied for $Gd^{3+}$ coordination and MRI applications. Several advantages or aims have motivated the design of DO3A-based ligands. Despite the very high inertness of [Gd(**DOTA**)($H_2O$)]$^-$ and most of the clinically used MRI agents, a small loss of free $Gd^{3+}$ occurs *in vivo* post-injection potentially causing the damages mentioned earlier. A way to reduce the amount of liberated $Gd^{3+}$ is to simply reduce the amount of injected CA, but this involves then to have more performant CAs. Of course, it would be of colossal work to cite all DO3A-based $Gd^{3+}$ complexes that have been studied; so in this chapter, we are only focusing on three important families.

It has been previously stated that ideally, at 20 MHz, the employed CAs should display a water residence lifetime ($\tau_M$) comprised between 10 and 30 ns. But most of the macrocyclic systems feature much longer $\tau_M$ preventing them from reaching optimum relaxivity [42–45]. For example, at 298 K and 20 MHz, the [Gd(**DOTA**)($H_2O$)]$^-$ complex shows a $\tau_M$ value of 243 ns [21]. Many works aimed at increasing the water exchange rate. Monohydrated $Gd^{3+}$ complexes undergo a dissociative exchange pathway that can be accelerated by adding more negative charges around the water molecule or by inducing steric crowding. This last point can be achieved by lengthening the carbon chain of the macrocyclic backbone (as it will be seen later) or the coordinating arm. Therefore, one acetate arm of $H_4$**DOTA** was replaced by a propionate one to give $H_4$**DO3AN**$_{\mathbf{prop}}$ (Figure 3) [46,47].

Other studies, focused on enhancing the water exchange rate, involve the incorporation of bulky substituents to create a steric constraint around the water

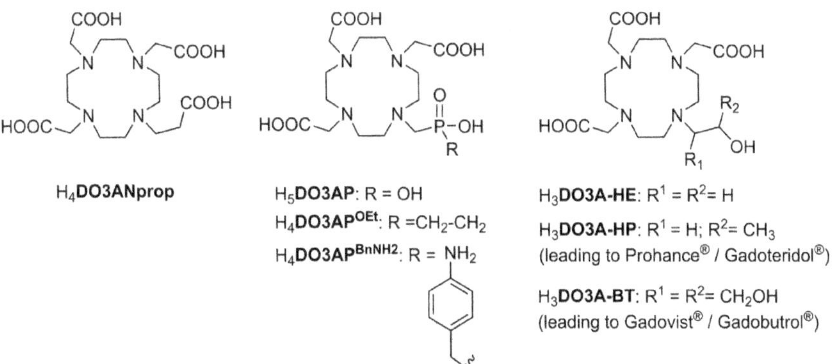


**FIGURE 3** DO3A-based chelates.

molecule within the inner coordination sphere. We have seen earlier that in solution, the DOTA Gd-chelate exists as two racemic diastereoisomers: Δδδδδ/Λλλλλ and Δλλλλ/Λδδδδ. The two diastereoisomers differ in the ω angle, which represents the mutual rotation of the O4 and N4 planes. In the Δλλλλ/Λδδδδ isomers, a rotation of ~40° leads to the SAP isomer. This diastereoisomer is more abundant in solution and is so denominated "major" (M) isomer [24]. In the Λλλλλ/Δδδδδ enantiomers pair, a rotation of ~–24° corresponds to the TSAP isomer by deduction named "minor" (m) isomer. It has been observed that SAP and TSAP isomers of [Gd(**DOTA**)($H_2O$)]$^-$ display different residence times ($\tau_M$) [9,25–29]. The steric crowding around the $H_2O$ molecule in the TSAP isomer leads to a much faster water molecule exchange (10–100 times faster than in the SAP isomer) [9,25,26]. Therefore, gadolinium(III) complexes with a higher proportion of the desired TSAP isomer are expected to exhibit higher relaxivity, particularly in high-molecular-weight molecules with very long $\tau_R$ values.

Thus, the introduction of mono-phosphonic acid functions on the DO3A-backbone has been proposed ($H_5$**DO3AP**, $H_4$**DO3AP**$^{BnNH2}$ and $H_4$**DO3AP**$^{OEt}$, among others, Figure 3) to benefit from the steric constraint imposed by bulky phosphorous groups to destabilize the coordinating water molecule and increase its exchange rate [48–54]. Noteworthy, phosphorous groups increase the negative charge around the water molecule and could potentially increase the second water coordination sphere (and so the overall relaxivity) as their phosphonate/phosphinate analogues [39,55–61].

Finally, the probably most promising DO3A-based derivatives in terms of MRI applications are those including a hydroxyalkyl chain since they led to the commercial ProHance® (or Gadoteridol) and Gadovist® (or Gadobutrol®) GBCAs. The $Gd^{3+}$ complexes formed from these chelators ($H_3$**DO3A-HE**, $H_3$**DO3A-HP** and $H_3$**DO3A-BT**, Figure 3) [8,18,62–64] have the specificity to be neutral, but highly hydrophilic, especially in the case of Gadobutrol®, which ensure a low protein binding and good biological tolerance [62].

Most of the chelators presented in Figure 3 are obtained through $H_3$**DO3A** or its ester derivative. The synthesis of these important building blocks will be discussed before to describe the formation of the DO3A-based ligands.

### 3.2.1 DO3A and DO3A-Based Derivatives Synthesis

Several strategies can be envisaged starting from **cyclen** for the preparation of $H_3$**DO3A** derivatives. A first method is based on the introduction of a coordinating arm on the cyclen backbone (directly on **cyclen** or on its di- or tri-diprotected form) prior to tri-functionalization with the three acetate pendants. A second strategy, which is more and more employed since the commercially availability of DO3A-building blocks, lies on the alkylation of $H_3$**DO3A** or its ester analogue **DO3A(O$^t$Bu)$_3$** where the acid functions are protected as *tert*-butyl esters. Despite their availability, it is interesting to remember the preparation of this very important DO3A-derivative precursors (Figure 4). **DO3A(O$^t$Bu)$_3$** can be prepared by direct alkylation of **cyclen** (between 0 and 25°C) using *tert*-butyl bromoacetate in the presence of sodium acetate, with dimethylacetamide as solvent [65,66].

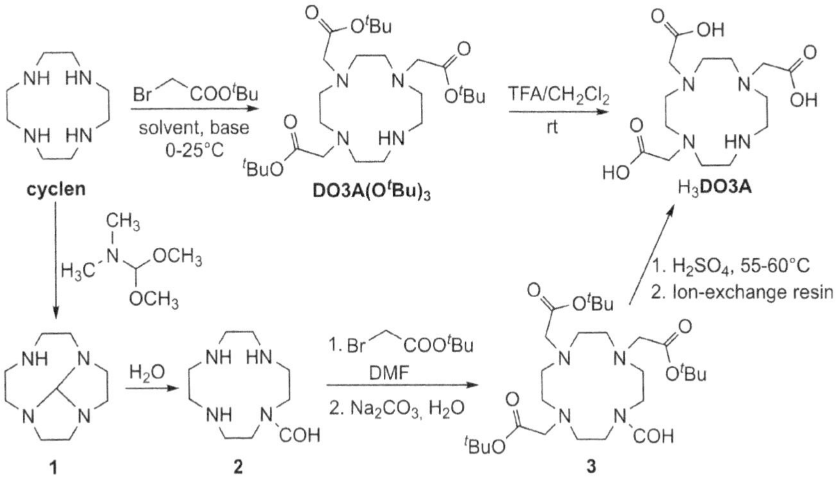

**FIGURE 4** Examples of synthetic routes for the preparation of **DO3A(O$^t$Bu)$_3$** and $H_3$**DO3A**.

However, the formation of the tri-alkylated desired compound **DO3A(O$^t$Bu)$_3$** is accompanied by the formation of the di-alkylated and tetra-alkylated side products. It has been established that the isolation of **DO3A(O$^t$Bu)$_3$** as a salt, especially a HBr salt, allows to obtain pure compounds with high yield. The addition of $NaHCO_3$.KBr to an aqueous solution of the crude obtained after the reaction of **cyclen** and *tert*-butyl bromoacetate (until pH 9) and the addition of diethyl ether leads to the formation of white crystals of **DO3A(O$^t$Bu)$_3$**.HBr [65]. Slightly modified conditions of base and solvent were also reported for the tri-functionalization reaction of **cyclen** with *tert*-butyl bromoacetate ($NaHCO_3$/ACN [67,68], $K_2CO_3$/ACN [69], $Et_3N/CH_3Cl$ [70], no base/$CH_3Cl$ [71] and NaOAc/DMF [72]). **DO3A(O$^t$Bu)$_3$** represents a tri-protected form of cyclen where only one macrocyclic secondary amine can be engaged in a functionalization reaction.

$H_3$**DO3A** can be obtained by a trifluoroacetic acid-mediated removal of *tert*-butyl groups of **DO3A(O$^t$Bu)$_3$** [73,74] or by direct alkylation of **cyclen** with chloroacetic acid [75]. Other synthetic pathways were described in the 1990s. For example, routes where **cyclen** was protected by benzyl or formyl groups were developed by Michael F. Tweedle et al. These groups have the advantage to be easily removed by hydrolysis or hydrogenolysis after the tri-alkylation with chloroacetic acid or with *tert*-butyl bromoacetate [76]. Alternatively, a method based on the formation of tricycle **1** from **cyclen** and *N,N*-dimethylformamide dimethyl acetal was also developed. This intermediate is subjected to hydrolysis to yield cyclen-formyl **2**, an interesting mono-protected precursor that allows tri-alkylation. Reaction of compound **2** with *tert*-butyl bromoacetate followed by an acid hydrolysis and an ion-exchange resin column leads to $H_3$**DO3A** [76]. However, due to the cost and toxicity of many reactants, this way is not industrially applicable.

$H_4$**DO3AN**$_{\mathbf{prop}}$ illustrates very well the two possibilities discussed above to offer a DO3A derivative. A propionate arm can be introduced prior to the tri-alkylation to insert the acetate pendants; alternatively, the propionate pendant can be attached on $H_3$**DO3A** or its *tert*-butyl ester analogue. Initially, Merbach et al. performed the mono-alkylation of **DO3A(O**$^t$**Bu)**$_3$ with *tert*-butyl 3-bromopropanoate that was followed by the acid-assisted removal of the *tert*-butyl esters to give $H_4$**DO3AN**$_{\mathbf{prop}}$ [46]. Two years later, Tóth et al. used the second strategy aiming to introduce the propionate arm prior to the acetate ones. **Cyclen** was therefore reacted with ethyl 3-bromopropanoate in ACN with $K_2CO_3$ as base. To favor the mono-alkylation, a large excess of **cyclen** was used (5 equiv.). Then, a tri-alkylation, this time with ethylbromoacetate (3 equiv.), in a mixture ACN/$K_2CO_3$ was performed before the hydrolysis in HCl 6M to yield $H_4$**DO3AN**$_{\mathbf{prop}}$ as its tetra-chlorhydride salt [47]. To avoid the use of a large excess of **cyclen**, an alternative way was described based on the bis-aminal tool. Cyclen glyoxal (**4**) was first synthesized from **cyclen** and glyoxal in MeOH [77]. As it will be further detailed later, cyclen-glyoxal is an important building block that, depending on the solvent or the alkylating agent, can be used for the selective mono-*N*-alkylation (mono-functionalization) or the *trans*-di-*N*-alkylation (di-functionalization) of **cyclen** [77–79]. In the present case, the selective mono-alkylation of cyclen glyoxal **4** was performed with iodopropanoic acid (1 equiv.) in water at 45°C for 4days to yield the ammonium salt **5**. The glyoxal bridge was removed in hydrazine hydrate. The three secondary amines of intermediate **6** were then functionalized with chloroacetic acid in an EtOH/$K_2CO_3$ system at 45°C to offer $H_4$**DO3AN**$_{\mathbf{prop}}$ (Figure 5) [47].

**DO3AP** derivatives were obtained by Mannich reactions either from **DO3A(O**$^t$**Bu)**$_3$ or directly from $H_3$**DO3A** (Figure 6). For example, the condensation of triethyl phosphite and **DO3A(O**$^t$**Bu)**$_3$ in the presence of paraformaldehyde led to intermediate **7**, whose *tert*-butyl groups were first removed with TFA in $CH_2Cl_2$ at room temperature before performing a basic hydrolysis to convert the diethyl phosphonate function into a monoester phosphate $H_4$**DO3AP**$^{\mathbf{OEt}}$ after purification on strong cation exchanger [53]. $H_4$**DO3AP** and $H_4$**DO3AP**$^{\mathbf{BnNH2}}$ were obtained by a Mannich reaction from $H_3$**DO3A**. $H_4$**DO3AP** could be obtained in one step from $H_3$**DO3A**, diethyl phosphonate and solid paraformaldehyde in acid medium (HCl/$H_2O$, 1:1) after a series of treatments with ion-exchange resins [49]. Finally, $H_4$**DO3AP**$^{\mathbf{BnNH2}}$ was prepared in two steps from $H_3$**DO3A**. This first step consisted of a Mannich reaction involving $H_3$**DO3A**, paraformaldehyde and phosphonic acid **8**. A hydrogenolysis catalysed by Pd/C led to the bifunctional ligand $H_4$**DO3AP**$^{\mathbf{BnNH2}}$ [54].

Since the approval of Gadovist® and ProHance® for clinical use, many groups aim to propose efficient high-scale synthesis for DO3A-ethanol derivatives, as proven by the high number of patents. In this chapter, we will mainly focus on the first established synthetic strategies.

$H_3$**DO3A-HE** can be obtained by the reaction of **DO3A(O**$^t$**Bu)**$_3$ with (2-bromoethoxy)tetrahydropyran [80] in a mixture $K_2CO_3$/ACN. Then, an acidic treatment (acetic acid followed by TFA) allowed the removal of the THP-protecting group and the acidolysis of the *tert*-butyl ester functions [81].

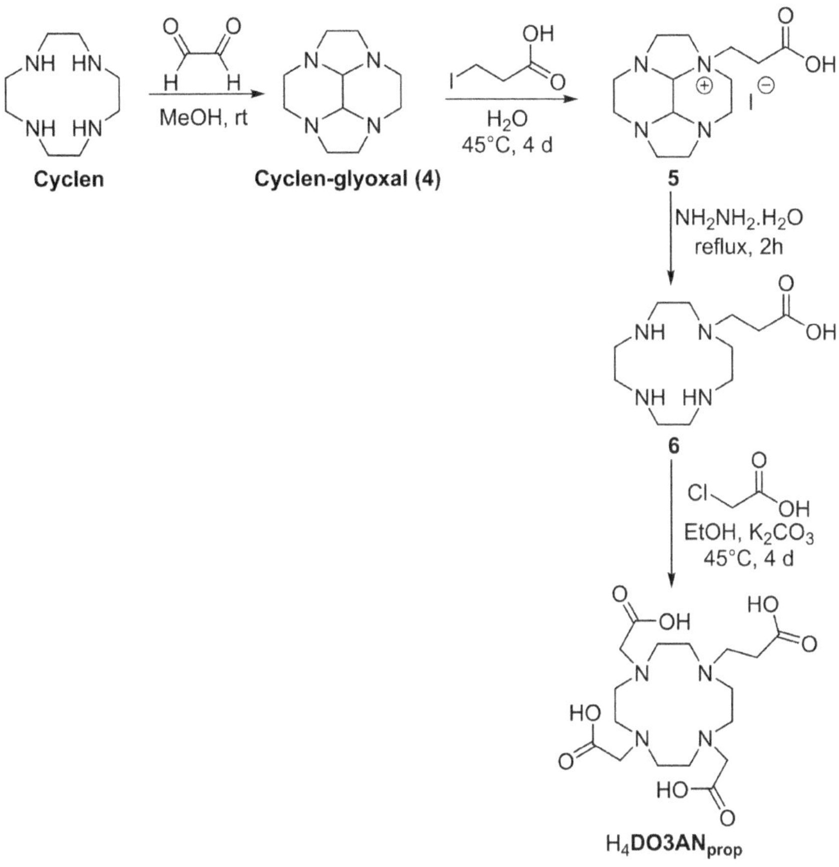

**FIGURE 5** Synthesis of $H_4$**DO3AN**$_{prop}$ according to Tóth et al. [47].

The synthesis of $H_3$**DO3A-HP** and $H_3$**DO3A-BT** relies on the alkylation of either $H_3$**DO3A** (or its *tert*-butyl ester analogue) or tri-protected **cyclen** with very reactive epoxide precursors.

The first published synthetic procedure of $H_3$**DO3A-HP** described its easy preparation by the reaction between $H_3$**DO3A** and propylene oxide in $H_2O$ at pH 12–12.5 (Figure 7). $H_3$**DO3A-HP** was obtained in its neutral form after the purification process on resins [76]. More recently, a slightly modified strategy used the protected **DO3A(O$^t$Bu)$_3$** for reaction with propylene oxide [82].

In 1997, Sülzel and colleagues described three routes to prepare $H_3$**DO3A-BT**. The first pathway involves isolating the triprotected cyclen intermediate (**9**) that was previously described in Figure 4. The latter was then mixed with a large excess of 4,4-dimethyl-3,5,8-trioxabicyclo[5.1.0]octane oil (**10**) at 120°C for 16h. Then, the mixture was treated with methanol/$H_2O$ (3:1) to give intermediate **11**. The formyl group was then removed by a basic treatment in a NaOH/MeOH/$H_2O$

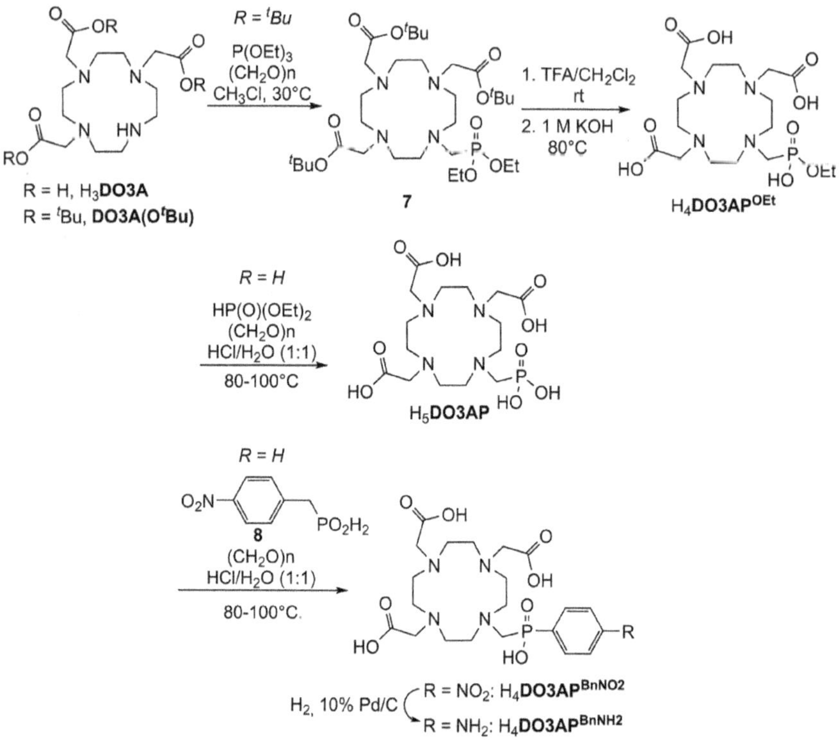

**FIGURE 6** Synthesis of various DO3AP derivatives based on a Mannich reaction.

mixture at 70°C for 12h to give compound **12**. Finally, the three remaining secondary amines of compound **12** were subjected to *N*-alkylation with chloroacetic acid in water at basic pH (9–10). After 18h at 70°C, concentrated HCl was added until pH 1 to open the seven-membered ring and offer $H_3$**DO3A-BT** after ion exchange column [62].

The second route described is quite similar but, according to the authors, it is more prone to high-scale synthesis. In this protocol, **cyclen** was first mixed with *N*,*N*-dimethylformamide dimethyl acetal in toluene. A methanol/toluene azeotrope was distilled off. This step probably allows the formation of tricycle **9** which was not isolated. After the concentration of the solvent, 4,4-dimethyl-3,5,8-trioxabicyclo[5.1.0]octane **13** was added and the mixture was heated to 110°C for 24h. Finally, concentrated HCl was added to isolate the cyclen derivative bearing the trihydroxy pendant arm (**14**). This intermediate was then functionalized with chloroacetic acid in classical conditions ($H_2O$, pH 9–10) [62]. These two synthetic pathways are detailed in Figure 7. The third route, implying the functionalization of a linear triamine with the trihydroxy precursor prior to cyclization, will not be further discussed. It is important to precise that $H_3$**DO3A-BT** is obtained as a racemic mixture. In Figure 7, only one of the two enantiomers is represented.

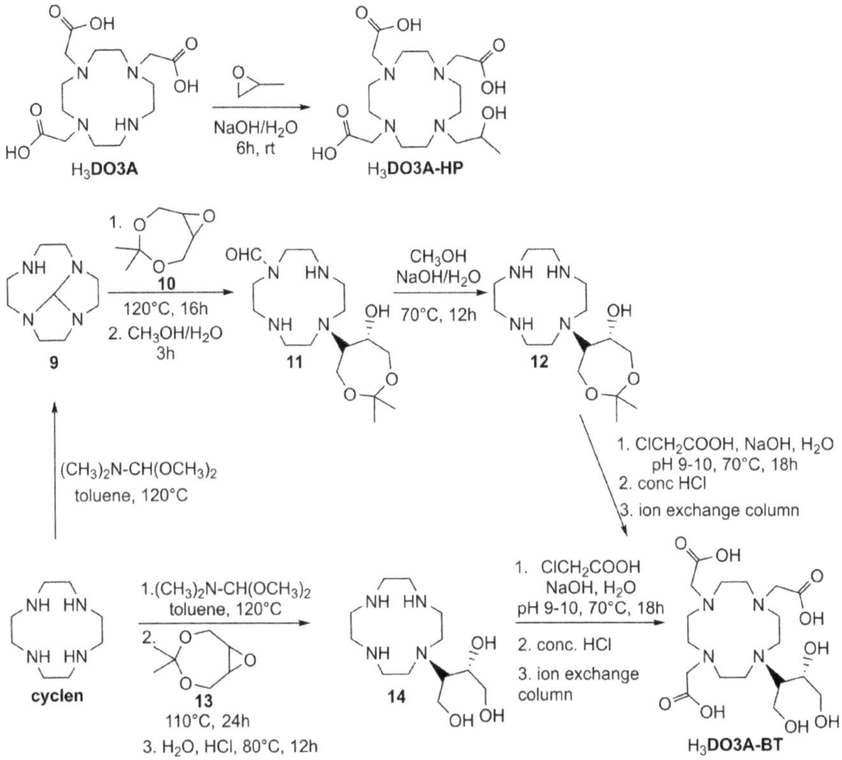

FIGURE 7 Initial synthesis of $H_3$**DO3A-HP** and $H_3$**DO3A-BT** [62,76].

Many patents were followed to improve the preparation or purification of $H_3$**DO3A-BT**, mainly to insure its production at the industrial scale [83].

### 3.2.2 DO3A and DO3A-Based Derivatives Properties

As explained earlier, $H_3$**DO3A** is clearly not a good candidate for application in MRI. Indeed, if the sacrifice of one acetate pendant to allow the insertion of 2 water molecules has a positive influence on the relaxivity, the thermodynamic stability and kinetic inertness of the $Gd^{3+}$ complex dramatically decrease, with a $t_{1/2}=9.6$ min in 0.1 M HCl and a log $K_{GdL}$ falling down to 21.0 [18,64].

Among various chelators cited here, some DO3A derivatives deserve discussion since they are currently clinically used for MRI exams.

Gadobutrol® is characterized by a stability constant log $K_{GdL}=21.8$ (measured in 0.1 M KCl) [84] which is considerably lower than the constant measured for [Gd(**DOTA**)]⁻ (log $K_{GdL}=24.7$ in 0.1 M KCl) [31]). The drop of stability when substituting an acetate pendant with a hydroxyalkyl chain has already been demonstrated with [Gd(**DO3A-HP**)] (ProHance®), which has a stability constant log $K_{GdL}=23.8$ [18]. This can easily be explained by the negatively charged oxygen donor of the carboxylate group, which binds more tightly to the $Gd^{3+}$ center than the neutral oxygen of the alcohol function. The further loss of stability of Gadobutrol® might be caused by the formation of intramolecular hydrogen interactions involving the additional hydroxy groups, which might reduce the electron density of coordinating oxygen atom [62]. However, as already stated, stability constant is not enough to validate the safe use of a GBCA; it must imperatively be accompanied by a very high kinetic inertness. The kinetic inertness of [Gd(**DO3A-HP**)] (ProHance®) and [Gd(**DO3A-BT**)] (Gadovist®) was compared, followed by spectrophotometry through which the exchange reaction with $Eu^{3+}$ between pH 3.2 and 5.3 was measured. The measured kinetic constant rates are $2.6\times10^{-4}\,M^{-1}s^{-1}$ and $2.8\times10^{-5}\,M^{-1}s^{-1}$ for [Gd(**DO3A-HP**)] and [Gd(**DO3A-BT**)], respectively, which means that the dissociation rate of [Gd(**DO3A-HP**)] is ~10 times faster than that of [Gd(**DO3A-BT**)] despite a lower stability constant for the latter [84]. Noteworthy, the exchange rates are linearly proportional to the proton concentration in the studied pH range. The kinetic inertness of [Gd(**DO3A-HP**)] was also evaluated in the presence of endogenous metal ions ($Zn^{2+}$ and $Cu^{2+}$) or anions (carbonate, phosphonate) at pH 7. No dissociation was observed after seven days proving the high inertness of [Gd(**DO3A-HP**)] [5]. The latter has also been involved in cell incubation experiments. Sylvio Aime et al. performed the cell labelling with [Gd(**DO3A-HP**)] (incubation of 4h, 37°C, [Gd**L**] = 1.6mM). The authors demonstrated that there was no release of free $Gd^{3+}$ occurred during the internalization process [85]. Later on, the same group published additional work proving that [Gd(**DO3A-HP**)] remains undissociated 96h after its endosomal internalization into both fibroblasts and macrophages [86].

Related to relaxivity, looking at various parameters, Gadovist® (or Gadobutrol) has a higher relaxivity than Dotarem® ($r_1 = 5.3\,mM^{-1}s^{-1}$ *vs* $4.2\,mM^{-1}s^{-1}$ at 20MHz, 25°C [52]), meaning it provides greater enhancement and image contrast, particularly in $T_1$-weighted sequences. As Dotarem®, Gadovist® has a predominant elimination through renal excretion; however, Dotarem® is eliminated through a lesser extent, with a higher percentage excreted through the hepatobiliary system (bile). Both contrast agents are approved for use in similar clinical indications.

The relaxivity of ProHance® ([Gd(**DO3A-HP**)($H_2O$)]) at 20MHz and 40°C is $3.7\,mM^{-1}s^{-1}$ [87] (*vs* $3.56\,mM^{-1}s^{-1}$ at 39°C and pH 7.3 for ([Gd(**DOTA**)($H_2O$)]$^-$ [35]). However, the relaxivity of ([Gd(**DO3A-HP**)($H_2O$)] is pH dependent. The $r_1$ value increases from pH 8 to pH 10.5 and then falls down to its initial value (298 K, 20MHz). The relaxivity gain observed in the pH range 8–10.5 is attributed to a base-catalysed exchange of the hydroxyl proton. At high pH, the alcohol function is deprotonated, quenching such effect; however, the initial relaxivity is recovered probably due to the interaction of the alkoxide group with water molecules in the second coordination sphere [45].

ProHance® has a renal clearance with an elimination lifetime of 1.57h [88]. It is recommended for whole body imaging and the typically injected dose is in the order of $0.1\,mmol\,kg^{-1}$ [52]. The water exchange rates of ProHance® and Dotarem® differ due to variations in their ligand structures, as proved by the analysis of the temperature dependence of water $^{17}O$ transverse relaxation rate. At 25°C, [Gd(**DOTA**)($H_2O$)]$^-$ displays a lower water exchange lifetime than

[Gd(**DO3A-HP**)($H_2O$)], with $\tau_M = 244$ and 350 ns, respectively [8]. No data concerning the water exchange rate of [Gd(**DO3A-BT**)($H_2O$)], nor any other parameters explaining the enhanced relaxivity, can be found in the literature.

The substitution of one acetate pendant by a propionate leads to the ligand $H_4$**DO3AN**$_{\mathbf{prop}}$. Compared to $H_4$**DOTA**, the lengthening of one coordinating arm has few impacts on the protonation steps and on the overall basicity of the ligand. However, the measures of the stability constants with a series of alkali earth and transition metals reveal the formation of less stable complexes with $H_4$**DO3AN**$_{\mathbf{prop}}$ than with $H_4$**DOTA**, probably due to the formation of a six-membered ring in the presence of a propionate arm, which leads to lower stability than a five-membered ring as is the case with acetate groups [47]. The literature lacks the log $K_{GdL}$ stability constant; however, in the case of [Gd(**DTPA**)($H_2O$)]$^{2-}$, the introduction of one propionate pendant causes a decrease of 2.7 log $K$ units; thus, it has been supposed that the stability constant of [Gd(**DO3AN**$_{\mathbf{prop}}$)]$^-$ is approximately log $K_{GdL} \approx 22$ [46]. The kinetic inertness of [Ce(**DO3AN**$_{\mathbf{prop}}$)]$^-$ towards acid-assisted dissociation has been evaluated at different [$H^+$] concentrations. The same trend as in the thermodynamic studies is observed, meaning a decrease of kinetic inertness when replacing one acetate pendant by a propionate one. The acid-catalysed dissociation rate constants of [Ce(**DO3AN**$_{\mathbf{prop}}$)]$^-$ are around one to two orders of magnitude higher than those of [Ce(**DOTA**)]$^-$ [47]. Some $^{17}O$ NMR experiments at 25°C on [Gd(**DO3AN**$_{\mathbf{prop}}$)($H_2O$)]$^-$ (with $q = 1$) demonstrated a faster water exchange rate constant than for [Gd(**DOTA**)($H_2O$)]$^-$ ($k_{ex} = 6.1 \times 10^7$ s$^{-1}$ and $0.46 \times 10^7$ s$^{-1}$, respectively) proving the efficiency to induce steric crowding by lengthening of one coordinating arm [46]. However, [Gd(**DO3AN**$_{\mathbf{prop}}$)($H_2O$)]$^-$ has not been further explored as GBCA.

As discussed before, the main aim of the DO3AP series is to take benefit of the bulky phosphorous group to increase the proportion of $Gd^{3+}$ complexes as TSA isomers, which exhibit faster water exchange rates and thus should provide higher relaxivities. So, intensive works about the determination of the structure of $Ln^{3+}$ complexes of DO3AP derivatives were conducted. Few thermodynamic and kinetic investigations were realized. One can however mention the paper of Lukeš and colleagues reporting the protonation constants of $H_5$**DO3AP** and $H_4$**DO3AP**$^{\mathbf{BnNH2}}$ as well as the stability constants and the dissociation kinetics of the associated $Gd^{3+}$ complexes [48]. The overall protonation constants of $H_5$**DO3AP** are higher than those of $H_4$**DO3AP**$^{\mathbf{BnNH2}}$ showing the lower basicity of macrocycles containing phosphinic acids compared to phosphonic acid-based macrocycles. Therefore, the stability constant of the $Gd^{3+}$ complex of $H_5$**DO3AP** is higher (log $K_{GdL} = 27.5$) than that of the $Gd^{3+}$ chelate formed with $H_4$**DO3AP**$^{\mathbf{BnNH2}}$ (log $K_{GdL} = 24.04$). The acid-assisted dissociation of the $Y^{3+}$ chelates was studied. In acidic medium ([$H^+$] = 0.01 M) and at 25°C, the dissociation half-lives of the two yttrium(III) complexes of DO3AP derivatives are quite similar ($t_{1/2} = 17.7$ h with $H_5$**DO3AP** and $t_{1/2} = 18.9$ h with $H_4$**DO3AP**$^{\mathbf{BnNH2}}$). These complexes have a better inertness than [Y(**DOTA**)]$^-$ that has, in the same conditions, a half-life of 0.96 h [48].

The lanthanide complexes formed with $H_5$**DO3AP** are structurally identical to those of $H_4$**DOTA**. However, it has been demonstrated that in solution, the ratio of TSAP/SAP isomers is highly pH dependent, probably influenced by different protonation states of the phosphonate moiety. At acidic pH, when the phosphonate group is fully protonated, the TSAP isomer is mostly present (Ln = Eu, Nd, Yb). Concerning the $Gd^{3+}$ complex of $H_5$**DO3AP**, $^{17}O$ NMR studies at variable pH revealed a constant TSAP/SAP ratio over the range 2.5–7. The TSAP isomer displays higher protonation constant values than the SAP one due to a longer distance between the $Gd^{3+}$ center and the coordinating phosphonate group.

From a simultaneous fitting of $^1H$ NMRD and $^{17}O$ NMR spectroscopic data, a much lower water residence time was found for [Gd(**DO3AP**)($H_2O$)]$^{2-}$ than for [Gd(**DOTA**)($H_2O$)]$^-$ ($\tau_M$ = 14 ns *vs* 244 ns, respectively at pH = 7 and 25°C), which leads to an enhanced relaxivity for [Gd(**DO3AP**)($H_2O$)]$^{2-}$ compared to [Gd(**DOTA**)($H_2O$)]$^-$ ($r_1$ = 4.37 mM$^{-1}$s$^{-1}$ *vs* 3.83 mM$^{-1}$s$^{-1}$, respectively, at 20 MHz, pH = 7 and 37°C). The short residence time, even at the border of the optimal theoretical value, was assigned to steric constraint imposed by the bulky phosphorous atom around the coordinated water molecule. A pH-dependence was also observed for the $\tau_M$ values those increase at lower pH [49].

As for its $H_5$**DO3AP** analogue, deep structural studies of Ln(III) complexes of $H_4$**DO3AP**$^{\mathbf{BnNH2}}$ were performed. The $^{17}O$ NMR on the Dy(III) complex confirmed the coordination of one water molecule to the metal center. The Eu(III) complex was studied by means of $^1H$ NMR and $^{17}P$ NMR. These experiments revealed the presence of three stereoisomers, one SAP isomer and two TSAP isomers. Compared to the [Eu(**DOTA**)($H_2O$)]$^-$ analogue, the SAP/TSAP ratio is higher which suggests that the bulky phosphorous moiety pushes the equilibrium towards the sterically demanding TSAP structure [54]. The relaxivity of [Gd(**DO3AP**$^{\mathbf{BnNH2}}$)($H_2O$)] was evaluated. A water proton relaxivity $r_1$ = 6.7 mM$^{-1}$s$^{-1}$ was measured at 37°C, 10 MHz and pH 7 which is higher than the value obtained for [Gd(**DOTA**)($H_2O$)]$^-$ under the same conditions ($r_1$ = 5.7 mM$^{-1}$s$^{-1}$). This enhancement can probably be explained by the shorter residence time of the water molecule ($\tau_M$ = 16.2 ns *vs* 244 ns for [Gd(**DOTA**)($H_2O$)]$^-$). Variable-temperature $^{17}O$ NMR transverse relaxation rates $R_{2r}$ measurements revealed that both SAP and TSAP isomers have similar $\tau_M$ values which means that in all cases, the bulkiness of the phosphonate group causes the lengthening of the Gd-$OH_2$ bond. Note that the presence of the aniline group of $H_4$**DO3AP**$^{\mathbf{BnNH2}}$ allows its conjugation to macromolecular entities, as it has been later demonstrated by the conjugation to dendrimers [89].

As its analogues, the $Gd^{3+}$ complex of $H_4$**DO3AP**$^{\mathbf{OEt}}$ contains one water molecule in the inner sphere and displays the same kind of isomerism than [Gd(**DOTA**)($H_2O$)]$^-$. However, due to the chiral phosphorous center in $H_4$**DO3AP**$^{\mathbf{OEt}}$, the $Gd^{3+}$ complex exhibits a total of eight isomers, four SAP (Λδδδδ-*R*, Λδδδδ-*S*, Δλλλλ-*R* and Δλλλλ-*S*) and four TSAP isomers (Λλλλλ-*R*, Λλλλλ-*S*, Δδδδδ-*R* and Δδδδδ-*S*). As for the other phosphorous-containing $Gd^{3+}$ chelates, the proportion of TSAP isomers is enhanced in this case. A report estimates 35% of TSAP isomers for [Gd(**DO3AP**$^{\mathbf{OEt}}$)($H_2O$)]$^-$ compared to 60% and 50% for [Gd(**DO3AP**)($H_2O$)]$^{2-}$ and [Gd(**DO3AP**$^{\mathbf{BnNH2}}$)($H_2O$)]$^-$, respectively [52]. The relaxivity $r_1$ at 20 MHz and 37°C of [Gd(**DO3AP**$^{\mathbf{OEt}}$)($H_2O$)]$^-$ is 3.49 mM$^{-1}$s$^{-1}$. This value is lower than the relaxivity of [Gd(**DOTA**)($H_2O$)]$^-$ ($r_1$ = 3.83 mM$^{-1}$s$^{-1}$)

**TABLE 2**
**Main Properties of the $Gd^{3+}$ Complexes of DO3A Derivatives**

| | $q$ | $\log K_{GdL}$ | Kinetic Data | $^{298}k_{ex}$ ($\times 10^6 s^{-1}$) | $r_1$ in $mM^{-1}s^{-1}$ (Conditions) |
|---|---|---|---|---|---|
| $[Gd(\mathbf{DO3AN_{prop}})]^-$ | 1 | ≈ 22[a] | $k_1 = 7\times10^{-3}$ $M^{-1}s^{-1}$[b] | 61[c] | n.d |
| $[Gd(\mathbf{DO3AP})]^{2-}$ | 1 | 27.05[d] | $t_{1/2} = 17.7$ h[e] | 71.4[f,g] | 4.57[g] (20 MHz, 37°C, pH 7) |
| $[Gd(\mathbf{DO3AP^{OEt}})]^-$ | 1 | n.d | n.d | 20[h] | 3.49[h] (20 MHz, 37°C) |
| $[Gd(\mathbf{DO3AP^{BnNH2}})]^-$ | 1 | 24.04[d] | $t_{1/2} = 18.9$ h[e] | 61.7[f,i] | 6.7[i] (10 MHz, 37°C, pH 7) |
| [Gd(**DO3A-HP**)] | 1 | 23.8[j] | $k_1 = 2.6\times10^{-4}$ $M^{-1}s^{-1}$[k] | 2.8[f,l] | 3.7[m] (20 MHz, 40°C) |
| [Gd(**DO3A-BT**)] | 1 | 21.8[k] | $k_1 = 2.8\times10^{-5}$ $M^{-1}s^{-1}$[k] | n.d | 5.3[n] (20 MHz, 25°C) |

[a] Approximation from Ref. [46].
[b] Kinetic dissociation at 25°C of $[Ce(\mathbf{DO3ANprop})]^-$ ($C = 5\times10^{-4}$ M) in acid ($[H^+] = 5\times10^{-3}$ to 0.2 M), from Ref. [47].
[c] From Ref. [46].
[d] From Ref. [48].
[e] Kinetic dissociation at 25°C in $[H^+] = 0.01$ M of the $Y^{3+}$ complexes, from Ref. [48].
[f] Calculated from $\tau_M$ (at 25°C).
[g] From Ref. [49].
[h] From Ref. [53].
[i] From Ref. [54].
[j] From Ref. [18].
[k] Exchange reaction with $Eu^{3+}$ between pH 3.2 and 5.3, from Ref. [84].
[l] From Ref. [8].
[m] From Ref. [87].
[n] From Ref. [52].

despite a faster water exchange rate at 25°C ($k_{ex} = 20\times10^6$ $s^{-1}$ and $4.1\times10^6$ $s^{-1}$, respectively) [53].

Different thermodynamic and relaxometric parameters of the discussed complexes of this section are summarized in Table 2.

## 3.3 DO2A or DO2PA Derivatives

DO2A-based chelators such as $H_6$**DO2A2P** [90] or $H_2$**DO2A-HE2** [91] and $H_2$**DO2A-HP2** [91] have been studied as structures with possible 8-coordination

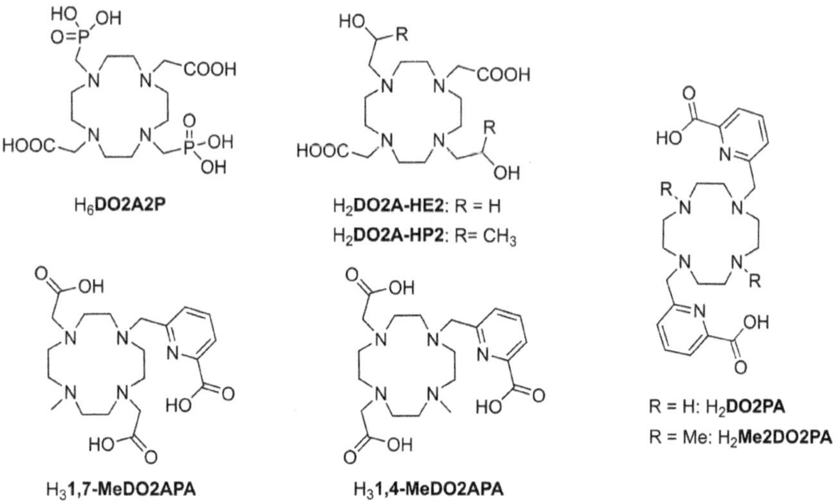

$H_6$**DO2A2P**

$H_2$**DO2A-HE2**: R = H
$H_2$**DO2A-HP2**: R= $CH_3$

R = H: $H_2$**DO2PA**
R = Me: $H_2$**Me2DO2PA**

$H_3$**1,7-MeDO2APA**

$H_3$**1,4-MeDO2APA**

**FIGURE 8** DO2A or DO2PA derivatives.

to combine the chelating properties of the acetate pendants, respectively, with phosphonic or propanol arms (Figure 8).

Another approach for manipulating the geometry of the coordination sphere of the ligand was to avoid the functionalization of one or two of the four amines of the macrocycle. To maintain an 8-coordination structure, bifunctional pendants are then necessary. Picolinate (pyridine-2-carboxylate, noted PA in the nomenclature employed) was then used as it proposed both an aromatic nitrogen and an acetate in a rigid structure (while maintaining the 5-membered chelate ring effect). $H_3$**1,7-MeDO2APA** and $H_3$**1,4-MeDO2APA** were then obtained to study the role of the nature of the capping bound of the $Gd^{3+}$ complexes (Figure 8).

To take advantage of the picolinate pendants and to still allow an 8-coordination structure, some DO2PA derivatives have been described as $H_2$**DO2PA** and $H_2$**Me2DO2PA** [92–94]. These $N_1$-$N_7$-difunctionalization gave interesting $Ln^{3+}$ complexes including $Gd^{3+}$ chelates (Figure 8).

### 3.3.1 DO2A or DO2PA Derivatives Synthesis

The direct $N_1$,$N_7$-di-functionalization of cyclen without prior protection is rarely the preferred option. This is primarily due to the limited reactivity on the distant secondary amines of the cyclen platform, especially when flexible pendant arms are present [94,95]. However, it has been demonstrated that selective $N_1$,$N_7$-acylation can be achieved by employing relatively bulky benzyl groups. The steric hindrance provided by benzyl chloroformate enhances their reactivity and prevents the formation of tri-substituted derivatives. Successful synthesis of $N_1$,$N_7$-protected cyclen has been reported by Sherry and colleagues in 2007 [96], as well as by Hopper and Allen [97], utilizing different solvents (without base), temperatures and reaction times, yielding comparable results

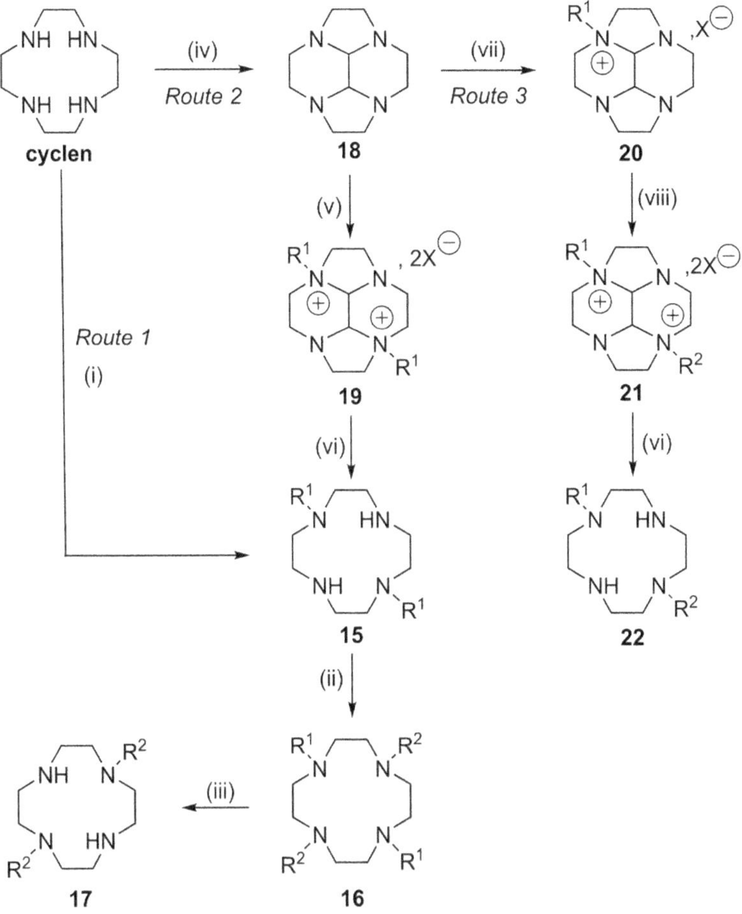


**FIGURE 9** Formation of *N*-functionalized cyclens. *Conditions*: (i) $BnOCOCl/CH_2Cl_2$ (or $CH_3Cl$); (ii) $BrCH_2C(O)R$, MeCN, $K_2CO_3$ or DIPEA ($R^1$ = CBz); (iii) $H_2$, Pd/C, EtOH, $R^2 = CH_2C(O)R$; (iv) glyoxal, MeOH; (v) 2 $R^1X$, DMF or $CH_3CN$, (NaI); (vi) $NH_2NH_2.H_2O$ or NaOH; or benzene-1,2-diamine; or ethylenediamine; (vii) $R^1X$, THF; (viii) $R^2X$, DMF or $CH_3CN$, (NaI).

(Route 1, Figure 9). This approach results in a di-protected cyclen that requires subsequent di-*N*-functionalization with the desired functions to produce compounds of type **16**. The resulting compounds are then deprotected through hydrogenolysis over a palladium-on-carbon catalyst, yielding ligands of type **17**. Further functionalization on one or two remaining secondary amines can be carried out afterwards [94]. Alternatively, $N_1$,$N_7$-di-functionalized cyclen derivatives can be obtained by reacting the macrocycle with two equivalents of different aldehydes under reductive amination conditions in the presence of $NaBH(OAc)_3$ [98].

Another approach for achieving highly selective $N_1,N_7$-protection of **cyclen** involves the use of bulky *N*-(*tert*-butoxycarbonyloxy)succinimide. This reagent allows for the introduction of *tert*-butoxycarbonyl (Boc) groups in nearly quantitative yields. The resulting diprotected cyclen can then be *trans*-di-alkylated to introduce additional substituents. The Boc groups can be easily removed using trifluoroacetic acid (TFA) in $CH_2Cl_2$ to obtain di-alkylated cyclen of type **17** [99]. This method is generally preferred over pH-controlled protection using different chloroformates [100].

A significant improvement in the $N_1,N_7$-di-*N*-alkylation procedure was achieved through bis-aminal chemistry [77–79], following a similar approach as mono-*N*-functionalization. However, in this case, a minimum of two equivalents of the desired electrophilic reagent are used to obtain the quaternary ammonium di-salts **19** (Route 2, Figure 9). To ensure successful reaction progression without premature precipitation of the mono-ammonium salt, a polar solvent like acetonitrile or DMF is employed [101]. Consequently, the di-*N*-alkylated derivatives **19** can be obtained in excellent yields without needing additional purification. However, it is worth noting that extended reaction times are often necessary for di-substitution. In certain instances, where the electrophile is an alkyl chloride or bromide, the reaction can be accelerated by the addition of NaI [102].

A noteworthy advancement in the *N*-functionalization method of cyclen-glyoxal was subsequently made by Archibald and colleagues. They introduced a highly efficient technique involving the direct grinding of the bis-aminal derivative with the alkyl- or allyl-halide derivative [103]. This approach dramatically improved the overall process, providing enhanced yields and simplifying the reaction conditions. The bis-aminal tool is also very useful for the design of non-symmetric di-$N_1,N_7$-alkylated cyclen compounds (Figure 9, Route 3), with the preparation of a mono-ammonium salt of type **20** followed by its reaction with a second electrophilic reagent leading to a quaternary ammonium di-salts **21** [77].

An example showcasing the preparation of disymmetrically tetra-*N*-functionalized cyclens is demonstrated through the synthesis of $H_3$**1,7-MeDO2APA** and $H_3$**1,4-MeDO2APA** (Figure 10) [94]. These ligands possess a distinct arrangement on the cyclen platform, consisting of a picolinate group, a methyl group and two acetate arms. To synthesize $H_3$**1,4-MeDO2APA**, the initial step involved the preparation of monomethyl cyclen **23**, achieved through bis-aminal chemistry. Conversely, for the synthesis of $H_3$**1,7-MeDO2APA**, an intermediate of type **16** was obtained by the cyclen-glyoxal strategy ($R^1 = Bn$, $R^2 = CH_2COO^tBu$, Route 2, Figure 9) that was subjected to hydrogenolysis to obtain a di-alkylated cyclen **25**. In both cases, the subsequent step involved the statistical *N*-functionalization/*N*-alkylation of one of the remaining secondary amine atoms. The mono-alkylation of intermediate **23** gives a dissymmetric di-*cis*-alkylated cyclen intermediate **24**, which was not isolated but directly reacted with 2 equivalents of *tert*-butyl bromoacetate. For the preparation of $H_3$**1,7-MeDO2APA**, one methyl substituent was introduced on cyclen **25** with iodomethane, followed by reaction with methyl 6-(chloromethyl)picolinate. The final stages encompassed the removal of ester

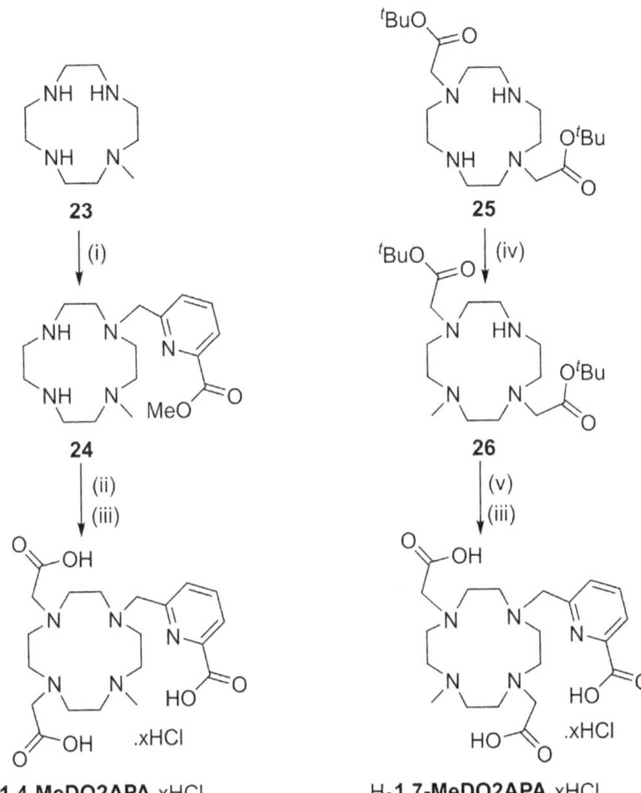


**FIGURE 10** Synthesis of $H_3$**1,4-MeDO2APA** and $H_3$**1,7-MeDO2APA**. *Conditions*: (i) Methyl 6-(chloromethyl)picolinate, $K_2CO_3$, $CH_3CN$; (ii) $BrCH_2COO^tBu$, $K_2CO_3$, $CH_3CN$; (iii) HCl, 6 M, Δ (iv) MeI, $K_2CO_3$, $CH_3CN$; (v) Methyl 6-(chloromethyl)picolinate, $K_2CO_3$, $CH_3CN$.

groups in acidic conditions, ultimately yielding the desired ligands as hydrochloride salts for subsequent gadolinium(III) complexation.

### 3.3.2 DO2A or DO2PA Derivative Properties

The series of ligands containing hydroxyl donor groups form $Gd^{3+}$ complexes with slightly lower stability than the parent $H_4$**DOTA** ligand (log $K_{GdL}=24.70$ and pGd = 18.9 [31]). The stability constants (0.1 M KCl) determined for the $DO2A^{2-}$ derivatives $H_2$**HE2-DO2A** and $H_2$**HP2-DO2A** are indeed, respectively, of log $K_{GdL}=21.10$, 22.50 and pGd = 17.15, 17.09 [91]. Furthermore, the complex with $H_2$**HE2-DO2A** has quite the same stability as the hydroxypropyl derivative $H_2$**HP2-DO2A**, suggesting that the presence of the electron-donating methyl group has a non-beneficial effect on thermodynamic stability when compared to the DO3A analogues [104]. The stability of the complex with $H_6$**DO2AP** (log $K_{GdL}=25.70$ and pGd = 17.40 [90]) is higher than that of the $H_4$**DOTA** analogue and the two previous DO2A derivatives determined using the same ionic strength (0.1 M KCl), which confirms that the substitution of carboxylate by phosphonate groups increases complex stability.

The incorporation of a water co-ligand in $Gd^{3+}$ complexes was investigated using $H_3$**1,4-MeDO2APA** and $H_3$**1,7-MeDO2APA** isomeric ligands. The stability constants of the $Gd^{3+}$ complexes with these two ligands are very similar, with log $K_{GdL}$ values of 16.98 and 16.33 for the complexes of $H_3$**1,4-DO2APA** and $H_3$**1,7-DO2APA**, respectively. These ligands were designed to lower the denticity of the **DO3APA**$^{4-}$ chelator, aiming to decrease the inertness of their $Ln^{3+}$ complexes [94,105] including gadolinium. Kinetic experiments were conducted to assess the rates of spontaneous dissociation and the involvement of the metal ion in reactions, revealing that the chelates were significantly labile. The $H_3$**1,7-MeDO2APA** ligand possesses a specific topology that leads to the coordination of the carboxylate group of the picolinate moiety at the capping position in the capped square antiprismatic coordination observed in the previous [Gd(**DO3APA**)]$^-$ complex. Consequently, the rates of acid-assisted dissociation were found to be 12.2 and 6.02 $M^{-1}s^{-1}$ for the $Gd^{3+}$ complexes of $H_3$**1,7-MeDO2APA** and $H_3$**1,4-MeDO2APA**, respectively. In terms of MRI applications, certain parameters need to be considered. These ligands are designed to provide eight-coordination to $Ln^{3+}$ ions, creating an available coordination position for a water molecule. This water molecule occupies a capping position in the twisted square antiprismatic polyhedron for $H_3$**1,4-MeDO2APA** or one of the positions in the square antiprism for $H_3$**1,7-MeDO2APA**. Luminescence measurements determined that [Gd(**1,7-MeDO2APA**)] has one water molecule ($q=1$) in its inner coordination sphere, while [Gd(**1,7-MeDO2APA**)] exhibits an equilibrium between a nine-coordinate species with one inner-sphere water molecule and an eight-coordinate species with $q=0.5$. Notably, the charge neutral [Gd(**1,7-MeDO2APA**)] complex demonstrates an exceptionally low water-exchange rate ($k_{ex}=8.8\times10^3\ s^{-1}$ at 298 K), whereas [Gd(**1,4-MeDO2APA**)] shows water exchange three orders of magnitude faster ($k_{ex}=6.6\times10^6\ s^{-1}$ at 298 K). These findings highlight the phenomenon of the labile capping bond, where a ligand occupying a capping position is hindered by its environment, rendering it inherently labile. The NMRD profiles of these complexes, which are characteristic of small $Gd^{3+}$ chelates, indicate that the relaxivity of [Gd(**1,4-MeDO2APA**)] is higher than that of [Gd(**1,7-MeDO2APA**)] across the entire range of proton Larmor frequencies. However, when examining the relaxivities ($r_1$) of aqueous solutions of the complexes at 20 MHz and 25°C, it becomes apparent that the relaxivity of [Gd(**1,7-MeDO2APA**)], measured in the pH range 5.1–7.2 (1.94 $mM^{-1}s^{-1}$), is considerably low compared to $Gd^{3+}$ complexes containing one inner-sphere water molecule like [Gd(**DOTA**)($H_2O$)]$^-$. Therefore, these complexes are far from being suitable for MRI applications.

The determination of stability constants of $Gd^{3+}$ complexes of $H_2$**DO2PA** and $H_2$**MeDO2PA** was measured by the relaxometric method that is proven to be very useful for very long equilibration times at room temperature (4 months in the case of $H_2$**DO2PA** and $H_2$**MeDO2PA**) [106]. Complexation constants measured for **DO2PA**$^{2-}$ and **MeDO2PA**$^{2-}$ for $Gd^{3+}$ (0.15 M NaCl) are log $K_{[Gd(DO2PA)]^+}=17.27$ and log $K_{[Gd(MeDO2PA)]^+}$ = 17.59, respectively. The [Gd(**DO2PA**)]$^+$ and [Gd(**MeDO2PA**)]$^+$ complexes display a fairly high kinetic inertness, as the rate constants of acid catalysed dissociation ($k_1$ = 2.5(4) × $10^{-3}$ and 8.3(4) × $10^{-4}$ $M^{-1}s^{-1}$ for [Gd(**DO2PA**)]$^+$ and [Gd(**MeDO2PA**)]$^+$, respectively) are smaller than the value reported for [Gd(**DO3A**)] ($k_1$ = 2.5 × $10^{-2}$ $M^{-1}s^{-1}$).

The calculated minimum energy conformation for the complex was found for Λ(λλλλ) stereoisomer. The luminescence lifetime measurements of $Eu^{3+}$ and $Tb^{3+}$ complexes in aqueous solution have revealed the hydration numbers $q$ of both complexes [92]. The DO2PA complexes exhibited $q = 1$, indicating the presence of one inner-sphere water molecule. In contrast, the methylated analogue $H_2$**Me-DO2PA** had $q = 0$, indicating the absence of inner-sphere water molecules; this was confirmed by the nuclear magnetic relaxation dispersion (NMRD) profiles recorded for [Gd(**MeDO2PA**)]$^+$. However, the NMRD profiles of [Gd(**DO2PA**)]$^+$ suggest the presence of both inner- and outer-sphere contributions to relaxivity. By fitting the NMRD profiles, variable temperature $^{17}$O NMR chemical shifts and transversal relaxation rates, the parameters governing relaxivity in [Gd(**DO2PA**)]$^+$ were determined. The results revealed a relatively fast water exchange rate, $k_{ex}$ = 58 × $10^6$ $s^{-1}$, for this system.

# 4 CYCLEN-BASED $Gd^{3+}$ CHELATORS WITH 4 IDENTICAL PENDANT ARMS OTHER THAN ACETATE

## 4.1 Synthesis

Many derivatives of cyclen with four identical pendant arms have been synthesized through *N*-functionalization of the macrocycle for coordinating lanthanides, including gadolinium(III). These ligands encompassed some amides, typically obtained by reacting the parent cyclen macrocycle with the corresponding α-haloacetamide in acetonitrile or ethanol. $Na_2CO_3$ or triethylamine were commonly used as bases in this process (Figure 11). Under these conditions, successful preparation of tetraamides such as **DOTAM** [107,108] and secondary amide derivatives containing methyl [109,110], ethyl [111], carboxyethyl [112] or methylphosphonate groups [113], among others [114], has been achieved. A similar procedure was employed to synthesize the dimethylacetamide analogue [115]. In some cases, the addition of catalytic amounts of KI promoted the formation of fully substituted tetraamide derivative [116]. Tetra ketone derivatives of cyclen have been obtained using a similar methodology by utilizing the appropriate bromo- or chloro-substituted ketone derivative [117]. The secondary amide derivative $H_4$**DOTAM$^{Gly}$**, which contains acetic acid groups, was obtained through the hydrolysis of the ethyl ester precursor under basic conditions [118]. Extensive researches have been conducted on the $Ln^{3+}$ complexes of the ligand $H_4$**DOTAM$^{Gly}$** and its derivatives in the context of PARACEST MRI contrast agents [119].

In the literature, cyclen-based ligands with four identical pendant arms have been reported, including the chiral $H_4$**DOTMA**, which was prepared by

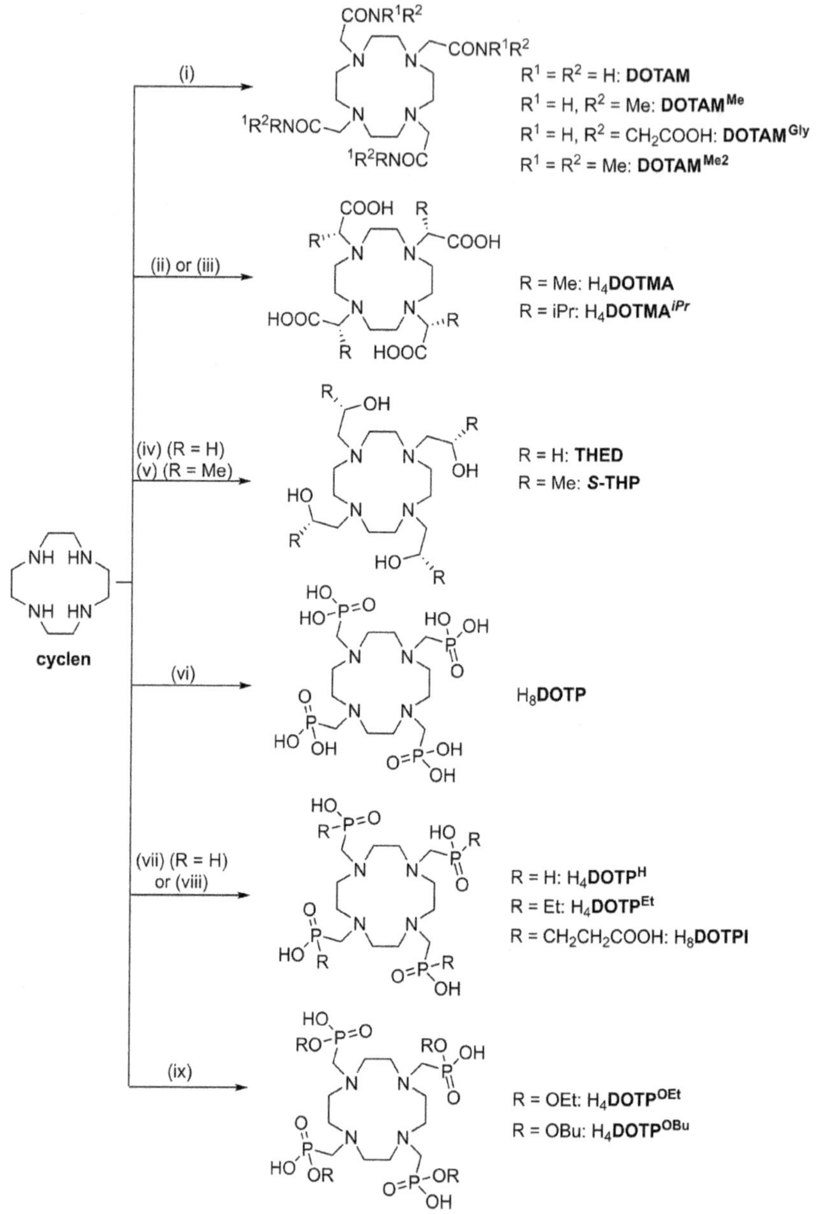

(i)

$R^1 = R^2 = H$: **DOTAM**
$R^1 = H$, $R^2 = Me$: **DOTAM$^{Me}$**
$R^1 = H$, $R^2 = CH_2COOH$: **DOTAM$^{Gly}$**
$R^1 = R^2 = Me$: **DOTAM$^{Me2}$**

(ii) or (iii)

R = Me: $H_4$**DOTMA**
R = iPr: $H_4$**DOTMA$^{iPr}$**

(iv) (R = H)
(v) (R = Me)

R = H: **THED**
R = Me: ***S*-THP**

cyclen

(vi)

$H_8$**DOTP**

(vii) (R = H) or (viii)

R = H: $H_4$**DOTP$^H$**
R = Et: $H_4$**DOTP$^{Et}$**
$R = CH_2CH_2COOH$: $H_8$**DOTPI**

(ix)

R = OEt: $H_4$**DOTP$^{OEt}$**
R = OBu: $H_4$**DOTP$^{OBu}$**

**FIGURE 11** Tetra-*N*-functionalization of cyclen. *Conditions*: (i) $XCH_2CONR^1R^2$ (X = Cl, Br, I), DIPEA or $K_2CO_3$, MeCN or DMF (then NaOH 1 M, EtOH/water for $H_4$DOTAM); (ii) (*S*)-2-chloropropanoic acid, $H_2O$, pH ~ 10; (iii) ethyl (*S*)-2-(trifluoromethylsulfonyl-oxy)propionate, $K_2CO_3$, $CHCl_3$, then NaOH; (iv) ethylene oxide, EtOH; (v) (*S*)-propylene oxide, EtOH; (vi) $(CH_2O)_n$, $H_3PO_3$, HCl; (vii) ethylphosphinic acid or (2-carboxyethyl) phosphinic acid, $(CH_2O)_n$, HCl; (viii) $(CH_2O)_n$, $RP(OEt)_2$, THF, then HCl; (ix) $(CH_2O)_n$, $(RO)_2P(O)H$, HCl or $CH_3COOH$, basic hydrolysis.

alkylation of **cyclen** with (*S*)-2-chloropropanoic acid in basic aqueous solution [120] (Figure 11). However, aqueous alkylation reactions often result in incomplete elimination of residual salts. An improved method for preparing $H_4$**DOTMA** involved using ethyl *L*-lactate triflate as the alkylating agent. The ester intermediate was subjected to saponification with sodium hydroxide and isolated as a sodium salt by crystallization of $Na_4$**DOTMA** [121].

The incorporation of hydroxyethyl pendant arms to obtain **THED** was achieved cleanly using ethylene oxide [122,123]. Similarly, 2-hydroxypropyl pendant arms were introduced using propylene oxide. The use of racemic propylene oxide led to the formation of multiple stereoisomers, whereas enantiomerically pure ***S*-THP** ligand was obtained using *S*-propylene oxide [124].

Various fully substituted cyclen derivatives carrying pendant arms containing phosphorus, such as phosphonic acids ($H_8$**DOTP**), phosphonic acid esters or phosphinic acids, have been prepared for lanthanide complexation, including $Gd^{3+}$. $H_8$**DOTP** was synthesized using a Mannich-type reaction with paraformaldehyde and phosphorous acid in relatively concentrated aqueous HCl solutions. Similarly, phosphinic acid derivatives such as $H_4$**DOTP$^H$**, $H_4$**DOTP$^R$** (R=Me, Et, Ph, Bn) [125,126] or related ligands (R=$CH_2OH$, $CH_2CH_2COOH$) [125,127] can be obtained using paraformaldehyde and hypophosphorous acid ($H_4$**DOTP$^H$**) [125]; or acidic formaldehyde with ethylphosphinic acid ($H_4$**DOTP$^{Et}$**) [126] or (2-carboxyethyl)phosphinic acid ($H_8$**DOTPI**, R=$CH_2CH_2COOH$) [127]. Alternatively, the condensation of paraformaldehyde and **cyclen** in the presence of an appropriately substituted alkyldiethoxyphosphine, $RP(OEt)_2$, yielded intermediate tetraphosphinate esters. These esters can be converted to phosphinic acid derivatives through acid hydrolysis, typically employing 6M HCl (i.e., $H_4$**DOTP$^{Bu}$** and $H_4$**DOTP$^{Bn}$**) [33]. Improved yields are generally achieved by increasing the overall concentration of the reagents, reducing the reaction temperature and selecting optimal acidic conditions.

The phosphonate ester ligands $H_4$**DOTP$^{OEt}$** and $H_4$**DOTP$^{OBu}$** were obtained using similar methods, employing paraformaldehyde and either diethylphosphite or dibutylphosphite [128,129]. These reactions typically required the use of benzene as a solvent for dissolving the dialkylphosphite. The syntheses resulted in dialkyl ester derivatives, which were then hydrolysed under basic conditions and isolated under sodium or potassium salts.

## 4.2 Properties

The thermodynamic constants of gadolinium complexes formed with some of the above described ligands are given in Table 3. [Gd(**DOTP**)]$^{5-}$ presents a remarkably high thermodynamic constant (log $K_{GdL}$=28.80) above that of [Gd(**DOTA**)]$^-$ [131]. The insertion of methyl substituents on the acetate pendants slightly impacts the stability of the resulting $Gd^{3+}$ complex (log $K_{GdL}$=23.60 and 24.70 for [Gd(**DOTMA**)] and [Gd(**DOTA**)]$^-$ respectively) despite the stronger basicity of the **DOTMA** ligand [121]. The lower stability of the lanthanide(III)–**DOTMA** complexes can be explained by the predominance of the TSAP geometry, which

**TABLE 3**
**Stability Constants of $Gd^{3+}$ Complexes with Cyclen-Based Ligands Containing Four Identical Pendant Arms**

| | *I* | log $K_{GdL}$ | pGd | Ref. |
|---|---|---|---|---|
| [Gd(**DOTA**)]$^-$ | 0.1 M KCl | 24.70 | 18.92 | [31] |
| [Gd(**DOTAM**)] | 1 M KCl | 13.12 | 12.34 | [130] |
| [Gd(**DOTAM$^{Me}$**)] | 1 M KCl | 13.54 | 12.31 | [130] |
| [Gd(**DOTAM$^{Et}$**)] | 1 M KCl | 14.83 | 13.99 | [130] |
| [Gd(**DOTAM$^{Gly}$**)] | 1 M KCl | 14.54 | 13.67 | [130] |
| [Gd(**DOTMA**)] | 0.1 M $Me_4NNO_3$ | 23.60 | 18.89 | [121] |
| [Gd(**DOTP**)]$^{5-}$ | 0.1 M $Me_4NCl$ | 28.80 | 22.37 | [131] |
| [Gd(**DOTP$^{Et}$**)]$^-$ | 0.1 M KCl | 16.50 | 13.02 | [126] |
| [Gd(**DOTP$^{MeOH}$**)]$^-$ | 0.1 M | 16.09 | 13.25 | [125] |
| [Gd(**DOTP$^{OEt}$**)]$^-$ | 0.1 M $Me_4NCl$ | 14.40 | 10.53 | [125] |

is less thermodynamically stable. In all the other cases, i.e., the substitution of the acetate coordinating arms by amide or phosphinate and phosphonate monoester groups, the stability of the resulting gadolinium(III) complexes decreases with pGd values between 10.5 and 14 (*vs* 18.92 for [Gd(**DOTA**)]$^-$), which is attributed to the lower overall basicity of these ligands [125]. However, even if the cyclen tetraamide $Gd^{3+}$ complexes are significantly less stable than the corresponding [Gd(**DOTA**)]$^-$ complex, it has been demonstrated that the protonation of the neutral amide side-chains is more difficult, which slows their rates of acid-catalysed dissociation [130]. In opposite ways, it has been proven that the proton-assisted dissociation of [Gd(**DOTP**)]$^{5-}$ and its derivatives is about one thousand times faster than that of [Gd(**DOTA**)]$^-$ [128].

To the best of our knowledge, no thermodynamic or kinetic data are available for **THED** nor ***S*-THP**.

Even if [Gd(**DOTMA**)]$^-$ mainly adopts a TSAP geometry with a higher water exchange rate, [Gd(**DOTMA**)]$^-$ presents a low-field relaxivity very close to that of [Gd(**DOTA**)]$^-$ [121].

Most of the systems described in this section have the particularity to have a relaxivity influenced by the second hydration sphere. In-depth studies aimed to determine the hydration number of $Gd^{3+}$-chelates formed with phosphorous cyclen derivatives. Various works agree that many of these systems present a $q$ value $< 1$ ([Gd(**DOTP**)]$^{5-}$, [Gd(**DOTP$^{Et}$**)]$^-$, [Gd(**DOTP$^{OEt}$**)]$^-$, [Gd(**DOTP$^{OBu}$**)]$^-$) [125,129]. However, these $Gd^{3+}$-chelates are characterized by the presence of water molecules in their second coordination sphere ($q_{ss} < 1$ for [Gd(**DOTP$^{Et}$**)]$^-$, [Gd(**DOTP$^{OEt}$**)]$^-$ and [Gd(**DOTP$^{OBu}$**)]$^-$, $q_{ss} = 1.3$ for [Gd(**DOTP$^{H}$**)]$^-$ and $q_{ss} = 3$ for ([Gd(**DOTP**)]$^{5-}$). Nevertheless, their relaxivities remain lower than those of [Gd(**DOTA**)]$^-$ ($r_1 = 2.1$–$2.8\,mM^{-1}s^{-1}$ at 20 MHz, 37°C for [Gd(**DOTP$^{R}$**)]$^-$ complexes and $r_1 = 3.5\,mM^{-1}s^{-1}$ for [Gd(**DOTP**)]$^{5-}$) [125].

The water exchange rate turns out to be particularly slow for the gadolinium(III) complexes of cyclen-tetraamide derivatives, which affects their relaxivity. For example, the $r_1$ value of [Gd(**DOTAM**$^{Gly}$)] is 2.1 mM$^{-1}$ s$^{-1}$ at neutral pH and 25°C. As for the phosphorous analogues, the relaxivity of [Gd(**DOTAM**$^{Gly}$)] is controlled by the outer-sphere hydration [132]. However, as mentioned earlier, [Gd(**DOTAM**$^{Gly}$)] and its derivatives were widely studied as PARACEST MRI contrast agents [119].

## 5 BACKBONE-MODIFIED DOTA DERIVATIVES

Desreux and colleagues proposed in the 1990s a new family of hepatobiliary GBCAs. The first examples are based on DOTA derivatives where the cyclen backbone fused with one or two cyclohexyl or 1,2,3,4-tetrahydronaphthalene rings (Figure 12) [133].

The synthesis of $H_4$**(CY)DOTA** is presented in Figure 13. It starts from the [1+1] condensation between *trans*-1,2-diaminocyclohexane and *N,N′*-dichlor oacetylethylenediamine (**27**) [134] in refluxing ACN with sodium carbonate. Noteworthy, the Richman and Atkins cyclization method was less efficient in this case. The absence of the activating tosyl group did not favour the cyclization. Many polymeric side products were observed despite the use of high dilution conditions [135]. In a second step, the amide functions of **28** were reduced using

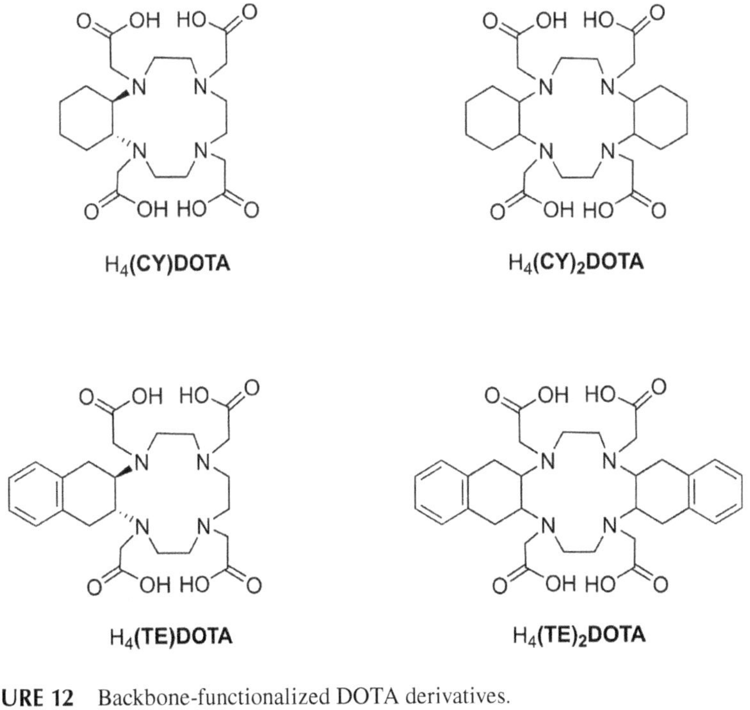

$H_4$(CY)DOTA

$H_4(CY)_2$DOTA

$H_4$(TE)DOTA

$H_4(TE)_2$DOTA

**FIGURE 12** Backbone-functionalized DOTA derivatives.

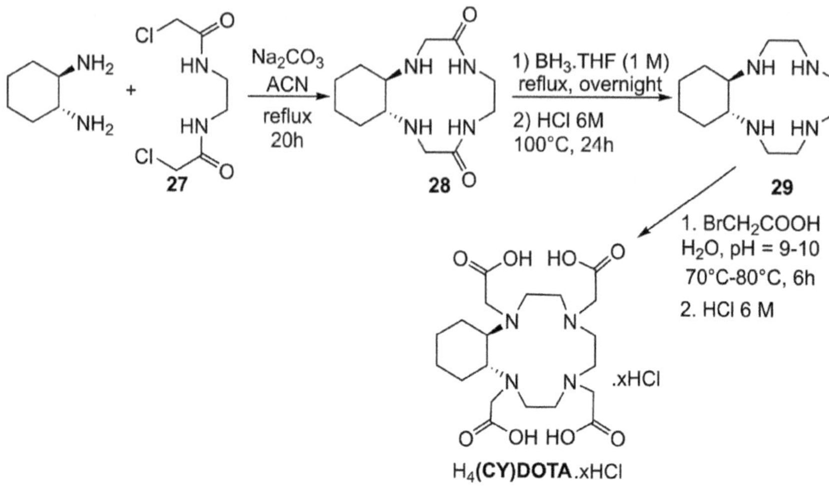

FIGURE 13 Synthesis of $H_4$**(CY)DOTA** by Desreux et al. [133].

a 1 M solution of $BH_3$ in THF and the borane complexes were hydrolysed with an acidic treatment (HCl 6 M for 24 h). Finally, the four acetate arms were introduced into macrocycle **29** through tetra-alkylation of the four secondary amine functions in the presence of bromoacetic acid in water (pH = 9–10). After the reaction, the pH was decreased to 3 with HCl 6 M. A precipitate of $H_4$**(CY)DOTA** in its hydrochloride form was obtained and purified by recrystallization (Figure 13). For the synthesis of $H_4$**(CY)$_2$DOTA**, the macrocyclization step was performed with *N,N′*-bis(chloroacetyl)-*trans*-1,2-diaminocyclohexane [136]. The anthracene analogues, $H_4$**(TE)DOTA** and $H_4$**(TE)$_2$DOTA**, were obtained using *trans*-2,3-diamino-1,2,3,4-tetrahydronaphthalene [137]. One can note that in the case of $H_4$**(CY)$_2$DOTA** and $H_4$**(TE)$_2$DOTA**, two stereoisomers can be obtained namely the *trans-anti-trans* and the *trans-syn-trans* isomers. In-depth NMR studies were performed on both stereoisomers showing that the *trans-anti-trans* ligand is extremely rigid and adopts the [3.3.3.3] conformation as observed for $H_4$**DOTA** [15,138].

The rigidity of this kind of chelators leads to $Gd^{3+}$ complexes with a particularly inertness towards acid-assisted dissociation. For example, the half-life of [Gd((**CY**)**DOTA**)]$^-$ in 1 M HCl is 50 h *vs* only 23 h for [Gd(**DOTA**)]$^-$ [138] despite the lower stability constant ($\log K_{GdL} = 23.6$ for [Gd((**CY**)$_2$**DOTA**)]$^-$ [139]).

Finally, neither the enhanced rigidity of the ligands (increase on the water exchange rate) nor the augmentation of the molecular weight (influence on $\tau_R$) do significantly ameliorate the relaxivity of these backbone-modified [Gd(**DOTA**)]$^-$ derivatives. The measured relaxivities are in the range of small monohydrated $Gd^{3+}$ chelates [138]. The principal interest in these compounds lies therefore on their very high kinetic inertness, which renders them particularly attractive for medical use and more specially interesting for diagnostic in lipophile areas such as liver due to the presence of the cyclohexyl and/or aromatic rings. This aim seems to be achieved since a high relaxivity was measured in liver for $[Gd((\mathbf{CY})_2\mathbf{DOTA})]^-$ ($r_1 = 17.2\,mM^{-1}\,s^{-1}$ at 20 MHz and 39°C) [139].

## 6 13-4-ANE AND CYCLAM DERIVATIVES

As already mentioned earlier, favoring the departure of the coordinating water molecule by inducing steric compression around it was often envisaged to increase its exchange rate and consequently the $r_1$ relaxivity. Attempts along this side have been carried out by elongating the size of the coordinating arm (propionate arm instead of acetate one) [46], as seen earlier, or by slightly increasing the size of the macrocyclic backbone (1 or maximum 2 additional carbon atoms in the chain) [47,140].

The modulation in the macrocycle's size has been studied by comparing $[Gd(\mathbf{DOTA})]^-$, $[Gd(\mathbf{TRITA})]^-$ and $[Gd(\mathbf{TETA})]^-$. $H_4$**TRITA** is based on a 13-membered cyclic platform that possesses a propyl bridge instead of an ethyl bridge compared to cyclen. $H_4$**TETA** is the direct cyclam-based equivalent of $H_4$**DOTA** (Figure 14).

The synthesis of $H_4$**TRITA** as published by Merbach and colleagues is presented in Figure 15 [140]. It started from the protection of tetraamine **30** with butanedione in acetonitrile to form a stable product of condensation **31** due to the formation of a maximum of six-membered rings as described by Handel et al. [141]. In this structure, the two methyl substituents adopt a *cis*-configuration. A ring-closure reaction was then performed between compound **31** and 1,2-dibromoethane in ACN with potassium carbonate as base to give the protected macrocycle that was then reacted with hydrazine to remove the bis-aminal bridge to obtain homocyclen **32** [142]. The latter was subjected to a tetra-alkylation reaction with ethyl bromoacetate in a $ACN/K_2CO_3$ mixture, and the ethyl ester functions were hydrolysed in acidic conditions. A final ion exchange process was performed to give the $H_4$**TRITA** ligand [140] (Figure 15).

To ensure its safe medical use, the thermodynamic stability of the resulting $Gd^{3+}$ complex was checked. A stability constant $\log K_{GdL} = 19.2$ was determined for $[Gd(\mathbf{TRITA})]^-$, which is lower than that of $[Gd(\mathbf{DOTA})]^-$ (see Table 1) [30]; at pH 7.4 and with $[L] = 10\ [Gd^{3+}] = 10 \times 10^{-6}$ M, pM values for

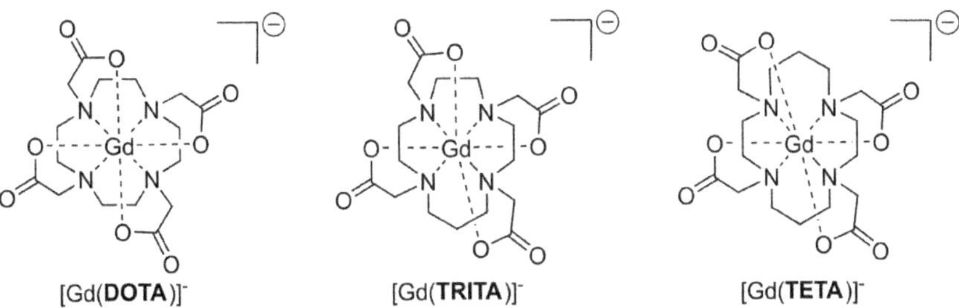


**FIGURE 14** Chemical structure of $[Gd(\mathbf{DOTA})]^-$, $[Gd(\mathbf{TRITA})]^-$ and $[Gd(\mathbf{TETA})]^-$.

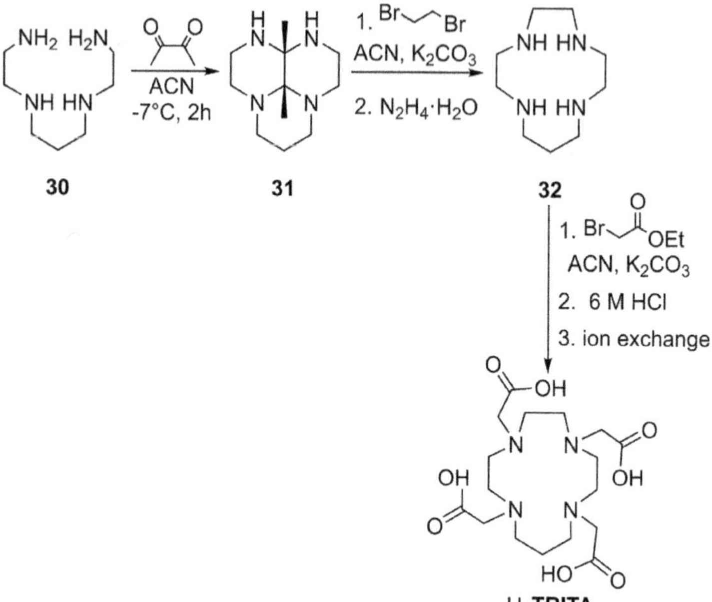

**FIGURE 15** Synthesis of $H_4$**TRITA** [140].

[Gd(**TRITA**)]$^-$ and [Gd(**DOTA**)]$^-$ are 14.6 and 19.2, respectively. Thus, the stability of [Gd(**TRITA**)]$^-$ is lower than that of [Gd(**DOTA**)]$^-$; however, it stays sufficiently reasonable for a medical use. The kinetic inertness of [Gd(**TRITA**)]$^-$ at pH 7.4 and 37°C, as well as its susceptibility to transmetallation by the endogenously most abundant $Zn^{2+}$, was later on investigated by Tóth and colleagues [143]. Going from the 12-membered ring $H_4$**DOTA** to the 13-membered ring $H_4$**TRITA**, the dissociation rate significantly increases. With an excess of $Zn^{2+}$, dissociation rate $k_0 = 5.0 \times 10^{-10}\,s^{-1}$ was measured for [Gd(**DOTA**)]$^-$ *vs* $k_0 = 4.2 \times 10^{-7}\,s^{-1}$ for [Gd(**TRITA**)]$^-$. At physiological pH (and 37°C), a dissociation half-life of 444 h was observed for [Gd(**TRITA**)]$^-$, which is remarkably high but lower than the $3.8 \times 10^5$ h observed for the very inert [Gd(**DOTA**)]$^-$. However, as expected, increasing the macrocyclic ring increases the water exchange rate. Compared to [Gd(**DOTA**)]$^-$, [Gd(**TRITA**)]$^-$ presents a higher water exchange rate which is about 65 times higher ($k_{ex} = 270 \times 10^6\,s^{-1}$) than its predecessor ($4.10 \times 10^6\,s^{-1}$), proving the efficiency of inducing steric compression around the water binding site [140].

All the observations done when going from the cyclen macrocycle to the homocyclen **32** are even more amplified when considering the cyclam 14-membered ring. In fact, both thermodynamic and kinetic stabilities of [Gd(**TETA**)]$^-$ fall down even more ($\log K_{GdL} = 15.75$ [144], $t_{1/2} = 0.23$ h for [Eu(**TETA**)]$^-$ at pH = 7.4 [143]), which illustrates that the larger sizes of the coordination sphere do not match so well with the $Gd^{3+}$ center. Finally, using the cyclam platform completely changes the ligand structure that loses its inner-sphere water molecule ($q=0$) and renders [Gd(**TETA**)]$^-$ useless for MRI application.

Efforts have been made to exploit cyclam-based chelators for the design of GBCAs. For example, the group of Tripier and colleagues proposed a reinforced cross-bridged cyclam macrocycle functionalized by two picolinate pendant arms ($H_2$**CB-TE2PA**) [145]. The associated $Ln^{3+}$ complexes are characterized by an exceptional kinetic inertness towards proton-assisted dissociation, and the X-ray structure of the [Gd(**CB-TE2PA**)]$^+$ chelate showed an exocyclic coordination resulting in a discrete structure with an eight-coordinated $Gd^{3+}$ center. However, once again, the absence of coordinating water molecule in the inner sphere was to be deplored, preventing any use in MRI [146].

To resume, increasing the size of the macrocyclic ring to induce steric compression around the water binding site is effective up to a point since a too important increase causes a too important steric compression that totally prevents the coordination of a water molecule in the inner sphere of $Gd^{3+}$. Other strategies to enhance proton relaxivity rate have been proposed, but these involve using smaller macrocycles, i.e. replacing cyclen by tacn (1,4,7-triazacyclononane), which should result in multi-hydrated $Gd^{3+}$ complexes. These results will be discussed in the coming section.

## 7 TACN-BASED GADOLINIUM(III) CHELATORS

The direct 9-membered ring of $H_4$**DOTA** was obviously studied as Gd(III) chelator, but, as it will been discussed below, it was not the best choice when dealing with tacn skeleton, and alternatives have been found such as substituting acetate by picolinate or HOPO pendants (Figure 16).

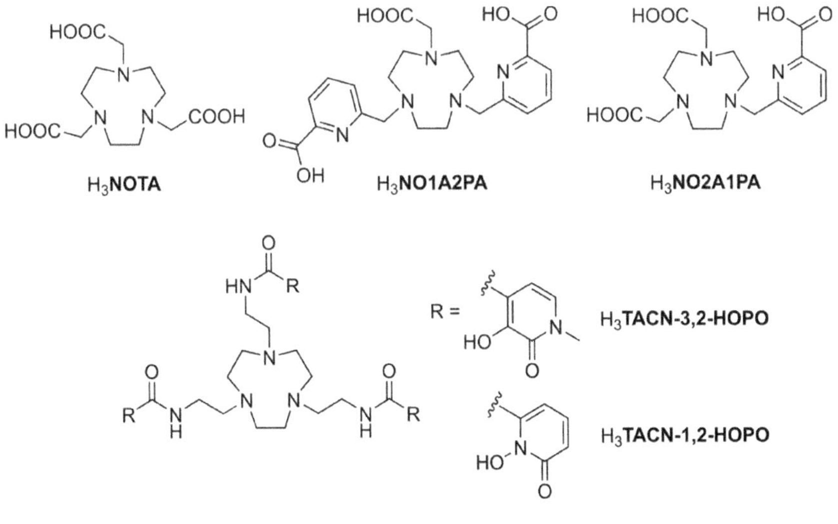

**FIGURE 16** Tacn-based ligands for the complexation of $Gd^{3+}$.

## 7.1 Synthesis

The $H_3$**NOTA** chelator is easily obtained from the direct alkylation of 1,4,7-triazacyclononane (**tacn**) with bromoacetic acid in refluxing water at basic pH (pH 10) according to a procedure developed by Takahashi et al. [147] and adapted by others [148].

In order to provide NOTA derivatives with more coordination sites than $H_3$**NOTA** and to allow the chelation of gadolinium(III), the substitution of one or two acetate arms by so many picolinate pendants has been investigated. The synthesis and studies of the tacn-containing picolinate ligands ($H_3$**NO1A2PA** and $H_3$**NO2A1PA**) were described by Mazzanti and colleagues [149,150].

The synthesis of both $H_3$**NO1A2PA** and $H_3$**NO2A1PA** started from the reaction of **tacn**.3HCl with 6-chloromethylpyridine-2-carboxylic acid ethyl ester in ACN with $K_2CO_3$ [151]. This reaction led to a mixture of mono- (11%, **33**), bi- (24%, **34**) and tri-alkylated (36%) products, which were separated by column chromatography. Note that the formation of the disubstituted derivative can be highly favoured in the following conditions: 2 equiv. of **tacn**.3HCl and 7.5 equiv. of diisopropylethylamine as base in ethanol [150]. After separation of different alkylated products, the second coordinating arm was introduced by reaction of either **33** or **34** with ethyl chloroacetate in heterogenous conditions ($K_2CO_3$/ACN). $H_3$**NO2A1PA**.xTFA was finally obtained after an ester hydrolysis reaction in acidic conditions, followed by purification on chromatography column ($C_{18}$, $H_2O$/TFA/$CH_3CN$) [149]. $H_3$**NO1A2PA** was formed through saponification with KOH (1 M) in ethanol and isolated as a hydrochloride salt after addition of a solution of HCl (Figure 17) [151].

In 1995, the group of Raymond proposed an interesting and thermodynamically stable $Gd^{3+}$ complex based on a TREN skeleton functionalized by three HOPO coordinating units. The resulting $Gd^{3+}$ chelate is bis-hydrated ($q=2$) and therefore showed a proton relaxation rate twice that of clinically used GBCAs [152]. However, this chelate and many other HOPO-TREN derivatives suffer from a two low water solubility [153]. Therefore, to maintain the thermodynamic and relaxivity properties of the previous TREN models while ensuring a better aqueous solubility and an easy synthetic approach, new HOPO ligands based on a tacn backbone were proposed [154].

The synthesis of $H_3$**TACN-3,2-HOPO** and $H_3$**TACN-1,2-HOPO** began from **tacn**.3HBr that was tri-substituted by three acetonitrile arms by action of chloroacetonitrile in the presence of an excess of triethylamine in ethanol. The nitrile functions of compound **35** were then reduced using a solution of $BH_3$·THF (1M). The hydrolysis of the borane complexes was operated in refluxing HCl solution [155] (Figure 18). Tacn-triethylamine (**36**) was then engaged in peptidic coupling reactions with the thiazolide-activated 3,2-HOPO moiety **37** or the thiazolide-activated 1,2-HOPO moiety **38** [156] in $CH_2Cl_2$ and $Et_3N$ as base. The final removal of the benzyl groups in a mixture $CH_3COOH$/HCl led to ligands $H_3$**TACN-3,2-HOPO** and $H_3$**TACN-1,2-HOPO** as hydrochloride salts (Figure 18) [154]. In this case, the associated $Gd^{3+}$ complexes were prepared

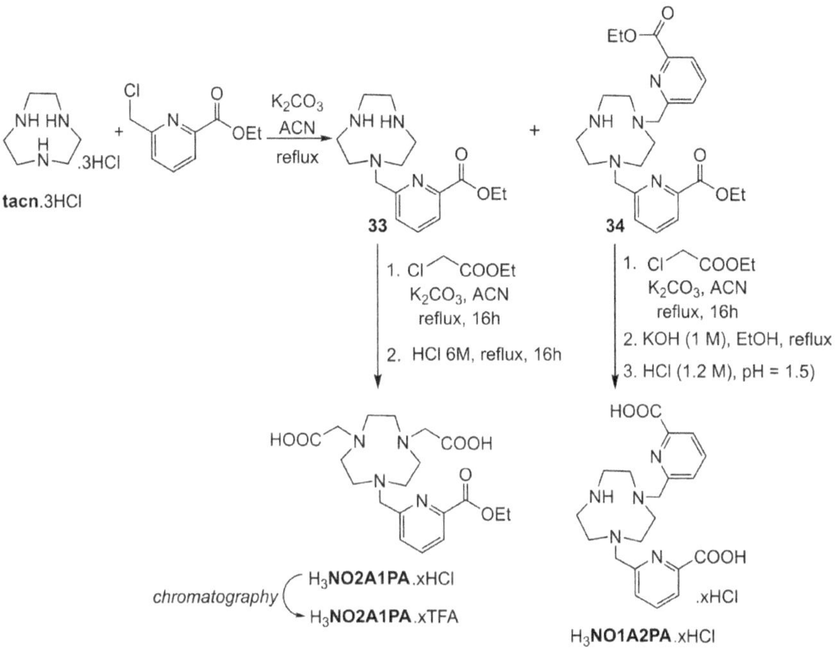

FIGURE 17 Synthesis of picolinate-tacn ligands by Mazzanti and colleagues [149,151].

in MeOH using a [Gd(**ACAC**)$_3$] salt [154]; however, the so obtained $Gd^{3+}$ chelates are about 100 times more soluble than their TREN analogues (solubility of [Gd(**L2**)] = 100 mM *vs* 1 mM for the TREN equivalent) [157].

## 7.2 Thermodynamic Stability

The use of the tacn platform rather than cyclen obviously diminishes the number of donor atoms available for the complexation of the metal center and increases the probability of having multi-aqua $Gd^{3+}$ complexes but at the detriment of the thermodynamic stability.

In an effort to make fairer comparisons, the pGd values (calculated with $[L] = 10^{-5}$ M and $[Gd] = 10^{-6}$ M at pH 7.4) will be directly discussed. For [Gd(**NOTA**)], the pGd value of 10.7 was reported, which constitutes an important decrease in stability compared to [Gd(**DOTA**)]$^-$ (pGd = 19.2 [31]), but this can easily be explained by the lower denticity of **NOTA**$^{3-}$ (CN = 6 *vs* 8 for **DOTA**$^{4-}$).

As already seen, picolinate coordinating groups are very interesting to obtain thermodynamically stable metal complexes, especially with metal with high coordination numbers. With H$_3$**NO2A1PA**, where one acetate arm has been substituted by a picolinate pendant arm, a more stable complex (compared to [Gd(**NOTA**)]) can therefore be obtained (pGd = 11.8), which is due to the presence of an additional N-donor atom [154].

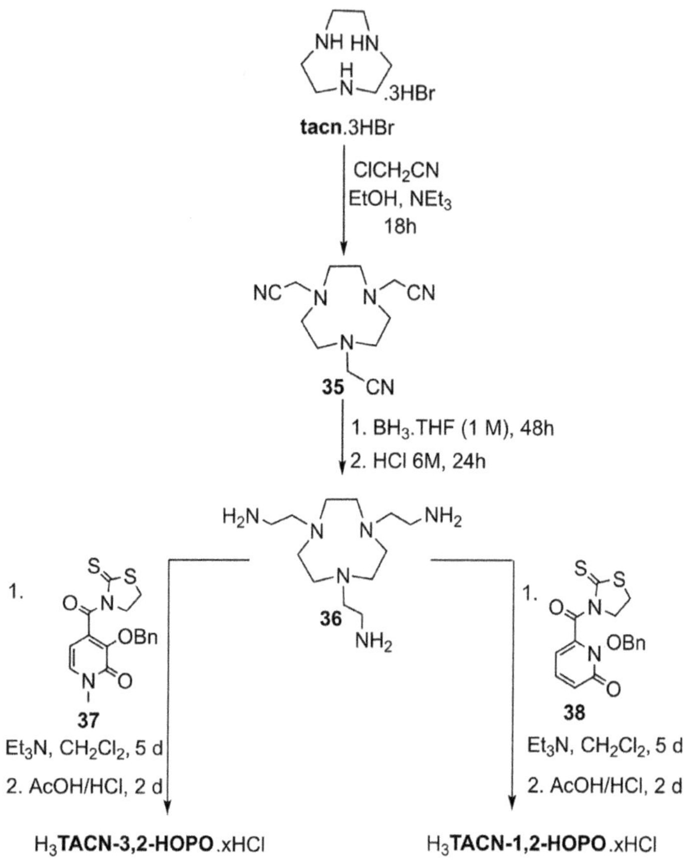

tacn.3HBr

$ClCH_2CN$
EtOH, $NEt_3$
18h

35

1. $BH_3$.THF (1 M), 48h
2. HCl 6M, 24h

36

1. 37
$Et_3N$, $CH_2Cl_2$, 5 d
2. AcOH/HCl, 2 d

1. 38
$Et_3N$, $CH_2Cl_2$, 5 d
2. AcOH/HCl, 2 d

$H_3$**TACN-3,2-HOPO**.xHCl

$H_3$**TACN-1,2-HOPO**.xHCl

**FIGURE 18** Tacn-HOPO ligands for $Gd^{3+}$ complexation by Raymond et al. [154].

## TABLE 4
## Thermodynamic Data of the $Gd^{3+}$ Complexes Formed with Tacn-Based Chelators

| | $H_3$NOTA[a] | $H_3$NO2A1PA[b] | $H_3$NO1A2PA[c] | $H_3$TACN-3,2-HOPO[d] |
|---|---|---|---|---|
| CN | 6 | 7 | 8 | 6 |
| log $K_{GdL}$ | 13.7[a] | 13.9 | 15.8 | - |
| pGd[e] | 10.7[b] | 11.8 | 13.6 | 18.7 |

[a] From Ref. [158].
[b] From Ref. [149].
[c] From Ref. [151].
[d] From Ref. [154].
[e] Calculated at pH 7.4 using $[L] = 10^{-5}$ M and $[Gd^{3+}] = 10^{-6}$ M.

[Gd(**NO1A2PA**)] is more thermodynamically stable than the previously cited analogues thanks to a higher coordination number leading to a better completion of the metal coordination sphere (CN = 7). The complexation constant, reported by Mazzanti and colleagues, reaches log $K_{GdL}$ = 15.8 and pGd = 13.6, which is still low compared to [Gd(**DOTA**)]$^-$.

As above mentioned, the tacn macrocycle was used to design HOPO-based $Gd^{3+}$ to enhance their solubility in water; however, according to structural and photophysical studies, the three amine functions of the macrocycle do not participate in the coordination of $Gd^{3+}$. Despite this, the hexacoordinated $Gd^{3+}$ complex formed with $H_3$**TACN-3,2-HOPO** presents a remarkably high pM value of 18.7, which is within the range of commercial contrast agents [154]. The high thermodynamic stability is the result of the rigidity brought by the macrocyclic tacn skeleton and the presence of hard oxygen donor atoms on the HOPO coordinating units. No data are available for [Gd(**TACN-1,2-HOPO**)]; however, one can also expect a high thermodynamic stability constant.

## 7.3 Relaxivity

As mentioned previously, the use of smaller macrocycles means reducing the denticity of the ligand and therefore gives access to multi-aqua $Gd^{3+}$ complexes those are probably characterized by very interesting relaxometric properties. These properties are given in Table 5.

Probably due to the low stability of [Gd(**NOTA**)], very few relaxometric studies were performed on this chelate. A report in the late 1980s compared the relaxation rates ($1/T_1$) for rat liver tissue of [Gd(**NOTA**)], [Gd(**DOTA**)]$^-$ and [Gd(**DTPA**)]$^{2-}$. A higher $1/T_1$ value observed for [Gd(**NOTA**)] compared to [Gd(**DOTA**)]$^-$ and [Gd(**DTPA**)]$^{2-}$ is not necessarily attributed to a greater efficiency of [Gd(**NOTA**)], but to a more important release of free $Gd^{3+}$ [159].

As said above, substitution of acetate arms by picolinate coordinating groups increases the stability of the associated $Gd^{3+}$ complex. In line with their number of coordinating atoms, the luminescent lifetimes of the $Eu^{3+}$ and $Tb^{3+}$ analogues predict hydration numbers of 2 ($q = 2$) for [Gd(**NO2A1PA**)] and 1 ($q = 1$) for [Gd(**NO1A2PA**)].

At 20 MHz, 25°C and pH 7.4, a relaxivity $r_1 = 6.6\,mM^{-1}s^{-1}$ has been measured for [Gd(**NO2A1PA**)] [151], which is in the same range as that of the bis-hydrated [Gd(**DO3A**)] chelate ($r_1 = 6.6\,mM^{-1}s^{-1}$ [39]). However, a faster water exchange rate was determined for [Gd(**NO2A1PA**)] ($k_{ex} = 60 \times 10^6 s^{-1}$) compared to [Gd(**DO3A**)] ($k_{ex} = 11 \times 10^6 s^{-1}$). Even if the relaxivities of $Gd^{3+}$ compounds are enhanced using heptadentate ligands, there is still concern about the relaxivity quenching of such compounds in BSA due to the substitution of coordinating water molecules by donor atoms of Asp or Glu proteins units [160]. In the present case of [Gd(**NO2A1PA**)], the relaxivity measured in bovine serum (4.5% BSA in $H_2O$) is more than twice than that of water ($r_{1,BSA} = 12.7\,mM^{-1}s^{-1}$), proving that the macromolecular system benefits from a slower rotational correlation time ($\tau_R$) without

**TABLE 5**
**Relaxometric Properties of $Gd^{3+}$ Tacn-Chelates at 20 MHz**

| | $q$ | $r_1^{298K}$ ($mM^{-1}s^{-1}$) | $k_{ex}$ ($10^6 s^{-1}$) |
|---|---|---|---|
| $H_3$**NO2A1PA**[a] | 2[b] | 6.6 | 60 |
| $H_3$**NO1A2PA**[c] | 1[b] | 4.59 (45 MHz) | 0.6 |
| $H_3$**TACN-3,2-HOPO**[d] | 3 | 13.1 | 500[e] |
| $H_3$**TACN-1,2-HOPO**[c] | 3 | 12.5 | 500[f] |

[a] From Ref. [151].
[b] Determined from photophysical data of the $Eu^{3+}$ and $Tb^{3+}$ complexes.
[c] From Ref. [150].
[d] From Ref. [154].
[e] Calculated from $\tau_M$.
[f] Fixed from $H_3$**TACN-3,2-HOPO**.

the substitution of one coordinated water molecule by the carboxylate groups of the lateral chains of the protein [151].

The relaxivity of [Gd(**NO1A2PA**)] was measured at 45 MHz (298 K) and the value rises to $r_1 = 4.59\,mM^{-1}s^{-1}$ which stays in the range of mono-hydrated $Gd^{3+}$ chelates with low molecular weight. From [Gd(**NO1A2PA**)], attempts were made to enhance the water exchange rate by synthesizing some analogues where the acetate pendant is replaced either by a propionate group or a phosphonate group, which increased steric compression around the coordinating water molecule [15x0]. Increased water exchange rates of $86 \times 10^6 s^{-1}$ for the propionate derivative and $34 \times 10^6 s^{-1}$ for the phosphonate one were measured, which led to relaxivities of $4.68\,mM^{-1}s^{-1}$ and $4.55\,mM^{-1}s^{-1}$ (45 MHz, 20°C), respectively. Associated with stabilities in the range than those determined for [Gd(**NO1A2PA**)], these tacn derivatives are very promising GBCAs.

Noteworthy, the nine-coordinated [Gd(**NO3PA**)] complex was also studied by the group of Mazzanti [161]. The complex obviously lacks a water molecule in the first coordination sphere of the Gd(III) center, but displays a particularly high relaxivity at low field ($5.3\,mM^{-1}s^{-1}$ at 0.02 MHz and 298 K) that can be attributed to the unusual coordination sphere containing six N donors associated with the complex $C_3$ symmetry.

The relaxivities measured at 20 MHz, 25°C and pH 7.0 for [Gd(**TACN-3,2-HOPO**)] and [Gd(**TACN-1,2-HOPO**)] are 13.1 and $12.5\,mM^{-1}s^{-1}$, respectively [154], which are the highest values measured for such low-molecular weight mononuclear $Gd^{3+}$ chelates. These high values fit well considering that [Gd(**TACN-3,2-HOPO**)] and [Gd(**TACN-1,2-HOPO**)] are tris-hydrated. The high proton relaxation rates may be also due to a very fast water exchange rate ($\tau_M = 2$ ns for [Gd(**TACN-3,2-HOPO**)]) as attested by variable temperature $^{17}O$ NMR. Later on, this kind of HOPO–tacn complexes has been covalently linked to polymeric and dendrimer carriers to go to macromolecular GBCAs with very high solubilities and proton relaxation rates [162].

The use of tacn-based ligands seems to be a good alternative to obtain multi-aqua $Gd^{3+}$ complexes with enhanced relaxivities. However, this is often in the detriment of their stability. By the way, even if thermodynamic constants were measured, no data about the kinetic dissociation of these complexes has been reported in the literature supposing a lack of inertness of this kind of $Gd^{3+}$ chelates where the sphere of coordination of the metal is not totally satisfied. However, another kind of macrocyclic backbone, the pyclen macrocycle, allows to reach very inert $Gd^{3+}$ chelates, as it will be discussed in the last part of this book chapter.

## 8 PYCLEN-BASED CHELATORS AS GBCAs

As mentioned in the introduction of this book chapter, in the challenge of designing most performant GBCAs, the modulation of the number of coordinating water molecules ($q$) was largely explored. However, if this task appears easy since it is accessible by simply reducing the number of coordination sites of the ligand, it is also risky since a direct consequence is the decrease of the thermodynamic and kinetic stability of the resulting $Gd^{3+}$ complexes. The best compromise seems to explore bis-hydrated $Gd^{3+}$ chelates ($q=2$) which obviously present enhanced relaxivity while keeping sufficient stability and inertness. In addition, the presence of two water molecules in the first coordination sphere of the $Gd^{3+}$ center is supposed to fasten the water exchange [39,163]. As discussed earlier, the passage from octacoordinated to heptacoordinated ligands was first accessed by removing one acetate coordinating arm from $H_4$**DOTA** to obtain the now well-known $H_3$**DO3A**. Many DO3A analogues followed [164]. However, with this type of structure, the relaxivity enhancement expected with the BSA is not observed and the opposite phenomenon of relaxivity quenching is being observed due to the substitution of coordinating water molecules by donor atoms of Asp or Glu protein units [160].

To avoid this undesirable effect, other heptacoordinated ligands have been investigated. From these studies, one particular ligand, offering the same coordination environment as $H_3$**DO3A** (i.e. three acetate groups and four macrocyclic nitrogen), emerged. The chelator in question, called $H_3$**PCTA** (3,6,9,15-tetraazabicyclo[9.3.1]pentadeca-1(15),11,13-triene-3,6,9,-triacetic acid), is based on a pyclen scaffold. Pyclen (3,6,9,15-tetraazabicyclo[9.3.1]pentadeca-1(15),11,13-triene), which cannot be strictly considered as an azamacrocycle, has nevertheless joined this family as an interesting alternative to the more common macrocycles previously discussed as cyclen, cyclam and tacn particularly for the complexation of $Ln^{3+}$ ions. The presence of an aromatic moieties into its macrocyclic backbone confers to pyclen interesting features for biomedical applications. Pyridine acts indeed as a chromophore to sensitize the luminescence of $Ln^{3+}$ ions for optical imaging [165,166,167], while bringing a certain rigidity that ensures fast chelation kinetics compared to the cyclen analogues [160], which is interesting for the complexation of radio-metals for nuclear medicine [165,166].

This last section of this book chapter will focus on GBCAs based on this particular macrocyclic skeleton offering very interesting properties in terms of thermodynamic stability and kinetics, as well as relaxivity properties for MRI. The archetype ligand $H_3$**PCTA** will first be described followed by different derivatives emanated from **PCTA**.

## 8.1 $H_3$PCTA: The Archetype of Pyclen-Based Chelators

### 8.1.1 Synthesis

The synthesis of $H_3$**PCTA** was described for the first time by Stetter et al. in 1981 [167]. The classical cyclization conditions from Richman and Atkins were employed [168]. Thus, in the first step, 2,6-bis(bromomethyl)pyridine (**39**) and the bis-sulfonamide sodium salt (**40**) were reacted in DMF at 100°C to form pyclen-tritosyl (**41**). The tosyl groups were removed with a concentrated solution of sulfuric acid and **pyclen** was produced with good yield. Finally, a tri-alkylation of **pyclen** with chloroacetic acid at pH 10 and 80°C followed by a treatment with HCl gives the desired $H_3$**PCTA** on its hydrochloride salt (Figure 19).

Modifications to protocols have been brought later on, specially concerning the cyclization step. It has been demonstrated that the success of this step depends on steric and conformational factors. The presence of bulky tosyl or nosyl groups,

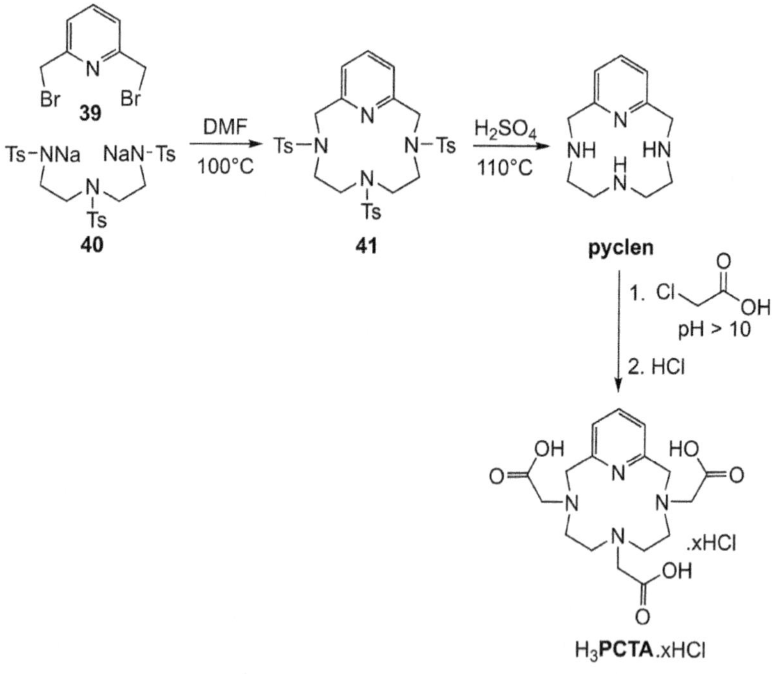

**FIGURE 19** First synthesis of $H_3$**PCTA** described by Stetter et al. [167].

in addition to playing the role of protecting/activating group, forces the triamine to adopt a conformation that may be more prone to cyclization.

In 1997, Aime and colleagues performed the crucial cyclization step from 2,6-bis(chloromethyl)-pyridine and 1,4,7-tritosyl-1,4,7-triazaheptane in heterogenous conditions with anhydrous acetonitrile as solvent and sodium carbonate as a base. After one night at reflux, the desired pyclen-tritosyl (**41**) was isolated with quantitative yield. The removal of the tosyl groups was realized in a mixture HBr/AcOH with phenol. Treatment with potassium hydroxide allowed to obtain **pyclen** with good yield. The alkylation step was performed in similar conditions than those described by Stetter et al.; **pyclen** was reacted at 80°C with chloroacetic acid in the presence of potassium hydroxide to maintain the pH of the solution about 10 [169].

In 2000, Siaugue et al. worked on the cyclization step of 1,4,7-nosyl-1,4,7-triazaheptane (**42**). The latter was reacted with 2,6-bis(bromomethyl)pyridine (**39**) in DMF and sodium carbonate to offer pyclen-trinosyl (**43**). Nosyl groups were chosen as an alternative to tosyl groups because of the smoother deprotection conditions. In fact, the removal of the nosyl groups of compound **43** was performed at room temperature with thiophenol in a $Na_2CO_3$/DMF mixture. **Pyclen** was isolated with good yield as its tri-hydrochloride salt and was subjected to alkylation with chloroacetic acid following the conditions previously described (pH > 10, 80°C). $H_3$**PCTA** was obtained pure after ion-exchange chromatography (Figure 20) [170].

More recently, the group of Picard published an alternative synthetic route where the acetate groups are installed on the triamine moiety prior to cyclization. This strategy avoids the deprotection reaction after the condensation step; however, the functionalization of the triamine necessitates many steps. One proposed

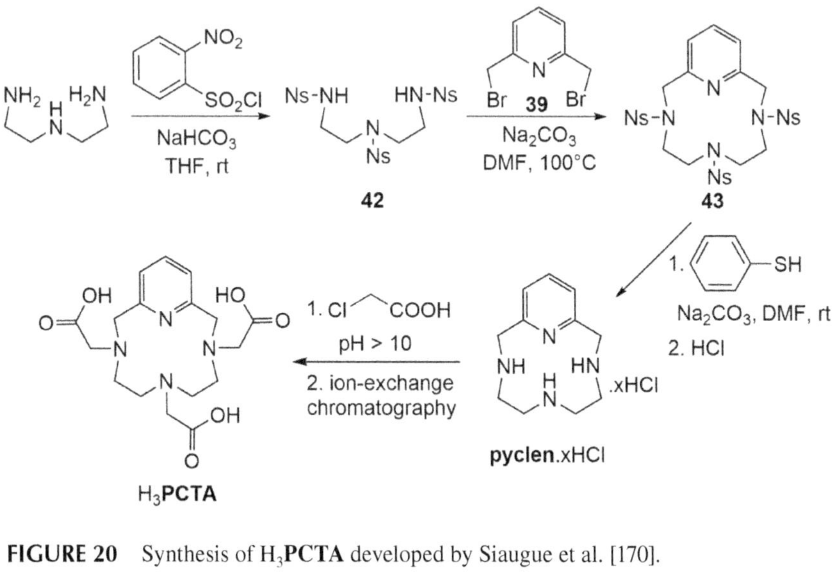

**FIGURE 20** Synthesis of $H_3$**PCTA** developed by Siaugue et al. [170].

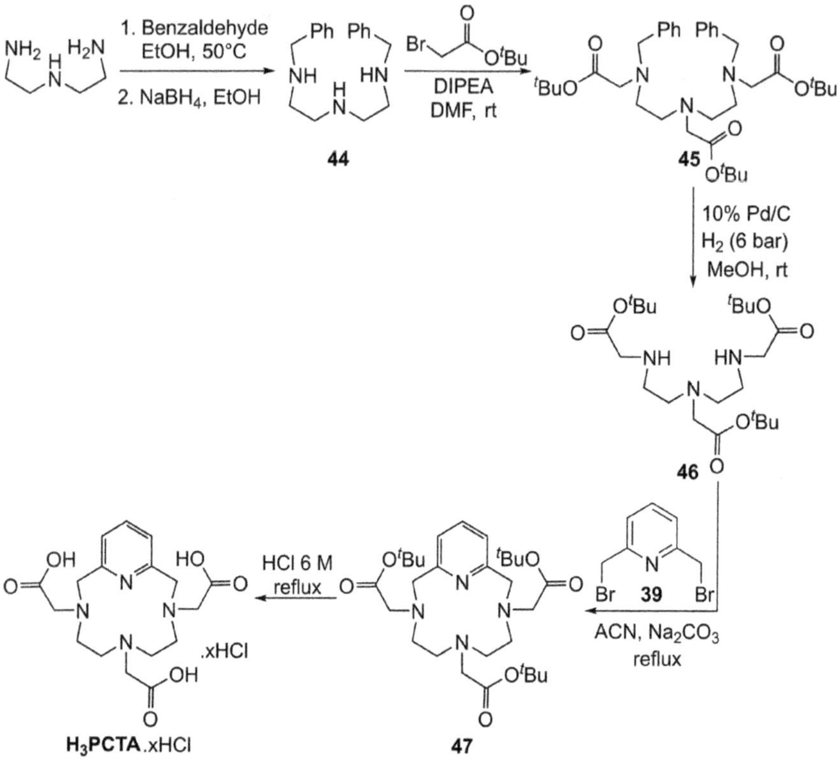

FIGURE 21 Synthesis of $H_3$**PCTA** developed by Picard et al. [171].

synthetic scheme was to start from diethyltriamine and to protect the primary amine functions using a reductive amination with benzaldehyde and sodium borohydride in a second time. The diprotected intermediate **44** was isolated and engaged in a tri-alkylation step with *tert*-butyl bromoacetate to obtain compound **45**. A (Pd/C)-catalysed hydrogenative deprotection of the *N*-benzyl-protecting group was realized in MeOH to form quantitatively the precursor of cyclization **46** (Figure 21). The condensation of **46** and 2,6-bis(bromomethyl)pyridine (**39**) in a mixture ACN/$Na_2CO_3$ offered pyclen-($^t$BuOAc)$_3$ **47**, whose *tert*-butyl ester functions were subjected to acidic hydrolysis in HCl 6M to obtain $H_3$**PCTA** on its hydrochloride salt (Figure 21) [171].

The list of the abovementioned synthetic routes is not exhaustive; slightly modified protocols have been reported. Noteworthy, other macrocyclization strategies have been attempted. For example, the [1+1] condensation based on the formation of imine bonds between 2,6-pyridinedicarbaldehyde and di-primary amines was studied using the *template effect* of metal cations. If this reaction allowed the formation of $H_3$**PCTA**[14] (the 14-membered ring $H_3$**PCTA** analogue), it never led to the formation of $H_3$**PCTA** despite many cations [Cu(II), Ag(I), Co(II), Ni(II)] were tested [164]. Other condensation reactions have been considered and will be treated later when dealing with aryl-functionalized PCTA derivatives.

### 8.1.2 Physicochemical and Relaxometric Properties of [Gd(PCTA)]

The [Gd(**PCTA**)] complex can be easily prepared by adding a stoichiometric amount of $GdCl_3$ salt to an aqueous solution of $H_3$**PCTA** at a pH of about 7 and at room temperature [169]. The formation kinetics of $Ln^{3+}$ complexes with macrocyclic ligands is usually slow; however, in the case of $H_3$**PCTA**, the formation of [Ln(**PCTA**)] complexes is about ten times faster than the analogous [Ln(**DOTA**)]$^-$ complexes [172]. The increase of complex formation rates can be explained by a preorganization of the pyclen macrocycle imposed by the pyridine ring.

The stability constant of the [Gd(**PCTA**)] complex was determined by Aime and colleagues using $H_5$**DTPA** as a competitive ligand. The equilibrium was measured by relaxometry and a value $\log K_{GdL} = 20.39$ was found [169]. More recently, Tirscó et al. also determined the thermodynamic stability of [Gd(**PCTA**)] with relaxometric methods by measuring the increase of proton relaxation times of [Gd(**PCTA**)] solutions between pH 0.7 and 4 due to its dissociation into $[Gd(H_2O)_8]^{3+}$. A value of $\log K_{GdL} = 18.28$ was obtained [173]. G. Tirscó et al. demonstrated that the acid-catalysed dissociation of [Ln(**PCTA**)] complexes is about one order of magnitude faster than for their [Ln(**DOTA**)]$^-$ analogues; however, it is still reasonable to consider the use of [Gd(**PCTA**)] as a GBCA [172].

The hydration number of [Gd(**PCTA**)], i.e. the number of inner-sphere water molecule, has been determined by correlation from the $q$ value calculated by luminescence lifetime measurements in $H_2O$ and $D_2O$ of the $Eu^{3+}$ and $Tb^{3+}$ analogues. These results confirmed that bishydrated complexes $Ln^{3+}$ complexes are formed with the heptadentate **PCTA**$^{3-}$ ligand [174]. Quite logically, an enhanced relaxivity $r_1 = 6.9\,mM^{-1}s^{-1}$ was measured for [Gd(**PCTA**)] at 20 MHz and 25°C (at pH $\approx$ 7) [39,170]. The relaxivity is mostly stable from pH 1.5 to 10 and starts to decrease above pH 10 probably because of the coordination of $OH^-$. [Gd(**PCTA**)] possesses a short average residence time of $\tau_M = 70$ ns (20 MHz, 25°C) that is in the optimal range (10–100 ns) and shorter than that of the bishydrated [Gd(**DO3A**)] ($\tau_M = 160$ ns) in the same conditions [39].

To resume, the neutral [Gd(**PCTA**)] complex is characterized by a quite high stability constants ($\log K_{GdL} = 20.39$) and a sufficient kinetic inert towards dissociation to be considered for an *in vivo* use. Added to its interesting relaxometric properties ($r = 6.9\,mM^{-1}s^{-1}$, $\tau_M = 70$ ns), [Gd(**PCTA**)] is definitively a good GBCA candidate, which is why more and more derivatives have been designed as it will be discussed in the next part of this chapter.

## 8.2 PCTA Derivatives

### 8.2.1 Synthesis of PCTA Derivatives

The in-depth investigation of new PCTA derivatives has several aims: increasing the stability of the Gd chelates while maintaining the relaxometric properties; on the contrary, increasing the $r_1$ value while keeping reasonable thermodynamic and kinetic parameters or playing on those aspects.

As for $H_4$**DOTA**, the modulation of the characteristics cited above lies on the modification of one or several of the coordination arms, which involves changing its nature or adding an additional group on the $sp^3$ carbon of the acetate pendant.

In 2003, Aime et al. synthesized a series of ligands ($H_3$**L1**–$H_5$**L3**, Figure 22) in which the acetate arm in the distal position is functionalized by a benzyl or a substituted benzyl group. In $H_5$**L3**, the two acetate groups adjacent to the pyridine moiety were substituted by phosphonate coordinating units. The objective of this work was to study the effect of the presence of an aromatic ring on the distal arm on the relaxivity as well as the importance of the type of coordinating arm [175].

The synthesis of these three ligands lies on the cyclization of 2,6-bis(chloromethyl)pyridine and a triamine (**48** or **50**) where the two primary amines are protected by tosyl groups and the secondary amine is substituted by the aryl group. After cyclization, the tosyl groups were removed in a HBr/AcOH mixture to give pyclen **53** or **54** (HBr/AcOH, phenol). Pyclen intermediate **53** was reacted either with chloroacetic acid (in $H_2O$, pH > 10) followed by HCl treatment or with $H_3PO_4/CH_2O$ in acid conditions (HCl 6 M) to yield $H_3$**L1** and $H_5$**L3**, respectively (Figure 23). The formation of $H_3$**L2** was more step-demanding because of the introduction of a functionalized aryl group. First, prior to cyclization, the phenol group of triamine **49** had to be protected to give silyl ester **50**. In a single step, methyl acetate groups were introduced on both the two secondary amines of the pyclen macrocycle **54** and the OH group of phenol by the action of bromomethyl acetate with $Ag_2CO_3$. A Kolbe-like carboxylation also took place, which led to the addition on a COOH group on the *ortho* position of phenol. A final saponification offered the ligand $H_3$**L2** (Figure 23) [176].

The substitution of the pendant acetate arms by phosphorus groups was largely studied in the case of cyclen and was also explored in the case of pyclen. In this section, we have selected three examples that are composed of a phosphinate coordinating group ($H_3$**PCP$^{Me}$2A**), one methylenephosphonic arm on the distal position ($H_4$**PCP2A**) or three methylenephosphonic units ($H_6$**PCTP**).

The synthesis of $H_6$**PCTP** is based on a Mannich-like reaction between pyclen, paraformaldehyde and phosphorous acid in acidic conditions (37% HCl) [39].

The preparation of ligand $H_3$**PCP$^{Me}$2A** is inspired from the synthesis of $H_3$**PCTA** according to the method developed by Picard and colleagues [171] meaning that the pendant coordinating groups are introduced on the triamine prior to cyclization. In the present case where the triamine is functionalized by two acetate groups on the primary amines and one diethyl methylphosphonite pendant on the central amine, four steps were necessary from *N*-(bromomethyl)phthalimide and diethyl methylphosphonite, as described in the literature [171]. Then, triamine **55** was engaged in a [1+1] condensation reaction with 2,6-bis(chloromethyl)pyridine in a mixture ACN/$Na_2CO_3$. The treatment of compound **56** with concentrated HCl (6 M) led to ligand $H_3$**PCP$^{Me}$2A** on its hydrochloride salt (Figure 24).

The synthesis of $H_4$**PCP2A** followed the route described earlier for the preparation of ligands $H_3$**L1**–$H_5$**L3**; in other words, the reaction of diethyl (aminomethyl)

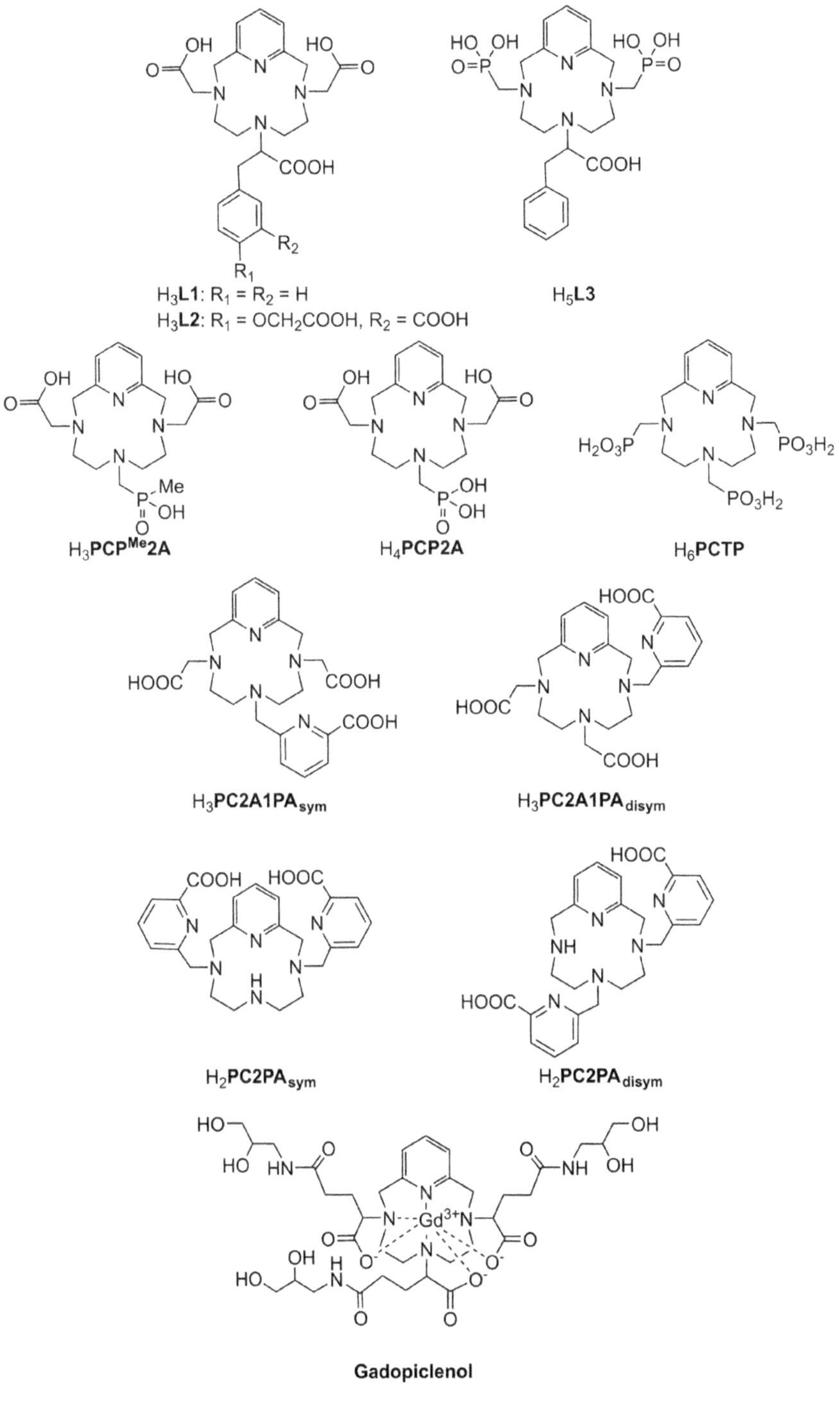

$H_3$**L1**: $R_1 = R_2 = H$

$H_3$**L2**: $R_1 = OCH_2COOH$, $R_2 = COOH$

$H_5$**L3**

$H_3$**PCP^Me^2A**

$H_4$**PCP2A**

$H_6$**PCTP**

$H_3$**PC2A1PA**$_{sym}$

$H_3$**PC2A1PA**$_{disym}$

$H_2$**PC2PA**$_{sym}$

$H_2$**PC2PA**$_{disym}$

**Gadopiclenol**

**FIGURE 22** PCTA derivatives discussed in this section.

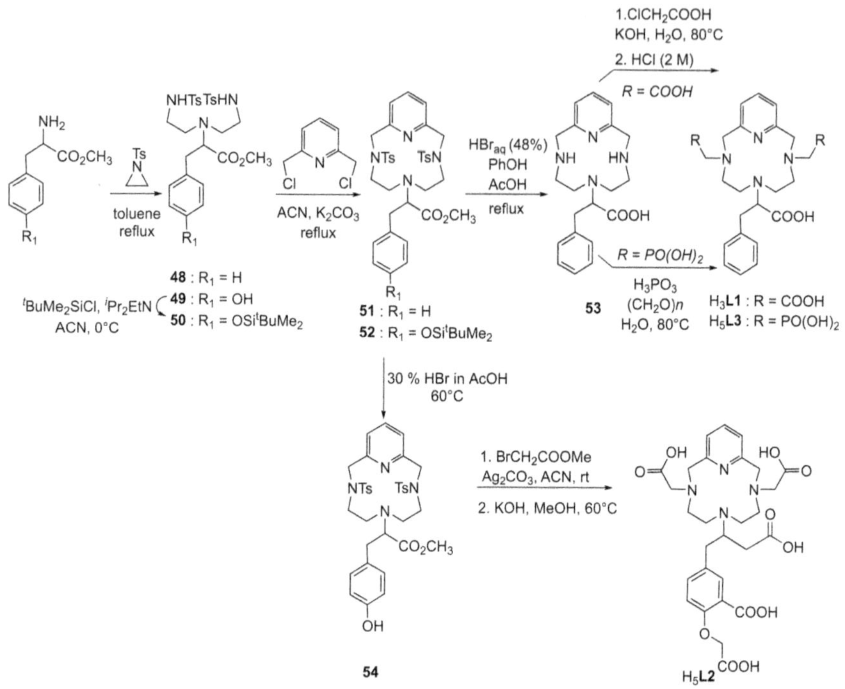

FIGURE 23 Synthesis of ligands $H_3$**L1**-$H_5$**L3** described by Aime et al. [176].

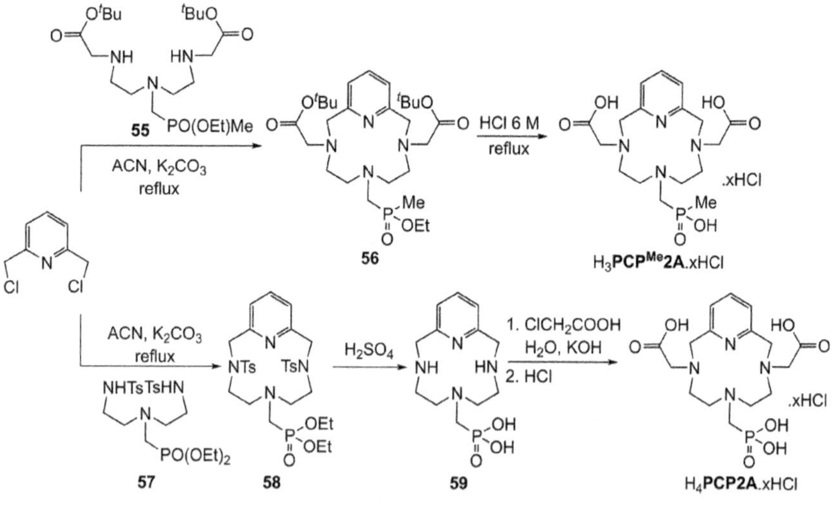

FIGURE 24 Synthesis of ligands $H_3$**PCP$^{Me}$2A** and $H_4$**PCP2A** described in the literature [171,175].

phosphonate with tosylaziridine allowed the obtention of triamine **57**, which was subjected to a cyclization reaction with 2,6-bis(chloromethyl)pyridine in heterogeneous conditions (ACN/$K_2CO_3$). The removal of the tosyl groups of pyclen **58** was performed in concentrated sulfuric acid. Finally, compound **59** was alkylated with chloroacetic acid in basic conditions and subjected to acidic treatment with HCl to give the hydrochloride salt of $H_4$**PCP2A** (Figure 24) [175].

As earlier mentioned in this chapter, picolinate bidentate moieties are really interesting for the formation of stable and inert lanthanide complexes. That is why the association of pyclen backbone plus picolinate coordinating groups was explored by the group of Tripier (Figure 25).

Two series of picolinate-pyclen chelators have been designed namely diacetate monopicolinate-pyclen ligands ($H_3$**PC2A1PA**$_{sym}$–$H_3$**PC2A1PA**$_{disym}$) and dipicolinate ligands ($H_2$**PC2PA**$_{sym}$–$H_2$**PC2PA**$_{disym}$) [177–179]. All these compounds are octadentate and each family is divided into two regioisomers. As will be discussed later, in-depth investigations proved that the arrangement of the coordinating atoms led to complexes with very different properties.

The synthesis of the symmetric ligands started from pyclen **60**, which had previously been described by Siaugue et al. and had a protecting Boc group on the distal amine [180]. The secondary amines were alkylated in ACN with potassium carbonate as base either with bromo *tert*-butylacetate or with methyl 6-(chloromethyl)picolinate to form pyclen **61** or **62**, respectively [181]. An acid hydrolysis allowed the removal of the Boc protecting group. This step permitted the obtention of $H_2$**PC2PA**$_{sym}$ as a chlorhydrate salt from intermediate **62** (Figure 25) [178]. After the hydrolysis of compound **61**, the re-esterification of the acetate groups of compound **63** was necessary before to perform a second alkylation to introduce the picolinate unit. The heating of pyclen **64** in HCl 6M gave the hydrochloride salt of the ligand $H_3$**PC2A1PA**$_{sym}$ (Figure 25) [177].

The synthesis of the dissymmetric regioisomers of $H_3$**PC2A1PA**$_{sym}$ and $H_2$**PC2PA**$_{sym}$, namely $H_3$**PC2A1PA**$_{disym}$ and $H_3$**PC2PA**$_{disym}$, required the development of a new synthetic route that allows to differentiate one of the secondary amine functions of pyclen that is adjacent to the pyridine ring. The procedure used for the acylation of **cyclam** with diethyl oxalate [182] was developed by R. Tripier and colleagues The acylation of **pyclen** was realized in MeOH in the presence of diethyl oxalate. Pyclen oxalate **65** was obtained in high yield (90%) [179]. The mono-functionalization of pyclen oxalate **65** could thus be performed with either the introduction of a coordinating arm thought a mono-alkylation or the introduction of an orthogonal protecting group. Then, the reaction of pyclen oxalate **65** with methyl 6-(bromomethyl)picolinate in ACN with a base gave pyclen **66**, while reaction of **65** with vinyl chloroformate allowed the introduction of an Alloc protecting group on the remaining secondary amine (**67**), which is resistant to both acidic and basic conditions [183]. The oxalate group of **66** and **67** was then removed in harsh acidic conditions to give **68** and **69**, respectively. Noteworthy, esterification conditions (MeOH/$H_2SO_4$) were applied to recover the ester function in pyclen **68**. From intermediate **68**, the second type of coordinating arm could be introduced, followed by an acidic hydrolysis of the ester functions to

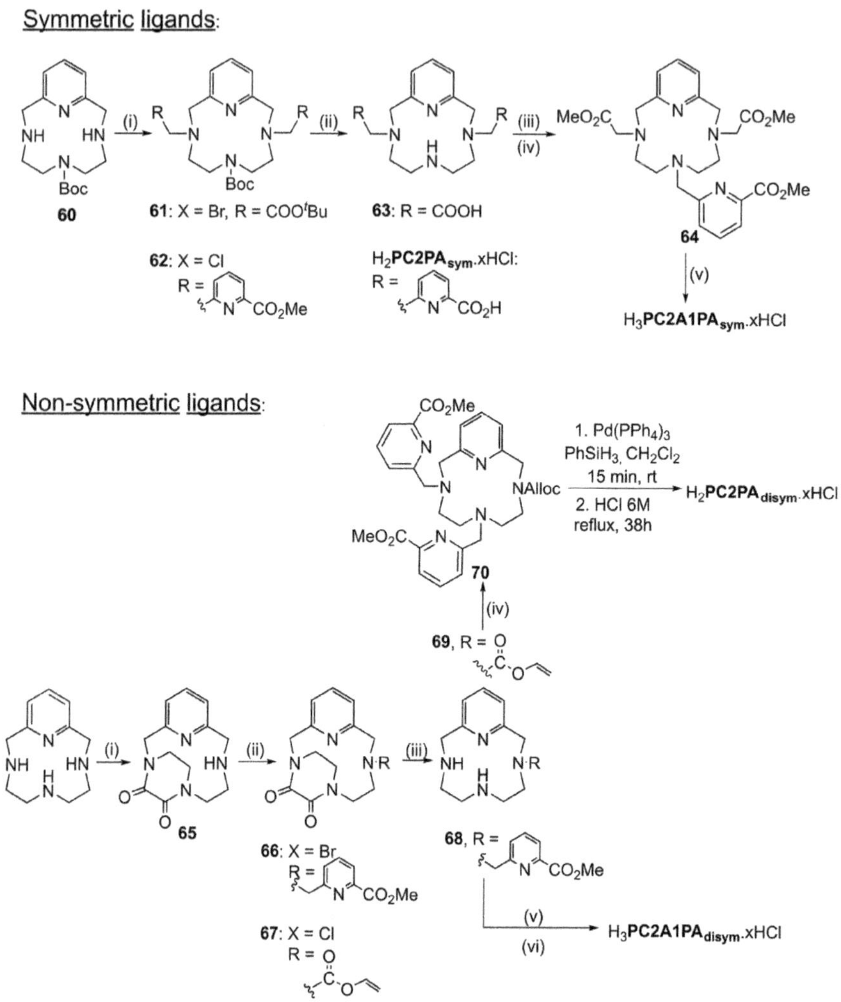

**FIGURE 25** Synthesis of picolinate-pyclen ligands by Tripier and colleagues [177–179]. *Conditions for symmetric ligands:* (i) R-$CH_2$X, ACN, $K_2CO_3$, reflux; (ii) HCl, reflux, 24h; (iii) MeOH/$H_2SO_4$, reflux, 24h; (iv) Methyl 6-(chloromethyl)picolinate, ACN, $K_2CO_3$, NaI, reflux, 7h; (v) HCl 6 M, reflux, 24h. *Conditions for non-symmetric ligands:* (i) diethyl oxalate, MeOH, rt, 24h; (ii) RX, ACN/base ($NaHCO_3$ or DIPEA), reflux; (iii) HCl 2M or $H_2$SO4/M$_e$OH, reflux; (vi) Methyl 6-(chloromethyl)picolinate, ACN, $K_2CO_3$, NaI, reflux, 24h. (v) Br$CH_2$COOMe, ACN, $K_2CO_3$, rt, 24h; (vi) HCl 3 M, 70°C, 20h.

lead to the formation of ligand $H_3$**PC2A1PA**$_{\mathbf{disym}}$.xHCl [177,178]. The two picolinate units were added to compound **69** and then the protecting Alloc group was removed with $Pd(PPh_3)_4$ and phenylsilane as a scavenger to offer the ligand $H_2$**PC2PA**$_{\mathbf{disym}}$.xHCl [178].

The performance of pyclen, and more precisely of PCTA derivatives, to form efficient GBCAs is confirmed by the recent marketing authorization of **Gadopiclenol** (Elucirem™) by the Food and Drug Administration (FDA). The synthesis of **Gadopiclenol** lies on the preparation of pyclen functionalized by three *N*-glutaryl pendant arms [184]. The synthesis of **Gadopiclenol** started from the tri-alkylation of **pyclen** with diethyl 2-bromopentanedioate **71** in classical conditions (ACN and $K_2CO_3$). The saponification of the ester functions of compound **72** was realized before its complexation to $Gd^{3+}$ to form complex **73**, which has four carboxylic acid functions available to increase the lateral chains. Thus, complex **73** was engaged in peptidic coupling reactions with isoserinol and EDCl/HOBT as coupling agents in water to obtain **Gadopiclenol** (Figure 26) [185]. Considering that both the alkylation with diethyl 2-bromopentanedioate **71** and

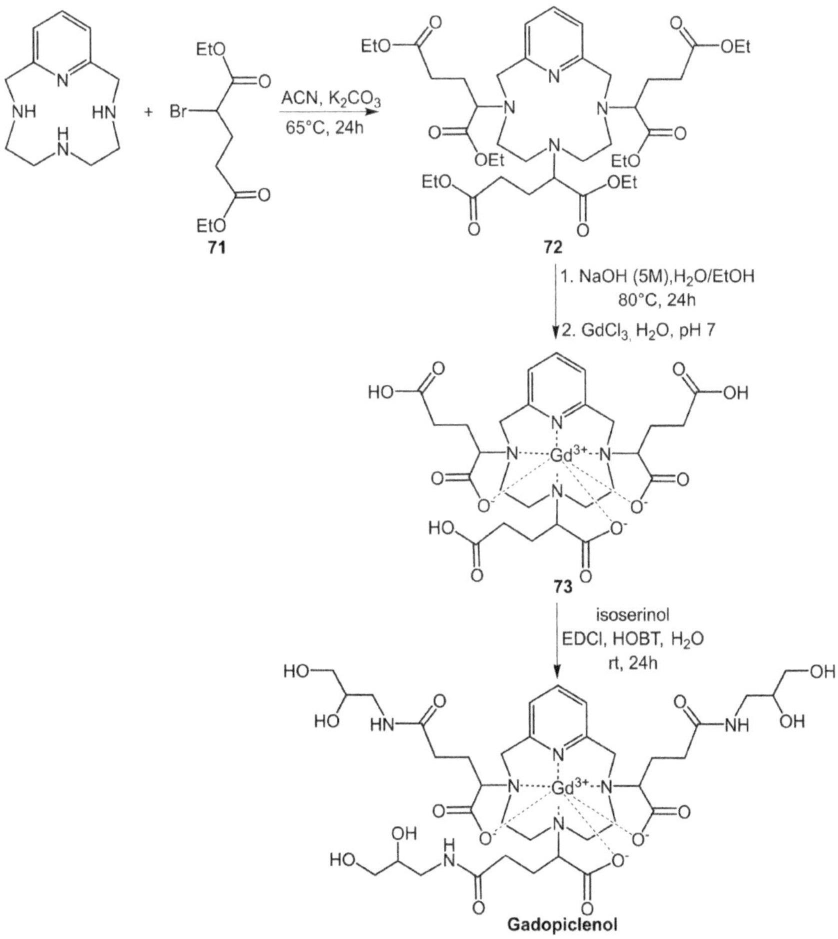

**FIGURE 26** Synthesis of **Gadopiclenol** [185,186].

the peptidic coupling with isoserinol were performed with racemic mixtures, **Gadopiclenol** is composed of 6 asymmetric carbons meaning that it exists as 64 stereoisomers [186]. However, giving that the isoserinol moieties are sufficiently distant from the metal center and are not involved in its coordination, the isomers coming from the isoserinol part can be ignored. Thus, one will only consider that **Gadopiclenol** forms 8 stereoisomers coming from the three asymmetric carbons of the glutaryl unit. The separation of the isomers by HPLC and/or UHPLC methods has been patented and it has been demonstrated that one group of isomers (containing the *RRR* and *SSS* isomers) distinguishes by its remarkable thermodynamic and relaxometric properties, which will be discussed later in this section [186].

The modification of the coordination environment (number, nature and arrangement of the coordinating atoms) has obviously an important effect on the resulting properties of the $Gd^{3+}$ chelates. The next sections aim to gather different thermodynamic, kinetic and relaxometric properties of the above-described ligands and their $Gd^{3+}$ complexes and to compare them to the archetype of the pyclen family, $H_3$**PCTA**.

### 8.2.2 Thermodynamic and Kinetic Stability

The protonation mechanism of pyclen derivatives is quite interesting. Initially, the first protonation of pyclen was assigned to take place on the nitrogen atom of the pyridine unit [163]. $^1$H NMR studies from S. Aime et al. demonstrated that the first protonation occurred on the N atom distal from the pyridine ring and that the addition of a second proton provokes a rearrangement resulting in the protonation of the two nitrogen atoms adjacent to the aromatic pyridine units [169]. This mechanism explains why the values of the first and second protonation of pyclen derivatives are very dependent on the nature of the coordinating arms. As shown in Table 6, $H_3$**PCTA** and its derivatives have quite similar basicity, except for ligand $H_4$**PCP2A**, which reveals to be particularly basic ($\Sigma \log K_i = 28.32$) probably due to the presence of a methylphosphonate group on the distal amine, which increases the first protonation constant ($\log K_1 = 12.8$) [175]. The protonation constants given for **Gadopiclenol** are in reality those reported for its precursor, ligand **72** (Figure 8) [184] and are quite similar to those reported for $H_3$**PCTA**. Noteworthy for the later, the initial protonation constants were determined with quite low concentrations of $H_3$**PCTA** [163,166] and were thus redetermined by Tirscó et al. using higher concentrations with different ionic strength [173]. The introduction of picolinate groups does also slightly increase the overall basicity of the resulting ligand ($\Sigma \log K_i \approx 23$) [177,180].

As discussed earlier, [Gd(**PCTA**)] is an interesting GBCA since it possesses a high stability constant $\log K_{GdL} = 20.39$, as reported by Aime et al. in 1997 [169], associated with a reasonable kinetic inertness ($t_{1/2} = 231$ min in $[H^+] = 1.0$ M) [173]. However, such data are always considered with caution; in fact, as also seen before, the stability constant of [Gd(**PCTA**)] has been more recently measured in other conditions and a lower value of $\log K_{GdL} = 18.28$ was determined [173].

**TABLE 6**
**Thermodynamic and Kinetic Data of the $Gd^{3+}$ Complexes Formed with Pyclen-Based Chelators**

| | $H_3PCTA$ | | $H_4PCP2A$ | $H_3PC2A1PA_{sym}$ | $H_3PC2A1PA_{disym}$ | $H_2PC2PA_{sym}$ | $H_2PC2PA_{disym}$ | Gadopiclenol |
|---|---|---|---|---|---|---|---|---|
| CN | 7 | | 7 | 8 | 8 | 8 | 8 | 7 |
| $\log K_1$ | 10.73[a] | 9.97[b,c] | 12.8[d] | 9.69[b,c] | 10.43[b,c] | 12.07[c,e] | 9.91[c,e] | 10.85[f] |
| $\log K_2$ | 7.52 | 6.73 | 6.7 | 7.63 | 6.47 | 5.18 | 8.37 | 6.86 |
| $\log K_3$ | 4.2 | 3.22 | 6.4 | 4.02 | 4.13 | 4.35 | 3.90 | 3.49 |
| $\log K_4$ | 2.4 | 1.4 | 2.42 | 2.35 | 2.71 | 2.02 | 2.77 | 1.92 |
| $\Sigma\log K_i$ | 24.86 | 21.32 | 28.32 | 23.69 | 23.74 | 23.62 | 24.95 | 23.12 |
| $\log K_{GdL}$ | 20.39[g] | 18.28[b,c] | 23.4[d,h] | 20.49[b,i] | 22.37[b,i] | 20.47[e,i] | 19.77[e,i] | 18.7[j] |
| pGd[k] | 18.05[l] | 16.62[b] | 18.9[l] | 17.74[b] | 20.25[b] | 16.76[e] | 17.20[e] | 16.09[m] |
| $t_{1/2}$ | 231 min[b,n] | | n.d | 167 min[b,n] | 542 min[b,n] | 5151 min[o] | 204 min[o] | 27 days[j,p] |

[a] From Ref. [166], 0.1 M KCl.
[b] From Ref. [173].
[c] $I = 0.15$ M NaCl.
[d] From Ref. [175].
[e] From Ref. [178].
[f] Protonation constants of ligand **72** (Figure 26) from Ref. [184], $I = 0.1$ M $NMe_4Cl$.
[g] From Ref. [172], $I = 1.0$ M KCl.
[h] Determined by relaxometry by competition with $H_5$**DTPA**.
[i] Determined by relaxometry.
[j] From Ref. [186].
[k] Calculated at pH 7.4 using $[L] = 10^{-5}$ M and $[Gd^{3+}] = 10^{-6}$ M.
[l] Calculated here from the protonation and stability constants.
[m] Calculated here from the protonation constants of ligand **72** and the stability constant of **Gadopiclenol**.
[n] $[H^+] = 0.1$ M, 25°C.
[o] $I = 1.0$ M $(Na^+ + H^+)Cl^-$, 25°C.
[p] pH = 1.2, 37°C.

Changing the nature of one or several coordination groups affects the overall basicity of the ligands, thereby having an impact on the stability constant of the resulting metal complexes. As it could be observed in Table 6, the substitution of the distal acetate arm of $H_3$**PCTA** by a methylphosphonate pendant group increases the global basicity of $H_4$**PCP2A** and gives complex [Gd(**PCP2A**)] that is characterized by log $K_{GdL}=23.4$ [175]. This result follows the trend already observed for $H_8$**DOTP** that, with equal number of coordination, forms more stable $Ln^{3+}$ complexes than $H_4$**DOTA** [131].

As seen earlier, the substitution of acetate by bidentate picolinate pendants has a beneficial effect on the stability of the corresponding $Ln^{3+}$ chelates. This is confirmed in the series of picolinate-containing ligands having either one picolinate and two acetate arms ($H_3$**PC2A1PA**$_{\mathbf{sym}}$–$H_3$**PC2A1PA**$_{\mathbf{disym}}$) or two picolinate groups ($H_2$**PC2PA**$_{\mathbf{sym}}$–$H_2$**PC2PA**$_{\mathbf{disym}}$), which are characterized by slightly higher stability constants compared to $H_3$**PCTA**; this is mainly the result of a higher denticity of the picolinate-pyclen ligands (CN=8 *vs* CN=7 for **PCTA**$^{3-}$). One can note that the comparison can be established between those $Gd^{3+}$ complexes and [Gd(**PCTA**)] since the constants were determined in the same conditions (measurement of the relaxivity of [GdL] solutions between pH 0.7 and 4) [177,180].

However, an interesting feature in this family of picolinate ligands is the importance of the arrangement of the coordinating atoms around the metal center that dictates the thermodynamic and kinetic properties, as well as the relaxometric ones, as it will be discussed later, of the resulting $Gd^{3+}$ chelates. The comparison of the regioisomers $H_3$**PC2A1PA**$_{\mathbf{sym}}$ and $H_3$**PC2A1PA**$_{\mathbf{disym}}$ shows that the non-symmetric ligand $H_3$**PC2A1PA**$_{\mathbf{disym}}$ leads to a more stable $Gd^{3+}$ complex than $H_3$**PC2A1PA**$_{\mathbf{sym}}$. The comparison of the half-lives of these two complexes reveals that [Gd(**PC2A1PA**$_{\mathbf{disym}}$)] is also more inert than [Gd(**PC2A1PA**$_{\mathbf{sym}}$)] ($t_{1/2}=542$ and 167 min, respectively) [173]. Theoretical studies have been undertaken to understand different behaviours of these two regioisomers. The calculation indicates that in [Gd(**PC2A1PA**$_{\mathbf{disym}}$)], the $Gd^{3+}$ center carries a water molecule in apical position, whereas this particular position is occupied by an oxygen atom of one acetate arm in [Gd(**PC2A1PA**$_{\mathbf{sym}}$)], which destabilizes the complex and render it more prone to dissociation [173]. A similar effect is observed when comparing [Gd(**PC2PA**$_{\mathbf{sym}}$)]$^+$ and [Gd(**PC2PA**$_{\mathbf{disym}}$)]$^+$. X-ray structures and DFT calculations point out that the coordinating water molecule is located on the labile apical position in [Gd(**PC2PA**$_{\mathbf{sym}}$)]$^+$, while for [Gd(**PC2PA**$_{\mathbf{disym}}$)]$^+$, the secondary amine of the pyclen macrocycle occupies this specific position [178]. Therefore, [Gd(**PC2PA**$_{\mathbf{sym}}$)]$^+$ is characterized by a slightly higher stability constant than its regioisomer, but also stands out by its remarkable inertness in $[H^+]=1.0$ M compared to the other picolinate-based complexes or [Gd(**PCTA**)] with a half-life of 5151 min [178].

In the previous part, we presented the synthesis of **Gadopiclenol** and mentioned that it can in reality be considered as a mixture of 8 stereoisomers. Without separation of the 8 stereoisomers, [Gd(**PCTA**)] has a stability constant log $K_{GdL}=14.9$. However, after the separation of groups of isomers, it has been established that the group containing the (*R*,*R*,*R*) and (*S*,*S*,*S*) isomers possesses a considerably higher constant which amounts to log $K_{GdL}=18.7$. This value is in the same range as the one measured for [Gd(**PCTA**)]. Also, the group of isomers (*R*,*R*,*R*) and (*S*,*S*,*S*) is notably inert since it provides a half-life of 27 days at pH 1.2 and 37°C. In comparison, $[Gd(\mathbf{DOTA})]^-$ has a half-life of 4 days under these conditions [186].

The comparison of the pGd values in Table 6, taken from the literature or calculated if not found (with $[L]=10^{-5}$ M and $[Gd]=10^{-6}$ M at pH 7.4), revealed that the octacoordinated $H_3$**PC2A1PA**$_{\mathbf{disym}}$ ligand forms the most stable $Gd^{3+}$ complex followed by the heptadentate $H_4$**PCP2A** chelator. Finally, considering their basicity, pyclen ligands with three acetate groups give the less stable $Gd^{3+}$ complexes.

### 8.2.3 Relaxometric Properties

The relaxivity data reported in Table 7 underline the enhancement of the relaxation rate ($r_1$) when going from monohydrated $Gd^{3+}$ chelates ($q=1$) to bis-hydrated ones ($q=2$). But they also illustrate the effect of the coordination environment on this parameter. In fact, one can observe that all PCTA derivatives with $q=2$ show a higher relaxivity. The functionalization of the distal acetate group of $H_3$**PCTA** with an aromatic aryl moiety confers interesting $r_1$ values at 298 K and 20 MHz to [Gd(**L1**)] and [Gd(**L2**)] ($r_1^{298K}=8.3$ and 10.5 $mM^{-1}s^{-1}$, respectively). These values are in the range expected for bis-hydrated $Gd^{3+}$ complexes of this molecular weight. As related earlier, the main drawback of GBCAs built with heptadentate ligands is the substitution of one exchanging water molecule by donor atoms contained in proteins, which provokes a phenomenon of relaxivity quenching [160]. The interaction of [Gd(**L1**)] and [Gd(**L2**)] with HSA was studied, and it showed that the aryl units introduced on the distal acetate group do not prevent the complexes from the substitution of the coordinating water molecules [176].

A relaxivity $r_1=12.5$ $mM^{-1}s^{-1}$ (20 MHz, 310 K) was reported for the (*R*,*R*,*R*) and (*S*,*S*,*S*) isomers group of **Gadopiclenol**. Noteworthy, all groups of isomers present quite similar proton relaxation properties [185,186]. The relaxivity does not significantly change with increasing fields and does not indicate any protein binding. The hydrophilic pendant arms aim to increase the global hydrodynamic size, while the presence of alcohol functions to increase the second hydration sphere. In a microreview, Botta evoked the possibility for hydrogen-bond acceptor groups, including phosphinate, phosphonate or carboxamide groups, to increase the number of water molecules in the second hydration sphere, which may contribute to enhance the relaxivity [187]. This phenomenon is probably occurring when the distal acetate group is substituted by a phosphinate or a phosphonate group, as proven by the relaxivity of the $Gd^{3+}$ complexes formed with ligands $H_3$**PCP**$^{\mathbf{Me}}$**2A** and $H_4$**PCP2A**, with measured $r_1$ values of 8.2 and 8.3 $mM^{-1}s^{-1}$, respectively (at 20 MHz, 298 K). However, the addition of phosphoryl groups can also affect the first coordination sphere. In fact, even if chelators $H_5$**L3** and $H_6$**PCTP** have a denticity of 7, they form monohydrated $Gd^{3+}$ complexes. It is postulated that the phosphonate groups increase the steric hindrance around the $Gd^{3+}$ center, reducing the number of coordinating water molecules ($q$). However,

## TABLE 7
## Relaxometric Properties of $Gd^{3+}$ Chelates at 20 MHz

| | $q$ | $r_1^{298K}$ ($mM^{-1}s^{-1}$) | $r_1^{310K}$ ($mM^{-1}s^{-1}$) | $k_{ex}$ ($10^6 s^{-1}$) |
|---|---|---|---|---|
| $H_3$**PCTA** | 2[a] | 6.9[b] | 5.1[c] | 14.3[b,d] |
| $H_3$**L1**[e] | 2 | 8.3 | n.d | 17.2[d] |
| $H_3$**L2**[e] | 2 | 10.5 | n.d | 12.7[d] |
| $H_5$**L3**[e] | 1 | 8.1 | n.d | 13.3[d] |
| $H_3$**PCP**$^{Me}$**2A**[f] | 2[g] | 8.2 | 5.8 | n.d |
| $H_4$**PCP2A**[h] | 2 | 8.3 | n.d | 16.7[d] |
| $H_6$**PCTP**[c] | 1 | 8.3 | 5.3 | 166.7[d] |
| $H_3$**PC2A1PA**$_{sym}$[i] | 1[j] | 4.74 | n.d | 1.08 |
| $H_3$**PC2A1PA**$_{disym}$[i] | 1[j] | 4.95 | n.d | 22.5 |
| $H_2$**PC2PA**$_{sym}$[k] | 1[l] | 4.46 | n.d | 87.1 |
| $H_2$**PC2PA**$_{disym}$[k] | 1[l] | 4.35 | n.d | 1.06 |
| **Gadopiclenol**[m] | 2[n] | n.d | 12.9 | n.d |

[a] From Ref. [174], determined from photophysical data of the $Eu^{3+}$ and $Tb^{3+}$ complexes.
[b] From Ref. [169].
[c] From Ref. [39].
[d] Calculated from $\tau_M$.
[e] From Ref. [39].
[f] From Ref. [171].
[g] Determined from photophysical data of the $Eu^{3+}$ and $Tb^{3+}$ complexes (tris buffer, pH 7.4, 298 K).
[h] From Ref. [175].
[i] From Ref. [173].
[j] From Ref. [174], determined from photophysical data of the $Eu^{3+}$ and $Tb^{3+}$ complexes.
[k] From Ref. [178].
[l] Determined from photophysical data of the $Eu^{3+}$ and $Tb^{3+}$ complexes in Tris/HCl 0.1 M, pH 6.8, 25°C.
[m] From Refs. [185,186].
[n] From Ref. [184], determined from photophysical data of the $Eu^{3+}$ and $Tb^{3+}$ analogues of complex 72.

the decrease of $q$ is compensated by the enhanced hydration of the second coordination sphere with shorter $Gd^{3+}$-$H_2O$ distances (3.8 Å for [Gd(**L3**)]). So finally, the [Gd(**L3**)] and [Gd(**PCTP**)] complexes present relaxivities in the same order as those of bis-hydrated $Gd^{3+}$ chelates ($r_1 = 8.1$ and $8.3\,mM^{-1}s^{-1}$, respectively, at 20 MHz and 298 K) [39,176].

The octadentate picolinate-pyclen ligands lead to $Gd^{3+}$ chelates with proton relaxation rate comparable or even slightly higher than the monohydrated [Gd(**DOTA**)]$^-$ compound. As for the thermodynamic and kinetic parameters, a different behaviour is observed between the two regioisomers of each series ($H_3$**PC2A1PA**$_{sym}$–$H_3$**PC2A1PA**$_{disym}$ and $H_2$**PC2PA**$_{sym}$–$H_2$**PC2PA**$_{disym}$) related to the differences in the structure of the $Gd^{3+}$ chelate as discussed earlier. For the regioisomers having the coordinating water molecule sitting on the more labile apical position, the exchange rate ($k_{ex}$) is considerably increased. In addition to this, a steric compression around the water binding site may also be considered. The exchange rate for [Gd(**PC2A1PA$_{disym}$**)] is about 20 times faster than for [Gd(**PC2A1PA$_{sym}$**)] [173], and 80 times faster for [Gd(**PC2PA$_{sym}$**)] than for [Gd(**PC2PA$_{disym}$**)] [178]. Interestingly, [Gd(**PC2A1PA$_{disym}$**)] and [Gd(**PC2PA$_{sym}$**)] present a $k_{ex}$ value higher than [Gd(**PCTA**)] and most of the $Gd^{3+}$ chelates presented in Table 7. Finally, one can observe that for $Gd^{3+}$ complexes (with $H_5$**L3** and $H_6$**PCTP** for example) whose relaxivities have been studied at higher temperature (310 K), a decrease of $r_1$ is observed, indicating that the relaxivity is limited by fast rotation.

## 9 CONCLUSION

Gadolinium(III) is a crucial element in medical imaging, particularly in MRI, owing to its unique magnetic properties. Chelation offers advantages beyond safety by influencing the pharmacokinetics of GBCAs. Variations in chelating agent structure allow for the customization of contrast agents, which has an impact on their physicochemical and magnetic properties. The design of $Gd^{3+}$ chelators involves expertise in organic chemistry, ligand design and coordination chemistry. This review chapter focuses on macrocyclic GBCAs, which are known for their higher thermodynamic and kinetic stability. Synthesis routes for azamacrocyclic GBCAs based on cyclen, cyclam, tacn and pyclen platforms have been highlighted. Physicochemical properties, including thermodynamics, inertness and relaxometry, play crucial roles in GBCAs' performance. This inventory delves into the intricate physicochemical factors that influence GBCA performance, providing insights into the design and synthesis of $Gd^{3+}$ chelators for optimal MRI contrast enhancement. In general, polyazacycloalkanes, including cyclen, cyclam, tacn and pyclen, provide suitable structures for gadolinium(III) coordination. Thermodynamic studies underscore the importance of stability in the complexes formed between $Gd^{3+}$ and these ligands. The results indicate that thermodynamic stability is crucial to prevent undesirable gadolinium release in the body. Additionally, kinetic inertness, evaluated through methods like transmetallation, is a crucial factor to ensure the persistence of complexes in the biological environment. The physicochemical properties of polyazacycloalkanes have a significant impact on the relaxivity of GBCAs. Relaxometric studies, particularly inner-sphere relaxivity ($r_1^{IS}$), demonstrate that ligand design directly influences the efficacy of contrast agents in terms of MRI signal enhancement. Modulation of parameters such as hydration number ($q$), exchange rate ($k_{ex}$) and rotational correlation time ($\tau_R$), has been explored to optimize these properties. Selected studies confirm the importance of designing ligands with high selectivity for $Gd^{3+}$ over other endogenous cations, such as $Ca^{2+}$, $Zn^{2+}$ and $Cu^{2+}$. This ensures the retention of gadolinium in the complex, minimizing the risks of toxicity and undesirable effects. The safety of contrast agents is crucial, especially considering potential risks in patients with kidney diseases. Customizing GBCAs by varying chelator structures offers significant advantages. The ability to adjust physicochemical and magnetic properties allows the development of contrast agents tailored to specific MRI applications. This customization can also influence distribution kinetics, circulation time and clearance of contrast agents. In conclusion, studies on the synthesis and coordination of polyazacycloalkanes for gadolinium(III) complexation have provided crucial insights for the development of effective MRI contrast agents. Even if there is no real tendency to find the best design for a good azamacrocyclic $Gd^{3+}$ chelate, these answers could offer perspectives on the design of optimal ligands, balancing stability, selectivity and efficacy for safe and effective clinical applications.

## ABBREVIATIONS AND DEFINITIONS

**ACAC** acetylacetone
**ACN** acetonitrile
**AcOH** acetic acid
**Alloc** *N*-allyloxycarbonyl
**Asp** asparagine
**BSA** bovine serum albumin
**Boc** *tert*-butoxycarbonyl
**CA** contrast agent
**CB** cross-bridge
**CN** coordination number
**cyclam** 1,4,8,11-tetraazacyclotetradecane
**cyclen** 1,4,7,10-tetraazacyclododecane
**DMF** dimethylformamide
**$H_3$DO3A** 1,4,7,10-tetraazacyclododecane-1,4,7-triacetic acid
**$H_4$DOTA** 1,4,7,10-tetraazacyclododecane-1,4,7,10-tetraacetic acid
**$H_5$DTPA** diethylenetriaminepentaacetic acid
**$Et_3N$** triethyl amine
**EDCl** 1-ethyl-3-(3-dimethylaminopropyl)carbodiimide
**FDA** U.S. Food and Drug Administration
**GBCA** gadolinium-based contrast agent
**Glu** glutamine
**HOBT** 1-hydroxybenzotriazole
**HOPO** hydroxypyridinone
**HSA** human serum albumin
**HSAB** hard and soft acid and base
**IS** inner sphere
**MRI** magnetic resonance imaging
**NMR** nuclear magnetic resonance
**NMRD** nuclear magnetic relaxation dispersion
**$H_3$NOTA** 2,2′,2″-(1,4,7-triazacyclononane-1,4,7-triyl)triacetic acid

| | |
|---|---|
| **NSF** | nephrogenic systemic fibrosis |
| **obs** | observed |
| **OS** | outer sphere |
| **PA** | pyridine 2-carboxylate |
| **PARACEST** | paramagnetic chemical exchange saturation transfer agents |
| **$H_3$PCTA** | 3,6,9,15-tetraazabicyclo[9.3.1]pentadeca-1(15),11,13-triene-3,6,9-triacetic acid |
| **PET** | positron emission tomography |
| **pyclen** | 3,6,9,15-tetraazabicyclo[9.3.1]pentadeca-1(15),11,13-triene |
| **SAP** | square-antiprismatic |
| **SPECT** | single-photon emission computed tomography |
| **SS** | second coordination sphere |
| **tacn** | 1,4,7-triazonane |
| **$H_4$TETA** | 1,4,8,11-tetraazacyclotetradecane-1,4,8,11-tetraacetic acid |
| **TFA** | trifluoroacetic acid |
| **THF** | tetrahydrofuran |
| **THP** | tetrahydropyran |
| **TREN** | 2,2′,2″-triaminotriethylamine |
| **$H_4$TRITA** | 2,2′,2″,2‴-(1,4,7,10-tetraazacyclotridecane-1,4,7,10-tetrayl) tetraacetic acid |
| **TSAP** | twisted-square-antiprismatic |

## REFERENCES


1. E. Kanal, *Magn. Reson. Imaging* **2016**, *34*, 1341–1345.
2. M. Rogosnitzky, S. Branch, *Biometals* **2016**, *29*, 365–376.
3. L. Blomqvist, G. F. Nordberg, V. M. Nurchi, J. O. Aaseth, *Biomolecules* 2022, *12*, 742.
4. J. Davies, P. Siebenhandl-Wolff, F. Tranquart, P. Jones, P. Evans, *Arch. Toxicol.* **2022**, *96*, 403–429.
5. M. Tweedle, J. Hagan, K. Kumar, S. Mantha, C. Chang, *Magn. Reson. Imaging* **1991**, *9*, 409–415.
6. S. Laurent, L. Vander Elst, R. N. Muller, *Contrast Media Mol. Imaging* **2006**, *1*, 128–137.
7. T. Frenzel, P. Lengsfeld, H. Schirmer, J. Hütter, H.-J. Weinmann, *Investig. Radiol.* **2008**, *43*, 817–828.
8. S. Aime, M. Botta, M. Fasano, E. Terreno, *Chem. Soc. Rev.* **1998**, *27*, 19–29.
9. M. Woods, Z. Kovacs, S. Zhang, A. D. Sherry, *Angew. Chem. Int. Ed.* **2003**, *42*, 5889–5892.
10. G. J. Stasiuk, N. J. Long, *Chem. Commun.* **2013**, *49*, 2732–2746.
11. M. R. Spirlet, J. Rebizant, J. F. Desreux, M.-F. Loncin, *Inorg. Chem.* **1984**, *23*, 359–363.
12. C. C. Bryden, C. N. Reilley, J. F. Desreux, *Anal. Chem.* **1981**, *53*, 1418–1425.
13. M. Magerstädt, O. A. Gansow, M. W. Brechbiel, D. Colcher, L. Baltzer, R. H. Knop, M. E. Girton, M. Naegele, *Magn. Reson. Imaging* **1986**, *3*, 808–812.
14. J. N. Caille, P. Kien, M. Allard, B. Bonnemain, *Am. J. Neuroradiol.* **1986**, *7*, 540–540.
15. J. F. Desreux, *Inorg. Chem.* **1980**, *19*, 1319–1324.

16. G. J. Stasiuk, H. Smith, M. Wylezinska-Arridge, J. L. Tremoleda, William Trigg, S. Kaur Luthra, V. Morisson Iveson, F. N. E. Gavins, N. J. Long, *Chem. Commun.* **2013**, *49*, 564–566.
17. E. E. Racow, J. J. Kreinbihl, A. G. Cosby, Y. Yang, A. Pandey, E. Boros, C. J. Johnson, *J. Am. Chem. Soc.* **2019**, *141*, 14650–14660.
18. K. Kumar, C. A. Chang, L. C. Francesconi, D. D. Dischino, M. F. Malley, J. Z. Gougoutas, M. F. Tweedle, *Inorg.Chem.* **1994**, *33*, 3567–3575.
19. X. Wang, T. Jin, V. Comblin, A. Lopez-Mut, E. Merciny, J. F. Desreux, *Inorg. Chem.* **1992**, *31*, 1095–1099.
20. P. Wedeking, K. Kumar, M. F. Tweedle, *Magn. Reson. Imaging* **1992**, *10*, 641–648.
21. D. H. Powell, O. M. N. Dhubhghaill, D. Pubanz, L. Helm, Y. S. Lebedev, W. Schlaepfer, A. E. Merbach, *J. Am. Chem. Soc.* **1996**, *118*, 9333–9346.
22. S. Hoeft, K. Roth, *Chem. Ber.* **1993**, *126*, 869–873.
23. S. Aime, L. Barbero, M. Botta, G. Ermondi, *Inorg. Chem.* **1992**, *31*, 4291–4299.
24. S. Aime, M. Botta, M. Fasano, M. P. M. Marques, C. F. G. C. Geraldes, D. Pubanz, A. E. Merbach, *Inorg. Chem.* **1997**, *36*, 2059–2068.
25. S. Zhang, Z. Kovacs, S. Burgess, S. Aime, E. Terreno, A. D. Sherry, *Chem. Eur. J.* **2001**, *7*, 288–296.
26. M. Woods, S. Aime, M. Botta, J. A. K. Howard, J. M. Moloney, M. Navet, D. Parker, M. Port, O. Rousseaux, *J. Am. Chem. Soc.* **2000**, *122*, 9781–9792.
27. F. A. Dunand, R. S. Dickins, D. Parker, A. E. Merbach, *Chem. Eur. J.* **2001**, *7*, 5160–5167.
28. F. A. Dunand, S. Aime, A. E. Merbach, *J. Am. Chem. Soc.* **2000**, *122*, 1506–1512.
29. S. Aime, A. Barge, M. Botta, A. S. De Sousa, D. Parker, *Angew. Chem. Int. Ed.* **1998**, *37*, 2673–2675.
30. E. T. Clarke, A. E. Martell, *Inorg. Chim. Acta* **1991**, *190*, 37–46.
31. W. P. Cacheris, S. K. Nickle, A. D. Sherry, *Inorg. Chem.* **1987**, *26*, 958–960.
32. E. Tóth, E. Brücher, *Inorg. Chim. Acta* **1994**, *221*, 165–167.
33. K. P. Pulukkody, T. J. Norman, D. Parker, L. Royle, C. J. Broan, *J. Chem. Soc., Perkin Trans. 2* **1993**, 605–620.
34. S. Aime, M. Botta, M. Panero, M. Grandi, F. Uggeri, *Magn. Reson. Chem.* **1991**, *29*, 923–927.
35. S. Aime, P. L. Anelli, M. Botta, F. Fedeli, M. Grandi, P. Paoli, F. Uggeri, *Inorg. Chem.* **1992**, *31*, 2422–2428.
36. R. B. Shukla, K. Kumar, R. Weber, X. Zhang, M. Tweedle, *Acta Radiologica* **1997**, *38*, 121–123.
37. K. Kumar, M. F. Tweedle, *Pure Appl. Chem.* **1993**, *65*, 515–520.
38. A. Takács, R. Napolitano, M. Purgel, A. C. Bényei, L. Zékány, E. Brücher, I. Tóth, Z. Baranyai, S. Aime, *Inorg. Chem.* **2014**, *53*, 2858–2872.
39. S. Aime, M. Botta, S. Geninatti Crich, G. Giovenzana, R. Pagliarin, M. Sisti, E. Terreno, *Magn. Reson. Chem.* **1998**, *36*, S200–S208.
40. E. Tóth, O. M. Ni Dhubhghaill, G. Besson, L. Helm, A. E. Merbach, *Magn. Reson. Chem.* **1999**, *37*, 701–708.
41. C. Feinle, P. Kunz, P. Boesiger, M. Fried, W. Schwizer, *Gut* **1999**, *44*, 106–111.
42. S. Aime, C. Cabella, S. Colombatto, S. G. Crich, E. Gianolio, F. Maggioni, *J. Magn. Reson.* **2002**, *16*, 394–406.
43. M. Woods, Z. Kovacs, A. D. Sherry, *J. Supramol. Chem.* **2002**, *2*, 1–15.
44. E. Tóth, D. Pubanz, S. Vauthey, L. Helm, A. E. Merbach, *Chem. Eur. J.* **1996**, *2*, 1607–1615.
45. S. Aime, M. Botta, M. Fasano, E. Terreno, *Acc. Chem. Res.* **1999**, *32*, 941–949.

46. Z. Jászberényi, A. Sour, E. Tóth, M. Benmelouka, A. E. Merbach, *Dalton Trans.* **2005**, 2713–2719.
47. E. Balogh, R. Tripier, P. Fousková, F. Reviriego, H. Handel, E. Tóth, *Dalton Trans.* **2007**, 3572–3581.
48. M. Försterová, I. Svobodová, P. Lubal, P. Táborský, J. Kotek, P. Hermann, I. Lukeš, *Dalton Trans.* **2007**, 535–549.
49. J. Rudovský, P. Cígler, J. Kotek, P. Hermann, P. Vojtíšek, I. Lukeš, J. A. Peters, L. Vander Elst, R. N. Muller, *Chem. Eur. J.* **2005**, *11*, 2373–2384.
50. P. Vojtíšek, P. Cígler, J. Kotek, J. Rudovský, P. Hermann, I. Lukeš, *Inorg. Chem.* **2005**, *44*, 5591–5599.
51. J. Kotek, J. Rudovský, P. Hermann, I. Lukeš, *Inorg. Chem.* **2006**, *45*, 3097–3102.
52. P. Hermann, J. Kotek, V. Kubíček, I. Lukeš, *Dalton Trans.* **2008**, 3027–3047.
53. P. Lebdušková, P. Hermann, L. Helm, E. Tóth, J. Kotek, K. Binnemans, J. Rudovský, I. Lukeš, A. E. Merbach, *Dalton Trans.* **2007**, 493–501.
54. J. Rudovský, J. Kotek, P. Hermann, I. Lukeš, V. Mainero, S. Aime, *Org. Biomol. Chem.* **2005**, *3*, 112–117.
55. F. Avecilla, J. A. Peters, C. F. G. C. Geraldes, *Eur. J. Inorg. Chem.* **2003**, 4179–4186.
56. C. F. G. C. Geraldes, A. D. Sherry, G. E. Keifer, *J. Magn. Reson.* **1992**, *97*, 290–304.
57. J. Rohovec, P. VojtÍšek, I. Lukeš, P. Hermann, J. LudvÍk, *J. Chem. Soc., Dalton Trans.* **2000**, 141–148.
58. S. Aime, A. S. Batsanov, M. Botta, R. S. Dickins, S. Faulkner, C. E. Foster, A. Harrison, J. A. K. Howard, J. M. Moloney, T. J. Norman, D. Parker, J. A. G. Williams, *J. Chem. Soc., Dalton Trans.* **1997**, 3623–3636.
59. S. Aime, A. S. Batsanov, M. Botta, J. A. K. Howard, D. Parker, K. Senanayake, J. A. G. Williams, *Inorg. Chem.* **1994**, 33, 4696–4706.
60. P. Caravan, M. T. Greenfield, X. Li, A. D. Sherry, *Inorg. Chem.* **2001**, *40*, 6580–6587.
61. S. Aime, M. Botta, E. Terreno, P. L. Anelli, F. Uggeri, *Magn. Reson. Med.* **1993**, *30*, 583–591.
62. J. Platzek, P. Blaszkiewicz, H. Gries, P. Luger, G. Michl, A. Müller-Fahrnow, B. Radüchel, D. Sülzle, *Inorg. Chem.* **1997**, *36*, 6086–6093.
63. S. Aime, S. Baroni, D. Delli Castelli, E. Brücher, I. Fábián, S. Colombo Serra, A. Fringuello Mingo, R. Napolitano, L. Lattuada, F. Tedoldi, Z. Baranyai, *Inorg. Chem.* **2018**, *57*, 5567–5574.
64. P. Caravan, J. J. Ellison, T. J. McMurry, R. B. Lauffer, *Chem. Rev.* **1999**, *99*, 2293–2352.
65. O. Axelsson, A. Olsson, Synthesis of Cyclen Derivatives. WO 2006112723 A1; CAN 145:419189, **2006**.
66. N. Raghunand, G. P. Guntle, V. Gokhale, G. S. Nichol, E. A. Mash, B. Jagadish, *J. Med. Chem.* **2010**, *53*, 6747–6757.
67. A. Dadabhoy, S. Faulkner, P. G. Sammes, *J. Chem. Soc., Perkin Trans.* **2002**, *2*, 348–357.
68. D. E. Prasuhn Jr., R. M. Yeh, A. Obenaus, M. Manchester, M. G. Finn, *Chem. Commun.* **2007**, 1269–1271.
69. M. Rami, A. Cecchi, J.-L. Montero, A. Innocenti, D. Vullo, A. Scozzafava, J.-Y. Winum, *ChemMedChem* **2008**, *3*, 1780–1788.
70. C. Li, W.-T. Wong, *Tetrahedron* **2004**, *60*, 5595–5601.
71. H.-S. Chong, S. Lim, K. E. Baidoo, D. E. Milenic, X. Ma, F. Jia, H. A. Song, M. W. Brechbiel, M. R. Lewis, *Bioorg. Med. Chem. Letters* **2008**, *18*, 5792–5795.
72. S. N. Aisyiyah Jenie, Z. Du, S. J. P. McInnes, P. Ung, B. Graham, S. E. Plush, N. H. Voelcker, *J. Mater. Chem. B* **2014**, *2*, 7694–7703.

73. L. Li, T. Olafsen, A.-L. Anderson, A. Wu, A. A. Raubitschek, J. E. Shively, *Bioconjug. Chem.* **2002**, *13*, 985–995.
74. T. Ke, Y. Feng, J. Guo, D. L. Parker, Z.-R. Lu, *Magn. Reson. Imaging* **2006**, *24*, 931–940.
75. S. J. Ratnakar, V. Alexander, *Eur. J. Inorg. Chem.* **2005**, 3918–3927.
76. D. D. Dischino, E. J. Delaney, J. E. Emswiler, G. T. Gaughan, J. S. Prasad, S. K. Srivastava, M. F. Tweedle, *Inorg. Chem.* **1991**, *30*, 1265–1269.
77. M. Le Baccon, F. Chuburu, L. Toupet, H. Handel, M. Soibinet, I. Déchamps-Olivier, J.-P. Barbier, M. Aplincourt, *New J. Chem.* **2001**, *25*, 1168–1174.
78. J. Kotek, P. Hermann, P. Vojtíšek, J. Rohovec, I. Lukeš, *Chem. Commun.* **2000**, 65, 243–266.
79. S. Develay, R. Tripier, F. Chuburu, M. Le Baccon, H. Handel, *Eur. J. Org. Chem.* **2003**, 3047–3050.
80. A. Anantanarayan, P. J. Dutton, T. M. Fyles, M. J. Pitre, *J. Org. Chem.* **1986**, 51, 752–755.
81. S. Aime, E. Terreno, D. Delli Castelli, D. L. Longo, F. Fedeli, F. Uggeri, WO2012059576.
82. Z. Zhang, Y. Liu, WO 2020/154892.
83. G. J. Lim, S. L. Chang, C. H. Byeon, H. J. Yoon, M. S. Kim, WO 2016/105172 Al; C. F. Viscardi, M. Ausonio, G. Dionisio, WO 98/56504.
84. E. Tóth, R. Király, J. Platzek, B. Radüchel, E. Brücher, *Inorg. Chim. Acta* **1996**, *249*, 191–199.
85. C. Cabella, S. Geninatti Crich, D. Corpillo, A. Barge, C. Ghirelli, E. Bruno, V. Lorusso, F. Uggeri, S. Aime, *Contrast Med. Mol. Imaging* **2006**, *1*, 23–29.
86. E. Di Gregorio, E. Gianolio, R. Stefania, G. Barutello, G. Digilio, S. Aime, *Anal. Chem.* **2013**, *85*, 5627–5631.
87. R. Shukla, M. Fernandez, R. K. Pillai, R. Ranganathan, P. C. Ratsep, X. Zhang, M. F. Tweedle, *Magn. Reson. Med.* **1996**, *35*, 928–931.
88. S.-P. Lin, J. J. Brown, *J. Magn. Reson. Imaging* **2007**, *25*, 884–899.
89. R. Esfand, D. A. Tomalia, *Drug Discov. Today* **2001**, *6*, 427–436.
90. F. K. Kálmán, Z. Baranyai, I. Tóth, I. Bányai, R. Király, E. Brücher, S. Aime, X. Sun, A.D. Sherry, Z. Kovács, *Inorg. Chem.* **2008**, *47*, 3851–3862.
91. J. Huskens, D.A. Torres, Z. Kovacs, J. P. André, C.F.G.C. Geraldes, A.D. Sherry, *Inorg. Chem.* **1997**, *36*, 1495–1503.
92. A. Rodríguez-Rodríguez, D. Esteban-Gómez, A. de Blas, T. Rodríguez-Blas, M. Fekete, M. Botta, R. Tripier, C. Platas-Iglesias, *Inorg. Chem.* **2012**, *51*, 2509–2521.
93. L. M. P. Lima, M. Beyler, R. Delgado, C. Platas-Iglesias, R. Tripier, *Inorg. Chem.* **2015**, *54*, 7045–7057.
94. A. Rodríguez-Rodríguez, M. Regueiro-Figueroa, D. Esteban-Gómez, T. Rodríguez-Blas, V. Patinec, R. Tripier, G. Tircsó, F. Carniato, M. Botta, C. Platas-Iglesias, *Chem. Eur. J.* **2016**, *23*, 1110–1117.
95. M. Suchý, R.H.E. Hudson, *Eur. J. Org. Chem.* **2008**, 4847–4865.
96. J. Vipond, M. Woods, P. Zhao, G. Tircso, J. Ren, S. G. Bott, D. Ogrin, G. E. Kiefer, Z. Kovacs, A. D. Sherry, *Inorg. Chem.* **2007**, *46*, 2584–2595.
97. L. E. Hopper, M. J. Allen, *Tetrahedron Lett.* **2014**, *55*, 5560–5561.
98. F. Chaux, F. Denat, E. Espinosa, R. Guilard, *Chem. Commun.* **2006**, 5054–5056.
99. L. M. De León-Rodríguez, Z. Kovacs, A. C. Esqueda-Oliva, A. D. Miranda-Olvera, *Tetrahedron Lett.* **2006**, *47*, 6937–6940.
100. Z. Kovács, A. D. Sherry, *Synthesis* **1997**, *1997*, 759–763.
101. J. Rohovec, R. Gyepes, I. Císařová, J. Rudovský, I. Lukeš, *Tetrahedron Lett.* **2000**, 411249–411253.

102. F. El Hajj, G. Sebki, V. Patinec, M. Marchivie, S. Triki, H. Handel, S. Yefsah, R. Tripier, C. J. Gómez-García, E. Coronado, *Inorg. Chem.* **2009**, *48*, 10416–10423.
103. B. H. Abdulwahaab, B. P. Burke, J. Domarkas, J. D. Silversides, T. J. Prior, S. J. Archibald, *J. Org. Chem.* **2016**, *81*, 890–898.
104. K. Kumar, T. Jin, X. Wang, J. F. Desreux, M. F. Tweedle, *Inorg. Chem.* **1994**, *33*, 3823–3829.
105. Z. Garda, V. Nagy, A. Rodríguez-Rodríguez, R. Pujales-Paradela, V. Patinec, G. Angelovski, É. Tóth, F. K. Kálmán, D. Esteban-Gómez, R. Tripier, C. Platas-Iglesias, G. Tircsó, *Inorg. Chem.* **2020**, *59*, 8184–8195.
106. A. Rodríguez-Rodríguez, Z. Garda, E. Ruscsák, D. Esteban-Gómez, A. de Blas, T. Rodríguez-Blas, L. M. P. Lima, M. Beyler, R. Tripier, G. Tircsó, C. Platas-Iglesias, *Dalton Trans.* **2015**, *44*, 5017–5031.
107. H. Maumela, R. D. Hancock, L. Carlton, J. H. Reibenspies, K. P. Wainwright, *J. Am. Chem. Soc.* **1995**, *117*, 6698–6707.
108. S. Amin, J. R. Morrow, C. H. Lake, M. R. Churchill, *Angew. Chem. Int. Ed.* **1994**, *33*, 773–775.
109. S. Aime, A. Barge, M. Botta, D. Parker, A. S. De Sousa, *J. Am. Chem. Soc.* **1997**, *119*, 4767–4768.
110. A. Bianchi, L. Calabi, C. Giorgi, P. Losi, P. Mariani, P. Paoli, P. Rossi, B. Valtancoli, M. Virtuani, *J. Chem. Soc. Dalton Trans.* **2000**, 697–705.
111. S. Shinoda, T. Nishimura, M. Tadokoro, H. Tsukube, *J. Org. Chem.* **2001**, *66*, 6104–6108.
112. S. Zhang, K. Wu, M. C. Biewer, A. D. Sherry, *Inorg. Chem.* **2001**, *40*, 4284–4290.
113. S. Zhang, K. Wu, A.D. Sherry, *Angew. Chem. Int. Ed.* **1999**, *38*, 3192–3194.
114. T. Mani, A. C. L. Opina, P. Zhao, O. M. Evbuomwan, N. Milburn, G. Tircsó, C. Kumas, A. D. Sherry, *J. Biol. Inorg. Chem.* **2014**, *19*, 161–171.
115. R. Kataky, K. E. Matthes, P. E. Nicholson, D. Parker, H.-J. Buschmann, *J. Chem. Soc. Perkin Trans.* **1990**, *2*, 1425–1432.
116. N. Cakić, B. Tickner, M. Zaiss, D. Esteban-Gómez, C. Platas-Iglesias, G. Angelovski, *Inorg. Chem.* **2017**, *56*, 7737–7745.
117. K. N. Green, S. Viswanathan, F. A. Rojas-Quijano, Z. Kovacs, A. D. Sherry, *Inorg. Chem.* **2011**, *50*, 1648–1655.
118. S. Aime, A. Barge, D. Delli Castelli, F. Fedeli, A. Mortillaro, F. U. Nielsen, E. Terreno, *Magn. Reson. Med.* **2002**, *47*, 639–648.
119. O. M. Evbuomwan, J. Lee, M. Woods, A. D. Sherry, *Inorg. Chem.* **2014**, *53*, 10012–10014.
120. H.G. Brittain, J. F. Desreux, *Inorg. Chem.* **1984**, *23*, 4459–4466.
121. S. Aime, M. Botta, Z. Garda, B. E. Kucera, G. Tircsó, V. G. Young, M. Woods, *Inorg. Chem.* **2011**, 50, 7955–7965.
122. S. Buoen, J. Dale, P. Groth, J. Krane, *J. Chem. Soc. Chem. Commun.* **1982**, 1172–1174.
123. B. F. Baker, H. Khalili, N. Wei, J. R. Morrow, *J. Am. Chem. Soc.* **1997**, *119*, 8749–8755.
124. K. O. A. Chin, J. R. Morrow, C. H. Lake, M. R. Churchill, *Inorg. Chem.* **1994**, *33*, 656–664.
125. Z. Kotková, G. A. Pereira, K. Djanashvili, J. Kotek, J. Rudovský, P. Hermann, L. Vander Elst, R. N. Muller, C. F. G. C. Geraldes, I. Lukeš, J. A. Peters, *Eur. J. Inorg. Chem.* **2009**, *2009*, 119–136.
126. I. Lazar, A. D. Sherry, R. Ramasamy, E. Brucher, R. Kiraly, *Inorg. Chem.* **1991**, *30*, 5016–5019.

127. J. Šimeček, P. Hermann, J. Havlíčková, E. Herdtweck, T. G. Kapp, N. Engelbogen, H. Kessler, H.-J. Wester, J. Notni, *Chem. Eur. J.*, **2013**, *19*, 7748–7757.
128. L. Burai, R. Király, I. Lázár, E. Brücher, *Eur. J. Inorg. Chem.* **2001**, *2001*, 813–820.
129. C. F. C. G. Geraldes, A. D. Sherry, I. Lázár, A. Miseta, P. Bogner, E. Berenyi, B. Sumegi, G. E. Kiefer, K. McMillan, F. Maton, R. N. Muller, *Magn. Reson. Med.* **1993**, *30*, 696–703.
130. A. Pasha, G. Tircsó, E. T. Benyó, E. Brücher, A. D. Sherry, *Eur. J. Inorg. Chem.* **2007**, *2007*, 4340–4349.
131. A. D. Sherry, J. Ren, J. Huskens, E. Brücher, É. Tóth, C. F. C. G. Geraldes, M. M. C. A. Castro, W. P. Cacheris, *Inorg. Chem.* **1996**, *35*, 4604–4612.
132. Z. Baranyai, E. Brücher, T. Iványi, R. Király, I. Lázár, L. Zékány, *Helv. Chim. Acta* **2005**, 88, 604–617.
133. J. F. Desreux, M. F. Tweedle, P. C. Ratsep, T. R. Wagler, E. R. Marinelli, US5358704 A, 1994.
134. W.C. Lin, J. A. D Fiqueira, H. G. Alt, *Monat. Chem.* **1985**, *116*, 217–221.
135. Q. N. Do, J. S. Ratnakar, Z. Kovács, G. Tircsó, F. K. Kálmán, Z. Baranyai, E. Brücher, I. Tóth, in *Contrast Agents for MRI: Experimental Methods*, Eds.: V. C. Pierre, M. J. Allen, The Royal Society of Chemistry, London, U.K. **2018**.
136. M. Saburi, S. Yoshikawa, *Bull. Chem. Soc. Jpn.* **1974**, *47*, 1184–1189.
137. T. Yano, H. Kobayashi, K. Ueno, *Bull. Chem. Soc. Jpn.* **1973**, *46*, 985–990.
138. V. Comblin, D. Gilsoul, M. Hermann, V. Humblet, V. Jacques, M. Mesbahi, C. Sauvage, J. F. Desreux, *Coord. Chem. Rev.* **1999**, *185–186*, 451–470.
139. K. Kumar, *J. Alloys Compd.* **1997**, *249*, 163–172.
140. R. Ruloff, É. Tóth, R. Scopelliti, R. Tripier, H. Handel, A. E. Merbach, *Chem. Commun.* **2002**, 2630–2631.
141. G. Hervé, H. Bernard, N. Le Bris, J.-J. Yaouanc, H. Handel, L. Toupet, *Tetrahedron Lett.* **1998**, *39*, 6861–6864.
142. G. Hervé, H. Bernard, N. Le Bris, M. Le Baccon, J.-J. Yaouanc, H. Handel, *Tetrahedron Lett.* **1999**, 40, 2517–2520.
143. E. Balogh, R. Tripier, R. Ruloff, E. Tóth, *Dalton Trans.* **2005**, 1058–1065.
144. M. F. Loncin, J. F. Desreux, E. Merciny, *Inorg. Chem.* **1986**, *25*, 2646–2648.
145. A. Rodríguez-Rodríguez, M. Regueiro-Figueroa, D. Esteban-Gómez, R. Tripier, G. Tircsó, F. K. Kálmán, A. Csaba Bényei, I. Tóth, A. de Blas, T. Rodríguez-Blas, C. Platas-Iglesias, *Inorg. Chem.* **2016**, *55*, 2227–2239.
146. A. Rodríguez-Rodríguez, D. Esteban-Gómez, R. Tripier, G. Tircsó, Z. Garda, I. Tóth, A. de Blas, T. Rodríguez-Blas, C. Platas-Iglesias, *J. Am. Chem. Soc.* **2014**, *136*, 17954–17957.
147. M. Takahashi, S. Takamoto, *Bull. Chem. Soc. Jpn* **1977**, *50*, 3413–3414.
148. V. Kubíček, Z. Böhmová, R. Ševčíková, J. Vaněk, P. Lubal, Z. Poláková, R. Michalicová, J. Kotek, P. Hermann, *Inorg. Chem.* **2018**, *57*, 3061–3072.
149. A. M. Nonat, C. Gateau, P. H. Fries, L. Helm, M. Mazzanti, *Eur. J. Inorg. Chem.* **2012**, 2049–2061.
150. A. Nonat, M. Giraud, C. Gateau, P. H. Fries, L. Helm, M. Mazzanti, *Dalton Trans.* **2009**, 8033–8046.
151. A. Nonat, C. Gateau, P. H. Fries, M. Mazzanti, *Chem. Eur. J.* **2006**, *12*, 7133–7150.
152. J. Xu, S. J. Franklin, D. W. Whisenhunt, K. N. Raymond, *J. Am. Chem. Soc.* **1995**, *117*, 7245–7246.
153. K. N. Raymond, V. C. Pierre, *Bioconjugate Chem.* **2005**, *16*, 3–8.
154. E. J. Werner, S. Avedano, M. Botta, B. P. Hay, E. G. Moore, S. Aime, K. N. Raymond, *J. Am. Chem. Soc.* **2007**, *129*, 1870–1871.

155. L. Tei, G. Baum, A. J. Blake, D. Fenske, M. Schröder, *J. Chem. Soc., Dalton Trans.* **2000**, 2793–2799.
156. J. Xu, B. Kullgren, P. W. Durbin, K. N. Raymond, *J. Med. Chem.* **1995**, *38*, 2606–2614.
157. A. Datta, K. N. Raymond, *Acc. Chem. Res.* **2009**, *42*, 938–947.
158. E. Brucher, S. Cortes, F. Chavez, A. D. Sherry, Inorg. Chem. 1991, 26, 2092–2097.
159. R. H. Knop, J. A. Frank, A. J. Dwyer, M. E. Girton, M. Naegele, M. Schrader, J. Cobb, O. Gansow, M. Maegerstadt, M. Brechbiel, L. Baltzer, J. L. Doppman, *J. Comp. Assisted Tomography* **1987**, *11*, 35–42.
160. S. Aime, E. Gianolio, E. Terreno, G. B. Giovenzana, R. Pagliarin, M. Sisti, G. Palmisano, M. Botta, M. P. Lowe, D. Parker, *J. Biol. Inorg. Chem.* **2000**, *5*, 488–497.
161. C. Gateau, M. Mazzanti, J. Pécaut, F. A. Dunand, L. Helm, *Dalton Trans.* **2003**, 2428–2433.
162. P. J. Klemm, W. C. Floyd, D. E. Smiles, J. M. J. Fréchet, K. N. Raymond, *Contrast Media Mol. Imaging* **2012**, *7*, 95–99.
163. R. Delgado, S. Quintino, M. Teixeira, A. Zhang, *J. Chem. Soc., Dalton Trans.* **1997**, 55–63.
164. S. M. Nelson, *Pure Appl. Chem.* **1980**, *52*, 2461–2476.
165. J.-M. Siaugue, F. Segat-Dioury, A. Favre-Réguillon, V. Wintgens, C. Madic, J. Foos, A. Guy, *J. Photochem. Photobiol. A* **2003**, *156*, 23–29.
166. J.-M. Siaugue, A. Favre-Réguillon, F. Dioury, G. Plancque, J. Foos, Charles Madic, C. Moulin, A. Guy, *Eur. J. Inorg. Chem.* **2003**, 2834–2838.
167. H. Stetter, W. Frank, R. Mertens, *Tetrahedron* **1981**, 37, 767–772.
168. J. E. Richman, T. J. Atkins, *J. Am. Chem. Soc.* **1974**, *96*, 2268–2270.
169. S. Aime, M. Botta, S. Geninatti Crich, G. B. Giovenzana, G. Jommi, R. Pagliarin, M. Sisti, *Inorg. Chem.* **1997**, *36*, 2992–3000.
170. J.-M. Siaugue, F. Segat-Dioury, A. Favre-Réguillon, C. Madic, J. Foos, A. Guy, *Tetrahedron Lett.* **2000**, *41*, 7443–7446.
171. M. Enel, N. Leygue, N. Saffon, C. Galaup, C. Picard, *Eur. J. Org. Chem.* **2018**, 1765–1773.
172. G. Tircsó, Zoltán Kovács, A. D. Sherry, *Inorg. Chem.* **2006**, *45*, 9269–9280.
173. M. Le Fur, E. Molnár, M. Beyler, F. K. Kálmán, O. Fougère, D. Esteban-Gómez, O. Rousseaux, R. Tripier, G. Tircsó, C. Platas-Iglesias, *Chem. Eur. J.* **2018**, *24*, 3127–3131.
174. W. D. Kim, G. E. Kiefer, F. Maton, K. McMillan, R. N. Muller, A. D. Sherry, *Inorg. Chem.* **1995**, *34*, 2233–2243.
175. S. Aime, M. Botta, L. Frullano, S. Geninatti Crich, G. Giovenzana, R. Pagliarin, G. Palmisano, F. Riccardi Sirtori, M. Sisti, *J. Med. Chem.* **2000**, *43*, 4017–4024.
176. S. Aime, E. Gianolio, D. Corpillo, C. Cavallotti, G. Palmisano, M. Sisti, G. B. Giovenzana, R. Pagliarin, *Helvetica Chim. Acta* **2003**, *86*, 615–632.
177. G. Nizou, C. Favaretto, F. Borgna, P. V. Grundler, N. Saffon-Merceron, C. Platas-Iglesias, O. Fougère, O. Rousseaux, N. P. van der Meulen, C. Müller, M. Beyler, R. Tripier, *Inorg. Chem.* **2020**, *59*, 11736–11748.
178. G. Nizou, E. Molnár, N. Hamon, F. K. Kálmán, O. Fougère, O. Rousseaux, D. Esteban-Gòmez, C. Platas-Iglesias, M. Beyler, G. Tircsó, R. Tripier, *Inorg. Chem.* **2021**, *60*, 2390–2405.
179. M. Le Fur, M. Beyler, E. Molnár, O. Fougère, D. Esteban-Gómez, G. Tircsó, C. Platas-Iglesias, N. Lepareur, O. Rousseaux, R. Tripier, *Chem. Commun.* **2017**, *53*, 9534–9537.
180. J.-M. Siaugue, F. Segat-Dioury, I. Sylvestre, A. Favre-Réguillon, J. Foos, C. Madic, A. Guy, *Tetrahedron* **2001**, *57*, 4713–4718.

181. M. Mato-Iglesias, A. Roca-Sabio, Z. Palinkas, D. Esteban-Gomez, C. Platas-Iglesias, E. Toth, A. de Blas, T. Rodriguez-Blas, *Inorg. Chem.* **2008**, *47*, 7840–7851.
182. F. Bellouard, F. Chuburu, N. Kervarec, L. Toupet, S. Triki, Y. Le Mest, H. Handel, *J. Chem. Soc., Perkin Trans. 1* **1999**, 3499–3505.
183. Z. Garda, E. Molnár, N. Hamon, J. L. Barriada, D. Esteban-Gómez, B. Váradi, V. Nagy, K. Pota, F. K. Kálmán, I. Tóth, N. Lihi, C. Platas-Iglesias, E. Tóth, R. Tripier, G. Tircsó, *Inorg. Chem.* **2021**, *60*, 1133–1148.
184. J. Moreau, J.-C. Pierrard, J. Rimbault, E. Guillon, M. Port, M. Aplincourt, *Dalton Trans.* **2007**, 1611–1620.
185. R. Napolitano, L. Lattuada, Z. Baranyai, N. Guidolin, G. Marazzi, World Intellectual Property Organization, WO2020030618 A1 2020-02-13.
186. S. Le Greneur, A. Chenede, M. Cerf, S. Decron, B. Francois, France, FR3091872 A1 2020-07-24.
187. M. Botta, *Eur. J. Inorg. Chem.* **2000**, 399–407.

# 2 PARASHIFT Systems
## *Including $^{19}F$ Effects*

*Peter Harvey*
Sir Peter Mansfield Imaging Centre – School of Medicine & School of Chemistry,
University of Nottingham,
Nottingham, NG9 2QL, UK
peter.harvey@nottingham.ac.uk

## CONTENTS



DOI: 10.1201/9781003374688-2

**ABSTRACT**

PARASHIFT refers to the use of paramagnetic metal ions to chemically shift the magnetic resonances of nearby spin-active nuclei. While originally used to probe complex structures, such as proteins, recent efforts have resulted in the emergence of a new type of magnetic resonance imaging and spectroscopy (MRI/S). Careful ligand design to place reported nuclei at an optimum location with respect to a paramagnetic ion can result in highly shifted resonances accompanied by rapid increases in longitudinal relaxation. The combination of improved sensitivity and chemical shift inequivalence opens a myriad of unique direct imaging approaches, including the design of responsive agents and multichannel imaging. This chapter surveys the progress of this relatively new and unexplored MRI approach.

## KEYWORDS

MRI; Contrast Agents; Lanthanide; Transition Metals; NMR; Paramagnetic; Responsive Probes; $^{1}H$; $^{19}F$; Ligand Design

## 1 INTRODUCTION

PARASHIFT magnetic resonance involves the use of paramagnetic metal ions to chemically shift magnetic resonances, with improved sensitivity resulting from increased longitudinal relaxation and the accompanying acceleration in scan times. The advancement of rapidly relaxing, highly shifted resonances that can be imaged directly by magnetic resonance imaging (MRI) and/or spectroscopy (MRS) opens the possibility of multichannel channel imaging, overcoming a limitation of standard MRI approaches – the traditional focus on the single $^{1}H$ resonance in water provides a robust and intense signal, yet ultimately limits most common MR techniques to "greyscale" imaging. PARASHIFT systems have so far focused on the use of small molecule complexes with an MR reporter group positioned within a set distance of the metal center. Careful positioning is required to finely tune various relaxation mechanisms at play in these systems. While the design of PARASHIFT agents can be complex and synthetically elaborate, the technique ultimately has the potential to overcome a number of challenges with more traditional MRS and molecular MRI approaches (Figure 1). The PARASHIFT term has traditionally been associated with recent $^{1}H$-based systems and less used with related heteronuclear approaches (e.g., $^{19}F$). As the basis for paramagnetic shift and relaxation is essentially the same across nuclei, PARASHIFT nomenclature will be applied across all relevant systems discussed throughout this chapter.

### 1.1 SENSITIVITY LIMITATIONS OF TRADITIONAL MR CHEMICAL SHIFT APPROACHES

MRI predominantly relies on the $^{1}H$ resonance of endogenous water molecules to form images, largely due to the insensitive nature of the technique. The very high concentration of water in our body (10s of M concentration) and the highest

**Traditional MRI contrast agent**

- indirect detection
- enhances local water molecule relaxation
- strong background water signal
- "greyscale" imaging

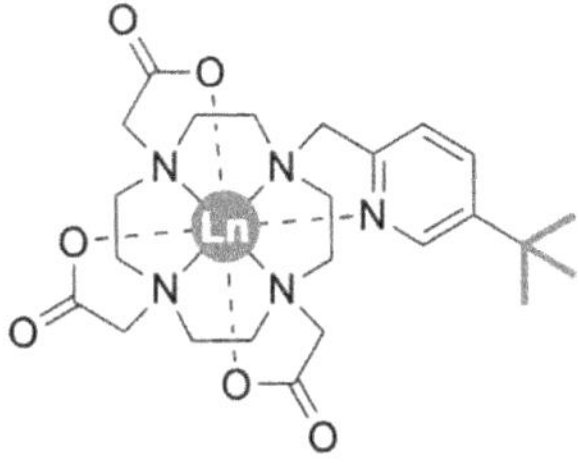

**PARASHIFT MR contrast agent**

- direct detection
- enhances internal reporter group relaxation
- zero background
- "multichannel" imaging capabilities

**FIGURE 1** Comparison of a standard Gd(III)-based magnetic resonance imaging (MRI) agent with a PARASHIFT agent. Traditional contrast agents rely on relaxation enhancement of nearby water molecules, limiting the signal to $^{1}H$ resonance of water. PARASHIFT uses a paramagnetic ion to shift the MR resonance of the reporter group (*t*Bu highlighted in this example) far from background biological signals. The accompanying relaxation increase increases sensitivity compared to diamagnetic magnetic resonance spectroscopy (MRS).

sensitivity of protons among all spin-active nuclei make the abundant water proton the standout choice for imaging. Most of the MR contrast agent methods to date are based on indirect detection via modulation of this water signal [1].

It is important, however, to remember that MRI is simply the imaging version of nuclear magnetic resonance (NMR) spectroscopy – any molecule that possesses a spin-active nuclei can, in theory, be directly imaged. The limitation of nuclei/molecule choice is embedded within the intrinsic shortcomings of NMR spectroscopy, namely that the energy levels involved are minute compared to most spectroscopic techniques. This low energy region, corresponding to radiofrequencies of the electromagnetic spectrum, infers very low differences between the excited and ground states of nuclear spins when placed in even a very powerful magnet. Such small energy differences lead to the intrinsically low sensitivity of magnetic resonance approaches, limiting applications to only the highest energy nuclei within ultra-high concentration molecules, such as $^{1}H$ nuclei of water found in our bodies.

Many other $^{1}H$- and/or heteronuclei-containing molecules have been directly detected using MR approaches, although very high concentrations are typically required (multi-mM to M). $^{1}H$ magnetic resonance spectroscopy (MRS) is often used to monitor the relative levels of naturally abundant sugars (e.g., glucose and lactate), but the narrow spectral window of standard $^{1}H$ NMR spectra also results in often overlapping signals with other endogenous biomolecules. Heteronuclei such as $^{2}H$, $^{13}C$, $^{31}P$, and $^{23}Na$ have been used to detect both endogenous and exogenous molecules, with often increased spectral windows available and/or reduced background competition. For direct detection of exogenous molecules, $^{19}F$ has been the go-to nuclei of choice, largely due to it having almost no natural abundance ($T_2$ of endogenous fluorine in bones/teeth generally too fast to image), spin of ½, and its relatively high sensitivity, with a gyromagnetic ratio second only to $^1$H [2,3]. However, even with these optimal properties, very high concentrations of fluorinated materials containing many copies of magnetically equivalent fluorine sites are typically required. Such materials typically take the form of perfluorocarbon (PFC) emulsions, with limitations in quantities required and frequent solubility issues.

## 1.2 Overcoming the MR Sensitivity Limitation

As sufficient recovery of $T_1$ relaxation is required between excitation pulses in MR experiments, a solution to overcome the sensitivity hurdle is to speed up the longitudinal relaxation time, $T_1$, of the observed nuclear resonance. By speeding up the relaxation, the time between excitations can be significantly reduced, allowing many more acquisitions in a given time window, thereby increasing the overall sensitivity. While relaxation times naturally vary according to environment and form the basis for the majority of MR image contrast, orders of magnitude increases in relaxation speed can be achieved through the use of unpaired electrons interacting with nuclear resonances. This mechanism is the basis behind standard water-based MRI contrast agents [1], but through careful molecular design, the unpaired electrons and signalling nuclear resonance can be co-located within the same compound, allowing fast relaxation and direct detection of the signalling resonance. As we will discover in this chapter, this approach can be used to dramatically lower the direct molecular detection limit from high mM to low μM, while also providing an added benefit of vastly increasing the spectral window of resonances observed. In the case of $^1$H, this increased spectral window can be expanded from the standard "biological window" of 0–10 ppm to beyond ±100 ppm.

Indeed, the use of paramagnetic metals to enhance the chemical shift of magnetic resonances predates the advent of MRI itself. An early application of lanthanide(III) complexes was to aid in NMR spectroscopy-based structure determination of complex molecules and proteins [4,5]. For proteins, Ln(III)-tags added at various locations on amino acid side chains result in highly shifted resonances that can be used to pinpoint the identity of the side chain and nearby groups. With the advent of higher field magnets, the increased use of 2-dimensional (2-D) techniques, and recent leaps in electron microscopy technology, this technique has largely fallen out of favour. The mechanism behind this protein determination technique, however, is the exact same mechanism behind PARASHIFT imaging.

This chapter will expand on the design principles and applications of using paramagnetic metal complexes to enable the direct visualisation of NMRs at increased sensitivity compared to more traditional MRI/S approaches. In theory, PARASHIFT works with any spin-active nuclei. The technique has, however, so far been mainly focused on $^1$H and $^{19}$F nuclei due to their increased sensitivity and natural abundance in comparison to other spin-active nuclei. This chapter will therefore focus on the theory and development of PARASHIFT with these two key nuclei, including interactions with both paramagnetic Ln(III) and transition metal (TM) ions.

It should be noted that there are other methods available for increasing sensitivity for the direct detection of spin-active nuclei, such as hyperpolarisation and dynamic nuclear polarisation [6], but they will not be covered in this chapter. PARASHIFT is also a distinct approach from chemical exchange saturation transfer (CEST) and its paramagnetic equivalent, PARACEST. This technique is discussed elsewhere in this volume.

## 2 EFFECTS OF PARAMAGNETIC IONS ON MAGNETIC RESONANCE

### 2.1 Chemical Shift

Unexpected shifts in the NMR spectra of solutions containing paramagnetic species were observed in the 1950s, with Bloemergen and Dickinson indicating that magnetic anisotropy must be present in some paramagnetic ions [7]. This magnetic anisotropy – the orientation-dependent unpaired electron-induced magnetic moment – occurs when there is a Boltzmann distribution across different low-lying energy states. For the *f*-block, this distribution would be across crystal-field splitting in the *J* manifold, while for *d*-block ions, zero-field splitting and spin–orbit coupling are involved [8]. These unpaired electrons can produce a large, temperature-dependent chemical shift well beyond that of its diamagnetic resonance. The observed shift ($\delta_{obs}$) is often described as follows:

$$\delta_{obs} = \delta_{dia} + \delta_{FCS} + \delta_{PCS} \tag{1}$$

where $\delta_{dia}$ is the diamagnetic term, $\delta_{FCS}$ is the Fermi contact term, and $\delta_{PCS}$ is the pseudocontact term [8]. The Fermi contact is important for atoms bound directly to a paramagnetic center, but it diminishes with increasing bond number. Due to the extreme line broadening usually associated with direct proximity to a paramagnetic center, this term is usually not relevant for PARASHIFT approaches. The pseudocontact shift, however, arises from through space dipolar interactions and is far more relevant to the chemical shifts observed. The physics of this shift in Ln(III) ions was described by Bleaney through which a series of coefficients or the so-called Bleaney constants can be derived for each individual Ln(III) ion [9]. For Ln(III) ions, these Bleaney constants are generally involved to describe the pseudocontact shift with respect to the orientation of the principle magnetic axis using a derivation of the McConnel equation [10]:

$$\delta_{PCS} = \frac{C_J \mu_B{}^2}{60(kT)^2}\left[\frac{(3\cos^2\theta - 1)B_0^2}{r^3} + \frac{(\sin^2\theta\cos^2\phi)B_2^2}{r^3}\right] \tag{2}$$

where $C_J$ is the Bleaney constant characteristic of a Ln(III) ion, $r$ is the distance of the nucleus from the Ln(III) ion, $\theta$ and $\phi$ are the angles between the nucleus and principle magnetic axis of the Ln(III) ion, $g$ is the Landé factor, $\mu_B$ is the Bohr magneton, and the $B$ parameters are the second-order ligand field terms [11]. These equations highlight the importance of Ln(III) identity and the effect of temperature ($T^{-2}$) on the chemical shift of the nucleus being observed, together with the orientation- ($\theta$, $\phi$) and distance-dependence ($r^3$) of the shift. It should be noted that Gd(III) typically does not display a paramagnetically induced shift due to the isotropic arrangement of its 7 unpaired electrons.

While the Bleaney theory gives good approximations in some cases, recent studies have shown some significant limitations [12–17]. Complexes do appear to follow a strict $T^{-2}$ dependency, yet the shift prediction provided by the traditional Bleaney method appears to fall down with certain ligand series. A classical assumption was that ligand field splitting of the ground $J$-multiplet is $<kT$ and that ligand field parameters are retained across the Ln(III) series of an isostructural ligand. For certain cases, these assumptions appear to be flawed. Indeed, ligand field splitting can be much higher than thermal energy even at room temperature. The Eu(III) luminescence spectrum provides a convenient handle for the measurement of $B_0^2$, and comparison of this value does not correlate with the observed pseudocontact shift in some cases [14]. It also appears that the position of the principal magnetic axis, often assumed to be invariant for Ln(III) across an isostructural series, may in fact vary from one Ln(III) to another [13]. The studies above are aiding to shape our understanding of a new generation of theoretical prediction approaches.

## 2.2 Relaxation

Bloch–Redfield–Wangsness theory is most commonly used to describe intramolecular nuclear relaxation rates. For nuclei not directly coordinated to the Ln(III) ion, the main input to relaxation is driven by the electron-nuclear dipolar interaction:

$$R_1 = \frac{2}{15}\left(\frac{\mu_0}{4\pi}\right)^2 \frac{\gamma_N{}^2 g_{Ln}{}^2 \mu_B{}^2 J(J+1)}{r^6}\left[3\frac{T_{1e}}{1+\omega_N{}^2 T_{1e}{}^2}+7\frac{T_{2e}}{1+\omega_N{}^2 T_{2e}{}^2}\right] + \frac{2}{5}\left(\frac{\mu_0}{4\pi}\right)^2 \frac{\omega_N{}^2 \mu_{eff}{}^4}{(3k_B T)^2 r^6}\left[2\tau_r \frac{3}{2}\frac{\tau_r}{1+\omega_N{}^2 \tau_r{}^2}\right] \tag{3}$$

$$R_2 = \frac{2}{15}\left(\frac{\mu_0}{4\pi}\right)^2 \frac{\gamma_N{}^2 g_{Ln}{}^2 \mu_B{}^2 J(J+1)}{r^6}\left[2T_{1e}\frac{3}{2}\frac{T_{1e}}{1+\omega_N{}^2 T_{1e}{}^2}+\frac{13}{2}\frac{T_{2e}}{1+\omega_N{}^2 T_{2e}{}^2}\right] + \frac{2}{5}\left(\frac{\mu_0}{4\pi}\right)^2 \frac{\omega_N{}^2 \mu_{eff}{}^4}{(3k_B T)^2 r^6}\left[2\tau_r \frac{3}{2}\frac{\tau_r}{1+\omega_N{}^2 \tau_r{}^2}\right] \tag{4}$$

where $\mu_0$ is the vacuum permeability, $\gamma_N$ is the magnetogyric ratio of the reporter nucleus, $g_{Ln}$ is the Landé factor of the fundamental multiplet $J$ of the free Ln(III) ion, $\mu_B$ is the Bohr magneton, $r$ is the distance between the Ln(III) and reporter nucleus (electron-nuclear distance), $\tau_r$ is the rotational correlation time, $\omega_N$ is the Larmor frequency of the reporter nuclei, $\omega_e$ is the electron Larmor frequency, $\mu_{eff}$ is the effective magnetic moment, and $T_{1e}$ is the longitudinal relaxation time of the electron spin (often assumed to be equal to $T_{2e}$) [11].

In terms of observed properties in PARASHIFT experiments, an increase in $R_1$ allows faster acquisition parameters due to the standard practice of waiting 4–5 times the $T_1$ ($T_1 = 1/R_1$) between repetitions to allow sufficient relaxation back to the ground state. An increased $R_1$ therefore translates to an improved sensitivity due to more rapid averaging possible in a given time period. An increase in $R_2$, however, is typically associated with line broadening, leading to a reduction in sensitivity due to a loss in signal amplitude. Due to the intrinsic nature of the relaxation rates, $R_2$ will always be faster than $R_1$. A key aspect of PARASHIFT design is, therefore, to aim for an $R_1$:$R_2$ ratio as close to the optimal 1 as possible.

While these equations have been used to model the designs of novel PARASHIFT complexes with the $r^{-6}$ dependence of the Ln(III)-reporter distance [11,18,19], recent studies have again found flaws in the existing calculations [16,20–23]. Again, ligand field splitting effects are much more important than classically assumed, and it appears that orientation of the reporter nuclei to the Ln(III) center, which has long been known to be relevant to chemical shift, is also highly important for relaxation. Anisotropy in the magnetic susceptibility tensor generates this orientation-dependence. Accounting for this effect has led to much more accurate calculations than solely relying on distance [21].

The focus of this section is on the work done to understand paramagnetic relaxation and chemical shift with Ln(III) ions due to the broader scope of work done in this field compared to TM ions. For a closer focus on the physical properties of PARASHIFT behavior with TMs, the reader is referred to recent discussions in the literature [24]. Understanding and prediction of MR properties is an evolving field, with significant need for improved models in both *f*- and *d*-block systems.

## 3 Ln(III)-BASED PARASHIFT AGENTS

### 3.1 Ln(III)-Based $^{19}$F PARASHIFT Systems

While some of the earliest work involving the use of paramagnetically shifted MR imaging focused on $^1$H-based systems, initial efforts into bespoke design and characterization of PARASHIFT agents were often based on $^{19}$F (though the term PARASHIFT was not directly used in these publications). Paramagnetic relaxation enhancement of $^{19}$F MR was demonstrated in the mid-1990s, with Gd(DTPA) used to enhance the fluorine signal from trifluoromethylsulfonate [25]. This approach generated a $^{19}$F signal enhancement factor of around four with an *in vivo* application in a rabbit brain abscess, demonstrating the potential of paramagnetic relaxation enhancement to enhance sensitivity. Later work demonstrated similar results with other fluorinated biosubstrates, such as fluoroacetate and trifluoroacetate [26,27]. It should be noted, however, that these approaches relied on intermolecular interactions between the fluorinated probe and the paramagnetic ion, with low likelihoods of interactions for most *in vivo* applications. Being based on the isotropic Gd(III), these approaches also do not take advantage of paramagnetic shift available with other ions – essentially the diamagnetic and paramagnetically enhanced $^{19}F$ resonances overlap, resulting in similar difficulties of distinguishing between background water and contrast-enhanced water in a standard, non-PARASHIFT $^{1}H$ MR image. A preferred approach would be to contain the $^{19}F$ reporter in the same molecule as the paramagnetic ion and to use Ln(III) other than isotropic Gd(III) to induce PARASHIFT responses.

An early example of the difficulty in designing such compounds, however, was demonstrated with NMR studies by Sherry and coworkers on a DOTP fluorinated ethyl ester analogue, F-DOTPME (**3-i**, Figure 2) [28]. Despite each of the four

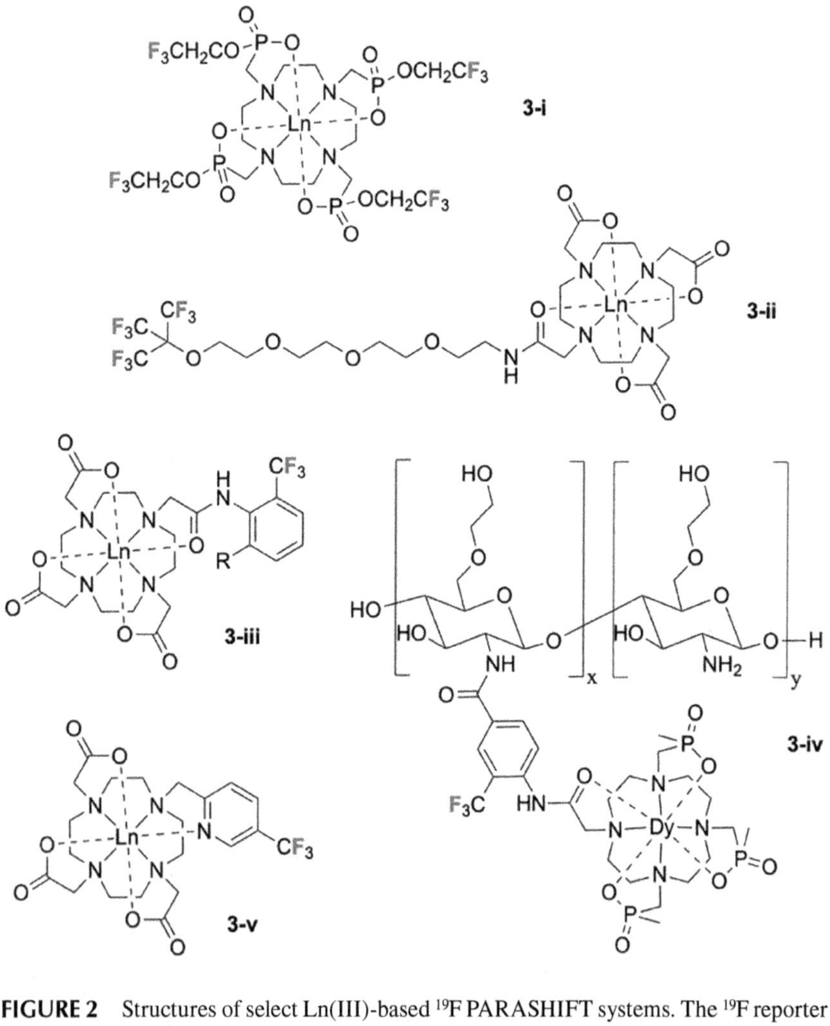

**FIGURE 2** Structures of select Ln(III)-based $^{19}F$ PARASHIFT systems. The $^{19}F$ reporter nuclei are highlighted in each case.

$CF_3$ groups appearing to be equivalent on paper, the $^{19}$F NMR spectra exhibit a complex mix of stereoisomers. While the hyperfine shifts generated by the paramagnetic ions aid in the assignment of some resonances in this case, splitting of the signal will result in reduced sensitivity for PARASHIFT approaches – a single, intense resonance is preferred for most applications. Development towards such a system was later shown by a DO3A-based ligand with a perfluoro-*tert*-butyl reporter group tethered via a PEG linker (**3-ii**, Figure 2) [29]. The encapsulation of various di- and trivalent paramagnetic metal ions led to distinct, single $^{19}$F resonances. While this complex demonstrated the potential for multicolor/channel approaches, the long and flexible linker, however, limits the chemical shift non-equivalence between ions to only 7–8 ppm. The distance between paramagnetic ion and reporter also results in modest enhancements of relaxation (Tb: $R_1$ = 14 Hz *versus* diamagnetic 0.75 Hz; pH 7, 298 K, 11.7 T), suggesting that the reporter group resides around 10 Å from the metal center.

Around the same time, Parker and coworkers developed a series of bespoke $^{19}$F-containing Ln(III) complexes that would pave the way for their $^{1}$H PARASHIFT systems to follow. These complexes were based on a $CF_3$-substituted aniline motif, coupled to a DOTA-type scaffold via an amide bond (**3-iii**, Figure 2). While a focus of the approach was to develop pH-responsive probes (see Section 3.3.2) [30,31], attention was also given to reduce the stereoisomeric distribution in solution and to generate as intense a $^{19}$F signal as possible [18,32]. The reporter $CF_3$ group was designed to be within an optimal distance to the Ln(III) ions – close enough to result in large paramagnetic chemical shifts and longitudinal relaxation, but not too close to lead to an excess of line broadening due to rapid transverse relaxation. In each of these systems, the $CF_3$ group lies around 5.8–6.3 Å from the Ln(III) center, with the optimal window calculated to be between 5 and 7 Å [18].

In addition to distance dependence, this series of complexes also demonstrated the importance of rigidity within the ligand scaffold. For example, the carboxylate DO3A-based ligand results in eight distinct isomeric species whose relative abundance changes slowly over time. The phosphinate analogue, however, gives rise to one dominant species in solution (>87%), with isomerism remaining fixed over time. The increased bulk of the phosphinate groups likely improves rigidity of the overall complex and inhibits isomeric exchange. This increased rigidity also infers improvements to the transverse relaxation properties of the phosphinate complexes *versus* the carboxylate counterparts – the more rigid the ligand scaffold, the less dynamic exchange and sharper linewidths will be present. This rigidity resulted in some of the most promising $^{19}$F PARASHIFT properties to date. The Dy(III) phosphinate system, for example, displayed a major resonance at −63.6 ppm with a relaxation rate enhancement of around 100-fold ($R_1$ = 185 Hz v diamagnetic 2.5 Hz; 295 K, pD 5.4, 9.4 T), while the $R_1/R_2$ ratio remained relatively close to parity (0.74) [32].

Even with these improvements to reporter sensitivity, efforts to demonstrate in intact *in vivo* models proved challenging, most likely due to rapid renal clearance of the complexes preventing sufficient build-up to image. In order to circumvent this issue, we appended the small molecular $^{19}$F PARASHIFT complexes to a larger molecule conjugate (**3-iv**, Figure 2). Glycol chitosan conjugates were prepared of the previously best performing $^{19}F$ system, with average molecular weights on the order of 14,000–16,000 (ca. 9–13 complexes conjugated) [33]. Administration of the Gd(III) version in colorectal tumor xenograft mice was used to confirm tumor uptake via standard $T_1$-weighted MR imaging, before the Dy(III)-conjugate was mapped using 2-D $^{19}F$ MR spectroscopy. While line broadening was observed compared to the *in vitro* phantom, this experiment demonstrated the first successful *in vivo* imaging using these paramagnetic $^{19}F$ reporters, at clinically relevant doses (0.037 $mmol^{-1}$ $kg^{-1}$) and scan times (<5 min).

We further explored the concept of increased rigidification by "locking" the reporter group in place via a pyridine moiety, removing the restricted rotation present in the previous aniline systems (**3-v**, Figure 2). While unfortunately no improvements in relaxation properties were observed over the aniline systems, an unprecedented range of chemical shifts was observed, with dipolar shifts of up to 100 ppm observed *versus* the diamagnetic complex and almost 200 ppm between the most shifted complex in each direction (Dy *versus* Tm) [13,34]. While these particular complexes were not explored further for direct imaging applications, they did form the inspiration for the $^1H$ PARASHIFT systems that followed – if similar dramatic increases in chemical shift could be designed with $^1H$, effective zero background imaging could be achieved with careful design of reporter group.

## 3.2 Ln(III)-Based $^1$H PARASHIFT Systems

The first $^1H$ PARASHIFT systems appeared in the mid-1990s, although the term would not be coined for another 15 years. Aime and coworkers suggested the use of Yb(III)-DOTMA as an alternative to the diamagnetic $^{19}F$ MRS probes that were being developed at the time [35]. The ligand possesses four equivalent methyl groups, with the twelve magnetically equivalent proton nuclei resonating at −14.2 ppm. While the phantom imaging experiment took 6 h 23 min to observe 3 mM of the complex, it demonstrated the potential of this approach to detect paramagnetically shifted $^1H$ resonances in the presence of water. Related work at the time from the same group and others also explored magnetically equivalent resonances in Yb(III), Tm(III), and Pr(III) complexes of DOTMA, DOTP, and MOE-DO3A (**3-vi to 3-viii**, Figure 3) [36–39].

Bansal and coworkers later continued the development of the DOTMA system, with the incorporation of Tm(III) increasing the methyl resonance chemical shift to −102 ppm [40]. This significant chemical shift, far removed from water protons, allowed improvements in MR parameters and image acquisition in less than a minute. The same group successfully demonstrated *in vivo* imaging using Tm(III)-DOTMA at doses of 0.5–0.6 mmol $kg^{-1}$ in nephrectomised rats [41]. Coman, Hyder, and coworkers also explored further applications of DOTP, focusing on the Tm(III) complex and its highly shifted resonances [42]. While the focus was on pH and temperature mapping (see Section 3.3.1 and 3.3.2), this work demonstrated the potential for imaging PARASHIFT resonances *in vivo* with scans that lasted less than 10 min. It should be noted, however, that these were demonstrated in renally ligated rats to prevent clearance of the agent and at a dose of 1 mmol $kg^{-1}$, around ten times higher than typical clinically approved doses of Gd(III) MR contrast agents.

These traditional ligand scaffolds present promising PARASHIFT properties, with their high symmetry leading to a range of magnetically equivalent proton resonances. This symmetry, however, provides a limitation as any modification to the ligand architecture will break the magnetic equivalency and decrease sensitivity. For example, to improve blood half-life and *in vivo* distribution, the same group appended Tm(III)DOTA analogues onto a PAMAM dendrimer [43]. While this was successfully achieved and allows for co-incorporation of bimodal reporters, the resulting $^{1}H$ NMR spectra presented far more speciation and line broadening compared to the complex alone.

Parker and coworkers, therefore, aimed to develop a series of complexes specifically dedicated to PARASHIFT imaging. These complexes were based on earlier, and previously discussed, work on $^{19}F$ PARASHIFT complexes, replacing the $CF_3$ group with a $^{1}H$-based *tert*-butyl reporter (**3-ix**, Figure 3). As with

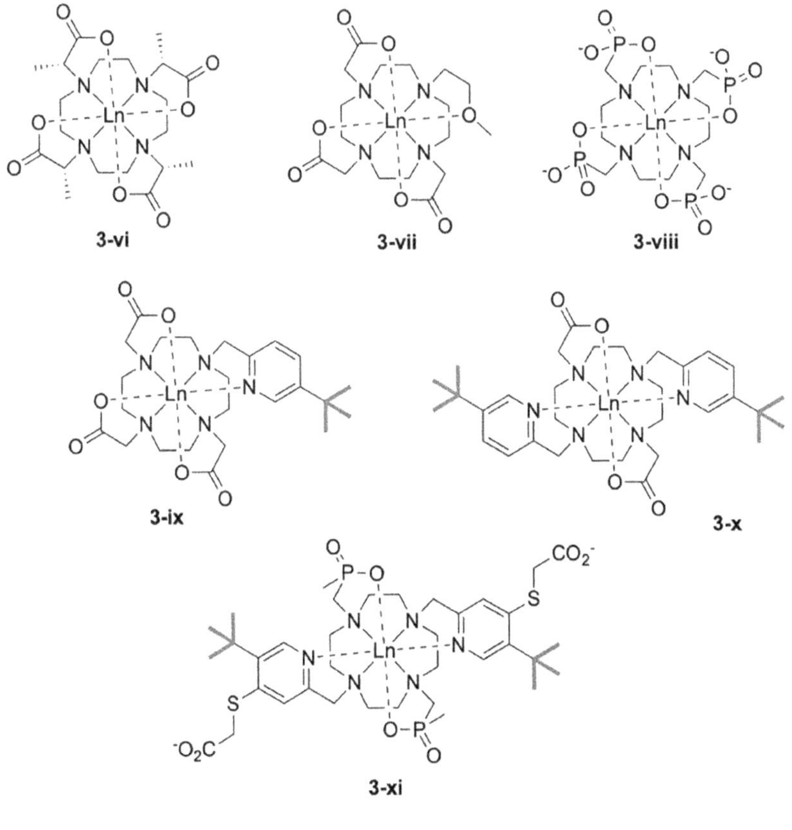

**3-vi** **3-vii** **3-viii** **3-ix** **3-x** **3-xi**

**FIGURE 3** Structures of select Ln(III)-based $^{1}H$ PARASHIFT systems. The $^{1}H$ reporter nuclei are highlighted in each case, unless multiple resonances are used across the ligand.

the previous $^{19}$F systems, the advantage of a dedicated reporter is that alterations to the ligand backbone do not break the magnetic equivalency of the reporter resonance, which has nine protons in this case, allowing a far broader range of potential chemical modifications. Our first systems focused on the use of a *t*Bu-substituted pyridine linker, taking advantage of the relatively sharp and highly shifted resonances that had been observed with the $CF_3$ analogues. The first-generation agents formed one major isomer in solution and displayed chemical shifts of up to 25 ppm from water (Dy; −20.5 ppm) [11]. For the bis-substituted version (**3-x**, Figure 3) with eighteen equivalent proton environments, the shift was slightly reduced (Dy; −17.8 ppm), but the paramagnetic relaxation enhancement was close to optimal ($R_1$ = 174 Hz, $R_1/R_2$ = 0.72; 295 K, pH 5.9, 9.4 T). Unlike with the related $^{19}$F systems, this complex was successfully observed *in vivo* without further modification or conjugation, following administration of 0.12 mmol$^{-1}$ kg$^{-1}$ dose of the agent. Anecdotally, from our experiences, the use of more standard $^1$H MR coils allowed far simpler fine tuning of the experimental parameters than the prior studies using more bespoke $^{19}$F coils.

While this is a promising initial experiment, increasing the chemical shift distance between water/fat and the reporter resonance would allow much wider sweep widths to be applied, facilitating ultra-fast TE values to be used [44]. This increased chemical shift was successfully achieved through functionalisation of the pyridine group and substitution of the carboxylic pendant arms for methyl phosphinates (**3-xi**, Figure 3) [45]. With this complex, the Dy(III) *t*Bu resonance was now shifted to −60 ppm, allowing a much wider sweep width to be applied. Successful imaging of this compound was achieved in mice with an estimated *in vivo* tissue detection concentration of 23 $\mu$M. This detection concentration approaches that of the indirect detection (i.e., influence on water $^1$H) of clinically used Gd(III) contrast agents [1], but with the potential added benefit of responsive and direct imaging afforded by the PARASHIFT technique.

## 3.3 Applications of Ln(III) PARASHIFT Systems

One of the most significant limitations of designing responsive probes using standard MRI contrast agents (e.g., Gd(III)) is their intrinsic reliance on concentration of the agent [1]. As the overall signal is directly proportional to the agent concentration, knowledge of the local concentration is required to interpret any response in the signal. For rapid dynamic processes, such as fluctuations in neurotransmitters or calcium(II) during neuronal firing, this problem is alleviated as the signal change is much faster than local pharmacokinetics [46]. For slower processes, however, it is a much more challenging limitation. There have been some attempts to overcome this issue with the co-injection of non-responsive agents, but this requires co-localisation of two chemically different compounds [47].

PARASHIFT overcomes this concentration reliance as chemical shift responses can be designed into the probe. While the overall signal is still concentration dependent, the frequency of the shift is instead dependent on the response being monitored. Disconnecting concentration to the response allows a single probe to be used to map physiological changes. With the combination of multiple probes or a single probe with multiple distinct resonances, the possibility for ratiometric sensing is unlocked by directly comparing variations at multiple frequencies. This multi-probe approach is particularly appealing with Ln(III) complexes as a single ligand structure can be used with varying Ln(III) ions. As the chemistry of the Ln(III) is largely unchanged across the series, different Ln(III) complexes of the same ligand scaffold could realistically be expected to possess similar pharmacokinetics and biodistribution.

Here, $^{1}H$ and $^{19}F$ will be discussed interchangeably due to the varying nature of work carried out with each nuclei to date. It should be noted that there are a number of reported systems, particularly with $^{19}F$, that use a paramagnetic ion (typically Gd(III)) to effectively quench the reporter resonance, which is only detectable when released [48,49]. While an interesting approach in their own right, they mostly rely on detection of the diamagnetic signal and so will not be discussed in detail here.

### 3.3.1 Temperature Sensitivity

The inverse square relationship of the pseudocontact shift with temperature provides interesting opportunities for the design of minimally invasive *in vivo* temperature probes for mapping disease states. With the development of thermal therapy techniques [50], in addition to clear neurological effects of hyperthermia [51], the ability to accurately monitor temperature in the body would be a powerful tool. Currently, temperatures in the body can be monitored with external probes (limited sampling, invasive) or with imaging thermometry methods, such as via $T_1/T_2$ measurements (low accuracy). The chemical shift of water can be used for thermal mapping using the proton resonance frequency shift method, but with very low sensitivity (0.01 ppm $K^{-1}$) [52].

The intrinsic temperature-dependence of PARASHIFT offers significant improvements to these standard approaches, retaining the minimally invasive nature of imaging but with the potential for far greater sensitivity. The increased chemical shift non-equivalence of PARASHIFT offers multiple orders of magnitude improvements in temperature-dependent chemical shift in comparison to diamagnetic water. A large focus of the early development of $^{1}H$ PARASHIFT approaches was focused on their temperature sensitivity [35,37,39], while most PARASHIFT reports to date have characterised the response of the chemical shift to variations in temperature. The temperature sensitivity of these PARASHIFT resonances is far superior to that of water. For example, the $^{1}H$ chemical shifts of Tm-DOTP$^{5-}$ display a temperature sensitivity of 0.89 ppm $K^{-1}$ *in vivo*, around a two orders of magnitude improvement over the water resonance alone [39].

Coman, Hyder, and coworkers further demonstrated this approach through which they term "biosensor imaging of redundant deviation in shifts", or BIRDS [42,53]. Despite different names, this approach is analogous to PARASHIFT and used $^{1}H$ resonances of Tm(III) DOTP and DOTMA complexes to map temperature in the living brain. The redundancy in the name arises from the presence of multiple resonances, each with a differential sensitivity to temperature.

The presence of this redundancy essentially creates an internal reference that can lead to increased temperature sensitivities, which was recently demonstrated using a tetra-amide Tm(III) complex [54]. By comparing the relative shift of two resonances on the same molecule with temperature-induced chemical shifts in opposite directions, the temperature sensitivity can be increased to 1.93 ppm $K^{-1}$ (compared to 0.74 ppm $K^{-1}$ for an individual resonance). Some groups have also demonstrated related temperature sensitivity using paramagnetically shifted $^{19}F$ systems [55], though applied work has so far been much more limited.

### 3.3.2 pH-Responsive Probes

Altered extracellular pH *in vivo* is associated with a range of acute and chronic conditions, including stroke and cancer, yet there are very few non-invasive methods available to measure pH deep in tissue. While temperature-sensitive changes in chemical shift are intrinsic to all PARASHIFT complexes, further responsive behavior requires careful, and often challenging, ligand design. In the case of pH-responsive PARASHIFT probes, this design typically involves either the use of intrinsic pH-sensitivity or the insertion of a pH-sensitive molecular group. It should be noted that relatively subtle pH-induced changes to the ligand can infer substantial PARASHIFT responses, with a potentially broader array of options compared to traditional Gd(III) contrast agent approaches [1]. A significant challenge in the design of pH-responsive probes is in tuning the response to biologically relevant changes. The $pK_a$ of any pH-responsive group therefore needs to be within a bio-relevant window (typically pH 6–7.5), though the inclusion of a highly charged metal center can lead to subtle changes in $pK_a$ of molecular groups compared to those measured in organic compounds.

A number of the early $^1H$ PARASHIFT systems, already discussed in the previous sections, display pH sensitivity of their chemical shifts over the desired biologically relevant pH window. While this can make analysis of spectra and temperature response more challenging, it also opens more possibilities for increased biomedical applications. Coman, Hyder, and coworkers have demonstrated a number of examples of this approach using their BIRDS method. They mapped both temperature and extracellular pH ($pH_e$) *in vivo* in their initial studies with Tm-DOTP [42], later expanding this approach to preclinical models of brain and liver tumours [56,57]. Through this work, they demonstrated the potential of PARASHIFT/BIRDS-type approaches for relatively high-resolution mapping of the intratumoural-peritumoural $pH_e$ gradient. These approaches can also be combined with other MR measurements. For example, comparison of the Tm-DOTP $pH_e$ measurement with $^{31}P$ MR measurements of the intracellular pH ($pH_i$) showed good agreement in healthy tissue but a clear discrepancy within glioma regions [57]. This approach has also been demonstrated in spatially resolved $pH_e$ mapping across 3-D *in vitro* tumor models [58] and to track tumor progression during chemotherapeutic intervention *in vivo* [59].

Due to the well dispersed nature of the shifts in Tm-DOTP, the influence of pH and temperature on the chemical shifts can be readily deconvoluted, improving accuracy in pH measurements. This work, and others, has demonstrated the advantages of chemical shift approaches over traditional contrast agent approaches – the pH/temperature response is independent of the probe concentration. It is also unaffected by overall changes in tissue relaxation, allowing measurements in otherwise challenging tissue environments where probe uptake can be achieved at sufficient localised levels.

Parker and coworkers early $^{19}$F approaches were largely focused on pH-responsive imaging systems, with pH-sensitivity introduced via the amide connecting the $CF_3$ reporter containing aromatic moiety to the Ln(III)–DO3A complex [18,30–32]. Fine tuning the functional groups on the reporter and/or amide linker allowed modification of the p$K_a$ of the response, with improved values observed in murine plasma and urine samples in some cases (p$K_a$ = 5.7 in water; 6.9 in urine/plasma) [30]. This latter example highlights the importance of measurements in biologically relevant media. The presence of stereoisomers in some of these systems actually provides a potential advantage in pH-sensing – the pH response of each isomer is distinct, allowing internal calibration between signals and an increased Δppm/pH unit where resonances are shifted in opposite directions [18]. While these systems largely focus on retaining the $^{19}$F reporter relatively close to the Ln(III) center and rely mostly on an alteration in rotation/orientation, a recent example has also demonstrated a pH-induced isomerisation that alters Ln(III)-to-$^{19}$F distance between 9 and 13 Å [60]. While this report focuses mainly on relaxation-enhancement with the Gd(III) analogue, it demonstrates the broad design possibilities available.

Parker and coworkers also incorporated pH-sensitivity to their previously reported $^{1}$H PARASHIFT systems through incorporation of a phosphonate group adjacent to the *t*Bu reporter group (**3-xii**, Figure 4) [61]. This design resulted in a pH-dependent chemical shift with a p$K_a$ between 6.7 and 7.1, depending on the media. They realised that co-administration of two different Ln(III) complexes would allow deconvolution of the effect of temperature from pH on the chemical shift while increasing chemical shift non-equivalence. This approach was demonstrated by co-administration of both the Dy(III) and Tm(III) complexes in healthy mice. Triple $^{1}$H imaging was achieved of the Dy(III), Tm(III), and baseline water proton resonances in the same experiment, using clinically relevant doses (0.04 mmol kg$^{-1}$ each of 1:1 Dy:Tm) in renally intact mice. This experiment allowed both temperature and pH to be measured across various organs and demonstrates the potential of PARASHIFT imaging – each "channel" (i.e., resonance) can be encoded with distinct information and acquired simultaneously.

### 3.3.3 Ion-Responsive Probes

Metal ions play an important role in cell signalling and function, while alterations in the concentrations of biologically relevant anions have been linked to a variety of disease. Methods to study ion concentration fluctuations *in situ* would provide routes for increased diagnostic and stratification of diseases, in addition to improved biological tools [62]. Traditional MR contrast methods to measure these analytes suffer from the same issues previously discussed – signal is proportional to both probe concentration and analyte presence. While this limitation can be circumvented where the dynamics are rapid on the MR timescale, such as recent successes in Ca(II) imaging *in vivo* [63,64], it is much more restrictive for slower measurements. While, to date, few PARASHIFT approaches have been demonstrated *in vivo*, the independence of signal and concentration suggests some intriguing opportunities.

In order to be effective, any ion-responsive MR probe requires specificity for the ion of interest and appropriate reversible binding kinetics. We demonstrated a Ca(II)-responsive $^{19}F$ system based on the calcium-binding ligand BAPTA tethered to a $CF_3$-containing Ln(III) complex (**3-xiii**, Figure 4) [65]. This complex displayed pH-independent Ca(II)-selectivity over other biologically relevant metal cations, with a distinct variation in $^{19}F$ chemical shift as a function of Ca(II) concentration. The chemical shift could be reversed back to the original position with the addition of EDTA to generate the "free" complex, while the $^{19}F$ relaxation rate remained essentially constant throughout (<4% variation). Fitting of the binding curve allowed Ca(II) concentration to be directly calculated from the $^{19}F$ resonance.

Removal of the Ca(II)-binding moiety to leave a free amine on the cyclen ring provided a route to an anion-responsive probe (**3-xiv**, Figure 4) [65]. While the resulting complex is pH sensitive, the $pK_a$ for the complexes was >8.5, resulting in a positively charged species at biologically relevant pH levels. Due to reduced coordination at the Ln(III) center, anionic species can bind to the metal centre and displace the labile water. Binding of a variety of anions (lactate, citrate, bicarbonate, and phosphate) led to distinct $^{19}F$ chemical shifts due to changes in the local Ln(III) ligand field. It should be noted, however, that with some Ln(III) ions, the resulting anion-bound $^{19}F$ resonances were extremely broad due to chemical exchange. This line broadening highlights a key challenge in designing responsive chemical shift systems, where the required speciation can impact with the desired spectroscopic properties if exchange falls into the NMR timescale. Both the Ho(III) and Tm(III) complexes displayed distinct $^{19}F$ resonances upon the addition of citrate, with selectivity over the other anions tested. A key advantage of Ln(III) systems is the interchangeability of the metal centre with minimal modification to overall chemical properties, allowing the citrate selectivity to be confirmed by luminescence with the analogous Eu(III) complex.

More recent work from the Parker group demonstrated proof-of-principle Zn(II) detection with a $^{1}H$ PARASHIFT complex [66]. Here, a *t*Bu-containing tris-pyridylamine (TPA) pendant arm was incorporated into the DO3A ligand (**3-xv**, Figure 4). Binding of Zn(II) resulted in significant chemical shift changes of the *t*Bu resonance, in some cases inverting direction of shift (i.e., positive to negative shifts, or vice versa, upon Zn(II) binding). This large change in shift suggests a dissociation of the pyridine arm from the Ln(III) centre upon Zn(II) binding, changing the position of the *t*Bu reporter with respect to the principle magnetic axis of the Ln(III) ion, which may itself be affected by an alteration in coordination. While the complex displayed high affinity and selectivity for Zn(II) over Mg(II) an Ca(II), the complex speciation observed in the $^{1}H$ NMR spectra and Zn(II) transmetallation of the DO3A macrocycle would likely need to be resolved to take this complex into *in vivo* studies. A related $^{19}F$-based Zn(II) probe was reported by Que and co-workers, where binding of Zn(II) to a $C(CF_3)_3$-containing TPA linker resulted in a "turn-on" response [67]. Prior to Zn(II)-binding, chemical exchange broadens the $^{19}F$ signal, while the more rigid nature of the Zn(II)-bound species modulates this chemical exchange and gives rise to a distinct and sharpened $^{19}F$ resonance. A similar system incorporated a much longer linker between a Ln(III)-DTPA complex and a $C(CF_3)_3$ reporter tag to induce large Ca(II)-induced changes in the distance between the $^{19}F$ group and the Ln(III) ion [68]. Binding of Ca(II) resulted in distinct alteration in the $^{19}F$ NMR spectra, but was accompanied by significant line broadening.

An alternative approach is to use an exogenous paramagnetic Ln(III) complex to modulate the magnetic properties of an endogenous cation. A number of studies have looked at the effect on endogenous sodium cations using $^{23}Na$ MR approaches [69,70]. Here, the shift agent is not being detected directly, but through its effect on the $^{23}Na$ MR spectra of the analyte itself. While Gd(III) complexes have been used to induce line broadening and relaxation effects, the use of Tm(III), for example, generates distinct and well-resolved $^{23}Na$ chemical shifts [69].

### 3.3.4 Enzyme-Responsive Probes

A number of Gd(III)-$^{19}F$ based systems have been designed for monitoring enzyme activity. Typically, this approach involves tethering a $^{19}F$ reporter to a Gd(III) complex by an enzyme-cleavable linker. Initially, the fluorine signal is effectively quenched by the presence of the Gd(III) and is not observable due to rapid $T_2$ relaxation. Upon enzyme activity and cleavage of the linker, the fluorine moiety is released and detectable as a diamagnetic entity with much longer $T_2$ values. While this approach uses paramagnetism to affect a $^{19}F$ chemical shift, it is effectively "turning off" the signal rather than utilising the properties of PARASHIFT.

We designed a lanthanide-based $^{19}F$ reporter that demonstrated the potential of PARASHIFT to report on enzyme activity with a fixed Ln(III)-$^{19}F$ distance before and after enzyme cleavage [65]. Here, we incorporated a self-immolative enzyme-triggerable tag on to the molecular group containing the $CF_3$ reporter group (**3-xvi**, Figure 4). Upon addition of α-chymotrypsin, cleavage of the pendant amide occurs, leading to self-immolation of the linker group and an accompanying $\Delta\delta_F$ of 4.5 (Dy(III)) or 12.1 ppm (Tm(III)). As the Ln(III)-$^{19}F$ distance is essentially unchanged during enzyme cleavage, the $R_1/R_2$ values are also unaffected during the process, allowing retention of imaging acquisition parameters throughout. Limitations with significant line broadening and an extremely slow response rate (10 days for full conversion) have precluded these particular systems from further applications, but they demonstrate the potential for direct monitoring of enzyme activity via PARASHIFT reporter groups.

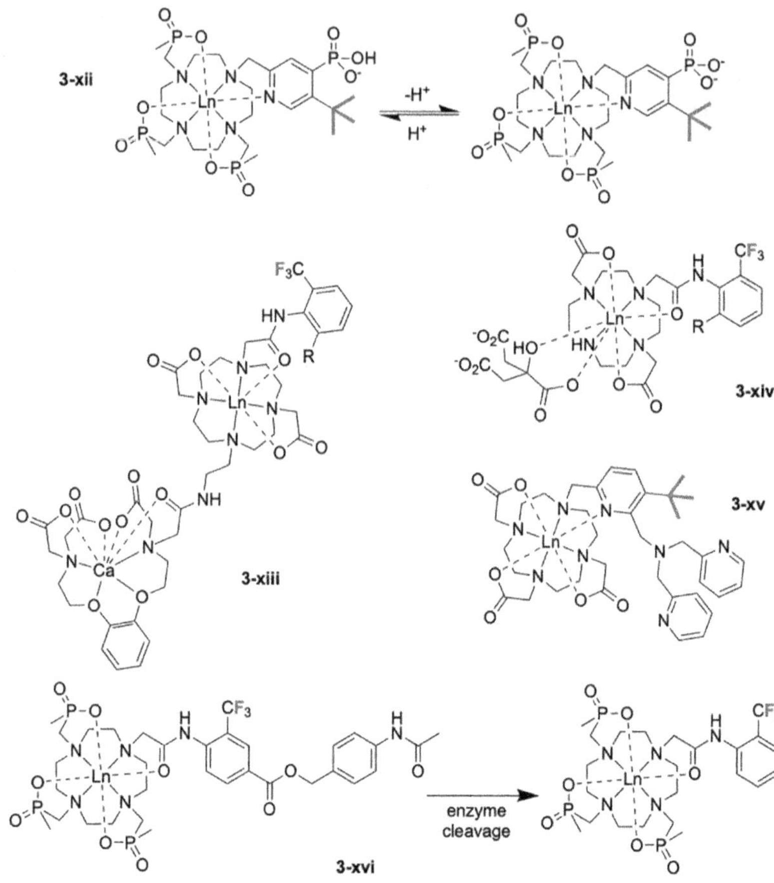

**FIGURE 4** Structures of select Ln(III)-based responsive PARASHIFT systems. The $^{1}H/^{19}F$ reporter nuclei are highlighted in each case. **3-xiii** includes bound Ca(II) while **3-xiv** indicates binding of citrate to the metal centre.

## 4 TRANSITION METAL-BASED PARASHIFT AGENTS

Issues around Gd(III) toxicity have led to some safety concerns over the use of currently applied clinical MRI contrast agents [71]. Of particular concern is nephrogenic systemic fibrosis, a serious effect that has been found to occur in patients with renal conditions. Reports have also demonstrated Gd(III) deposits in the brains of patients following administration of contrast agents and even environmental build-up of the ion over decades due to the widespread use of MR imaging [72]. While the majority of studies have focused on Gd(III) due to its common usage, it is likely that similar concerns may arise through the use of other Ln(III) ions. There has, therefore, been considerable recent interest in the development of alternative forms of contrast, ranging from TMs [73] to organic radicals [74].

TM complexes also have the additional benefit of freedom of design – less restrictive ligand requirements open the potential of broader applications, such as tuneable lipophilicity [75] and intracellular targeting [76]. The wider variety in oxidation and spin states also opens intriguing opportunities for redox [77] and spin crossover [78] responsive applications.

While the bulk of work around TMs for MRI have focused on traditional $T_1$ small molecule contrast agents or the far more studied nanoparticles for $T_2$-weighted imaging, there has been some recent interest in the development of TM-based approaches for both $^1H$ and $^{19}F$ paramagnetic chemical shift imaging. The requirements and physical properties of TM PARASHIFT compounds have been less well characterised, though there have been some attempts to group and classify them alongside related Ln(III) systems [79]. It should be noted that there have also been considerable related efforts in the development of TM-based complexes for PARACEST applications, with potential for overlap in future directions. While these systems will not be discussed further here, they are detailed elsewhere in this volume.

The bulk of work in chemical shift approaches using paramagnetic TM ions has involved $^{19}F$ nuclei as the reporter group, though there has been growing recent interest in the development of systems with distinct $^1H$ reporters. The general design principles for TM-based PARASHIFT complexes are the same as for Ln(III) – multiple equivalents of spin-active nuclei in close proximity to a paramagnetic ion. In comparison to *f*-electron driven chemistry, the more varied nature of *d*-block metal chemistry simultaneously complicates design while providing for new possible approaches. A range of TM ions have been explored for use as PARASHIFT probes, with the only requirement being the presence of unpaired electron(s). While Ln(III) ligand design largely focuses on stable thermodynamic coordination, TM ligand design needs to additionally account for stabilising the desired oxidation and spin states of the ion. Variation in either of these states will have dramatic impact on the observed PARASHIFT behavior, though this variation can also be taken advantage of in the design of responsive agents. Initial ligand designs have often focused on similar architectures to those used with Ln(III), including cyclen scaffolds and the related cyclam derivative.

## 4.1 TM-Based $^{19}F$ PARASHIFT Systems

There have been a number of cyclam derivatives studied containing $CF_3$-capped pendant arms, including the complexation of Co(II), Cu(II), and Ni(II) to generate paramagnetic shift behavior. Blahut, Hermann, and coworkers compared the paramagnetic Co(II) and diamagnetic Co(III) complexes of *trans*-$CF_3$ labelled cyclam ligands (**4-i**, Figure 5) with accompanying X-ray diffraction and *ab initio* and DFT calculations [24]. The accompanying calculations, alongside the characterisation of paramagnetic NMR spectroscopic behavior, are one of the few contributions to understanding the basis behind paramagnetic NMR shift and relaxation in macrocyclic TM complexes towards MR applications. Similar to the previously discussed related Ln(III) complexes, the most promising observations

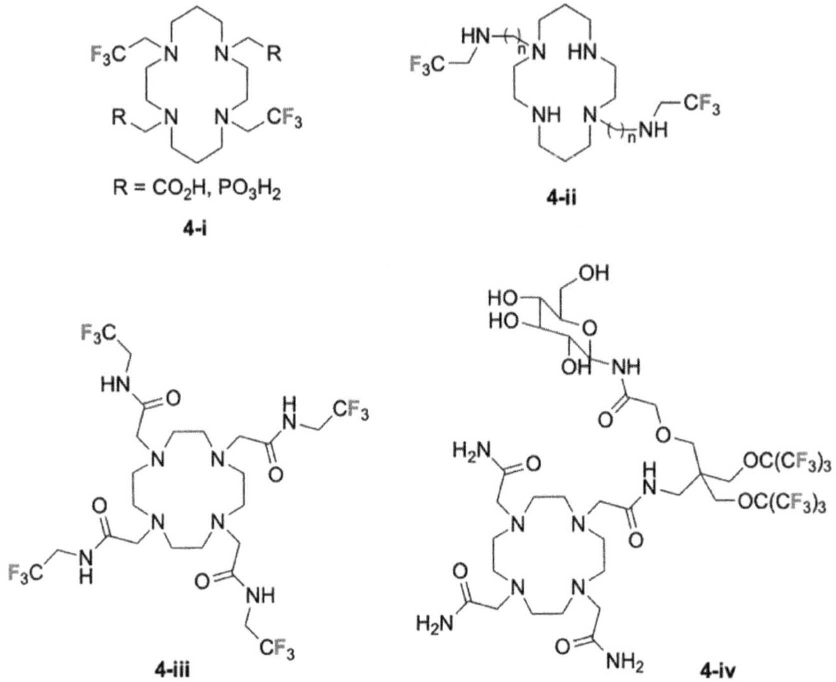

**FIGURE 5** Structures of select TM-based $^{19}$F PARASHIFT systems. The $^{19}$F reporter nuclei are highlighted in each case.

were with the bulkier phosphonate containing complex. The fluorine atoms in these systems are located around 5.3 Å from the Co(II). Fast longitudinal relaxation of the $^{19}$F nuclei (ca. 80 $s^{-1}$) was observed with excellent $T_2^*/T_1$ ratios (>0.78). The $^{19}$F NMR spectra were well resolved, with shifts of around 35 ppm from the diamagnetic analogues. Most strikingly, the $^{1}$H and $^{13}$C NMR spectra were also highly resolved with the presence of large paramagnetic shifts, indicating potential utility in $^{1}$H PARASHIFT studies as well. After careful manipulation of the pulse sequence, the $^{13}$C resonance of the $CF_3$ carbon could be detected at 936 ppm, which is >1000 ppm from any other $^{13}$C resonance in the molecule. The accompanying computational simulations suggest a delocalisation of single-occupied molecular orbitals from Co(II) to ligand atoms, with the resulting contact spin density leading to this extreme $^{13}$C shift. These simulations also suggest the importance of Fermi contact relaxation mechanisms in TM systems, indicating that these effects cannot be assumed negligible as they often are for Ln(III).

The same group has previously demonstrated the potential of Ni(II) cyclam analogues for *in vitro* use in cell labelling studies with ultrashort TE sequences [80]. They have also recently expanded studies with Co(II) [81], including comparative analysis alongside Ni(II), Co(II), and Zn(II) [82]. Comparison of these related approaches, including understanding the effect of TM to $CF_3$ distance through variable pendant arm length (**4-ii**, Figure 5) [81], is aiding to build up the knowledge of TM-based paramagnetic shift towards imaging applications.

Other groups have focused on cyclen-based ligands for TM-based $^{19}F$ PARASHIFT approaches. Srivastava et al. directly compared the paramagnetic shift properties of Fe(II) against the analogous Ln(III) complexes on an eight-coordinate DOTAM ligand (**4-iii**, Figure 5) [83]. While some of the Ln(III) ions demonstrated significantly enhanced paramagnetic shift compared to the Fe(II) system, the relaxation properties of the $^{19}F$ nuclei were found to be superior in the TM complex. In water, the best performing Ln(III) ions (Ho/Tm/Yb) displayed comparative sensitivity to Fe(II). In blood, however, significant increases in the transverse relaxation rates were observed with Ln(III) complexes, impacting the sensitivity and rendering the signal unobservable in some cases. The Fe(II) complex retained a $T_2/T_1$ ratio close to unity (0.98 in water, 0.57 in blood) and overall provided the fastest longitudinal relaxation (175 $s^{-1}$) and the highest estimated sensitivity (300 μM) of the ions tested.

The Que group designed a cyclen-based ligand containing 18 magnetically equivalent fluorine atoms, the highest reported in a stable, small molecule complex (**4-iv**, Figure 5) [84]. The $^{19}F$ resonances of subsequent Fe(II), Co(II), and Ni(II) complexes were not highly shifted in comparison to the diamagnetic ligand, likely due to a relatively long distance and weak through through-space interactions. Longitudinal relaxation rates, however, were increased over 30-fold (0.7–22 $s^{-1}$ at 7 T) with $R_2/R_1$ ratios near parity. Interestingly, the Ni(II) displayed a highly field-dependent longitudinal relaxation, with an increase from 16 $s^{-1}$ at 7 T to almost 70 $s^{-1}$ at 9.4 T (>400% $\Delta R_1$ compared to maximum 45% increase observed with other metal ions). The authors attribute this observation to the relatively longer electronic relaxation time of Ni(II), the effect of which is reduced at higher field. Variable concentration phantom experiments with these complexes in aqueous buffer estimated the detection sensitivity at 40–60 μM. These values are among the highest reported for $^{19}F$ MR systems, though assessment in more biologically relevant settings is required to directly compare with related systems. The presence of the amide pendant arms of the DOTAM scaffold also provides a suitable handle for PARACEST experiments, providing routes for dual-imaging approaches.

## 4.2 TM-Based $^1H$ PARASHIFT Systems

Early work on $^1H$-based TM PARASHIFT arose from complexes previously investigated for use as PARACEST agents [85,86]. Morrow and coworkers identified that common macrocyclic amine scaffolds (e.g., cyclen and TACN) can stabilize the high-spin state of Fe(II), in addition to providing protection against common biological reactivity and oxidation. They also recognized that the high symmetry of these complexes is favorable for increasing the number of chemically equivalent non-exchangeable protons, with distinct paramagnetically induced shifts [87]. However, difficulties were observed in the dynamic nature of several of the complexes, leading to significant broadening of certain resonances.

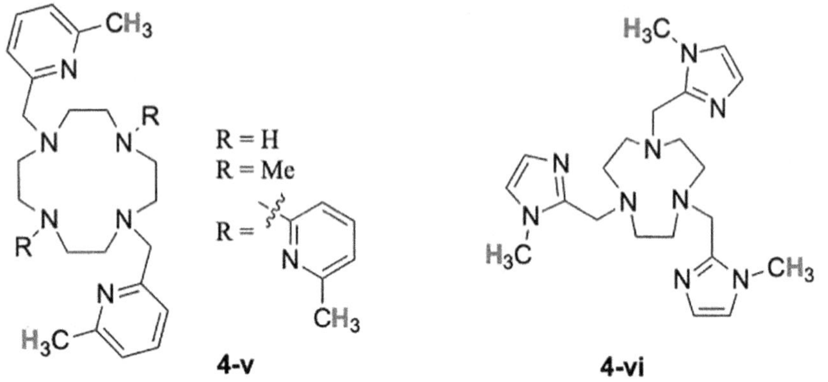

**4-v** **4-vi**

**FIGURE 6** Structures of select TM-based [1]H PARASHIFT systems. The [1]H reporter nuclei are highlighted in each case.

More recent work from the same group has focused on expanding the use of these scaffolds with further metal ions. While Ni(II) complexes of cyclen, cyclam, and TACN scaffolds were predominantly investigated for PARACEST, they also demonstrated significant hyperfine shifted proton resonances of up to 250 ppm [88]. Cyclen-based ligands with purposefully designed pendant arms were explored with Fe(II) and Co(II), with methyl substituents at various distances and orientations from the metal centre explored as potential [1]H PARASHIFT reporters (**4-v**, Figure 6) [89]. In addition to highly shifted resonances (~20–160 ppm), these Me reporters also demonstrate remarkably rapid longitudinal relaxation (>1000 $s^{-1}$) with ideal $T_2/T_1$ ratios of 0.8–1. Related TACN-based complexes (**4-vi**, Figure 6) have even demonstrated that the single unpaired electron in low-spin Fe(III) can be used to generate moderately shifted proton resonances, though inducing a large enough shift in the desired reporter group is likely challenging [90].

## 4.3 Applications of TM-Based PARASHIFT Systems

### 4.3.1 Temperature Sensitivity

As with Ln(III) complexes, temperature sensitivity is an intrinsic property of TM-based PARASHIFT. For example, variable temperature NMR spectroscopy of the cyclen-based [1]H reporters discussed in the previous section displayed temperature-dependent chemical shifts of up to 0.52 ppm $K^{-1}$ for the methyl reporters and up to 1.10 ppm $K^{-1}$ for the most highly shifted resonances in Co(II) complexes [91]. These resonances displayed negligible line broadening up to 60°C and were shown to be independent of pH (5.6–8.6) and Ca(II) concentrations, indicating promise for use in MR thermometry.

### 4.3.2 pH-Responsive Probes

Similar to examples with Ln(III)-based systems, TM PARASHIFT complexes have also been proposed for applications in pH mapping and imaging, as well as in diagnosis of cancer and other diseases, though relatively few systems have been successfully demonstrated to date. The Que group designed a Ni(II) dioxocyclam derivative with a pH-sensitive intramolecular pyridine donor [92]. While the pyridine aided in stabilizing the high-spin Ni(II) and generating a distinct $^{19}F$ chemical shift (compared to the diamagnetic analogue without the pyridine), lowering the pH destabilised the complex, and the pH-response could not be determined.

The use of $^{19}F$ resonances in Fe(II) complexes has also been demonstrated to act as an internal calibration for measuring pH by other methods, in this case PARACEST [93]. A fundamental challenge with most MR-based pH sensing approaches is that the signal is proportional to both concentration and pH, as in this case where the magnitude of the CEST response is dependent on both pH and complex concentration. With a pH-insensitive internal reference – the $^{19}F$ resonance in this case – the effect of concentration can then be corrected to extrapolate the direct contribution of pH to the CEST signal.

### 4.3.3 Redox Active Probes

While temperature and pH applications have been demonstrated more widely for Ln(III) complexes, redox activity is an area primed for exploitation with TMs. Understanding of redox processes is crucial in elucidating physiological and pathological processes. While this area is well explored with luminescence approaches [94], approaches using MR probes have been far more limited. The highly stable +3 oxidation state of the lanthanides renders direct metal-center redox processes improbable in a biological setting, outside of a few outlier ions such as Eu(II/III) [95]. TMs, on the other hand, are known to exist in a broad range of oxidation states. Careful ligand design can lead to systems capable of stabilising various oxidation states of the same metal [96], providing routes to biologically triggered redox processes. Due to the change in electron configuration of the metal ion during redox, any incorporated PARASHIFT resonance would be expected to undergo significant modification.

Morrow and coworkers demonstrated an example of redox-triggered PARASHIFT in a simple and elegant Co complex (**4-vii**, Figure 7) – so simple, in fact, that it formed the basis of undergraduate research experiments [97]. In this experiment, a classical diamagnetic Co(III) cage complex is reduced to paramagnetic Co(II), with an accompanying dramatic increase in the chemical shift range of the $^{1}H$ resonances of the cage from $<1.5$ ppm to $>400$ ppm. The high symmetry of the cage leads to relatively simple to elucidate spectra, making them ideal for student interpretation.

The Que group have focused on the development of biologically relevant redox processes. Initial studies demonstrated a NOTA-based $CF_3$-containing scaffold (**4-viii**, Figure 7) with complexation of paramagnetic Co(II) [98]. Oxidation with hydrogen peroxide led to the formation of the diamagnetic Co(III), with distinct changes to the relaxation rate and chemical shift of the $^{19}F$ resonance. While the focus here was on modulation of relaxation, there was clear chemical shift inequivalence between the two oxidation states. This chemical shift difference was enhanced with the subsequent incorporation of a $C(CF_3)_3$ reporter group

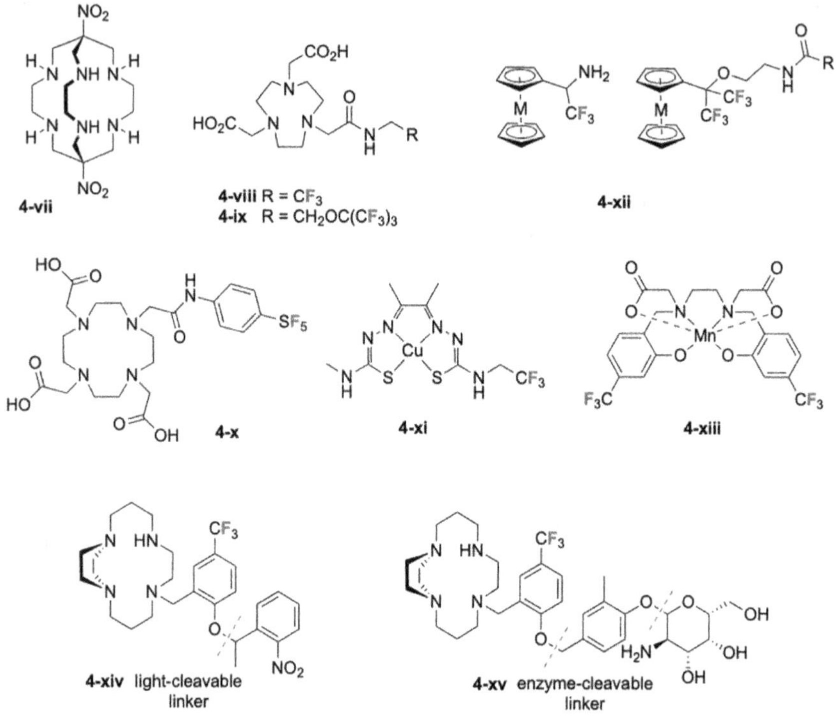

**FIGURE 7** Structures of select TM-based responsive PARASHIFT systems. The $^{19}F$ reporter nuclei are highlighted in each case. For **4-vii**, $^{1}H$ resonances throughout the molecule are used to track redox processes.

(**4-ix**, Figure 7), with variation in the ligand architecture providing different reactivity between $H_2O_2$ alone or in the presence of peroxidase enzymes [99].

Recent work in the same group demonstrated the potential for redox switching between two paramagnetic oxidation states, in this case Fe(II/III) [100]. Upon cysteine addition to an $SF_5$-containing DO3A ligand (**4-x**, Figure 7), high-spin Fe(III) ($S$ = 5/2) was reduced to high-spin Fe(II) ($S$ = 2) with an accompanying "turn on" of a paramagnetically enhanced $^{19}F$ signal (signal suppressed in high-spin state). This redox state could be interchangeably reversed with increasing additions of $H_2O_2$ and cysteine. To demonstrate potential for redox imaging in hypoxic environments, the complex was combined with their previously reported $CF_3$-tagged hypoxia probe (**4-xi**, Figure 7) [101], which itself is a Cu(I/II) redox switch. The differing frequencies of the $^{19}F$ resonances of the two complexes demonstrate the potential for multiplexed imaging approaches [100]. Related recent examples also include the demonstration of redox-dependent paramagnetic $^{19}F$ switches in polymeric fluorinated ferrocene complexes (**4-xii**, Figure 7) [102].

A recent Mn(II/III) example showed promise for dual $^{1}H/^{19}F$ redox imaging [103]. The Mn(II) complex (**4-xiii**, Figure 7) displays a significantly broadened $^{19}F$ resonance with enhanced relaxation of the $^{1}H$ water signal, while oxidation to the

Mn(III) reduces the $^{1}H$ MR contrast but "turns on" a paramagnetically enhanced $^{19}F$ MR signal. This redox process could be controlled by biologically relevant levels of glutathione and ascorbic acid. The dual imaging approach was successfully demonstrated *in vitro* following incubation in a human cell line.

### 4.3.4 Spin State-Responsive Probes

The ability of certain d-block metals to exist in either their high- or low-spin state unlocks intriguing possibilities for spin crossover PARASHIFT probes. The number of unpaired electrons is intrinsically linked to the PARASHIFT properties – if the metal spin state is triggered by physiological conditions, it would provide a route to a unique class of responsive imaging probes. The Que group have demonstrated an elegant example of this approach, triggering spin state switch of a $^{19}F$-containing cyclam Ni(II) complex by either light (**4-xiv**, Figure 7) or enzyme (**4-xv**, Figure 7) activity [78]. The light-activated or enzyme-induced cleavage of a deliberately designed self-immolative moiety results in distinct changes in the Ni(II) coordination, triggering conversion from a low-spin ($S$ = 0) to high-spin ($S$ = 1) species. The paramagnetic high-spin Ni(II) induces a large $^{19}F$ chemical shift change compared to the initial low-spin system ($\Delta\delta_F > 35$ ppm). Demonstration of this spin state sensing of light and enzyme activity was performed *in vitro* (HeLa and HEK 293T cells, respectively).

Fe(II) spin crossover systems have also been demonstrated to show potential for improved thermometry properties [104]. Related work also demonstrated linking pH to Fe(II) spin state population to create a highly sensitive $^{19}F$-based pH sensor [105].

## 5 CONCLUDING REMARKS AND FUTURE DIRECTIONS

Despite almost 30 years of study, the area of PARASHIFT imaging remains an untapped potential. Various groups have demonstrated its application in living systems, yet no commercial biological tool or clinical agent exists. An increased awareness of the potential of PARASHIFT imaging may drive the next generation of agents, with the majority of studies to date limited to a handful of research groups. Demonstration in real-world approaches and addressing unmet challenges may expand the current horizons.

One of the most significant advantages of PARASHIFT complexes, in comparison to more traditional methods of MR contrast, is the ability to perform multichannel imaging. Each chemical shift is a discrete frequency that can be simultaneously acquired in the same imaging acquisition. With careful choice of PARASHIFT agents – combining similar relaxation properties ($R_1/R_2$ values) with large chemical shift separation (e.g., Tm *versus* Dy) – each frequency can then be used as a different imaging channel. This multichannel approach can be thought of as akin to the common blue/green/red imaging applied in confocal microscopy and similar techniques, where each channel typically reports on a different molecular marker. While only a handful of studies have demonstrated and highlighted this approach [61,100,106], incorporation of this multichannel/multiplex methodology to the much deeper tissue imaging capabilities of MRI will result in powerful new tools for mapping biological response to disease and therapy.

## ABBREVIATIONS AND DEFINITIONS

| | |
|---|---|
| **BAPTA** | 1,2-bis(2-aminophenoxy)ethane-N,N,N′,N′-tetraacetic acid |
| **BIRDS** | biosensor imaging of redundant deviation in shifts |
| **CEST** | chemical exchange saturation transfer |
| **DFT** | density functional theory |
| **DOTA** | 1,4,7,10-tetraazacyclododecane-1,4,7,10-tetraacetic acid |
| **DOTAM** | 1,4,7,10-tetrakis(carbamoylmethyl)-1,4,7,10-tetraazacyclododecane |
| **DOTMA** | (1R,4R,7R,10R)-$\alpha,\alpha',\alpha'',\alpha'''$-tetramethyl-1,4,7,10-tetraazacyclododecane-1,4,7,10-tetraacetic acid |
| **DOTP** | 1,4,7,10-tetraazacyclododecane-1,4,7,10-tetra(methylenephosphonic acid) |
| **DO3A** | 1,4,7,10-tetraazacyclododecane-1,4,7-triacetic acid |
| **DTPA** | diethylenetriaminepentaacetic acid |
| **EDTA** | ethylenediaminetetraacetic acid |
| **Ln** | lanthanide |
| **MRI** | magnetic resonance imaging |
| **MRS** | magnetic resonance spectroscopy |
| **NMR** | nuclear magnetic resonance |
| **NOTA** | 1,4,7-triazacyclononane-1,4,7-triacetic acid |
| **PAMAM** | poly(amidoamine) |
| **PARACEST** | paramagnetic chemical exchange saturation transfer |
| **PARASHIFT** | paramagnetically induced chemical shift |
| **PFC** | perfluorocarbon |
| $\mathbf{R_1}$ | longitudinal relaxation rate |
| $\mathbf{R_2}$ | transverse relaxation rate |
| $\mathbf{T_1}$ | longitudinal relaxation time |
| $\mathbf{T_2}$ | transverse relaxation time |
| **TACN** | 1,4,7-triazacyclononane |
| **TE** | echo time |
| **TM** | transition metal |
| **TPA** | tris(2-pyridylmethyl)amine |

## REFERENCES


1. J. Wahsner, E. M. Gale, A. Rodríguez-Rodríguez, P. Caravan, *Chem. Rev.* **2019**, *119*, 957.
2. J. Ruiz-Cabello, B. P. Barnett, P. A. Bottomley, J. W. M. Bulte, *NMR Biomed.* **2011**, *24*, 114.
3. I. Tirotta, V. Dichiarante, C. Pigliacelli, G. Cavallo, G. Terraneo, F. B. Bombelli, P. Metrangolo, G. Resnati, *Chem. Rev.* **2015**, *115*, 1106.

4. C. C. Hinckley, *J. Am. Chem. Soc.* **1969**, *91*, 5160.
5. G. Otting, *Ann. Rev. Biophys.* **2010**, *39*, 387.
6. J. Eills, D. Budker, S. Cavagnero, E. Y. Chekmenev, S. J. Elliott, S. Jannin, A. Lesage, J. Matysik, T. Meersmann, T. Prisner, J. A. Reimer, H. Yang, I. V. Koptyug, *Chem. Rev.* **2023**, *123*, 1417.
7. N. Bloembergen, W. C. Dickinson, *Phys. Rev.* **1950**, *79*, 179.
8. A. C. Harnden, D. Parker, N. J. Rogers, *Coord. Chem. Rev.* **2019**, *383*, 30.
9. B. Bleaney, *J. Magn. Reson.* **1972**, *8*, 91.
10. H. M. McConnell, *J. Chem. Phys.* **1957**, *27*, 226.
11. P. Harvey, A. M. Blamire, J. I. Wilson, K.-L. N. A. Finney, A. M. Funk, P. K. Senanayake, D. Parker, *Chem. Sci.* **2013**, *4*, 4251.
12. G. Castro, M. Regueiro-Figueroa, D. Esteban-Gómez, P. Pérez-Lourido, C. Platas-Iglesias, L. Valencia, *Inorg. Chem.* **2016**, *55*, 3490.
13. A. M. Funk, K.-L. N. A. Finney, P. Harvey, A. M. Kenwright, E. R. Neil, N. J. Rogers, P. K. Senanayake, D. Parker, *Chem. Sci.* **2015**, *6*, 1655.
14. A. C. Harnden, E. A. Suturina, A. S. Batsanov, P. K. Senanayake, M. A. Fox, K. Mason, M. Vonci, E. J. L. McInnes, N. F. Chilton, D. Parker, *Angew. Chem. Int. Ed.* **2019**, *58*, 10290.
15. K. Mason, A. C. Harnden, C. W. Patrick, A.W. J. Poh, A. S. Batsanov, E. A. Suturina, M. Vonci, E.J. L. McInnes, N. F. Chilton, D. Parker, *Chem. Commun.* **2018**, *54*, 8486.
16. E. A. Suturina, K. Mason, M. Botta, F. Carniato, I. Kuprov, N. F. Chilton, E. J. L. McInnes, M. Vonci, D. Parker, *Dalton Trans.* **2019**, *48*, 8400.
17. E. A. Suturina, K. Mason, C. F. G. C. Geraldes, I. Kuprov, D. Parker, *Angew. Chem. Int. Ed.* **2017**, *56*, 12215.
18. K. H. Chalmers, E. De Luca, N. H. M. Hogg, A. M. Kenwright, I. Kuprov, D. Parker, M. Botta, J. I. Wilson, A. M. Blamire, *Chem. Eur. J.* **2010**, *16*, 134.
19. P. Harvey, I. Kuprov, D. Parker, *Eur. J. Inorg. Chem.* **2012**, *2012*, 2015.
20. A. M. Funk, P. Harvey, K.-L. N. A. Finney, M. A. Fox, A. M. Kenwright, N. J. Rogers, P. K. Senanayake, D. Parker, *Phys. Chem. Chem. Phys.* **2015**, *17*, 16507.
21. D. Parker, E.A. Suturina, I. Kuprov, N. F. Chilton, *Acc. Chem. Res.* **2020**, *53*, 1520.
22. N. J. Rogers, K.-L. N. A. Finney, P. K. Senanayake, D. Parker, *Phys. Chem. Chem. Phys.* **2016**, *18*, 4370.
23. E. A. Suturina, K. Mason, C. F. G. C. Geraldes, N. F. Chilton, D. Parker, I. Kuprov, *Phys. Chem. Chem. Phys.* **2018**, *20*, 17676.
24. J. Blahut, L. Benda, J. Kotek, G. Pintacuda, P. Hermann, *Inorg. Chem.* **2020**, *59*, 10071.
25. H. Lee, R. R. Price, G. E. Holburn, C. L. Partain, M. D. Adams, W. P. Cacheris, *J. Magn. Reson. Imaging* **1994**, *4*, 609.
26. C. S. Bonnet, P. H. Fries, *ChemPhysChem* **2010**, *11*, 3474.
27. E. Terreno, M. Botta, W. Dastrù, S. Aime, *Contrast Media Mol. Imaging* **2006**, *1*, 101.
28. W. D. Kim, G. E. Kiefer, J. Huskens, A. D. Sherry, *Inorg. Chem.* **1997**, *36*, 4128.
29. Z.-X. Jiang, Y. Feng, Y. B. Yu, *Chem. Commun.* **2011**, *47*, 7233.
30. A. M. Kenwright, I. Kuprov, E. D. Luca, D. Parker, S. U. Pandya, P. K. Senanayake, D. G. Smith, *Chem. Commun.* **2008**, 2514–2516.
31. P. K. Senanayake, A. M. Kenwright, D. Parker, S. K. van der Hoorn, *Chem. Commun.* **2007**, 2923.
32. K.H. Chalmers, M. Botta, D. Parker, *Dalton Trans.* **2011**, *40*, 904.
33. E. De Luca, P. Harvey, K. H. Chalmers, A. Mishra, P. K. Senanayake, J. I. Wilson, M. Botta, M. Fekete, A. M. Blamire, D. Parker, *J. Biol. Inorg. Chem.* **2014**, *19*, 215.

34. A. M. Funk, P. H. Fries, P. Harvey, A. M. Kenwright, D. Parker, *J. Phys. Chem. A* **2013**, *117*, 905.
35. S. Aime, M. Botta, M. Fasano, E. Terreno, P. Kinchesh, L. Calabi, L. Paleari, *Magn. Reson. Med.* **1996**, *35*, 648.
36. S. Aime, M. Botta, L. Milone, E. Terreno, *Chem. Commun.* **1996**, 1265.
37. T. Frenzel, K. Roth, S. Koßler, B. Radüchel, H. Bauer, J. Platzek, H.-J. Weinmann, *Magn. Reson. Med.* **1996**, *35*, 364.
38. K. Roth, G. Bartholomae, H. Bauer, T. Frenzel, S. Kossler, J. Platzek, H.-J. Weinmann, *Angew. Chem. Int. Ed. Engl.* **1996**, *35*, 655.
39. C. S. Zuo, J. L. Bowers, K. R. Metz, T. Nosaka, A. D. Sherry, M. E. Clouse, *Magn. Reson. Med.* **1996**, *36*, 955.
40. S. K. Hekmatyar, P. Hopewell, S. K. Pakin, A. Babsky, N. Bansal, *Magn. Reson. Med.* **2005**, *53*, 294.
41. S. K. Pakin, S. K. Hekmatyar, P. Hopewell, A. Babsky, N. Bansal, *NMR Biomed.* **2006**, *19*, 116.
42. D. Coman, H. K. Trubel, R. E. Rycyna, F. Hyder, *NMR Biomed.* **2009**, *22*, 229.
43. Y. Huang, D. Coman, F. Hyder, M. M. Ali, *Bioconjugate Chem.* **2015**, *26*, 2315.
44. F. Schmid, C. Höltke, D. Parker, C. Faber, *Magn. Reson. Med.* **2013**, *69*, 1056.
45. P. K. Senanayake, N. J. Rogers, K.-L. N. A. Finney, P. Harvey, A. M. Funk, J. I. Wilson, D. O'Hogain, R. Maxwell, D. Parker, A. M. Blamire, *Magn. Reson. Med.* **2017**, *77*, 1307.
46. S. Ghosh, P. Harvey, J. C. Simon, A. Jasanoff, *Curr. Opin. Neurobiol.* **2018**, *50*, 201.
47. M. L. Garcia-Martin, G. V. Martinez, N. Raghunand, A. D. Sherry, S. Zhang, R. J. Gillies, *Magn. Reson. Med.* **2006**, *55*, 309.
48. A. Li, X. Luo, L. Li, D. Chen, X. Liu, Z. Yang, L. Yang, J. Gao, H. Lin, *Anal. Chem.* **2021**, *93*, 16552.
49. H. Lin, X. Tang, A. Li, J. Gao, *Adv. Mater.* **2021**, *33*, 2005657.
50. P. Wust, B. Hildebrandt, G. Sreenivasa, B. Rau, J. Gellermann, H. Riess, R. Felix, P. M. Schlag, *Lancet Oncol.* **2002**, *3*, 487.
51. E. J. Walter, M. Carraretto, *Crit. Care* **2016**, *20*, 199.
52. H. Odéen, D. L. Parker, *Prog. Nucl. Magn. Reson. Spectrosc.* **2019**, *110*, 34.
53. D. Coman, H. K. Trubel, F. Hyder, *NMR Biomed.* **23**, 277 (2010).
54. A. B. M. Zakaria, Y. Huang, D. Coman, S. K. Mishra, J. M. Mihailovic, S. Maritim, F. A. Rojas-Quijano, P. Jurek, G. E. Kiefer, F. Hyder, *NMR Biomed.* **2022**, *35*, e4687.
55. F. Mysegaes, P. Voigt, P. Spiteller, I. Prediger, J. Bernarding, M. Plaumann, *Chem. Commun.* **2023**, *59*, 9340–9343.
56. D. Coman, D. C. Peters, J. J. Walsh, L. J. Savic, S. Huber, A. J. Sinusas, M. Lin, J. Chapiro, R. T. Constable, D. L. Rothman, J. S. Duncan, F. Hyder, *Magn. Reson. Med.* **2020**, *83*, 1553.
57. D. Coman, Y. Huang, J. U. Rao, H. M. De Feyter, D. L. Rothman, C. Juchem, F. Hyder, *NMR Biomed.* **2016**, *29*, 309.
58. L. J. Savic, I. T. Schobert, C. A. Hamm, L. C. Adam, F. Hyder, D. Coman, *NMR Biomed.* **2021**, *34*, e4465.
59. J. U. Rao, D. Coman, J. J. Walsh, M. M. Ali, Y. Huang, F. Hyder, *Sci. Rep.* **2017**, *7*, 7865.
60. D. Janasik, K. Jasiński, J. Szreder, W. P. Węglarz, T. Krawczyk, *ACS Sens.* **2023**, *8*, 1971.
61. K.-L. N. A. Finney, A. C. Harnden, N. J. Rogers, P. K. Senanayake, A. M. Blamire, D. O'Hogain, D. Parker, *Chem. Eur. J.* **2017**, *23*, 7976.
62. D. J. Hare, E. J. New, M. D. de Jonge, G. McColl, *Chem. Soc. Rev.* **2015**, *44*, 5941.

63. A. Barandov, B. B. Bartelle, C. G. Williamson, E. S. Loucks, S. J. Lippard, A. Jasanoff, *Nat. Commun.* **2019**, *10*, 897.
64. G. D. Thiabaud, M. Schwalm, S. Sen, A. Barandov, J. Simon, P. Harvey, V. Spanoudaki, P. Müller, J. L. Sessler, A. Jasanoff, *ACS Sens.* **2023**, *8*, 3855.
65. P. Harvey, K. H. Chalmers, E. De Luca, A. Mishra, D. Parker, *Chem. Eur. J.* **2012**, *18*, 8748.
66. A. C. Harnden, A. S. Batsanov, D. Parker, *Chem. Eur. J.* **2019**, *25*, 6212.
67. M. Yu, D. Xie, R. T. Kadakia, W. Wang, E. L. Que, *Chem. Commun.* **2020**, *56*, 6257.
68. G. Gambino, T. Gambino, R. Pohmann, G. Angelovski, *Chem. Commun.* **2020**, *56*, 3492.
69. M. H. Khan, S. K. Mishra, A. B. M. Zakaria, J. M. Mihailović, D. Coman, F. Hyder, *Anal. Chem.* **2022**, *94*, 2536.
70. A. D. Sherry, C. R. Malloy, F. M. H. Jeffrey, W. P. Cacheris, C. F. G. C. Geraldes, *J. Magn. Reson.* **1988**, *76*, 528.
71. E. Lancelot, J.-S. Raynaud, P. Desché, *Investig. Radiol.* **2020**, *55*, 578.
72. V. Hatje, K. W. Bruland, A. R. Flegal, *Environ. Sci. Technol.* **2016**, *50*, 4159.
73. A. Gupta, P. Caravan, W. S. Price, C. Platas-Iglesias, E. M. Gale, *Inorg. Chem.* **2020**, *59*, 6648.
74. H. V.-T. Nguyen, Q. Chen, J. T. Paletta, P. Harvey, Y. Jiang, H. Zhang, M. D. Boska, M. F. Ottaviani, A. Jasanoff, A. Rajca, J. A. Johnson, *ACS Cent. Sci.* **2017**, *3*, 800.
75. E. A. Kras, E. M. Snyder, G. E. Sokolow, J. R. Morrow, *Acc. Chem. Res.* **2022**, *55*, 1435.
76. A. Barandov, B. B. Bartelle, B. A. Gonzalez, W. L. White, S. J. Lippard, A. Jasanoff, *J. Am. Chem. Soc.* **2016**, *138*, 5483.
77. J. R. Morrow, J. J. Raymond, M. S. I. Chowdhury, P. R. Sahoo, *Inorg. Chem.* **2022**, *61*, 14487.
78. D. Xie, M. Yu, Z.-L. Xie, R. T. Kadakia, C. Chung, L. E. Ohman, K. Javanmardi, E. L. Que, *Angew. Chem.* **2020**, *132*, 22712.
79. M. Zalewski, D. Janasik, A. Wierzbicka, T. Krawczyk, *Inorg. Chem.* **2022**, *61*, 19524.
80. J. Blahut, K. Bernášek, A. Gálisová, V. Herynek, I. Císařová, J. Kotek, J. Lang, S. Matějková, P. Hermann, *Inorg. Chem.* **2017**, *56*, 13337.
81. Z. Kotková, F. Koucký, J. Kotek, I. Císařová, D. Parker, P. Hermann, *Dalton Trans.* **2023**, *52*, 1861.
82. F. Koucký, J. Kotek, I. Císařová, J. Havlíčková, V. Kubíček, P. Hermann, *Dalton Trans.* **2023**, *52*, 12208.
83. K. Srivastava, E. A. Weitz, K. L. Peterson, M. Marjańska, V. C. Pierre, *Inorg. Chem.* **2017**, *56*, 1546.
84. M. Yu, B. S. Bouley, D. Xie, E. L. Que, *Dalton Trans.* **2019**, *48*, 9337.
85. S. J. Dorazio, P. B. Tsitovich, K. E. Siters, J. A. Spernyak, J. R. Morrow, *J. Am. Chem. Soc.* **2011**, *133*, 14154.
86. S. J. Dorazio, J. R. Morrow, *Eur. J. Inorg. Chem.* **2012**, *2012*, 2006.
87. P. B. Tsitovich, J. R. Morrow, *Inorganica Chim. Acta* **2012**, *393*, 3.
88. S. M. Abozeid, E. M. Snyder, A. P. Lopez, C. M. Steuerwald, E. Sylvester, K. M. Ibrahim, R. R. Zaky, H. M. Abou-El-Nadar, J. R. Morrow, *Eur. J. Inorg. Chem.* **2018**, *2018*, 1902.
89. P. B. Tsitovich, T. Y. Tittiris, J. M. Cox, J. B. Benedict, J. R. Morrow, *Dalton Trans.* **2018**, *47*, 916.
90. P. B. Tsitovich, F. Gendron, A. Y. Nazarenko, B. N. Livesay, A. P. Lopez, M. P. Shores, J. Autschbach, J. R. Morrow, *Inorg. Chem.* **2018**, *57*, 8364.

91. P. B. Tsitovich, J. M. Cox, J. B. Benedict, J. R. Morrow, *Inorg. Chem.* **2016**, *55*, 700.
92. D. Xie, L. E. Ohman, E. L. Que, *Magn. Reson. Mater. Phy.* **2019**, *32*, 89.
93. K. Srivastava, G. Ferrauto, V. G. Jr. Young, S. Aime, V. C. Pierre, *Inorg. Chem.* **2017**, *56*, 12206.
94. A. Kaur, J. L. Kolanowski, E. J. New, *Angew. Chem. Int. Ed.* **2016**, *55*, 1602.
95. L. A. Basal, M. D. Bailey, J. Romero, M. M. Ali, L. Kurenbekova, J. Yustein, R. G. Pautler, M. J. Allen, *Chem. Sci.* **2017**, *8*, 8345.
96. E. M. Gale, C. M. Jones, I. Ramsay, C. T. Farrar, P. Caravan, *J. Am. Chem. Soc.* **2016**, *138*, 15861.
97. P. J. Burns, P. B. Tsitovich, J. R. Morrow, *J. Chem. Educ.* **2016**, *93*, 1115.
98. M. Yu, D. Xie, K. P. Phan, J. S. Enriquez, J. J. Luci, E. L. Que, *Chem. Commun.* **2016**, *52*, 13885.
99. M. Yu, B. S. Bouley, D. Xie, J. S. Enriquez, E. L. Que, *J. Am. Chem. Soc.* **2018**, *140*, 10546.
100. R. T. Kadakia, R. T. Ryan, D. J. Cooke, E. L. Que, *Chem. Sci.* **2023**, *14*, 5099.
101. D. Xie, T. L. King, A. Banerjee, V. Kohli, E. L. Que, *J. Am. Chem. Soc.* **2016**, *138*, 2937.
102. P. Švec, O. V. Petrov, J. Lang, P. Štěpnička, O. Groborz, D. Dunlop, J. Blahut, K. Kolouchová, L. Loukotová, O. Sedláček, T. Heizer, Z. Tošner, M. Šlouf, H. Beneš, R. Hoogenboom, M. Hrubý, *Macromolecules* **2022**, *55*, 658.
103. H. Chen, X. Tang, X. Gong, D. Chen, A. Li, C. Sun, H. Lin, J. Gao, *Chem. Commun.* **2020**, *56*, 4106.
104. A. E. Thorarinsdottir, A. I. Gaudette, T. D. Harris, *Chem. Sci.* **2017**, *8*, 2448.
105. A. I. Gaudette, A. E. Thorarinsdottir, T. D. Harris, *Chem. Commun.* **2017**, *53*, 12962.
106. Y. Jiang, X. Luo, L. Chen, H. Lin, J. Gao, *Fundamental Res.* **2022**.

# 3 Targeted MRI Contrast Agents

*Alexander G. Sertage*[*], *Md. Sydul Islam*[*], and *Matthew J. Allen*[†]
Department of Chemistry, Wayne State University, 5101 Cass Avenue,
Detroit, MI 48202, USA
alexandersertage@wayne.edu,
sydul.islam@wayne.edu, mallen@chem.wayne.edu

## CONTENTS



[*] These authors contributed equally.
[†] Corresponding author.

DOI: 10.1201/9781003374688-3

**Abstract**

Targeted contrast agents for magnetic resonance imaging (MRI) are contrast agents that have been modified to accumulate in specific areas. Targeted contrast agents have the potential to enable the use of small amounts of contrast agents because of localized accumulation, which could greatly assist in the early diagnosis and monitoring of diseases. The development of targeted contrast agents is challenging because they must be specific and retain the ability to enhance contrast. This task is especially difficult to translate from *in vitro* and cell-based studies to *in vivo* systems. In this chapter, we present examples from the past ten years of targeted contrast agents for MRI that have been demonstrated to function *in vivo*.

## KEYWORDS

Targeted Contrast Agents; Magnetic Resonance Imaging; *In vivo* Imaging; Relaxivity; Specificity; Tumor Targeting

## 1 INTRODUCTION

Magnetic resonance imaging (MRI) is a non-invasive technique that uses applied magnetic fields and radiofrequency radiation to generate images of objects, including tissues and organs, with high spatial resolution. Contrast agents administered during MRI enhance contrast most commonly by quickly relaxing the nuclear spins of protons of nearby water molecules. To increase the information provided by contrast-enhanced MRI, contrast agents have been designed to target organs and receptors to enhance contrast in specific regions, and these probes are referred to as targeted contrast agents or probes. This chapter contains examples of targeted probes for MRI reported from 2013 to 2023. For reviews covering targeted contrast agents reported prior to 2013, readers are referred elsewhere [1–4]. This chapter is not exhaustive, but it features examples that are grouped based on the location of the target and have demonstrated efficacy *in vivo*. The chapter is limited to *in vivo* studies because *in vitro* and cell studies do not always translate to successful targeting *in vivo*. Some of the probes described in this chapter are multimodal, but we focus on the part of the probes relevant to MRI in this chapter. For reviews of multimodal probes, readers are referred elsewhere [5–8].

# 2 TUMORS

Tumors are primary diagnostic signs of cancer, making them targets for imaging. Imaging tumors is necessary because the location, size, and spread of tumors are relevant to the selection of treatment options. With respect to contrast agents for MRI, gadolinium-based agents are the most used clinically, but other types of imaging probes have demonstrated promise in targeting tumors preclinically. In this chapter, tumor targeting with a variety of contrast agents is organized based on the strategy used for tumor targeting. The selected examples highlight recently reported *in vivo* targeting strategies.

## 2.1 General Tumor Targeting

### 2.1.1 Enhanced Permeability and Retention Effect

The enhanced permeability and retention (EPR) effect is a physiological mechanism in which macromolecular compounds with molecular weights greater than roughly 40kDa accumulate in solid tumors [9, 10]. One class of contrast agents that can be modified to be compatible with the EPR effect is manganese-containing contrast agents. For example, $Mn^{II}$-diethylenetriaminepentaacetic acid (DTPA) conjugated to copper sulfide nanoparticles (**1**, Figure 1) was examined as a potential contrast agent for tumor-targeted imaging [11]. The copper sulfide nanoparticles enhance contrast in MRI due to the presence of manganese. The nanoparticles exhibited a longitudinal relaxivity ($r_1$) of 7.10 $mM^{-1}s^{-1}$ and accumulated in tumor tissue due to the EPR effect, enhancing contrast in $T_1$-weighted MRI of a tumor (Figure 2) [11].

### 2.1.2 Epidermal Growth Factor Receptor

Epidermal growth factor receptor (EGFR) is overexpressed in several types of cancer, has a key role in oncogenesis, and is a prognostic indicator. Further, human epidermal growth factor receptor 2 (HER2) is another member of the EGFR family of receptor tyrosine kinases [12]. Overexpression of HER2 is associated with malignant tumors, including breast, pancreatic, colorectal, and bladder cancers [13]. In one study targeting EGFR and HER2, tumor-specific protein cage nanoparticle-based $T_1$ contrast agents were developed. $Gd^{III}$-containing chelates were conjugated to the surface of a lumazine synthase by genetically fusing the SpyTag peptide (AHIVMVDAYKPTK) to the C-terminus of the protein cage subunits. Later, two different targeting affibody molecules of EGFR (**2**) or HER2 (**3**) were covalently linked to the surface of the Gd complex using the SpyTag-SpyCatcher bacterial superglue network (Figure 1) [14]. Affibodies mimic the specific binding of antibodies, but with smaller molecules than antibodies. In this study, affibodies selectively identified and attached to EFGR and HER2 on the surface of targeted tumor cells. The $r_1$ values of **2** and **3** were found to be 25.2 and 27.0 $mM^{-1}s^{-1}$ at 7 T, respectively. $T_1$-weighted *in vivo* MRI on tumor-bearing mice confirmed target-specific contrast enhancement by **2** and **3**.

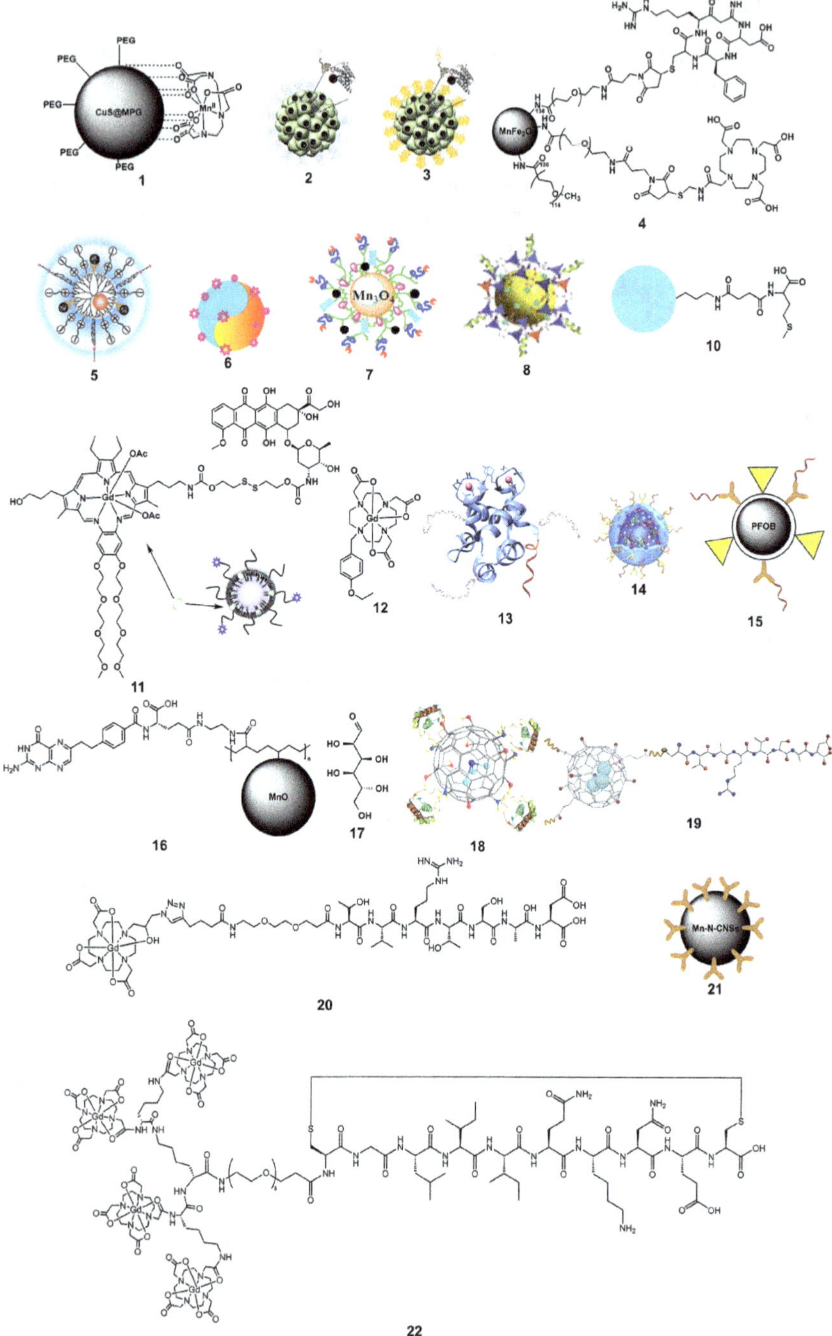


**FIGURE 1** Structures of targeted MRI contrast agents for tumor imaging. For **2** and **3**, green shapes represent lumazine synthase protein cage nanoparticles, black circles represent

(*Continued*)

**FIGURE 1 (*Continued*)** $Gd^{III}$-containing complexes, light blue spirals represent EGFR affibodies, and yellow spirals represent HER2 affibodies. (Contrast agents **2** and **3** are reprinted with permission from [14], Copyright 2021, Elsevier.) Contrast agent **5** is a generation 5 poly(amidoamine) dendrimers with amine termini functionalized with 1,3-propane sultone, PEG-RGD, and $Gd^{III}$. (Contrast agent **5** is adapted with permission from [16]. Copyright 2019, American Chemical Society.) For contrast agent **6**, the yellow and blue portions represent gold clusters and gadolinium oxide integrated nanoparticles, respectively, and pink stars represent folic acid. (The image of **6** is reprinted with permission from [17]. Copyright 2017, Royal Society of Chemistry), For contrast agent **7**, the $Mn_3O_4$ core is functionalized with polyethyleneimine represented by green lines and PEGylated folic acid represented by blue lines with a red tip. (Reprinted with permission from [18], Copyright 2015, Elsevier.) For contrast agent **8**, Mn-ZIF-90 is represented by the chains of compounds between the tetrahedra, and the $Y_1R$ ligand APT-NPY is represented by yellow helices. (Reprinted with permission from Ref. [19]. Open source link: http://creativecommons.org/licenses/by/4.0/.) For contrast agent **10**, the blue circle represents $Gd^{III}$-containing mesoporous silica. For contrast agent **11**, blue stars represent 1,2-distearoyl-sn-glycero-3-phosphoethanolamine-N-[folate(polyethylene glycol)-2000] (folate-PEG-DSPE). (Contrast agent **11** is adapted with permission from Ref. [23], Copyright 2016, American Chemical Society.) For contrast agent **13**, the pink circles represent $Gd^{III}$, the red helix represents a targeting peptide, and the gray lines represent PEG. (Contrast agent **13** is minimally adapted from [14], Copyright 2019, Elsevier.) Contrast agent **14** is mesoporous organosilica nanoparticles containing $Gd^{III}$ that are functionalized with the RGD peptide. (Contrast agent **14** is reprinted with permission from [16] Copyright 2023, Elsevier.) For contrast agent **15**, the gray circle represents the perfluorooctylbromide (PFOB) nanoparticles, the white circle represents a phospholipid layer, orange shapes represent PEG, and red helices represent the RGD peptide. Contrast agent **18** is a $Gd^{III}$-containing metallofullerene functionalized with the cytokine interleukin-13 peptide. (Contrast agent **18** is adapted with permission from [30], Copyright 2015, American Chemical Society.) Contrast agent **19** is a tri-gadolinium nitride metallofullerene functionalized with the ZD2 peptide. (The image of contrast agent **19** is adapted from the open access article [31].) Contrast agent **21** is a manganese-nitrogen doped carbon nanosheet (Mn-N-CNSs) functionalized with Anti-$HE_4$ monoclonal antibodies.

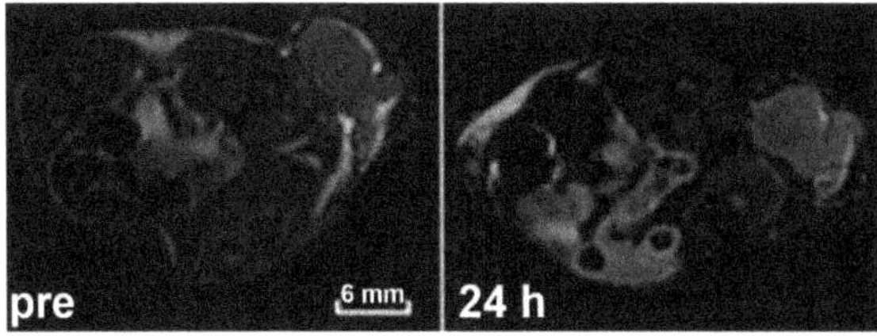


**FIGURE 2** $T_1$-weighted MRI of MDA-MB-231 tumor xenografts in nude mice before and 24 h after injection with $Mn^{II}$-DTPA conjugated to copper sulfide nanoparticles (**1**). Contrast in the tumor (circled in red) is enhanced following accumulation of the nanoparticles due to the EPR effect. Image obtained from the open access article [11].

### 2.1.3 Integrin $\alpha_v\beta_3$ Receptors

Another target for tumor imaging is integrin $\alpha_v\beta_3$ that is present in cell angiogenesis and helps degrade the extracellular matrix connecting the cytoskeleton and cells. Integrin $\alpha_v\beta_3$ contributes to increased cell migration and proliferation in tumors, enhancing tumor progression [15]. In one example targeting integrin $\alpha_v\beta_3$, $MnFe_2O_4$ nanoparticles were labeled with the peptide arginine-glycine-aspartic acid (RGD) (**4**, Figure 1). The nanoparticles effectively targeted integrin $\alpha_v\beta_3$ receptors in tumors because of the presence of arginine-glycine-aspartic acid that binds to integrin $\alpha_v\beta_3$. The functionalized nanoparticles exhibited a transverse relaxivity ($r_2$) of 267.5 mM$^{-1}$ s$^{-1}$, bound to integrin $\alpha_v\beta_3$, and were used in MRI of U87MG glioblastoma tumors expressing integrin $\alpha_v\beta_3$ *in vivo* [15].

Another integrin-targeting contrast agent includes zwitterionic gadolinium-based dendrimers entrapped gold nanoparticles modified with 1,3-propane sultone and RGD peptide (**5**, Figure 1) [16]. Agent **5** has a $r_1$ of 13.17 mM$^{-1}$ s$^{-1}$ and targets $\alpha_v\beta_3$ integrin-expressing cancer cells due to the RGD peptide. With the aforementioned features, **5** endows enhanced contrast in MRI of a B16 lung cancer metastasis *in vivo* (Figure 3).

### 2.1.4 Folate Receptors

Folate receptors are overexpressed on cancer cell surfaces, and this overexpression enables tumor targeting using folic acid as the folate receptor ligand. In one study, gold cluster and gadolinium oxide conjugated nanoparticles were modified with folic acid (**6**) for tumor-targeting imaging by contrast-enhanced MRI (Figure 1) [17]. The $r_1$ of **6** is 12.39 mM$^{-1}$ s$^{-1}$ at 7 T. *In vivo* MRI of KB tumor-bearing mice confirmed the ability of **6** to enhance contrast in folate-expressing tumors *in vivo*.

$Mn_3O_4$ nanoparticles were functionalized with folic acid to target cancer cells overexpressing the folic acid receptor (**7**, Figure 1) [18]. The functionalized nanoparticles were synthesized using solvothermal decomposition and exhibited an $r_1$ of 0.57 mM$^{-1}$ s$^{-1}$. The nanoparticles accumulated in cancer cells overexpressing the folic acid receptor via a receptor-mediated process, and the accumulation enabled $T_1$-weighted MRI with contrast enhancement from the nanoparticles in the tumor.

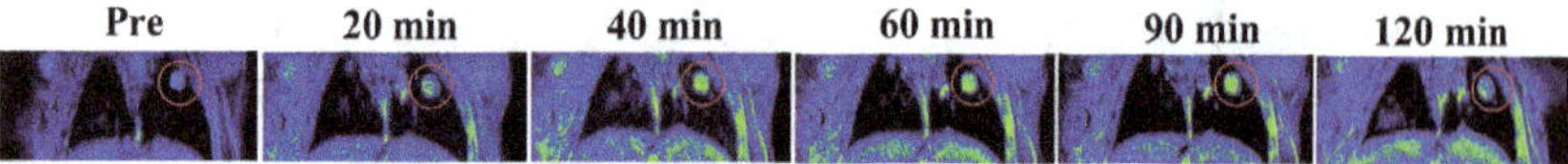


**FIGURE 3** *In vivo* $T_1$-weighted MRI of B16 lung cancer metastasis in mice at different time points after injection of **5**. Red circles indicate the tumor, demonstrating that the tumor exhibits increased contrast enhancement after injection of **5**. The increased contrast enhancement is due to the targeting with $\alpha_v\beta_3$ integrin. (Adapted with permission from [16], Copyright 2019, American Chemical Society.)

### 2.1.5 Other Receptors

$Y_1$ receptor molecules promote targeting of the $Y_1$ receptor in MCF-7 breast tumor cells. $Y_1$ receptor molecules were conjugated to zeolitic imidazolate frameworks as contrast agents for tumor targeting [19]. The specific agent was a manganese-zeolitic imidazolate framework (Mn-ZIF-90) (**8**, Figure 1). The zeolitic imidazolate framework exhibited a $r_1$ of $3.16\,mM^{-1}\,s^{-1}$. *In vivo* MRI using human breast cancer cell line MCF-7 in murine models demonstrated that the nanoparticles enhanced contrast in the tumors [19].

Glucose transporter-1, a cellular transmembrane receptor, has overexpression in many tumor cell lines and tissues [20]. In one study, a tumor-specific MRI contrast agent consisting of $Gd_2O_3$-modified polycyclodextrin conjugated with glucose (**9**) was investigated for targeting the glucose transporter-1 [21]. The $r_1$ value for **9** was $4.86\,mM^{-1}\,s^{-1}$. *In vivo* MRI performed on 4T1 tumor-bearing mice demonstrated enhancement in the tumor caused by **9**.

In another study, a tumor-targeted MRI contrast agent based on methionine conjugation to Gd-based mesoporous silica nanospheres (**10**, Figure 1) was employed to evaluate the ability of the contrast agent to detect tumors by enhancing contrast in $T_1$-weighted images [22]. Agent **10** targets methionine receptors that are expressed in different tumor cells because proliferating tumor cells absorb more methionine than normal cells. Results from *in vivo* studies demonstrated that accumulation of **10** in tumors results in an increase in signal intensity in MRI.

## 2.2 Liver Tumor Targeting

### 2.2.1 Folate

As described earlier, folate receptors are overexpressed in some tumors; consequently, a folate-targeted system containing Gd-texaphyrin was investigated for liver tumor imaging (**11**, Figure 1) [23]. The texaphyrin was loaded into folate-containing liposomes resulting in particles with $r_1$ of 11.8 and $7.1\,mM^{-1}\,s^{-1}$ at 0.47 and 4.7 T, respectively. With contrast agent **11**, $T_1$-weighted contrast enhancement was observed during *in vivo* MRI and aided in early-stage detection of metastatic liver cancer (Figure 4).

### 2.2.2 Organic Anion-Transporting Polypeptide 1

Organic anion-transporting polypeptide 1 proteins are expressed in hepatocytes and are a target for imaging the liver. In a study targeting these proteins, a Gd-chelate bearing an ethoxybenzyl moiety was employed as a hepatobiliary-specific MRI contrast agent for imaging of hepatocellular carcinoma (**12**, Figure 1) [24]. Contrast agent **12** had an $r_1$ of $8.07\,mM^{-1}\,s^{-1}$. From the *in vivo* MRI study of hepatocellular carcinoma in mice, differences were observed in uptake of contrast agents between normal and liver tumor hepatocytes due to abnormal expression of organic anion-transporting polypeptides in tumor cells (Figure 5).

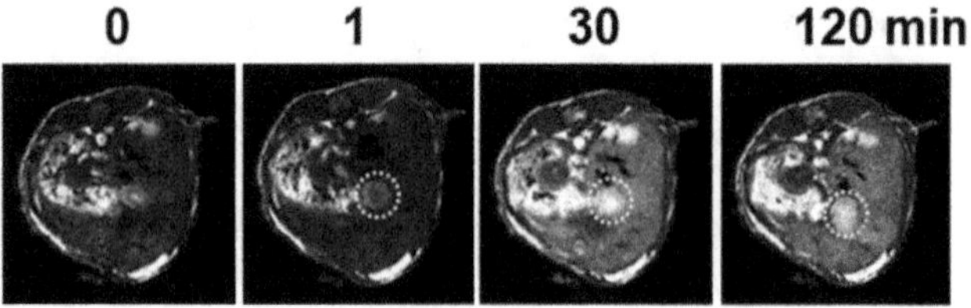


**FIGURE 4** $T_1$-weighted MRI of **11** in a metastatic liver cancer mouse model. Dashed circles indicate the metastatic tumor regions that were revealed by the accumulation of **11**. (Adapted with permission from [23], Copyright 2016, American Chemical Society.)

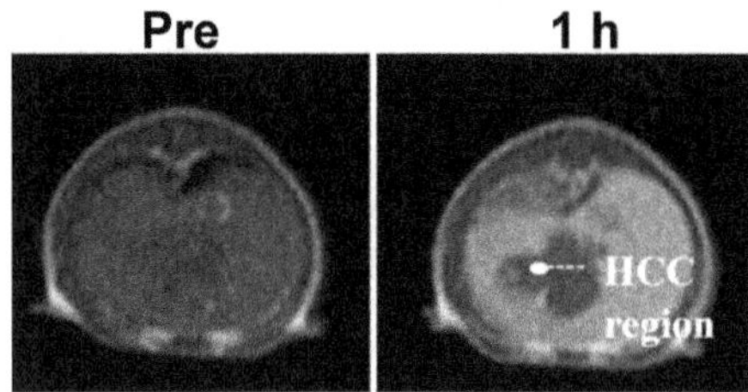


**FIGURE 5** Axial $T_1$-weighted MRI of hepatocellular carcinoma in mice. Images were acquired before and one hour after injection of **12** that targets liver cancer. Regions of hepatocellular carcinoma are marked with a white dashed line. In contrast to normal hepatocytes, liver cancer cells typically have abnormal expression of organic anion-transporting polypeptide 1, resulting in inhibition of the uptake of **12** and, consequently, darker contrast compared with normal liver tissue. (Adapted with permission from Ref. [24], Copyright 2017, American Chemical Society.)

### 2.2.3 Chemokine Receptor 4

Chemokine receptor 4, a G-protein-integrated transmembrane receptor, is an imaging biomarker due to its overexpression in liver metastases. In one study, gadolinium-containing contrast agent **13** was modified to contain a targeting moiety (Figure 1) for chemokine receptor 4 [25]. The targeted contrast agent **13** displays relaxivities of $r_1=30.9\,\text{mM}^{-1}\text{s}^{-1}$ and $r_2=43.2\,\text{mM}^{-1}\text{s}^{-1}$ at 1.5 T and $r_1=23.5\,\text{mM}^{-1}\text{s}^{-1}$ and $r_2=98.6\,\text{mM}^{-1}\text{s}^{-1}$ at 7.0 T. Using **13**, contrast was enhanced in MRI enabling detection of uveal melanoma hepatic metastases in mice.

## 2.3 Brain Tumor Targeting

Imaging the brain with contrast agents is more challenging than the other organs because the blood–brain barrier inhibits the transport of exogenous molecules into the brain. However, many brain tumors have weak blood–brain barriers that enable the delivery of contrast agents. Beyond passive targeting, contrast agents have been labeled with targeting groups to aid in imaging brain tumors.

### 2.3.1 Integrin $\alpha_v\beta_3$ Receptors

As mentioned in Section 2.1.3, the transmembrane glycoprotein $\alpha_v\beta_3$ integrin is overexpressed in some tumors and tumor-induced angiogenic vasculature. Consequently, several $\alpha_v\beta_3$ integrin-targeted gadolinium-based contrast agents for MRI have been reported to image brain tumors. One example of a contrast agent specific for $\alpha_v\beta_3$ integrin receptors is Gd-hybridized mesoporous organosilica nanoparticles conjugated to the RGD peptide (**14**, Figure 1) [26]. After injecting **14** to tumor-bearing mice, MRI revealed increased contrast enhancement in the tumor region. The increased signal was due to the accumulation of **14** at the tumor due to the affinity of the RGD peptide for $\alpha_v\beta_3$ integrin.

In another example of targeting the integrin $\alpha_v\beta_3$ receptor in brain tumors, perfluorooctylbromide nanoparticles functionalized with RGD (**15**) served as $^{19}F$-MRI probes targeted to the integrin $\alpha_v\beta_3$ receptor (Figure 1) [27]. *In vivo* studies were performed using a U87 glioblastoma murine model, and $^{19}F$ signal was observed due to specific binding of the nanoparticles to the integrin $\alpha_v\beta_3$ receptor.

### 2.3.2 Folic Acid Receptors

As described in the liver-tumor targeting section, folate receptors are overexpressed on the surface of cancer cells, enabling tumor targeting. With respect to the use of folate in targeting of brain tumors, one study involved folic acid-conjugated MnO nanoparticles coated with polyacrylic acid (**16**, Figure 1) [28]. The nanoparticles were functionalized with folic acid to target cancer cells overexpressing the folic acid receptor, and polyacrylic acid provides colloidal stability for the nanoparticles. The nanoparticles exhibited a $r_1$ of $9.3\,mM^{-1}s^{-1}$ and enhanced contrast in U87 MG cancer murine models because of the selective targeting to folic acid receptors along cancerous tissue sites (Figure 6).

### 2.3.3 D-Glucose

D-Glucose (**17**, Figure 1) has also been investigated as a potential contrast agent for targeting brain tumors using dynamic glucose-enhanced chemical exchange saturation transfer (CEST) MRI. Tumors generally display elevated levels of

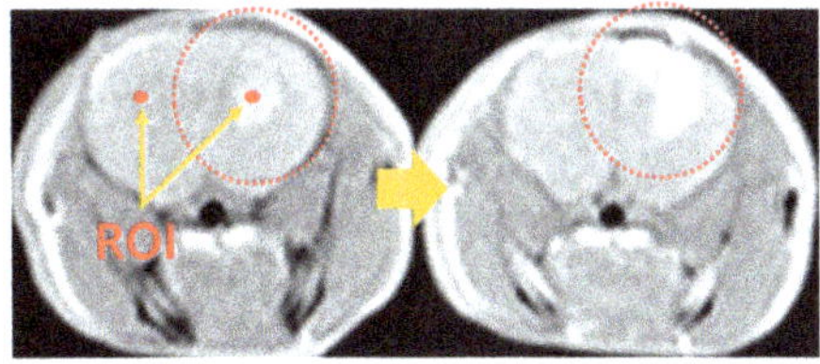


**FIGURE 6** $T_1$-weighted MRI (left) before and (right) 4h after intravenous injection of folic acid-conjugated MnO nanoparticles (**16**). The red circles indicate the region-of-interest (ROI) of brain cancer. Contrast enhancement is observed after 4h, indicating nanoparticle accumulation in the cancerous tissue due to the targeting with folic acid. Image adapted from open access source [28].

glucose uptake, and D-Glucose was previously shown to enhance contrast using CEST [29]. In this study, 20% D-glucose was injected through the cubital vein in newly diagnosed glioblastoma patients to examine healthy white matter tissue compared to cancerous tissue [29]. The glucose-enhanced tumor region displayed a median signal intensity of 2.02%, while the healthy white matter tissue displayed a lower median signal intensity of 0.08%, indicating that the tumor and region surrounding the tumor increased glucose uptake and could be used to image tumor environments in the brain.

### 2.3.4 Interleukin-13 Receptor Subunit Alpha-2

Another brain-tumor-targeting contrast agent is a functionalized gadolinium nitride endohedral metallofullerene nanoplatform conjugated with interleukin-13 peptide (**18**, Figure 1) [30]. The cytokine interleukin-13 peptide enables enhanced targeting of glioblastoma multiforme cells that overexpress the interleukin-13 receptor subunit alpha-2. This peptide delivers nanoparticles to interleukin-13 receptor subunit alpha-2. *In vivo* MRI studies in mice models with orthotopically implanted U-251 glioblastoma tumors exhibited enhanced contrast in mouse brain tumors (Figure 7).

## 2.4 Breast Tumor Targeting

Extradomain B fibronectin is expressed in the extracellular matrix of aggressive cancers, especially in malignant breast cancers. This expression permits easy access and targeted binding of contrast agents for effective MRI. The ZD2 peptide (TVRTSAD) labels extradomain B fibronectin; consequently, targeted contrast

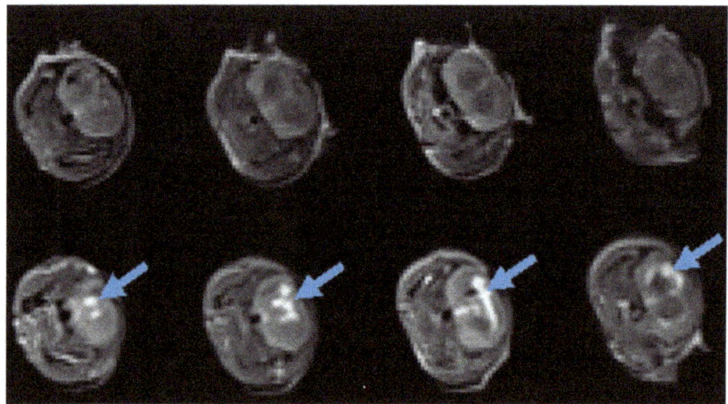

**FIGURE 7** *In vivo* imaging following intravenous delivery of **18**. (top row) MRI of a mouse brain without tumor and (bottom row) MRI acquired 15 min after intravenous injection of **18**. The bright contrast (labeled with blue arrows) is due to the presence of **18** that targets interleukin-13 receptor subunit alpha-2 overexpressed glioblastoma multiforme cell. (Adapted with permission from [30], Copyright 2015, American Chemical Society.)

agents for MRI have been developed using this peptide, including tri-gadolinium nitride metallofullerene (**19**) and gadolinium 1,4,7-triscarboxymethyl-1,4,7,10-tetraazacyclododecane (**20**) (Figure 1) [31,32]. Gadolinium-containing chelating moieties are conjugated to the targeting ZD2 peptide to synthesize the contrast agents. The per-Gd $r_1$ of **19** and **20** are 75 (1.5 T) and 3.7 (7 T) $mM^{-1}s^{-1}$, respectively. In $T_1$-weighted *in vivo* MRI studies, enhanced contrast was observed in aggressive triple negative breast tumor model in mice due to the targeting of the ZD2 peptide.

### 2.5 Ovarian Tumor Targeting

One example of ovarian-tumor-targeting contrast agents includes $Mn^{II}$-doped carbon nanosheets conjugated with Anti-$HE_4$ monoclonal antibodies (**21**, Figure 1) [33]. The antibody was selected for its ability to bind to human epididymis protein 4 ($HE_4$) that is overexpressed in ovarian cancer [32]. $Mn^{II}$ enhances MRI, and the carbon nanosheets are a biocompatible system for coupling $Mn^{II}$ and Anti-$HE_4$ monoclonal antibodies. *In vivo* studies confirm positive enhancement when the HO-8910 cell line was imaged in murine models because of binding to human epididymis protein 4.

### 2.6 Prostate Tumor Targeting

Early detection of prostate cancer is key to developing effective treatment strategies and clinical management of prostate cancer. Consequently, different targeted contrast agents have been explored for imaging of prostate cancer. One example of a peptide-specific MRI contrast agent for prostate cancer is a gadolinium chelate conjugated to CLT1 peptide sequence (CGLIIQKNEC) that targets clotted plasma proteins (**22**, Figure 1) [34]. Clotted plasma protein is highly expressed in cancer stroma and provides a cancer-specific molecular target. The per-Gd $r_1$ of **22** is 10.1 $mM^{-1}s^{-1}$ at 1.5 T. In contrast-enhanced MRI with **22**, contrast enhancement was observed after injection of mice bearing orthotopic PC3 prostate tumors.

## 3 LIVER

The liver is responsible for filtering blood, removing toxins, and metabolizing proteins and carbohydrates. Common liver diseases include cancer, hepatitis, and cirrhosis. Targeted imaging of the liver is potentially important to determine patient prognosis. This section is organized based on specific targets within the liver. Section 3 differs from Section 2.2 by focusing on imaging of pathological conditions of the liver as opposed to tumors within the liver.

### 3.1 Organic Anion-Transporting Polypeptide 1

The organic anion-transporting polypeptide 1 is a transporting protein that mediates the uptake of different substrates in the liver. $Mn^{II}$-ethylenediaminetetraacetic acid conjugated to benzothiazole aniline (**23**, Figure 8) was investigated as a

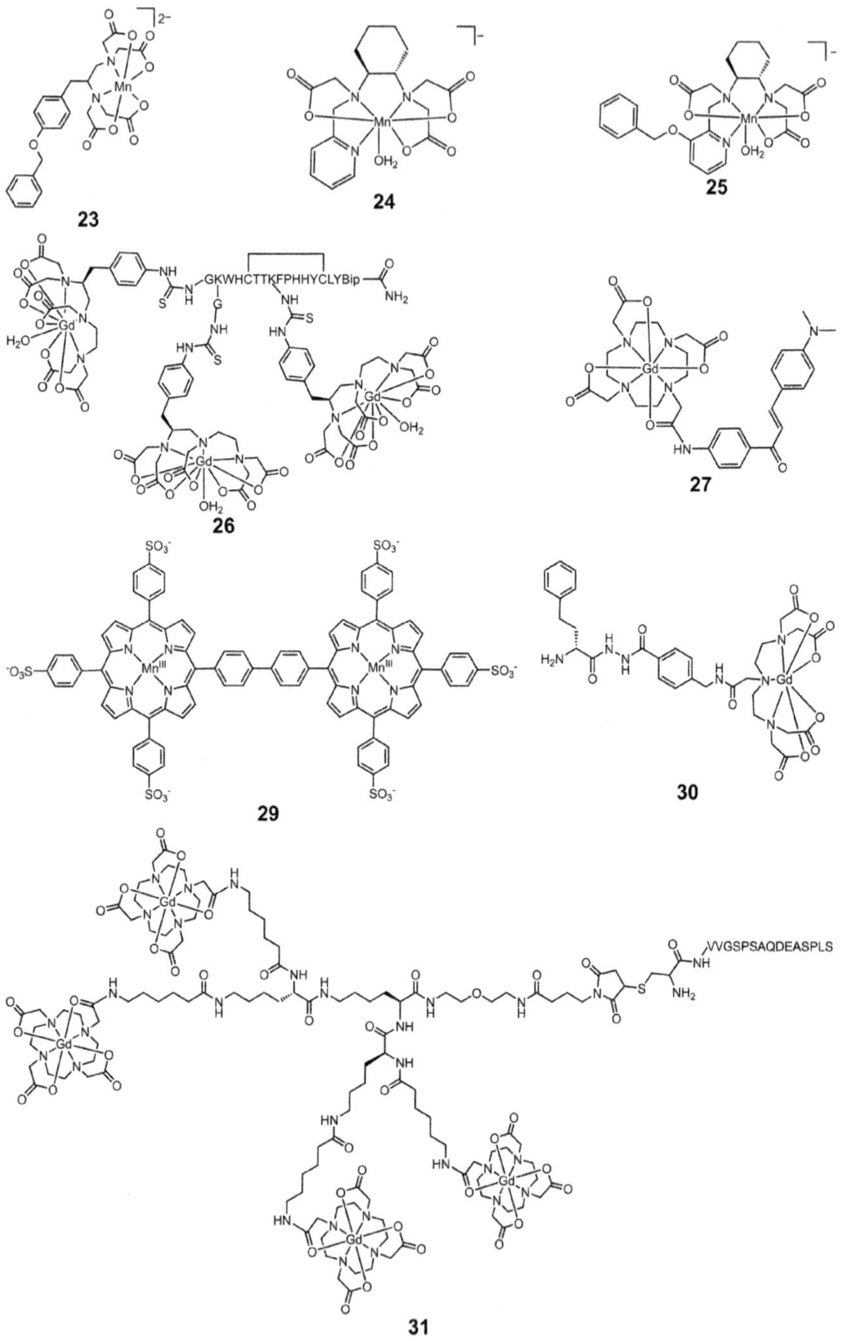

23

24

25

GKWHCTTKFPHHYCLYBip

26

27

29

30

VVGSPSAQDEASPLS

31

**FIGURE 8** Structures of targeted MRI contrast agents for imaging of liver, brain, lung, and other targets described in this chapter.

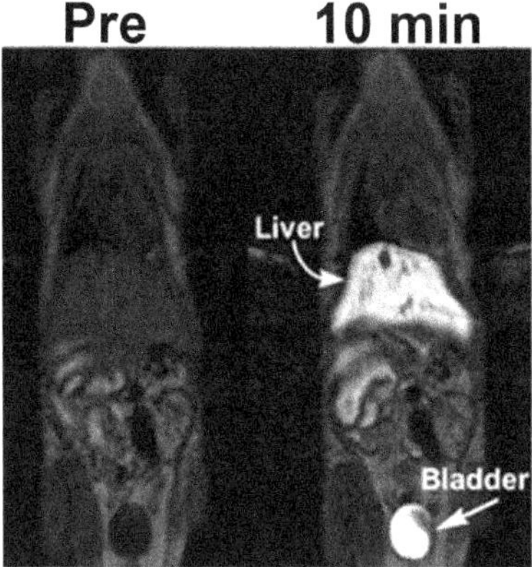


**FIGURE 9** $T_1$-weighted MR images (left) before and (right) 10 min after intravenous injection of **23**. Contrast enhancement is observed after injection, indicating that the contrast agent experienced uptake via organic anion-transporting polypeptide 1. (Adapted with permission from [35], Copyright 2023, American Chemical Society.)

liver-targeting contrast agent that associates with hepatocytes in the liver through uptake via organic anion-transporting polypeptide 1 (Figure 8) [35]. Relaxivity ($r_1$) was measured in aqueous bovine serum albumin to be $15.81 \pm 1.72\,mM^{-1}\,s^{-1}$ at 0.47 T. The contrast agent experienced hepatocellular uptake via the organic anion-transporting polypeptide 1 proteins on the membranes of hepatocytes from interactions with benzothiazole [35]. *In vivo* studies confirmed that the contrast agent experienced uptake via organic anion-transporting polypeptide 1 (Figure 9).

## 3.2 Fibrin

In another attempt to target the liver, nine derivatives of fibrin-targeting $Mn^{II}$ *N*-picolyl-*N,N′,N′*-*trans*-1,2-cyclohexenediaminetriacetate ($PyC_3A$) (**24**, Figure 8) were synthesized to examine the effects of increasing the lipophilicity of the initial complex [36]. Fibrin can be used as an imaging marker for liver inflammation because fibrin concentrations increase in inflamed areas of the liver due to endothelial disruption. The initial complex **24** targeted fibrin but did not feature rapid blood clearance required for liver-targeting agents [37]. One derivative from the study, $Mn^{II}$-$PyC_3A$-3-OBn (**25**), exhibited a $r_1$ of $9.0 \pm 1.4\,mM^{-1}\,s^{-1}$ in human blood plasma at 1.4 T [36]. This complex was promising due to its large relaxivity and hepatocellular uptake, enabling liver targeting due to the lipophilicity of the complex (Figure 10).

## 3.3 Collagen

Hepatic fibrosis is characterized by an excess accumulation of extracellular type-I collagen matrix protein; therefore, type-I collagen was investigated as a target for

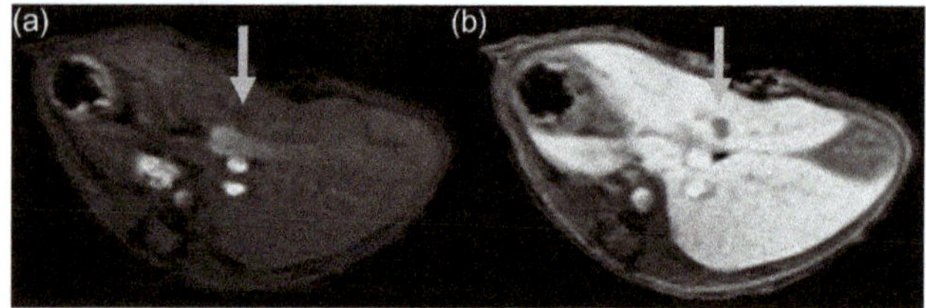


**FIGURE 10** $T_1$-weighted MRI of colorectal liver metastasis of an orthotopic mouse model. The images are (a) before and (b) after injection of **25**. The arrows point to a hepatocyte-devoid lesion. Hepatocyte-filled areas of the liver show contrast enhancement (brightness) due to presence of the contrast agent. (Adapted with permission from [36], Copyright 2023, American Chemical Society.)

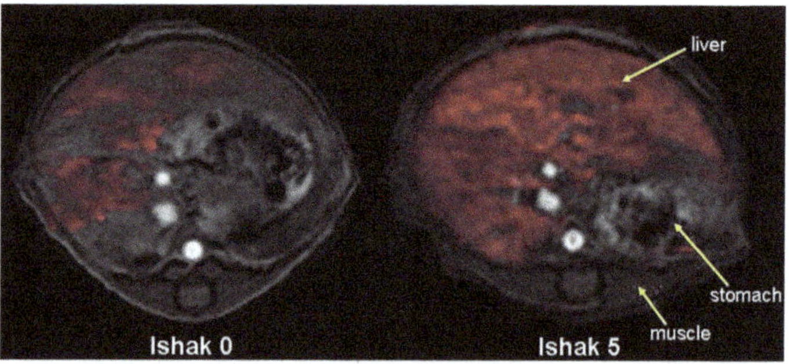


**FIGURE 11** Staging of liver fibrosis using **26** and MRI. Representative axial MRI of a control (Ishak 0) and severe fibrotic (Ishak 5) mouse showing the liver, stomach, and dorsal muscle. False color overlay is the difference image between post-injection and the image acquired prior to injection. (Reprinted with permission from [38], Copyright 2013, Elsevier.)

MRI contrast agents. In one example of targeting collagen, a collagen-binding peptide was conjugated to three units of $Gd^{III}$-DTPA (**26**) to monitor the progression of fibrosis (Figure 8) [38]. Using contrast agent **26** for staging of liver fibrosis with collagen-enhanced MRI, enhanced contrast was observed in the liver of fibrotic mice (Figure 11). Additional studies for targeted detection of liver fibrosis by MRI using **26** can be found elsewhere [39,40].

## 4 BRAIN

Common brain-related diseases are cancer, Parkinson's disease, stroke, cerebral palsy, epilepsy, and Alzheimer's disease. Targeted imaging of the brain is potentially important to monitor disease progression and minimize damage to healthy brain tissue from treatments. Although Section 2.1.3 describes tumor targeting in the brain, this section differs by focusing on imaging of other aspects of the brain. In all brain-specific imaging, the blood–brain barrier is a critical challenge. For example, some targeted agents focus on myelin but cannot cross an intact blood–brain barrier; therefore, the status of the blood–brain barrier and the ability of a contrast agent to cross it must be considered when designing targeted contrast agents for imaging in the brain.

For diagnosing Alzheimer's disease, insoluble $\beta$-amyloid peptides are an important imaging biomarker. However, to address this limitation, in one study, gadolinium 1,4,7,10-tetraazacyclododecane-1,4,7-triacetic acid was conjugated with chalcone to synthesize the $\beta$-amyloid-targeted contrast agent (**27**, Figure 8) [41]. The chalcone moiety binds to $\beta$-amyloid oligomers and crosses the blood–brain barriers because of its lipophilicity. Relaxivity values ($r_1$) of **27** are 4.95±0.22 at 3 T and 4.99±0.19 $mM^{-1}s^{-1}$ at 9.4 T. *In vivo* $T_1$-weighted MRI studies of a transgenic mouse brain using **27** revealed contrast enhancement in regions containing $\beta$-amyloid with specificity confirmed using histology and fluorescence imaging.

Oligomeric $\beta$-amyloid-specific detection and imaging was also investigated using another Gd-based contrast agent (**28**) [42]. In **28**, the surface of Gd-based nanoparticles was encapsulated with $\beta$-amyloid oligomer-specific cyanine dye (fluorophore, F-SLOH). Incorporating $\beta$-amyloid-selective fluorophore facilitates targeting of $\beta$-amyloid oligomers, biocompatibility, and blood–brain barrier permeability [42,43]. The $r_1$ of **28** is 5.56 $mM^{-1}s^{-1}$ [42]. The nanoparticle contrast agent **28** showed blood–brain barrier penetration, which is confirmed by biodistribution studies. From *in vivo* MRI studies of transgenic Alzheimer's disease mice, contrast agent **28** was able to target $\beta$-amyloid and produce contrast enhancement in regions containing $\beta$-amyloid species. The strong binding interactions of the blood–brain barrier permeable agent **28** with $\beta$-amyloid species permit a prolonged retention time for the nanoparticles to obtain contrast enhancement.

# 5 LUNG

Common lung diseases include cancer, asthma, and chronic obstructive pulmonary disease. Targeted imaging of the lungs is potentially important to assess disease states for selection of treatments and determination of patient prognosis. Although Section 2.1.6 mentions tumor targeting in the lungs, this section differs by describing non-tumor imaging of the lungs.

A collagen-targeted enhanced MRI probe (**26**) was used in assessing idiopathic pulmonary fibrosis in the lung of a bleomycin mouse model [44]. Idiopathic pulmonary fibrosis is a progressive lung disease that causes fibrosis of the lungs. Because the overexpression of collagen is a characteristic of fibrosis, **26** enabled imaging of collagen and detected pulmonary fibrosis by showing enhanced contrast in MRI of the lungs of bleomycin mice.

# 6 OTHER TARGETS

## 6.1 Blood Targeting

Blood is responsible for transporting oxygen, nutrients, and waste throughout the body. Targeted imaging of the blood helps detect blocked blood vessels and signs of heart attack and stroke. One agent studied for targeted imaging of blood is a dimeric $Mn^{III}$ porphyrin that features a biphenyl bridge (**29**, Figure 8) [45]. The incorporation of a biphenyl bridge promotes interaction with human serum albumin, the most abundant protein in blood. The dimeric porphyrin exhibited a per-Mn $r_1$ of 10.0 $mM^{-1} s^{-1}$ in the presence of human serum albumin. *In vivo* studies were performed in murine models to demonstrate that the complex interacts with serum albumin and enhances contrast post-injection.

## 6.2 Elastin and Tropoelastin

Elastin is an extracellular matrix protein that endows elasticity in connective tissue and organs. Monomeric tropoelastin is the soluble precursor of elastin and forms elastic fibers along with microfibrils. Elastin and tropoelastin are present in greater than normal amounts in cells after damage or plaque formation. Hence, elastin and tropoelastin are interesting choices of imaging biomarkers because of the potential to enable non-invasive longitudinal monitoring of progressive diseases.

One example of an elastin targeting contrast agent consists of Gd-DTPA bearing an elastin-binding small molecule via amide linkage (**30**, Figure 8). Contrast agent **30** was used in detecting and monitoring of elastin composition in vascular remodeling post-myocardial infarction, as well as rupture of aortic aneurysms [46–48]. *In vivo* MRI using **30** was performed at 3, 7, and 21 days post-injury in a mouse model of myocardial infarction [46]. Two hours after injection, signal enhancement was detected in the infarcted regions because of the presence of elastin or tropoelastin.

Additionally, to visualize dysfunctional elastogenesis in atherosclerotic plaques in mice, a tropoelastin-targeted 15-residue peptide sequence VVGSPSAQDEASPLS was conjugated to four Gd-containing complexes (**31**, Figure 8) [49]. The per-Gd $r_1$ of **31** is $8.5 \pm 0.2$ $mM^{-1} s^{-1}$ at 0.7 T. MRI after injection of **31** in an apolipoprotein-E-deficient mouse model of progressive atherosclerosis revealed circumferential contrast of the plaque surrounding the lumen.

# 7 CONCLUSIONS

Progress toward designing and synthesizing targeted contrast agents for MRI has grown over the past ten years, with several agents being translated to preclinical *in vivo* studies. Several mechanisms are used to advance the ability to image specific physiological regions or molecular events. However, many of the targeting strategies work across multiple targets (for example, folate or integrins being used to target multiple tumor types), so designing targeted contrast agents must be done carefully to account for potential undesired targets that could lead to false positives or poor signal-to-noise ratios. Contrast agents for targeted imaging that function *in vivo* have the potential to improve the diagnoses and monitoring of cancer and other pathological conditions; however, additional research is needed to continue improving the efficacy and safety of targeted contrast agents for MRI for clinical translation of this class of probes.

## ACKNOWLEDGMENT

A.G.S. gratefully acknowledges the National Institutes of Health (T32GM142519) for support.

## ABBREVIATIONS AND DEFINITIONS

| | |
|---|---|
| **CEST** | chemical exchange saturation transfer |
| **CuS@MPG** | CuS nanoparticle functionalized with $Mn^{II}$-containing chelates and PEG |
| **DTPA** | diethylenetriaminepentaacetic acid |
| **EGFR** | epidermal growth factor receptor |
| **EPR** | enhanced permeability and retention effect |
| **HE4** | human epididymis protein 4 |
| **HER2** | human epidermal growth factor receptor 2 |
| **Mn-N-CNSs** | manganese-nitrogen doped carbon nanosheet |
| **MRI** | magnetic resonance imaging |
| **PEG** | polyethylene glycol |
| **PFOB** | perfluorooctylbromide |
| **$PyC_3A$** | *N*-picolyl-*N,N',N'-trans*-1,2-cyclohexenediaminetriacetate |
| $r_1$ | longitudinal relaxivity |
| $r_2$ | transverse relaxivity |
| **RGD** | arginine-glycine-aspartic acid |
| **ROI** | region-of-interest |

## REFERENCES


1. A. M. Morawski, G. A. Lanza, S. A. Wickline, *Curr. Opin. Biotechnol.* **2005**, *16*, 89–92.
2. S. M. Vithanarachchi, M. J. Allen, *Curr. Mol.* Imaging **2012**, *1*, 12–25.
3. S. Lacerda, *Inorganics* **2018**, *6*, 129.
4. J. Wahsner, E. M. Gale, A. Rodríguez-Rodríguez, P. Caravan, *Chem. Rev.* **2019**, *119*, 957–1057.
5. S. A. A. S. Subasinghe, R. G. Pautler, M. A. H. Samee, J. T. Yustein, M. J. Allen, *Bionsensors* **2022**, *12*, 478.
6. A. Louie, *Chem. Rev.* **2010**, *110*, 3146–3195.
7. D.-E. Lee, H. Koo, I.-C. Sun, J. H. Ryu, K. Kim, I. C. Kwon, *Chem. Soc. Rev.* **2012**, *41*, 2656–2672.
8. A. Walter, P. Paul-Gilloteaux, B. Plochberger, L. Sefc, P. Verkade, J. G. Mannheim, P. Slezak, A. Unterhuber, M. Marchetti-Deschmann, M. Ogris, K. Bühler, D. Fixler, S. H. Geyer, W. J. Weninger, M. Glösmann, S. Handschuh, T. Wanek, *Front. Phys.* **2020**, *8*, 47.
9. J. Wu, *J. Pers. Med.* **2021**, *11*, 771.

10. H. Maeda, *J. Pers. Med.* **2021**, *11*, 229.
11. R. Liu, L. Jing, D. Peng, Y. Li, J. Tian, Z. Dai, *Theranostics* **2015**, *5*, 1144–1153.
12. Z. Mitri, T. Constantine, R. O'Regan, *Chemother. Res. Pract.* **2012**, 2012, 743193.
13. C. Lohrisch, M. Piccart, *Semin. Oncol.* **2001**, *28*, 3–11.
14. H. Kim, S. Jin, H. Choi, M. Kang, S. G. Park, H. Jun, H. J. Cho, S. Kang, *J. Control. Release* **2021**, *335*, 269–280.
15. X. Shi, L. Shen, *J. Inorg. Biochem.* **2018**, *186*, 257–263.
16. J. Liu, Z. Xiong, J. Zhang, C. Peng, B. Klajnert-Maculewicz, M. Shen, *X. Shi. ACS Appl. Mater. Interfaces* **2019**, *11*, 15212–15221.
17. C. Xu, Y. Wang, C. Zhang, Y. Jia, Y. Luo, X. Gao, *Nanoscale* **2017**, *9*, 4620–4628.
18. Y. Luo, J. Yang, J. Li, Z. Yu, G. Zhang, X. Shi, M. Shen, *Colloids Surf. B* **2015**, *136*, 506–513.
19. Z. Jiang, B. Yuan, N. Qiu, Y. Wang, L. Sun, Z. Wei, Y. Li, J. Zheng, Y. Jin, Y. Li, S. Du, J. Li, A. Wu, *Nano-Micro Lett.* **2019**, *11*, 61.
20. K. C. Carvalho, I. W. Cunha, R. M. Rocha, F. R. Ayala, M. M. Cajaíba, M. D. Begnami, R. S. Vilela, G. R. Paiva, R. G. Andrade, F. A. Soares, *Clinics* **2011**, *66*, 965–972.
21. T. Mortezazadeh, E. Gholibegloo, A. N. Riyahi, S. Haghgoo, A. E. Musa, M. Khoobi, *J. Biomed. Phys. Eng.* **2020**, *10*, 25–38.
22. B. Mehravi, M. S. Ardestani, M. Damercheli, H. Soltanghoraee, N. Ghanaldarlaki, A. M. Alizadeh, M. A. Oghabian, M. S. Shirazi, S. Mahernia, M. Amanlou, *Mol. Imaging Biol.* **2014**, *16*, 519–528.
23. M. H. Lee, E.-J. Kim, H. Lee, H. M. Kim, M. J. Chang, S. Y. Park, K. S. Hong, J. S. Kim, J. L. Sessler, *J. Am. Chem. Soc.* **2016**, *138*, 16380–16387.
24. A. R. Baek, H.-K. Kim, S. Park, G. H. Lee, H. J. Kang, J.-C. Jung, J.-S. Park, H.-K. Ryeom, T.-J. Kim, Y. Chang, *J. Med. Chem.* **2017**, *60*, 4861–4868.
25. S. Tan, H. Yang, S. Xue, J. Qiao, M. Salarian, K. Hekmatyar, Y. Meng, R. Mukkavilli, F. Pu, O. Y. Odubade, W. Harris, Y. Hai, M. L. Yushak, V. M. Morales-Tirado, P. Mittal, P. Z. Sun, D. Lawson, H. E. Grossniklaus, J. J. Yang, *Sci. Adv.* **2020**, *6*, eaav7504.
26. J. Zhang, X. Su, L. Weng, K. Tang, Y. Miao, Z. Teng, L. Wang, *J. Colloid Interface Sci.* **2023**, *633*, 102–112.
27. C. Giraudeau, F. Geffroy, S. Mériaux, F. Boumezbeur, P. Robert, M. Port, C. Robic, D. Le Bihan, F. Lethimonnier, J. Vallette, *Angiogenesis* **2013**, *16*, 171–179.
28. S. Marasini, H. Yue, S.-L. Ho, J.-A. Park, S. Kim, J.-U. Yang, H. Cha, S. Liu, T. Tegafaw, M. Y. Ahmad, A. K. A. Al Saidi, D. Zhao, Y. Liu, K.-S. Chae, Y. Chang, G.-H. Lee, *Appl. Sci.* **2021**, *11*, 2596.
29. D. Paech, P. Schuenke, C. Koehler, J. Windschuh, S. Mundiyanapurath, S. Bickelhaupt, D. Bonekamp, P. Bäumer, P. Bachert, M. E. Ladd, M. Bendszus, W. Wick, A. Unterberg, H.-P. Schlemmer, M. Zaiss, A. Radbruch, *Radiology* **2017**, *285*, 914–922.
30. T. Li, S. Murphy, B. Kiselev, K. S. Bakshi, J. Zhang, A. Eltahir, Y. Zhang, Y. Chen, J. Zhu, R. M. Davis, L. A. Madsen, J. R. Morris, D. R. Karolyi, S. M. LaConte, Z. Sheng, H. C. Dorn, *J. Am. Chem. Soc.* **2015**, *137*, 7881-7888.
31. Z. Han, X. Wu, S. Roelle, C. Chen, W. P. Schiemann, Z. R. Lu, *Nat. Commun.* **2017**, *8*, 692.
32. Z. Han, H. Cheng, J. G. Parvani, Z. Zhou, Z.-R. Lu, *Magn. Reason. Med.* **2018**, *79*, 3135–3143.
33. C. Han, T. Xie, K. Wang, S. Jin, K. Li, P. Dou, N. Yu, K. Xu, *J. Nanobiotechnol.* **2020**, *18*, 175.

34. X. Wu, S. M. Burden-Gulley, G.-P. Yu, M. Tan, D. Lindner, S. M. Brady-Kalnay, Z.-R. Lu, *Bioconjugate Chem.* **2012**, *23*, 1548-1556.
35. K. Chen, P. Li, C. Zhu, Z. Xia, Q. Xia, L. Zhong, B. Xiao, T. Cheng, C. Wu, C. Shen, X. Zhang, J. Zhu, *J. Med. Chem.* **2021**, *64*, 9182–9192.
36. J. Wang, H. Wang, I. A. Ramsay, D. J. Erstad, B. C. Fuchs, K. K. Tanabe, P. Caravan, E. M. Gale, *J. Med. Chem.* **2018**, *61*, 8811–8824.
37. E. M. Gale, I. P. Atanasova, F. Blasi, I. Ay, P. Caravan, *J. Am. Chem. Soc.* **2015**, *137*, 15548–15557.
38. B. C. Fuchs, H. Wang, Y. Yang, L. Wei, M. Polasek, D. T. Schühle, G. Y. Lauwers, A. Parkar, A. J. Sinskey, K. K. Tanabe, P. Caravan, *J. Hepatol.* **2013**, *59*, 992–998.
39. C. T. Farrar, D. K. DePeralta, H. Day, T. A. Rietz, L. Wei, G. Y. Lauwers, B. Keil, A. Subramaniam, A. J. Sinskey, K. K. Tanabe, B. C. Fuchs, P. Caravan, *J. Hepatol.* **2015**, *63*, 689–696.
40. B. Zhu, L. Wei, N. Rotile, H. Day, T. Rietz, C. T. Farrar, G. Y. Lauwers, K. K. Tanabe, B. Rosen, B. C. Fuchs, P. Caravan, *Hepatology* **2017**, *65*, 1015–1025.
41. G. Choi, H.-K. Kim, A. R. Baek, S. Kim, M. J. Kim, M. Kim, A. E. Cho, G.-H. Lee, H. Jung, J.-u. Yang, T. Lee, Y. Chang, *J. Ind. Eng. Chem.* **2020**, *83*, 214–223.
42. C. Wang, X. Wang, H.-N. Chan, G. Liu, Z. Wang, H.-W. Li, M. S. Wong, *Adv. Funct. Mater.* **2020**, *30*, 1909529.
43. Y. Li, D. Xu, A. Sun, S.-L. Ho, C.-Y. Poon, H.-N. Chan, O. T. W. Ng, K. K. L. Yung, H. Yan, H.-W. Li, M. S. Wong, *Chem. Sci.* **2017**, *8*, 8279–8284.
44. P. Caravan, Y. Yang, R. Zachariah, A. Schmitt, M. Mino-Kenudson, H. H. Chen, D. E. Sosnovik, G. Dai, B. C. Fuchs, M. Lanuti, *Am. J. Respir. Cell Mol. Biol.* **2013**, *49*, 1120-1126.
45. W. Cheng, T. Ganesh, F. Martinez, J. Lam, H. Yoon, R. B. Macgregor, Jr., T. J. Scholl, H.-L. M. Cheng, X.-a. Zhang, *J. Biol. Inorg. Chem.* **2014**, *19*, 229–235.
46. A. Protti, B. Lavin, X. Dong, S. Lorrio, S. Robinson, D. Onthank, A. M. Shah, R. M. Botnar, *J. Am. Heart Assoc.* **2015**, *4*, e001851.
47. R. M. Botnar, A. J. Wiethoff, U. Ebersberger, S. Lacerda, U. Blume, A. Warley, C. H. P. Jansen, D. C. Onthank, R. R. Cesati, R. Razavi, M. S. Marber, B. Hamm, T. Schaeffter, S. P. Robinson, M. R. Makowski, *Circ. Cardiovasc. Imaging* **2014**, *7*, 679–689.
48. M. Wildgruber, I. Bielicki, M. Aichler, K. Kosanke, A. Feuchtinger, M. Settles, D. C. Onthank, R. R. Cesati, S. P. Robinson, A. M. Huber, E. J. Rummeny, A. K. Walch, R. M. Botnar, *Circ. Cardiovasc. Imaging* **2014**, *7*, 321–329.
49. F. Capuana, A. Phinikaridou, R. Stefania, S. Padovan, B. Lavin, S. Lacerda, E. Almouazen, Y. Chevalier, L. Heinrich-Balard, R. M. Botnar, S. Aime, G. Digilio, *J. Med. Chem.* **2021**, *64*, 15250–15261.

# 4 Responsive MRI Contrast Agents

*Luke A. Marchetti, Manon Isaac, and Célia S. Bonnet**
Centre de Biophysique Moléculaire,
CNRS UPR 4301, Université d'Orléans,
Rue Charles Sadron,
F-45071 Orléans Cedex 2, France
celia.bonnet@cnrs-orleans.fr

## CONTENTS



### Abstract

Molecular imaging has gained increasing interest over the last years, especially in the area of MRI. This has prompted the design of so-called "responsive" or "smart" contrast agents, for which the efficacy is modulated by a biomarker of interest. This, in principle, allows for early diagnosis as these changes will happen prior to the morphological changes that they trigger. This chapter will focus on $Ln^{3+}$-based MRI contrast agents that are responsive to relevant biomarkers such as temperature, ions, pH, neurotransmitters, enzymes, and redox state. Representative examples are chosen to illustrate the strategies used depending on the biomarker as well as the advantages and limitations. The problem of quantification of the biomarker is also discussed at the end of the chapter.

* Corresponding author.

DOI: 10.1201/9781003374688-4

## KEYWORDS

Gadolinium; Europium; Biomarkers; Early Diagnosis; Quantification

Since the approval of the first $Gd^{3+}$-based MRI contrast agent (CA) in 1988, tremendous efforts have been made to improve their efficacy, which is called relaxivity. Relaxivity is defined as the paramagnetic increase of the relaxation rate of water proton per unit of the CA. The relationship between relaxivity and the microscopic parameters of $Gd^{3+}$ complexes is now well-known (see Chapter 1 for the theory of relaxation). Chemists have cleverly played with the number of water molecules directly coordinated to $Gd^{3+}$ ($q$), their water exchange rate with the bulk ($k_{ex}$), and the rotational correlation time of the CA ($\tau_R$), among other parameters. In the past twenty years, much effort has been dedicated to render those CA specific to a given biomarker, such as pH, ions, neurotransmitters, and enzymes. Imaging those important biomarkers non-invasively *in vivo* is very valuable for biomedical research to better understand their physiological and pathological role, as well as for early diagnosis of pathologies. Indeed, those changes are thought to happen prior to the morphological changes that they trigger, and that can be tracked by clinically approved MRI CA. For such molecular imaging applications, the presence of the CA is necessary and its relaxivity has to be influenced by the biomarker of interest, leading to the so-called smart or responsive CA. The last few years have seen tremendous efforts in the development of responsive CA with encouraging successful *in vivo* applications. In this chapter, we will concentrate on molecular $Ln^{3+}$-based $T_1$-responsive CA (mainly $Gd^{3+}$ and $Eu^{2+}$). Examples of ParaCEST, Parashift, transition metal-based CA, and nanoparticles will be described in other chapters. The biomarkers that can be imaged by $T_1$-responsive CA should be extracellular as it remains very challenging to load enough $Gd^{3+}$ complexes intracellularly and in sufficient quantity to be adapted to the low sensitivity of MRI. We will survey some representative examples of responsive probes to sense temperature, metal ions, pH, neurotransmitters, enzymatic activity, or redox state. Finally, the quantification of the biomarker will be discussed as it is of prime importance for successful *in vivo* application.

## 1 TEMPERATURE SENSING

The employment of thermal therapies, e.g., thermal ablation and hyperthermia, in the treatment of various diseases has gained traction in recent years [1–3]. The methods to measure the applied temperature during these therapies often display disadvantages such as being limited to surface temperature measurement or being invasive in nature [4]. Using magnetic resonance (MR) thermometry in combination with CA circumvents these drawbacks by allowing the clinician to non-invasively obtain real-time information with high spatial resolution.

A method that dominates the literature to date is the encapsulation of a contrast agent within a thermosensitive liposome (TSL), which then creates a paramagnetic nanoparticle with temperature-responsive properties while also granting the

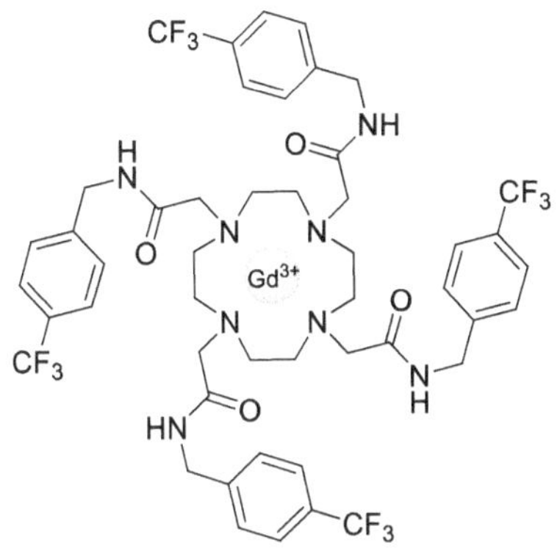

**T1**

**FIGURE 1** Structure of a temperature-responsive contrast agent.

ability to obtain absolute temperature values *in vivo* by the use of MRI. However, as nanoparticles fall outside the scope of this chapter, we will focus on an example of a molecular CA. There is, in fact, a distinct lack of CA that can act as MR thermometers of their own accord. A recent example reports a fluorinated Gd complex, **T1** (Figure 1), in which the conformation of the complex forms a cage-like structure around the $Gd^{3+}$ ion, inhibiting a fast exchange of water molecules coordinated to the metal center with the bulk water and making $k_{ex}$ the determining factor of relaxivity. An increase in temperature results in a sufficient change in conformation to allow the bulk water greater access to the $Gd^{3+}$ ion, leading to a relaxivity increase [5]. The temperature range examined was between 25°C and 44°C, giving this thermosensitive complex possible clinical application in the future.

## 2 ION-RESPONSIVE CONTRAST AGENTS

Despite the physiological importance of anions, the development of CA responsive to anions remains limited as the interactions mostly occur at the $Gd^{3+}$ center where the anions replace the water molecules. It is therefore very difficult (if not impossible) to selectively detect a given anion. Moreover, the displacement of the water molecule is detrimental for the relaxivity. We can nevertheless cite the development of a DTPA-bisamide $Gd^{3+}$ complex substituted with Zn-dipicolyl moieties, which are known to bind strongly to phosphates. The system is selective for polyphosphorylated species [6].

The development of cation-responsive CA has been a very active area of research these past decades. Cations are important biological targets that are implicated in numerous physiological and pathological processes. It is possible to target the extracellular pool of "free" cations (loosely bound to proteins) when this pool is in sufficient quantity (μM at least). Historically, $Ca^{2+}$ was the first metal ion to be detected by MRI due to its important extracellular concentration and its prime role in neuroimaging. $Zn^{2+}$ has also been successively imaged *in*

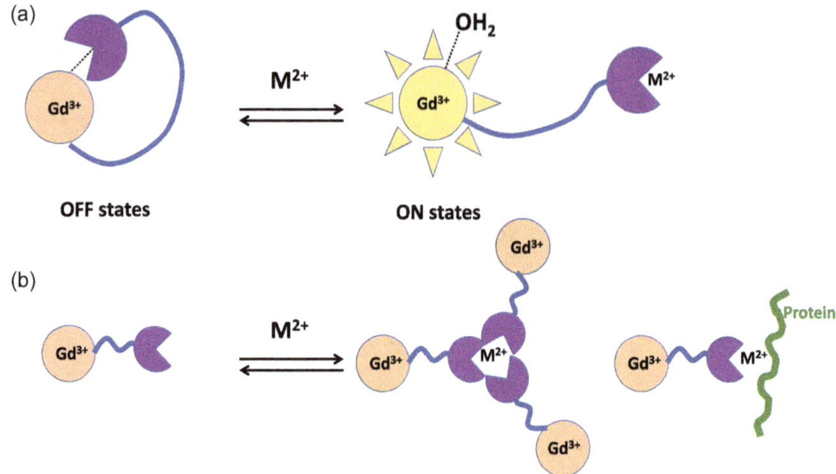


**FIGURE 2** Schematic representation of the two main routes to achieve relaxivity increase of molecular $T_1$-based CA upon cation binding: (a) through a change in the hydration number ($q$): in the absence of the metal to detect, the cation binding part weakly coordinates $Gd^{3+}$, and this arm flips upon cation complexation, leaving space for one water molecule; (b) through a change in the rotational correlation time ($\tau_R$), either via self-assembly or modified interaction for a given protein.

*vivo* recently, and examples of $Cu^{2+}$ sensing are appearing in the literature. Most of the designs involve a Gd-binding unit (MRI active part), which is thermodynamically stable and kinetically inert, a $Zn^{2+}$-binding unit, which is selective and has an adapted affinity, and a linker between the two. The two main mechanisms of response rely on a change in $Gd^{3+}$ hydration number ($q$) or on the rotational correlation time of the system ($\tau_R$) (see Figure 2).

In addition to obtaining a maximum change in relaxivity upon cation binding, there are two main challenges: (1) the system needs to be selective for a given cation and even for a given oxidation state in the case of Cu and (2) the affinity for the cation should be adapted to physiological conditions in order not to perturb homeostasis. Given that the concentration of cations is dependent on the body localisation, this last point is not easy to solve and has been rarely addressed in *in vivo* experiments. Since there are many reviews on cation-responsive CA in the literature [7–9], we will survey briefly representative examples and concentrate on the most recent examples to illustrate different strategies used.

## 2.1 Calcium

The labile pool of $Ca^{2+}$ is particularly important in the brain where it acts as a carrier of electrical current and secondary messenger. The extracellular resting concentration is ~1.25 mM, which is detectable by MRI, and this concentration varies during normal and pathological brain activity [10]. The development of a CA to visualise fluctuations in $Ca^{2+}$ concentrations has attracted much attention to help understand normal and pathological brain activity. One clear challenge to overcome is the delivery of the CA to the brain.

The first $Ca^{2+}$-responsive CA, described in the late 1990s, is composed of two $Gd^{3+}$ complexes linked through a BAPTA (1,2-bis(*o*-aminophenoxy)ethane-*N,N,N′,N′*-tetraacetic acid) chelator [11], which is known to selectively bind $Ca^{2+}$ (Figure 3, **Ca1**). The relaxivity increase upon $Ca^{2+}$ binding is based on a change in the hydration number of $Gd^{3+}$ ($q$=0.47–1.05) [12]. The dissociation constant in the μM range (0.96 μM) is more suitable for the detection of intracellular $Ca^{2+}$. Therefore, the carboxylate functions of BAPTA were esterified to obtain a more lipophilic complex with better cell internalisation (**Ca2**, Figure 3). Moreover, the complex remains in an "off-state" for $Ca^{2+}$ detection until the ethyl ester functions are hydrolysed by the esterase present inside the cells [13].

Detection of extracellular $Ca^{2+}$ is easier to achieve as cell internalisation can be neglected, and the mM $Ca^{2+}$ concentration is more likely to induce changes in the CA present in sub-mM concentration. In a first attempt, the BAPTA linker was successfully modified in **Ca3** (Figure 3) to decrease the affinity for $Ca^{2+}$ by 4 orders of magnitude, while keeping a good selectivity against $Mg^{2+}$ [14]. The relaxivity increase remained however modest (15%). Other chelators, such as modified EDTA (ethylenediamine-tetraacetic acid) (**Ca4**, Figure 3), DTPA (diethylenetriamine pentaacetic acid) (**Ca5**, Figure 3), and EGTA (ethylene glycol-bis(β-aminoethylether)-*N,N,N′,N′*-tetraacetic acid) (**Ca6**, Figure 3) [15,16],

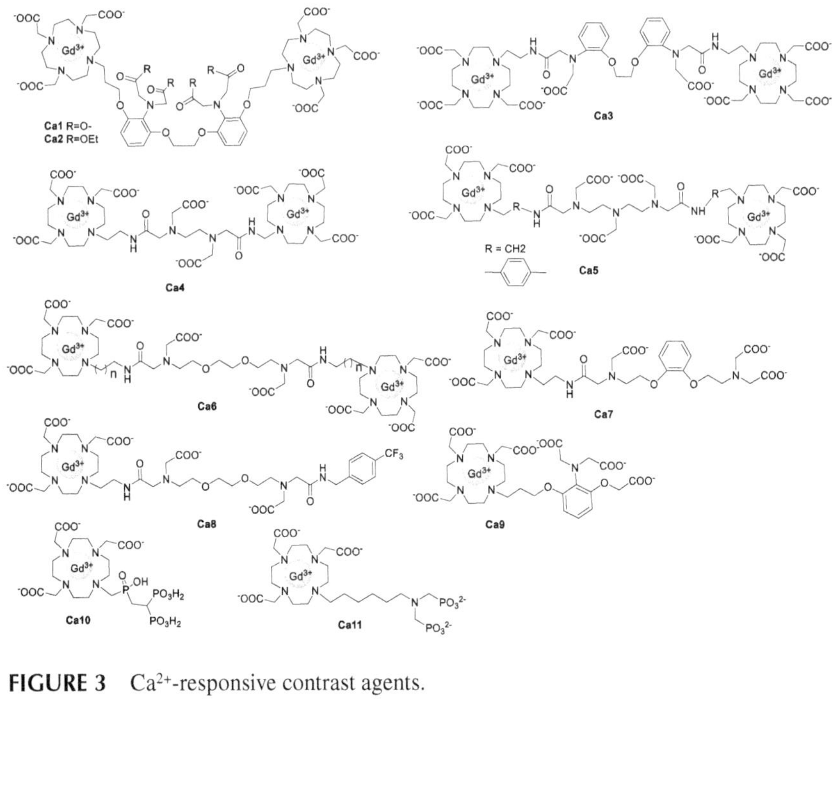

**FIGURE 3** $Ca^{2+}$-responsive contrast agents.

enabled to vary $K_d$ for $Ca^{2+}$ in the mM to sub-mM range. All complexes show relaxivity increase upon $Ca^{2+}$ binding, which could be ascribed to an increase in the hydration number of $Gd^{3+}$, and to some extent to the rigidification of the complex. The nature and length of the linker between the GdDO3A unit and the $Ca^{2+}$ chelator is crucial to determine the relaxivity response upon $Ca^{2+}$ binding. For example, in **Ca5**, when a rigid benzyl linker is used, the CA is insensitive to $Ca^{2+}$, while a 15% relaxivity increase is obtained with the ethyl linker [16]. The most successful complex was certainly **Ca6**, which shows a good selectivity against $Mg^{2+}$ and the highest relaxivity increase (83%) upon $Ca^{2+}$ binding in this series. Importantly, this increase remains sizeable (50%) in the brain extracellular medium (BEM) [15]. **Ca6** was successfully used recently to detect cerebral ischemia through real-time monitoring of $Ca^{2+}$ fluctuations [17]. This allows the early detection of ischemia compared to the existing techniques which visualizes the affected area minutes or hours after the initiation of ischemia.

Monomeric $Gd^{3+}$ complexes based on EGTA derivatives were also studied in detail and evidenced the importance of small ligand differences in the response by modification of the length of the alkyl linker between the $Gd^{3+}$ complex and the EGTA [18], or by the addition of an extra coordinating function to tune the selectivity from $Zn^{2+}$ to $Ca^{2+}$ (**Ca7**, Figure 3) [19]. Such systems were also derivatized to append a $^{19}F$ moiety (**Ca8**, Figure 3) and obtain a dual MRI CA which responds to the presence of $Ca^{2+}$ both in $^{1}H$ and $^{19}F$ MRI [20].

Other $Ca^{2+}$ chelating moieties such as APTRA (*o*-amino-phenol-*N*,*N*,*O*-triacetate) (**Ca9**, Figure 3) or bisphosphonates (**Ca10–11**, Figure 3) were also used. **Ca9** shows a relaxivity increase of almost 100% upon $Ca^{2+}$ addition in buffered solution, 37% in artificial cerebral spinal fluid (ACSF) and 25% in artificial extracellular matrix (AECM) [21]. Evidently, the selectivity against $Zn^{2+}$ and $Mg^{2+}$ is not as good as it is with EGTA for example. **Ca10** oligomerises in the presence of $Zn^{2+}$ at pH 5–7, and $Ca^{2+}$ and $Mg^{2+}$ above pH 8, with a concomitant relaxivity increase [22]. Finally, **Ca11** responds to $Ca^{2+}$ by an unexpected relaxivity decrease of 38% due to aggregation phenomena. It was however not successful to detect neuronal activity in the brain [23].

## 2.2 Zinc

Zinc is the second most abundant transition metal ion in the body after iron. It is implicated in various biological processes including enzymatic activity, signalling, protein synthesis, and transcription [24]. Zinc concentrations can reach μM to mM, making it a realistic target for MRI detection. Since the first example of GdDTPA disubstituted by two BPEN (*N*,*N*-bis-(2-pyridyl-methyl)ethylene diamine) more than twenty years ago (**Z1**, Figure 4) [25], many examples have appeared in the literature [26]. **Z2** (Figure 4) is certainly the most successful example of $Zn^{2+}$-responsive CA developed so far. It responds to $Zn^{2+}$ upon interaction with human serum albumin (HSA), an abundant protein in the blood [27]. **Z2** has been used to detect $Zn^{2+}$ release from the pancreas upon glucose stimulation [28,29] and for the early detection of prostate cancer through differential $Zn^{2+}$

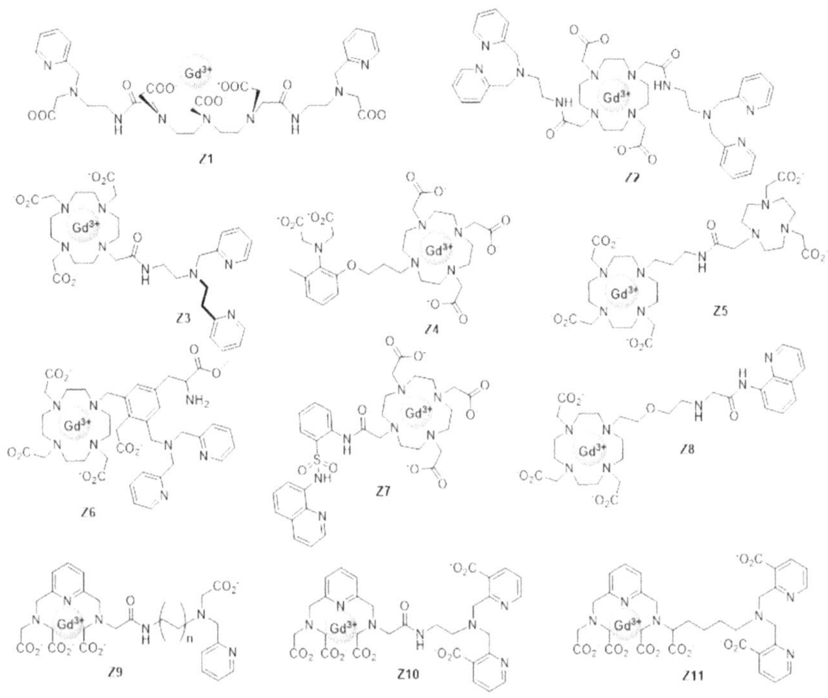

**FIGURE 4** Representative molecular $Gd^{3+}$ complexes for $Zn^{2+}$ sensing.

release after glucose stimulation [30]. Modifications of the $Gd^{3+}$ binding unit of **Z2** to modify the water exchange rate resulted in an amplification of the response [31]. It was also shown that the modification of the $Zn^{2+}$ binding unit to decrease its affinity (**Z3**, Figure 4) improves the imaging of $Zn^{2+}$ release (therefore insulin secretion) from mouse pancreas [32]. This type of CA based on a $\tau_R$ change gives the best relaxivity response at intermediate fields (20–80 MHz), which are often not the imaging field. However, these nice examples illustrate that it is possible to image $Zn^{2+}$ at 9.4 T, a field at which the response is not optimal. The intrinsic entanglement between the $Gd^{3+}$-binding unit, the $Zn^{2+}$-binding unit, and the linker renders the prediction of the relaxivity response upon $Zn^{2+}$-binding difficult. **Z4** (Figure 4) responds to $Zn^{2+}$ by a modification of the hydration number of $Gd^{3+}$; however, the replacement of one carboxylate function on the benzene moiety by a pyridine or a proton, or the modification of the length of the linker, is detrimental for relaxivity modification, illustrating the difficulty of such a design [33–35]. Similarly, **Z5** shows a relaxivity increase upon $Zn^{2+}$ binding, which was ascribed to a change in the coordination number of the $Gd^{3+}$ and not a direct coordination of the $Zn^{2+}$-binding unit. This relaxivity increase is lost when a more preorganised linker is used [36]. Recently, the use of a modified tyrosine (with an additional carboxylate function) as a linker between the GdDO3A unit and the $Zn^{2+}$-binding unit (**Z6**, Figure 4) was successfully used to obtain a *q*-switch and an unprecedented 400% relaxivity increase upon $Zn^{2+}$ binding at 7 T [37]. This increase was explained not only by the increase in inner-sphere water molecules but also by some aggregation processes in solution that are highly dependent on small ligand modification [38]. The $Zn^{2+}$-responsive CA could be rendered bimodal (optical imaging and MRI) by the use of chromophores such as quinoline derivatives for $Zn^{2+}$-binding (**Z7–8**, Figure 4) [39,40].

Bishydrated $Gd^{3+}$ complexes based on a pyridinic platform, which show good thermodynamic stability and kinetic inertness, high relaxivity, and no *in vivo* toxicity [41], were also derivatised to detect $Zn^{2+}$. The first generation of CA (**Z9**, Figure 4) showed a response to $Zn^{2+}$ through the formation of a dimeric $Gd^{3+}$ complex around $Zn^{2+}$ as evidenced by $^{1}H$ NMR and diffusion coefficient measurements [42]. This behavior is not optimal for quantitative $Zn^{2+}$ detection; therefore, the design was modified to limit the formation of dimers (**Z10**, Figure 4). Further modifications concerned the $Gd^{3+}$-binding moiety, increasing the thermodynamic stability and kinetic inertness (**Z11**, Figure 4). This second generation of complex responds to $Zn^{2+}$ in the presence of HSA, with an increased affinity for the protein in the presence of $Zn^{2+}$ by twofold, at least. It was shown by PRE (paramagnetic relaxation enhancement) and competitive fluorescence measurements that the affinity for HSA is driven by the $Zn^{2+}$-binding unit. However, the presence of the $Gd^{3+}$ complex seems important to obtain an affinity increase in the presence of $Zn^{2+}$ [43]. These systems function on the basis of a $\tau_R$ change (more rigid systems in the presence of $Zn^{2+}$) and therefore give maximum response at intermediate field. Fast-field cycling MRI, where the derivative of the relaxivity as a function of the field rather than the relaxivity itself, can be used at 3 T to recover a $Zn^{2+}$ response otherwise absent [44]. Interestingly, at 9.4 T, these systems give a small negative response in the presence of $Zn^{2+}$ in mouse serum *in vitro*. Surprisingly, *in vivo*, after glucose stimulation to release $Zn^{2+}$ from the pancreas, an increase of the MRI intensity in the pancreas is observed. This could be correlated to a local accumulation of $Gd^{3+}$ in $Zn^{2+}$-rich tissues as demonstrated by ICP [45]. This contradictory *in vitro* and *in vivo* behavior showed for the first time that, for such protein-bound $Zn^{2+}$-responsive CA, the biodistribution of the CA in the absence and in the presence of $Zn^{2+}$ is more important than relaxivity variations.

In a very different approach, bioinspired systems based on classical ββα zinc-fingers (ZF) scaffolds were used for $Zn^{2+}$ detection by MRI. Classical ββα ZF are 20–30 amino acid-peptides with the following sequence: (Tyr/Phe)-Xaa-Cys-$Xaa_{2/4}$-Cys-$Xaa_3$-Phe-$Xaa_5$-Leu-$Xaa_2$-His-$Xaa_3$-His, where Xaa can be any amino acid. These peptides change from an undefined conformation in the absence of $Zn^{2+}$ to a compact ββα fold in the presence of $Zn^{2+}$. The addition of a monohydrated DOTA derivative to a lateral chain of a lysine (Figure 5, top) rendered possible $Zn^{2+}$ detection by MRI for the $Gd^{3+}$ complex and by luminescence for the $Tb^{3+}$ complex [46]. Indeed, Trp is able to sensitize $Tb^{3+}$ luminescence by the so-called "antenna effect". In the absence of $Zn^{2+}$, Trp is far away for $Tb^{3+}$ preventing an efficient energy transfer. In the presence of $Zn^{2+}$, the two entities come in close proximity, which results in a 30-fold enhancement of $Tb^{3+}$ luminescence intensity. The relaxivity of the corresponding $Gd^{3+}$ complex is increased by 40% in the presence

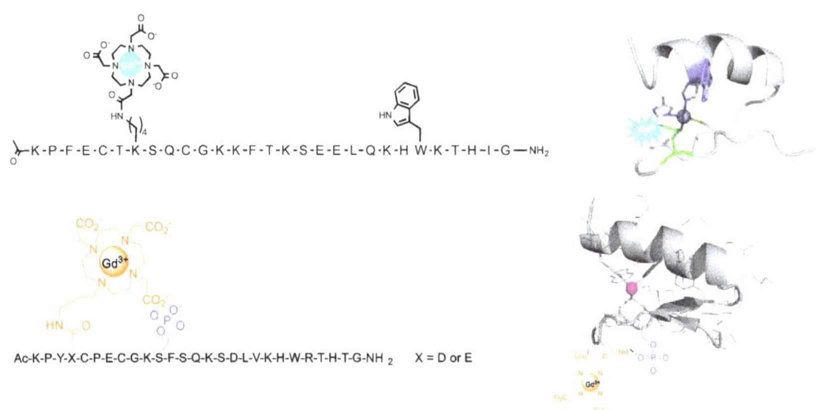


**FIGURE 5** Schematic representation of the ZF peptides appended with a $Gd^{3+}$ complex for MRI detection of $Zn^{2+}$ based on a change in $\tau_R$ (top) and $q$ and $\tau_R$ (bottom).

of $Zn^{2+}$ at 40 MHz, 37°C, due to an increased rotational correlation time, explained by the more rigid confirmation. Interestingly, in the absence of $Zn^{2+}$, relaxivity is limited by $\tau_R$, while in the presence of $Zn^{2+}$, the water exchange rate starts to limit relaxivity. The peptide was also phosphorylated to obtain $q$-change (Figure 5, bottom). In the absence of $Zn^{2+}$, the phosphate is partially bound to $Gd^{3+}$ (exchange between a $q=0$ and $q=1$ species), while one water molecule is present in the first coordination sphere of $Gd^{3+}$ in the presence of $Zn^{2+}$. Due to strong H-bonding network, the system is highly rigid in the presence of $Zn^{2+}$, leading to relaxivity increase at 25°C of 150% at 40 MHz and 60% at 400 MHz. Once again, small amino acid modifications resulted in dramatic changes in the relaxation properties. When glutamate is replaced by aspartate, no $q$-change is observed [47].

## 2.3 Copper

Detection of $Cu^{2+}$ is only at its infancy compared to $Ca^{2+}$ or $Zn^{2+}$ detection. Firstly, the labile pool of $Cu^{2+}$ is at the lower limit of the MRI detection (ca. 5 μM in the blood for example), and secondly, the development of selective probes for $Cu^{2+}$, particularly over $Zn^{2+}$ is highly challenging. As the labile pool of $Zn^{2+}$ is always more important, this is not problematic for $Zn^{2+}$ detection, but it becomes crucial when it comes to reliable $Cu^{2+}$ detection. This is well illustrated by the first example of $Cu^{2+}$-responsive probe (**Cu1**, Figure 6), which is very similar to a $Zn^{2+}$-responsive probe and functions on a basis of $q$-change [48]. To circumvent the selectivity problem, sulfur donor atoms were added [49], and the resulting probes respond to the softer metal ion $Cu^{+}$, which is more present intracellularly. It was shown that the addition of 8 arginines enables the delivery of a sufficient payload of CA *in cellulo* for $Cu^{+}$ detection in a Menkes disease model [50]. By combining sulfur atoms and carboxylate functions as in **Cu2** (Figure 6), CAs respond equally to $Cu^{+}$ and $Cu^{2+}$, with an affinity for $Cu^{2+}$ of log $K_{Cu}=15$ [49].

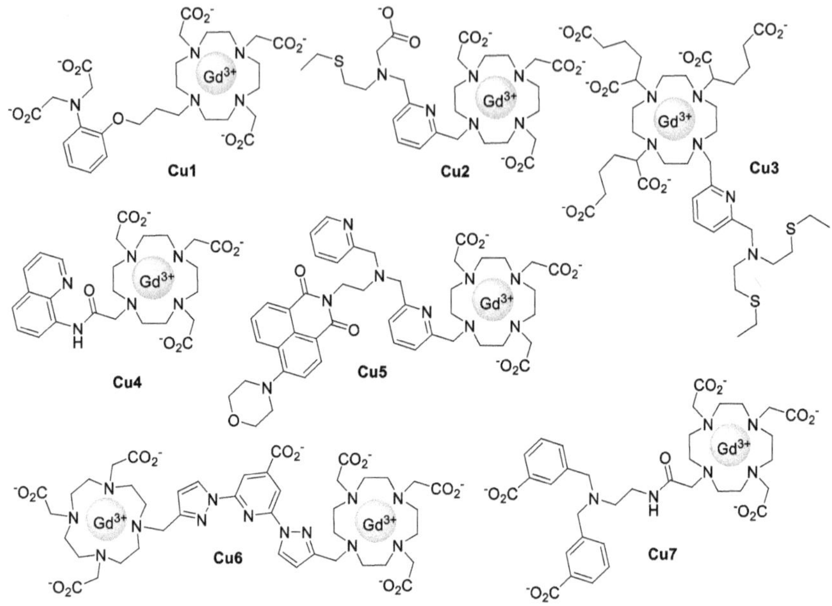

**FIGURE 6** Copper-responsive contrast agents.

$Cu^{2+}$ complexation is accompanied by an increase in the hydration number of $Gd^{3+}$ ($q$), which results in a relaxivity increase. The main drawback of this system, similar to all DO3A-based complexes, is its sensitivity to physiological anions in the presence of $Cu^{2+}$, which renders the practical detection of $Cu^{2+}$ problematic. It was shown, however, with **Cu3** (Figure 6) responding to $Cu^{+}$ with the same mechanism, that this problem can be overcome by the addition of negatively charged pendant carboxylate functions, while still retaining a good relaxivity increase upon $Cu^{+}$ complexation [51].

Other $Cu^{2+}$-binding moieties have been appended to DO3A systems, such as amidoquinoline (**Cu4,** Figure 6) [52] or DPA (**Cu5,** Figure 6) [53]. Since these binding moieties are similar to those proposed for $Zn^{2+}$, the selectivity is not expected to be good. In the case of DPA, a naphthalimide was appended which decreased the affinity for metal ions and enabled a good selectivity for $Cu^{2+}$ *vs* $Zn^{2+}$. The affinity for $Cu^{2+}$ is low (log $K_{Cu}$=4.9), which is problematic in displacing loosely bound $Cu^{2+}$ ions, e.g., in the blood where $Cu^{2+}$ is bound to HSA (log $K_{CuHSA}$=13) [54].

The dinuclear $Gd^{3+}$ CA **Cu6** (Figure 5) shows a nice relaxivity enhancement of 75% upon $Cu^{2+}$-binding due to a change in the hydration number of $Gd^{3+}$, with a good selectivity over other physiological cations including $Zn^{2+}$, even if the selectivity was only tested up to 2 eq. of $Zn^{2+}$. The relaxivity response drops down to 26% in the presence of a mixture of anions simulating the extracellular media [55].

Similarly with $Zn^{2+}$-responsive systems, a CA based on the variation of $\tau_R$ was recently developed (**Cu7,** Figure 5). This system responds to $Cu^{2+}$ with a 43% relaxivity increase, which is amplified to 270% in the presence of HSA. This is due to the formation of a ternary complex at the ATCUN site of the HSA, where $Cu^{2+}$ is naturally bound. A response to $Zn^{2+}$ is also observed in the presence of HSA, which can be problematic for practical detection. The CA was however used *in vivo* to image mouse liver [56].

## 3 pH-RESPONSIVE PROBES

pH plays a pivotal role in human biology, influencing important processes such as enzymatic activity and cellular function. The pH of healthy tissues is maintained at 7.1–7.4 by homeostasis, and a deviation from this range can be indicative of various pathological states such as bacterial infection, kidney disease, and cancers like solid tumors [57–59]. The ability to measure tissue pH has been achieved through various modalities, such as fluorescent probes or CEST probes [60,61], and in recent years, $T_1$-weighted Gd-based probes have garnered significant attention [62]. Generally, these pH-responsive CAs feature a chelating unit for $Gd^{3+}$ with an appended pH-responsive moiety. Similar to the principles of ion-responsive CA discussed in the previous section (See Figure 2), the relaxivity is then effectively modulated by the surrounding pH due to either (1) a change in the number of coordinated water molecules, $q$, or (2) a change in the rotational correlation time, $\tau_R$.

Early examples of pH-responsive CA, **P1–3** (Figure 7), feature a Gd-DO3A chelate with a sulfonamide moiety acting as the pH-responsive unit [63]. By varying the $R$ groups on the sulfonamide aryl ring, the protonation constants of the NH could be modulated to better fit physiological pH. This design of complex would see the sulfonamide NH deprotonated at basic pH, resulting in low relaxivity values due to the coordination of the nitrogen to the $Gd^{3+}$ center therefore blocking water access to the coordination sphere. However, in acidic media, the sulfonamide NH would become protonated and would no longer bind to the metal center, thereby increasing $q$ and thus the relaxivity of the complex. Complexes **P1a–c** demonstrated such a "pH-switch" at pH 7, where relaxivity increased from $2.0\,mM^{-1}s^{-1}$ to $7.5\,mM^{-1}s^{-1}$ (25°C, 20 MHz). Also, in this family, **P2** and **P3** (Figure 7) contained additional carboxylalkyl groups to inhibit intermolecular anion binding and provide greater stability. These complexes saw a greater increase in relaxivity, displaying $r_1$ values corresponding to $q=2$ complexes ($10.9\,mM^{-1}s^{-1}$; 25°C, 20 MHz).

Sulfonamides are also useful pH-responsive handles for changing the relaxivity of a CA through the mechanism of altering the rotational correlation time. HSA is often a target for stimuli-responsive contrast agents, as seen previously. Briefly, upon interaction of the contrast agent with HSA, the rotational correlation time vastly increases and thus so does the relaxivity of the CA. **P4** (Figure 7) is such a contrast agent that features a HSA binding moiety linked to the DO3A-type

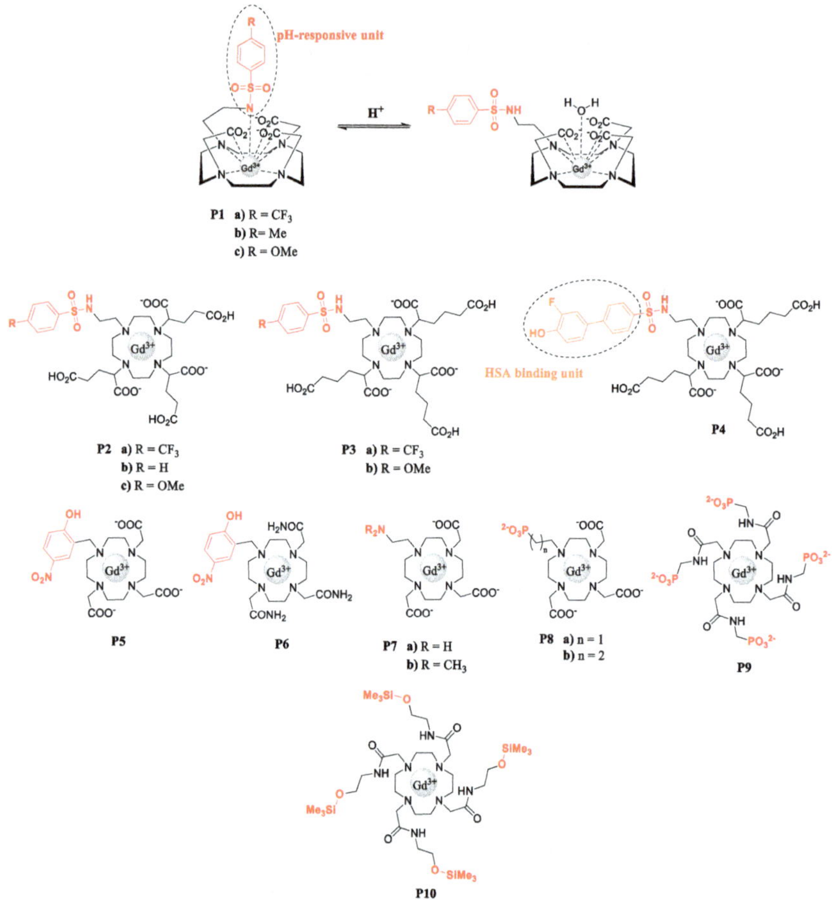

**FIGURE 7** pH-sensitive probes.

macrocyle *via* pH-sensitive sulfonamide [64]. As pH decreases, the relaxivity of **P4** increases due to an increase in $q$ as the sulfonamide NH becomes protonated, a behavior we have seen multiple times thus far. However, when in the presence of HSA, the changes in relaxivity were not as expected. The $r_1$ values were lower than expected at low pH, and the authors postulated that this could be due to inner-sphere water molecules being displaced by HSA once binding occurs or the exchange rate of inner-sphere water molecules is slowed by the protein.

A study comparing **P5** and **P6** (Figure 7) highlighted the importance of the choice of ligating groups [65]. While **P5** employs a DO3A-type chelate and **P6** employs a DO3Am-chelate, they both feature a pendant nitrophenolic arm which acts as a pH-responsive moiety. The relaxivity of **P5** increases by 71% when the pH passes from high to low pH (4.1 $mM^{-1}s^{-1}$ to 7.0 $mM^{-1}s^{-1}$; 25°C, 20 MHz), consistent with a $q$ change from 1 to 2, due to dissociation of the nitrophenolic

OH from the $Gd^{3+}$ center, once protonated. However, **P6** does not display such behaviour. While in acidic media, the relaxivity increases by 25%, and luminescence measurements show that there is no significant change in hydration number between acidic and basic pH ($q$=1.1 and 1.2). This would indicate that the nitrophenolic arm remains coordinated to the metal center, even at low pH. The authors attribute this due to the decreased electron-donating ability of the amides to the $Gd^{3+}$ ion, resulting in the $Gd^{3+}$ metal center of **P6** being less electron-rich than that of **P5**. This has a consequent result in which the nitrophenolic OH cannot dissociate from the lanthanide center, preventing a $q$ change.

The p$K_a$ values can be tuned conveniently by small chemical modifications as exemplified by **P7a** and **P7b** (Figure 7). The addition of two methyl groups on an aminoethyl moiety increased p$K_a$ values of ca. 7.5–9 [66]. Another interesting difference exists between the two complexes: while the $r_1$ values in acidic media are similar, 5.7 and 5.8 $mM^{-1}s^{-1}$ (25°C, 20 MHz), in basic conditions, the relaxivity is significantly different between the two, 4.1 and 3.1 $mM^{-1}s^{-1}$ (25°C, 20 MHz) for **P7a** and **P7b**, respectively. This is owed to the steric hindrance of the dimethyl group of **P7b**, which produces an equilibrium between $q$=0 and $q$=1.

Phosphonates have been successfully employed as pH-sensitive moieties in $T_1$ relaxation contrast agents. Similar to previously discussed examples, protonation of the pH-sensitive phosphonate at acidic pH leads to a dissociation from the $Gd^{3+}$ center, resulting in an increase of $q$ value. Examples of this include **P8** (Figure 7), where the molecules feature a DO3A-type chelate and a phosphonate arm, differing by the distance between the metal center and the phosphonate [67]. The $r_1$ values of **P8a** and **P8b** increased by 50% and 60% respectively, when pH moved from neutral to acidic, while $r_1$ values of esterified analogues that were also reported did not change at all in this pH range.

Similar systems have been successfully applied in *in vivo* settings for the measuring of pH in various tissues. **P9** (Figure 7), also referred to as GdDOTA-4AmP$^{5-}$, creates a hydrogen-bonding network that catalytically increases the exchange of bound water protons with protons of bulk water, thereby increasing relaxivity [68]. It was shown to be capable of measuring the extracellular pH (pH$_e$) in mice kidneys [69] and the ability to map the pH of various structures inside the kidney [70], alongside measuring the pH$_e$ in rat gliomas [71]. In order to map the pH, **P9** was co-injected with the corresponding pH-insensitive CA. When **P9** was incorporated onto a PAMAM (polyamidoamine) dendrimer, the relaxivity response in accordance to pH more than doubled [72].

Recently, a new class of pH-responsive CA has been reported which involves a change in $q$ value due to the removal of an acid-labile silyl ether moiety which sterically blocks water access to the $Gd^{3+}$ center [73]. The silyl ether groups of **P10** (Figure 7) are readily cleaved at acidic pH, revealing the metal center to the bulk water. The $T_1$ relaxation time of the CA steadily decreases as the silyl ether groups are hydrolysed from the CA, i.e., the complex was tested *in vivo* where a mouse was injected with a lactic acid buffer (pH 5.5). Here, **P10** displayed the capacity to detect a change in pH with an increase in $T_1$ signal intensity in the injected region, but also saw an increase in signal intensity in acidic tissues like the stomach and thyroid gland.

## 4 NEUROTRANSMITTER DETECTION

The detection of neural activity by MRI is appealing in neuroimaging sciences, and one way to achieve this is to track the release of neurotransmitters with responsive CA. Paramagnetic engineered metalloproteins ($Fe^{3+}$ metal center) were mutated to detect dopamine or serotonin and were successfully used *in vivo* to track the release of those neurotransmitters [74,75]. Superparamagnetic iron oxide nanoparticles appended with neurotransmitter analogues and protein-binding neurotransmitters were also used as $T_2$ contrast agents for neurotransmitter monitoring [76]. In a different approach, the zwitterionic properties of neurotransmitters were exploited for their detection by appending a crown-ether to a macrocyclic $Gd^{3+}$ complex. A double interaction occurs between the amine of the neurotransmitter and the crown ether, as well as between the carboxylate and the $Gd^{3+}$ ion. This results in a change in the number of water molecules coordinated upon neurotransmitter interaction and a subsequent change in relaxivity (see Figure 8). The complex has been used to follow neural activity *ex vivo* in mouse brain slices [77,78]. Further modification allows for the introduction of a benzophenone chromophore to sensitize $Nd^{3+}$ or $Yb^{3+}$ luminescence in the near-infrared range [79].

The ditopic systems were also conjugated to a generation 4 PAMAM dendrimer to increase its relaxivity [80]. After pegylation to ensure sufficient solubility in physiological conditions, the dendrimer was able to detect acetylcholine and glutamate over hydrogenocarbonate.

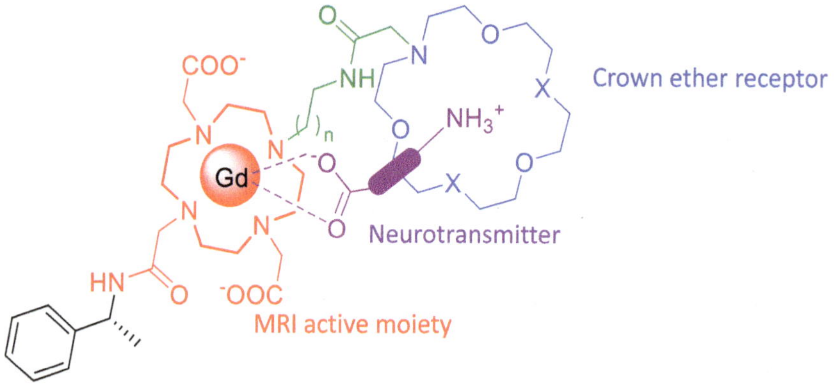

**FIGURE 8** Schematic representation of the macrocyclic $Gd^{3+}$ complex with a crown ether receptor for sensing zwitterionic neurotransmitters.

## 5 ENZYMATICALLY RESPONSIVE CONTRAST AGENTS

Enzymes are a diverse class of biomolecules that play crucial roles in human metabolism and homeostasis. They also contribute to pathological processes and conditions such as malignant tumour growth, inflammation, and neurodegenerative diseases [81,82]. There are many obstacles to face when attempting to sense enzymes *in vivo*. Certain enzymes are promiscuous and will interact with multiple chemical species, enzyme activity closely relies on its environment – pH, temperature, presence of cofactors, etc., and enzymes are usually present in relatively low concentrations of approximately 10 pM to 10 nM [83]. While MRI is not without its own challenges when attempting to detect biomarkers *in vivo*, MRI has proven effective in the detection of enzymes multiple times over. The typical low detection sensitivity of MRI can be counterbalanced by an enzyme's ability to catalytically convert a high concentration of a substrate. The classic design of enzymatically responsive CA features a chelating unit for $Gd^{3+}$ for $T_1$ relaxivity, and a substrate for a specific enzyme connected *via* a linker moiety. Upon interaction with the desired enzyme, several mechanisms can take place which results in an increase in $T_1$ relaxivity, such as increasing $H_2O$ access to the $Gd^{3+}$ center thereby effectively increasing $q$, or a change in $\tau_R$ due to self-assembly or further interaction with a biomacromolecule.

The first set of examples that will be examined in this section will focus on CAs which have been designed to demonstrate a change in $q$ after enzymatic interaction. The use of carbohydrates as enzyme substrates is a frequent theme with this strategy, as they allow for the detection of glycoside hydrolases. Glycoside hydrolases, enzymes which hydrolyse glycosidic bonds, such as $\beta$-galactosidase and $\beta$-glucuronidase, are perceived as biomarkers for tumours due to their upregulation in malignant tumours [84,85] and are as such prime targets for the imaging applications. $\beta$-Galactosidase can also be exploited for the indirect detection of gene expression, e.g., in biopsies and in tissue sections [86]. Indeed, the seminal paper from Meade et al. in 1997, which describes the first enzymatically responsive $T_1$ CA, features a $\beta$-galactose moiety linked to a $Gd^{3+}$-containing DO3A macrocycle, **E1** (Figure 9) [87]. The $\beta$-galactose ligand obstructs water access to the metal center by coordinating to the ninth coordination site of the $Gd^{3+}$, thereby limiting relaxivity. Upon cleavage of the galactopyranose from the chelating unit by $\beta$-galactosidase, the $Gd^{3+}$ ion is then free to coordinate to water molecules in the surrounding environment (demonstrated by a $q$ change from 0.7 to 1.2). This resulting relaxivity increase of 20% was a landmark result – opening the door for a new class of imaging agents, stimulating a novel avenue of research, and allowing for the design of CAs with specific applications and further diagnostic applicability.

Further work from this group focused on optimising the structure of the CA to obtain greater relaxivity changes [88]. By installing a methyl group on the linker, the $q$ values of the CA, **E2** (Figure 9), were proposed to be 0 and $1.3\pm0.1$ for pre-enzymatic and post-enzymatic cleavage, respectively. When assessed *in vivo* with *Xenopus laevis* embryos for the detection of the *lacZ* gene, the gene

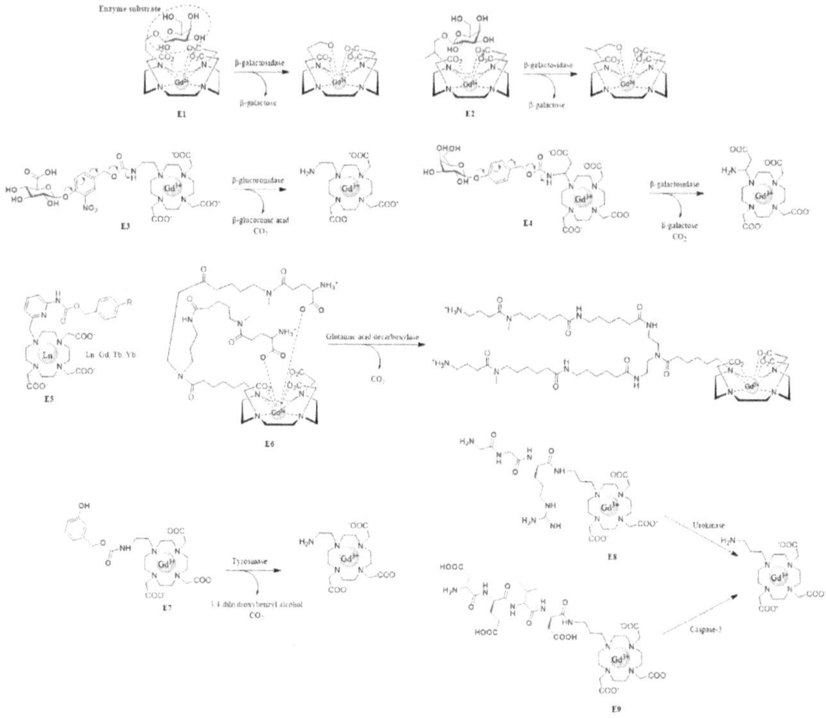

**FIGURE 9** Enzyme-responsive probes that feature an increase in $q$ post-enzymatic interaction.

which encodes $\beta$-galactosidase, an increase in MRI signal intensity of 57% was observed [89]. The slight change in the overall structure of the CA resulted in a significant change in behavior, as highlighted by the reported $q$ values alone. This emphasises that in the design of CAs, the conformation and the overall sterics of the agent are significant elements, alongside the selection of an enzyme substrate to append to the CA. It is also important to note that the efficacy of a CA *in vitro* does not necessarily reflect its behaviour *in vivo.* This was observed with the *X. laevis* embryos mentioned previously and also observed with **E3** (Figure 9). In this example, the authors describe the behaviour of **E3**, an enzyme-responsive CA with a $\beta$-glucuronic acid moiety appended to the macrocycle *via* a self-immolative linker [90]. Upon cleavage of the sugar due to $\beta$-glucuronidase, there is a resulting electron cascade, further fragmenting the linker from the molecule, releasing the contrast agent. While *in vitro* experiments in solutions mimicking extracellular anion composition demonstrated a 17% increase in relaxivity of the released CA, the same experiments repeated in human blood serum revealed that the relaxivity of the released CA, in fact, decreased by 27%. While the authors were unable to ascertain the exact mechanism behind this decrease in relaxivity due to the complex composition of human serum, they postulated that the viscosity of human serum or the potential interaction with macromolecules could have an effect on the rotational correlation time.

Similar behavior was observed with **E4** (Figure 9), which did not exhibit a significant change in relaxivity after enzymatic conversion [86]. This CA featured $\beta$-galactosidase as the enzyme substrate, connected to a GdDO3A macrocycle *via* a self-immolative linker. The $q$ values pre- and post-enzymatic cleavage were reported to be 0.8 (±0.3), which was reflected with the limited change in relaxivity (3.68 $mM^{-1} s^{-1}$ *vs* 3.15 $mM^{-1} s^{-1}$ at 37°C, 60 MHz). This decrease in relaxivity was ascribed to a decrease of rotational correlation time due to a decrease in complex size post-enzymatic cleavage. While it was deemed insufficient for *in vivo* enzymatic detection, the $Yb^{3+}$ analogues displayed more promising results for CEST imaging. Indeed, the authors continued on with the avenue of research and presented a trimodal probe, **E5** (Figure 9), for $T_1$-MRI, PARACEST, and visible/near-infrared luminescence, based on the structure of **E4** [91].

Other enzymes which have been the focus of MRI detection include glutamate decarboxylase (GAD), which is of particular relevance in the central nervous system and in neurological disorders such as epilepsy and Parkinson's disease [92,93]. The design of the GAD-responsive CA is based on a GdDO3A macrocycle with a long, flexible branched substituent ending with two glutamate residues, **E6** (Figure 9) [94]. The two glutamate residues coordinate to the $Gd^{3+}$ metal center, blocking water access. At basic pH, water is completely precluded from $Gd^{3+}$, with $q=0$. When shifted towards acidic media, $q=0$ species still dominate; however, an increase in relaxivity is observed which is attributed to strong contributions from outer-sphere water molecules. Upon enzymatic decarboxylation of **E6**, the relaxivity displayed a 60% increase relaxivity (~7 $mM^{-1} s^{-1}$ *vs* ~11.5 $mM^{-1} s^{-1}$) due to an increase to $q=2$. This CA was applied *in vivo* for the detection of inhibitory neurons in mice, where the monitoring of differentiation of stem cell grafts into GABAergic neurons was achieved [95].

Tyrosinase has become an enzyme of interest in recent years due to its possible implication in Parkinson's disease [96]; however, its exact role is not well understood. In order to provide a toolset to elucidate the role of tyrosinase, scientists developed **E7** (Figure 9) to measure tyrosinase activity in the brain [97]. This CA features *m*-hydroxyphenyl group as the enzyme substrate, which is appended to a GdDO3A *via* a carbamate linker, similar to the design strategy of **E3** and **E4**. Relaxivity studies were performed in ACSF, to mimic the environment in the brain, where the relaxivity of **E7** was found to be 1.88 $mM^{-1} s^{-1}$ (37°C, 60 MHz). This was attributed to the high concentration of carbonate in ACSF, which is known to coordinate to $Gd^{3+}$ and limit relaxivity. When the same experiments were repeated in phosphate-buffered saline (PBS), the relaxivity of **E7** was observed to be 4.6 $mM^{-1} s^{-1}$ (37°C, 60 MHz), once again highlighting the impact of the chemical environment on a CA. Upon interaction with tyrosinase, **E7** gains an additional hydroxyl functional group in the 4-position. This additional –OH results in an electron cascade releasing Gd-aminoethyl-DO3A which features improved water access to the $Gd^{3+}$ center. This increase in $q$ led to a subsequent increase in the relaxivity by 21%, granting a dark to bright signal change in the presence of tyrosinase.

Despite the plethora of research on enzyme-responsive CA and various DO3A-type chelates, not only does it remain difficult to predict their efficacy *in vivo* as discussed previously, but the coordination mode of the complexes and their interactions with enzymes also remains difficult. This is illustrated by a recent example which features two Gd-DO3A macrocycles, **E8** and **E9** (Figure 9), bearing peptide arms as substrates for urokinase and caspase-3, respectively [98]. Despite both complexes displaying a $q$ of 0, their relaxivities were higher than expected, 3.27 and 3.65 $mM^{-1}s^{-1}$ (pH 7.5, 25°C, 20 MHz) for **E8** and **E9**, respectively. The low $q$ values were attributed to the peptide arms coordinating to the $Gd^{3+}$ center *via* the amino acid residues, precluding water from accessing the inner sphere. The authors postulated that the high relaxivities of **E8** and **E9** were due to second-sphere mechanisms, brought about by the hydrophilic peptides retaining $H_2O$ molecules close to the metal center. Both complexes were shown to be inactive towards their respective enzymes, with no increase in relaxivity observed when treated with a large excess of paramagnetic complex. Once again, the authors propose that the binding of the amino acid residues to the $Gd^{3+}$ center is the root cause. They suggest that if the binding of the peptide to the $Gd^{3+}$ is relatively strong, it would effectively prevent the enzyme's access to the substrate and effectively hinder enzymatic conversion.

Another common mechanism that is exploited in the design of enzyme-responsive CA is to affect an increase in the rotational correlation time, $\tau_R$, after enzymatic conversion. This can be achieved in two ways: (1) commonly known as receptor-induced magnetisation enhancement (RIME), enzymatic conversion causes a subsequent binding of the CA to a large biomacromolecule, e.g., HSA, thereby slowing the molecular rotation, increasing $\tau_R$ and thus $r_1$ and (2) enzymatic conversion of the CA stimulates either a polymerisation or a supramolecular self-assembly process, effectively increasing the molecular weight and overall size of the paramagnetic species, again resulting in an increase of $\tau_R$ and consequently $r_1$.

Just as with the "$q$ change" examples examined previously, hydrolases feature as common targets for pro-RIME CAs. Early examples include **E10** (Figure 10) for the detection of $\beta$-galactosidase [99]. Here, the authors describe a Gd–DPTA complex appended to a biphenyl moiety, which exhibits a high affinity for HSA, alongside a $\beta$-galactose group acting as a "masking group", preventing interaction with HSA. Upon enzymatic hydrolysis of the carbohydrate, the complex becomes more hydrophobic thereby restoring the biphenyl moiety's affinity for albumin. This is reflected in the 57% increase in relaxivity in the presence of both $\beta$-galactosidase and HSA (4.06 $mM^{-1}s^{-1}$ *vs* 9.51 $mM^{-1}s^{-1}$; 37°C, 20 MHz).

A similar example exists with **E11** (Figure 10), which exhibits an even greater relaxivity change once interacting with HSA [100]. This pro-RIME CA contains $\beta$-glucuronic acid as the masking group of the albumin binding site, a 2-difluoromethylphenyl moiety. Relaxivity values of **E11**, *in vitro*, begin at 3.90 $mM^{-1}s^{-1}$

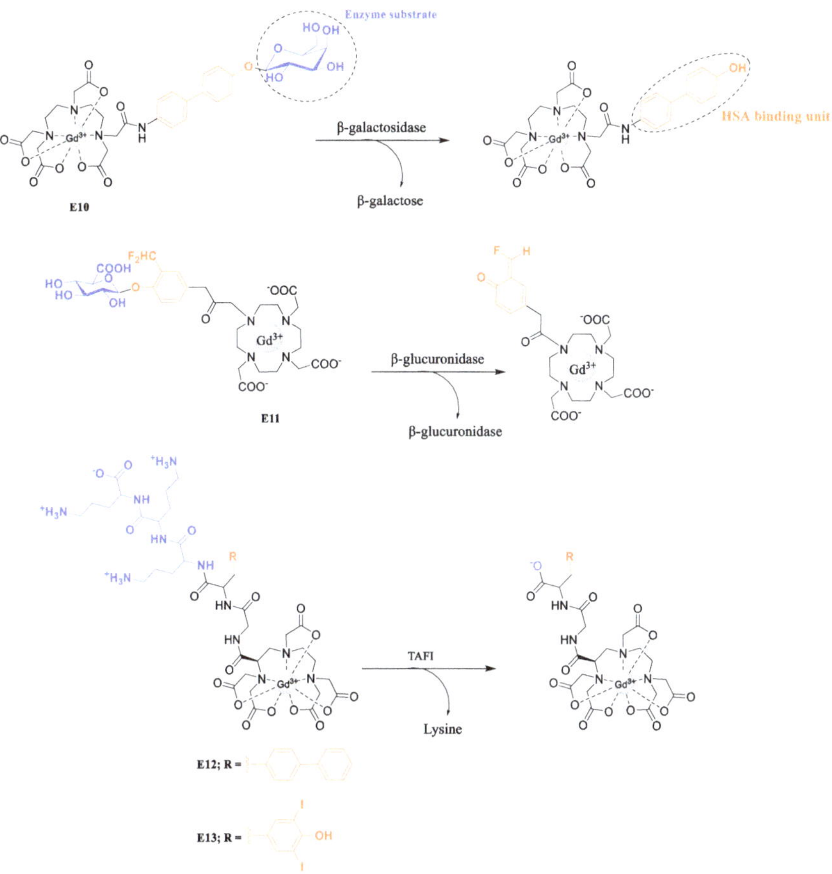


**FIGURE 10** Enzyme-responsive probes that feature an increase in $\tau_R$ post-enzymatic interaction.

under the same conditions (37°C, 20MHz) but in the presence of HSA, $r_1$ increases to 7.76mM$^{-1}$s$^{-1}$, indicating interaction with HSA even with the intact masking group. When the environment contains both $\beta$-glucuronidase and HSA, the relaxivity displays a remarkable increase to 14.3mM$^{-1}$s$^{-1}$. *In vivo* experiments studying tumor xenografts in BALB/c mice also saw significant enhancement of the MR signal at multiple time points.

Other targeted enzymes for pro-RIME CAs include thrombin-activatable fibrinolysis inhibitor (TAFI), an enzyme which is implicated in thrombotic diseases. **E12** and **E13** (Figure 10) vary by their HSA binding groups, but they retain their enzyme substrates (three lysine residues), their MR signal generator, Gd-DPTA, and a glycine linker between the $Gd^{3+}$ complex and the HSA binding groups [101]. The three lysine groups act as efficient blocking groups for HSA due to charged moieties having a low affinity for HSA, while the albumin binding groups, biphenylalanine and 3,5-diiodotyrosine, display albumin binding affinity due to their hydrophobic nature. Upon enzymatic-cleavage of the three lysine groups to reveal the aryl HSA binding motifs, there was a stark increase in relaxivity. The relaxivity of **E12** increased from 11.1 $mM^{-1}s^{-1}$ to 24.5 $mM^{-1}s^{-1}$ (37°C, 20 MHz), an increase of 220%. Likewise, **E13** saw an increase in its relaxivity from 9.8 $mM^{-1}s^{-1}$ to 24.5 $mM^{-1}s^{-1}$ (37°C, 20 MHz), an increase of approximately 239%.

Oligomerisation of a contrast agent to induce an increase in relaxivity has been a common tactic in the detection of oxidoreductases, namely peroxidase and myeloperoxidase. The mechanism of action for the detection of peroxidases relies on the enzyme using the CA as an electron donor in the reduction of hydrogen peroxide. The oxidised, free radical form of the CA would then self-polymerise into a higher relaxivity oligomers (Figure 11a). This mechanism of increasing relaxivity is exemplified by **E14** (Figure 11), where a DOTA-type chelate has an appended dopamine moiety that acts as the electron donor for the enzymatic reduction of $H_2O_2$ [102]. In the presence of horseradish-peroxidase and $H_2O_2$, relaxivity increases from 3.75 $mM^{-1}s^{-1}$ to 11.5 $mM^{-1}s^{-1}$ (40°C, 20 MHz). To verify self-polymerisation, reaction products were subjected to size-exclusion chromatography and HRMS, which revealed oligomerisation of up to 12 degrees. A continuation of this work, **E15** (Figure 11), changed the dopamine moiety to 5-hydroxytryptamide in order to selectively target myeloperoxidase (MPO), an enzyme implicated in inflammation and atherosclerotic diseases [103]. Reaction with MPO resulted in an increase of relaxivity by 70% (40°C, 20 MHz) and in *in vitro* MR experiments using model tissue systems, while a 40% signal increase was observed with those containing MPO and $H_2O_2$.

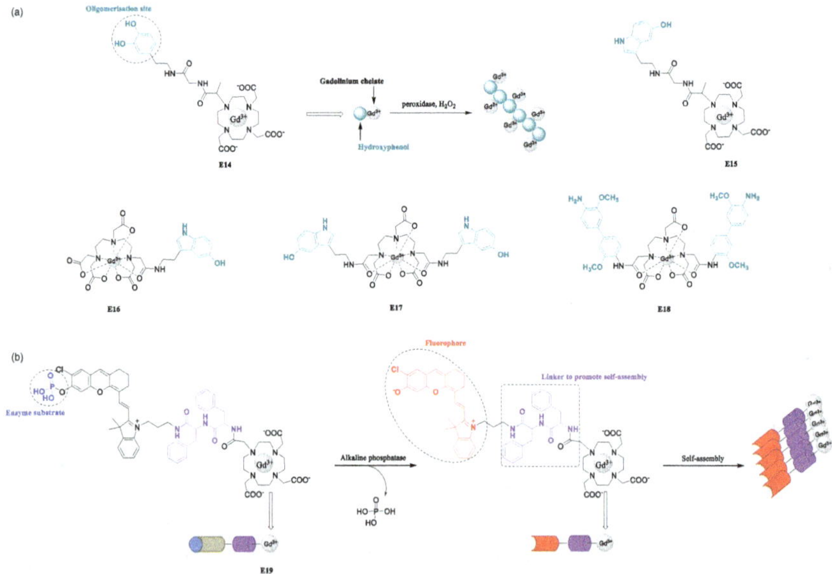


**FIGURE 11** (a and b) Enzyme-responsive probes that self-assemble post-enzymatic interaction.

Enzyme-responsive CAs sensitive to MPO have also been shown to increase their relaxivity through a combination of two mechanisms – oligomerisation and RIME. Such is the case with **E16–18** (Figure 11) which, after interaction with MPO, sees an increase in relaxivity between 50% and 80% [104]. This is attributed to the polymerisation of the CAs, with the degree of relaxivity increase being directly related to the extent of polymerisation. When **E16–18** are exposed to MPO and polypeptides with tyrosine residues, there is a further, concomitant increase in relaxivity of 22%–59% in $r_1$ values. These systems were successfully applied *in vivo* for the detection of inflammation in mice, where all three agents displayed a 100% enhancement of MR signal.

The supramolecular self-assembly of CA into larger paramagnetic nanoparticles has been demonstrated as an effective strategy to detect the presence of enzymes. The assembly of small molecule into a larger structure, *via* non-covalent interactions, effectively results in an increase of $\tau_R$. A prime example of this is **E19** (Figure 11b), a bimodal probe that features a merocyanine fluorophore linked to paramagnetic Gd–DOTA complex *via* a Phe-Phe linker which increases the propensity for self-assembly through π–π stacking and hydrophobic interactions [105]. The fluorophore is equipped with a phosphate moiety which acts as a fluorescent quencher and a substrate for alkaline phosphatase. Enzymatic cleavage of the phosphate moiety sees an increase of fluorescence alongside self-assembly of the probe into nanoparticles, which results in an increase in $r_1$ from $8.9 \pm 0.3\,mM^{-1}s^{-1}$ to $20.1 \pm 0.5\,mM^{-1}s^{-1}$ (37°C, 20 MHz). These paramagnetic nanoparticles demonstrated their MR utility *in vivo* where they demonstrated the ability to diffuse into HeLa tumors in mice and detect ALP activity, with a signal enhancement of ~58%.

## 6 REDOX-RESPONSIVE CONTRAST AGENTS

The redox environment in biological systems refers to the balance between reduction and oxidation reactions and is essential for various cellular processes such as maintaining cellular homeostasis [106], cell proliferation, differentiation, and cell death [107]. A disturbance of redox equilibrium is associated with cancers [108,109], autoimmune disorders [110], and schizophrenia [111]. As such, the development of tools such as imaging agents to monitor the redox environment has garnered great interest in recent years and has been the topic of many reviews [112–115]. This section will focus on a few key examples of redox-responsive CAs and discuss the main strategies for examining the redox environment using MRI contrast agents. The two main ways to design redox CA are either to use a redox active ligand or metal ion. In the first case, the $Ln^{3+}$ ion is $Gd^{3+}$, while in the second, the reduction of $Eu^{3+}$ to $Eu^{2+}$ is mainly exploited. Redox coupling of transition metal ions such as $Mn^{2+}/Mn^{3+}$ or $Cu^{+}/Cu^{2+}$ has also been exploited but will not be discussed here.

For ligand center redox activity, the main design strategy is to induce a structural change in the chelate which consequentially increases the relaxivity of the CA – either through an increase in $q$ value or an increase in $\tau_R$. The first example

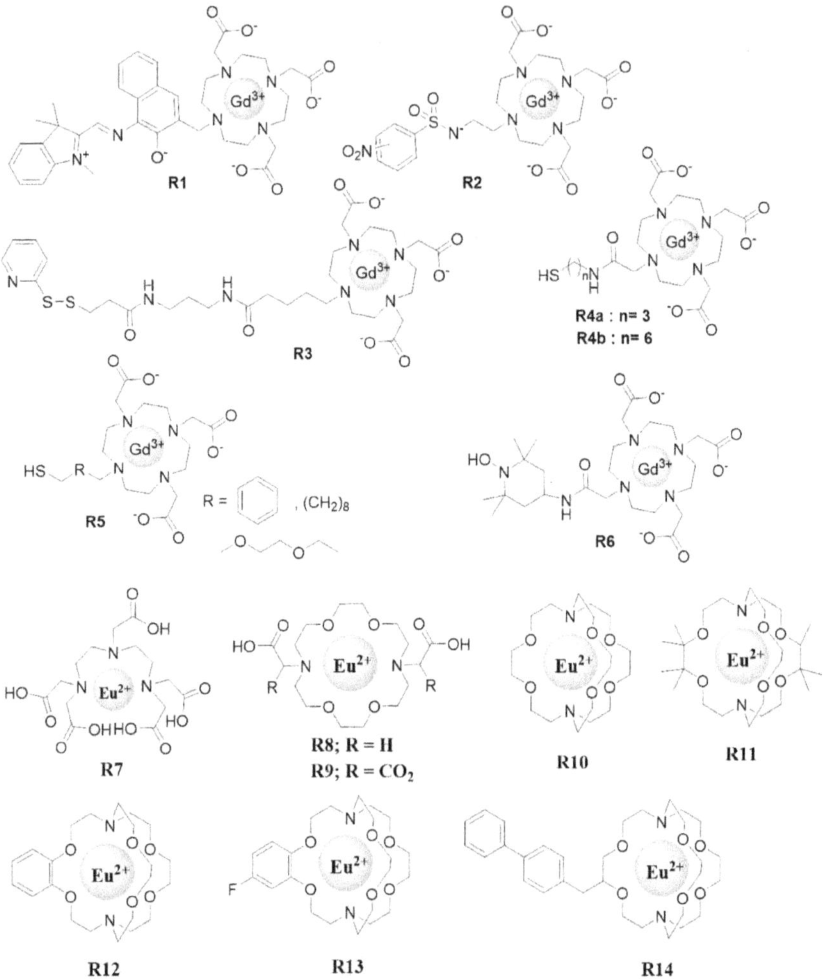

FIGURE 12 Ligand redox active $Gd^{3+}$-based CA and $Eu^{2+}$ redox active CA.

of such a CA, **R1** (Figure 12), employs a Gd-DO3A chelate with a merocyanine moiety as the redox-sensitive unit [116]. Merocyanines and spirooxazines are known to interconvert when exposed to photoirradiation [117]; however, in this instance, the isomerisation from merocyanine to spirooxazine was caused by the oxidation of NADH to $NAD^+$. Upon conversion to the spirooxazine form, one of the coordination bonds to the $Gd^{3+}$ center is broken, resulting in an increase in relaxivity from $5.56\,mM^{-1}s^{-1}$ to $8.60\,mM^{-1}s^{-1}$, due to an increase in $q$ value of the complex from 1 to 2. It is important to note that the exact mechanism of this isomerisation is unclear due to no electron transfer occurring, and while the spirooxazine could be partially converted back to merocyanine by $H_2O_2$, **R1** did not undergo isomerisation in the presence of other reducing agents such as L-cysteine and L-ascorbic acid.

Hypoxia is characterised by its reductive environment, and it occurs in a multitude of pathologies, including cancer, and heart and kidney diseases [118]. A small library of redox-responsive contrast agents for the detection of hypoxia was reported, where the CAs displayed both redox-sensitive and pH-sensitive behaviours [119]. These CAs employed arylsulfonamides bearing a nitro group as the redox-responsive trigger linked to a DO3A-type chelate. Taking **R2** (Figure 12) as an example, the nitrogen of the arylsulfonamide is deprotonated and coordinates to the $Gd^{3+}$ center. In hypoxic environments, the nitro moiety is reduced to an amino group thereby changing the $pK_a$ of the sulfonamide NH. As seen previously with examples in Section 3, a decrease in pH leads to the nitrogen to become protonated, dissociating the atom from the lanthanide core and leading to a consequent increase in relaxivity (2.1 $mM^{-1}s^{-1}$ *vs* 3.6 $mM^{-1}s^{-1}$; 37°C, 20 MHz). When in the hypoxic environment of rat microsomes, **R2** granted significantly increased MR signal enhancement.

Glutathione (GSH) is a vital reducing agent with roles in reducing oxidative stress [120] and in the innate immune system [121], and its dysregulation is implicated in many disorders such as neurodegenerative diseases [122] and liver disease [123]. **R3** (Figure 12) demonstrated the ability to detect GSH through a decrease in relaxivity [124]. Displaying a Gd-DO3A chelate with an appended 2-pyridyl-dithio moiety *via* a flexible linker, in the presence of GSH, the complex reacts to form mixed disulfides. These mixed disulfides can form intramolecular coordination bonds to the $Gd^{3+}$ metal center through the carboxyl group in the case of GSH, leading to a decrease in relaxivity (8.1 $mM^{-1}s^{-1}$ to 4.1 $mM^{-1}s^{-1}$; 25°C, 20 MHz) due to a decrease in the number of water molecules coordinated to the lanthanide. When an inclusion complex between **R3** and β-cyclodextrin or poly-β-cyclodextrin was examined, relaxivity differences of up to 60% were observed.

Thiol moieties appended to paramagnetic chelates can also affect relaxivity through modulating the rotational dynamics of the complex post-redox interaction. **R4** (Figure 12), bearing either a simple propyl or hexyl chain terminating with a primary thiol, demonstrated an increase in relaxivity in the presence of HSA (**R4b** –2.3 $mM^{-1}s^{-1}$ to 4.5 $mM^{-1}s^{-1}$; 37°C, 200 MHz) [125]. This increase in relaxivity was attributed to the complexes binding to HSA, resulting in an increase in $\tau_R$. Further studies including complexes **R5** (Figure 12), which are Gd-DO3A chelates with various linkers ending with a free thiol, revealed that the DO3A complexes exhibited lower relaxivity values than the Gd-DOTA chelates of **R4** [126]. Direct and competition binding assays indicated that the increase in $\tau_R$ for these CAs arises from the formation of a disulphide bridge with Cys34 of HSA. **R4b** was successfully employed *in vivo*, where the CA, in combination with glutathione synthesis inhibitor or thiol-oxidising cancer therapeutic Imexon, provided high MRI contrast enhancement in NCI-N87 and Mia-PaCa-2 tumor xenografts in mice [127,128]. The degradation of perthiolated β-cyclodextrin-based nanocapsules incorporating $Gd^{3+}$ complexes under reducing conditions was used to transition from a high relaxivity state to a low relaxivity state [129].

The TEMPO radical has also been recently used for such purposes in **R6** (Figure 12) with a different mechanism of detection [130]. The relaxivity increase of the $Gd^{3+}$ complex from the hydroxylamine form to the nitroxide form by 26% at 30 MHz, 25°C, is explained by the generation of the paramagnetic center in the vicinity of $Gd^{3+}$, which accelerates spin relaxation.

While $Gd^{3+}$-based probes have been proven to be effective redox-responsive CAs through various mechanisms and redox-sensitive ligands, they do no exhibit a complete "OFF/ON" switch, i.e., the CA still displays an MR signal before interaction with its chosen stimulus. By changing the metal center from Gd to Eu, such an "OFF/ON" switch can be observed. $Eu^{2+}$ and $Gd^{3+}$ are isoelectronic so it can be expected to have similar proton relaxivities, while $Eu^{3+}$ shows poor relaxation properties due to its small magnetic moment, thus making the $Eu^{3+}/Eu^{2+}$ redox couple a prime candidate for such redox-responsive CAs.

The first examples of $Eu^{2+}$ complexes as contrast agents, **R7–9** (Figure 12), were reported in 2000 with $r_1$ values of 2.7, 3.49, and 3.57 $mM^{-1} s^{-1}$ (25°C, 20MHz), respectively [131]. However, the thermodynamic stability of these complexes was lower than that of their $Gd^{3+}$ analogues, limiting their *in vivo* applications. To increase the stability of $Eu^{2+}$ complexes, the design of the chelate shifted towards cryptates, e.g., **R10** (Figure 12) [132].

Many structural modifications were made to the simple $Eu^{2+}$ cryptate **R10** to optimise its stability by modulating several important factors such as the metal's exposure to its environment and its propensity for $Eu^{2+}$ over $Eu^{3+}$. **R11** bears additional methyl groups on the cryptand to limit $Eu^{2+}$ interactions with the environment; **R12** and **R13** feature a phenyl ring with varying substituents to alter the Lewis basicity to favor $Eu^{2+}$ and $Eu^{3+}$ [132].

These $Eu^{2+}$-based redox-responsive CAs can also display similar behavior to $Gd^{3+}$ examples we have examined previously. For instance, **R14** (Figure 12) with its appended biphenyl moiety can bind to HSA and demonstrate a 170% increase in $r_1$ due to an increase in $\tau_R$ [133]. Despite initial struggles in obtaining a $Eu^{2+}$-based CA, they have been successfully applied *in vivo.* **R13** was tested in a 4T1 mammary carcinoma xenograft in a mouse model [133]. Immediately post-injection of the CA, a contrast enhancement was observed throughout the tumor. After 2 h, the contrast became localised to the necrotic core of the tumor, demonstrating the potential of such agents.

## 7 QUANTIFICATION METHODS

Intrinsically, the MRI signal observed depends not only on the presence of the biomarker but also on the local concentration of the CA, and both contributions cannot be separated, especially for $T_1$-based CA, as "averaged" signal is measured. For some biomarkers such as enzymes, the quantification is not crucial as the detection of the activity is more important; however, it is not the case for other biomarkers such as pH or cations. Therefore, strong quantification methods must be developed. Early work on pH quantification exploited an *in vitro* $R_2/R_1$ method; however, it remains difficult to use in practice due to the large inherent $R_2$ in living tissues [134]. Co-injection of pH-independent and pH-dependent probes has been used *in vivo* to map values in brain tumour models [71]. However, different biodistribution of the probes remain problematic.

The use of bimodal probes is a promising way to achieve quantification. MRI can be combined with a complementary technique such as nuclear imaging. Several approaches have been used to introduce PET or SPECT reporters. A $^{18}F$ PET reporter was appended to a pH-sensitive CA (**P9** derivative, Figure 7) and diluted in $^{19}F$ to respect the sensitivity of each technique to successfully measure pH *in vitro* [135]. Metal surrogates are also easy to use in a chemical point of view. Recently, $^{64}Cu^{2+}$ and $^{68}Ga^{3+}$ were used as $Gd^{3+}$ surrogates to measure pH *in vivo* [136]. However, the PET agents were not stable enough *in vivo*, and the charge difference and coordination difference of the metal might also be a limitation. $Ln^{3+}$ surrogates seem appealing to avoid those difficulties, and the recent use of $^{86}Y$ to quantify whole-body distribution of $Gd^{3+}$ CA supports this idea [137]. $^{166}Ho$, a SPECT reporter, was used to quantify pH *in vitro* [138]. $^{166}Ho$ is produced from $^{165}Ho$ and cannot be separated from its cold parent. It is therefore possible to quantify pH using this isotope, but not realistic for cation detection. Indeed, cations are present in the μM concentration range and the high quantity of cold$^{166}Ho$ will limit the MRI response and give unreliable quantification information.

We recently proposed to use $^{165}Er$ as a SPECT surrogate. $^{165}Er$ can be produced from a cyclotron and purified from its parent compound $^{165}Ho$ for a final $^{165}Ho$ concentration less than 1000-fold that of $Gd^{3+}$ not to hamper cation detection. Ligand of **Z11** (Figure 4) was radiolabelled with $^{165}Er^{3+}$, and a cocktail of the $Gd^{3+}$ and $^{165}Er^{3+}$ complex was used *in vitro* in the presence of HSA to successfully quantify $Zn^{2+}$, as demonstrated by ICP measurements [139].

Finally, $^{19}F$ MRI can also be used in complement with $^{1}H$ MRI for quantification. It was cleverly used for pH quantification *in vitro* through the formation of supramolecular adducts. **P1** (Figure 7) was modified to add an adamantane, known to form supramolecular adducts with poly-β-cyclodextrins [140]. A similar adamantane was substituted by a $CF_3$ group, which was used to report on the local concentration of the $Gd^{3+}$ complex. Micelles composed of $Ca^{2+}$-responsive CA, surfactant, and perfluoro-15-crown-5-ether were also used for $Ca^{2+}$ quantification *in vitro* and *in vivo* as the $^{19}F$ MRI signal was shown insensitive to $Ca^{2+}$, while the $^{1}H$ MRI signal was [141]. In a different approach, a $Ca^{2+}$-responsive CA labelled with 9 magnetically equivalent F shows modified distances between the F and a paramagnetic centre ($Dy^{3+}$) upon $Ca^{2+}$ complexation. When $Dy^{3+}$ is replaced by diamagnetic $Y^{3+}$, the signal remains unchanged allowing for *in vitro* quantification [142].

To conclude, recent years have seen tremendous developments in the design and applications of responsive $Ln^{3+}$-based probes, both *in vitro* and in preclinical studies. Contrast agents, which are enzymatically activated, pH sensing, and $Ca^{2+}$ and $Zn^{2+}$ sensing, have all been successfully employed in preclinical models, showing the full potential of such agents to revolutionise diagnostic imaging and medical diagnoses. However, amidst these achievements, it is vital to note that there are still obstacles yet to be overcome – the delivery of contrast agents to the brain, crucial for neurological applications, is yet to be presented and robust quantification methods are still in their infancies, to name but two challenges. The success over these obstacles would undoubtedly push *in vivo* applications into brand new territory. To achieve this, the development of bimodal agents is a promising avenue of exploration, and it should be accompanied with technological development to allow single imaging of the patient and to reduce the dose of probe administered. At the heart of these advancements in responsive contrast agents, coordination chemistry provides the basis of the development of such successful contrast agents which are the future of diagnostic imaging and diagnosis.

## ABBREVIATIONS AND DEFINITIONS

**ACSF** artificial cerebral spinal fluid
**AECM** artificial extracellular matrix
**APTRA** o-amino-phenol-N,N,O-triacetate
**BAPTA** 1,2-bis(o-aminophenoxy)ethane-N,N,N′,N′-tetraacetic acid)
**BEM** brain extracellular medium
**BPEN** N,N-bis-(2-pyridyl-methyl)ethylenediamine)
**CA** contrast agent
**DOTA** 1,4,7,10-tetraazacyclododecane-1,4,7,10-tetraacetic acid
**DTPA** diethylenetriamine pentaacetic acid
**EDTA** ethylenediamine tetraacetic acid
**EGTA** (ethyleneglycol-bis(β-aminoethylether)-N,N,N′,N′-tetraacetic acid
**GAD** glutamate decarboxylase
**GSH** glutathione
**HSA** human serum albumin
**MPO** myeloperoxidase
**PRE** paramagnetic relaxation enhancement
**TSL** thermosensitive liposome
**ZF** zinc finger

## REFERENCES

1. B. S. H. Chia, S. Z. Ho, H. Q. Tan, M. L. K. Chua, J. K. L. Tuan, *Cancers* **2023**, *15*, 346.
2. L. H. Lindner, H. M. Reinl, M. Schlemmer, R. Stahl, M. Peller, *Int. J. Hyperthermia.* **2005**, *21*, 575–588.
3. D. Long, T. Liu, L. Tan, H. Shi, P. Liang, S. Tang, Q. Wu, J. Yu, J. Dou, X. Meng, *ACS Nano.* **2016**, *10*, 9516–9528.
4. M. Peller, A. Schwerdt, M. Hossann, H. M. Reinl, T. Wang, S. Sourbron, M. Ogris, L. H. Lindner, *Invest. Radiol.* **2008**, *43*, 877–892.
5. S. A. A. S. Subasinghe, J. Romero, C. L. Ward, M. D. Bailey, D. R. Zehner, P. J. Mehta, F. Carniato, M. Botta, J. T. Yustein, R. G. Pautler, M. J. Allen, *Chem. Commun.* **2021**, *57*, 1770–1773.

6. A. J. Surman, C. S. Bonnet, M. P. Lowe, G. D. Kenny, J. D. Bell, É. Tóth, R. Vilar, *Chem. Eur. J.* **2011**, *17*, 223–230.
7. K. P. Malikidogo, H. Martin, C. S. Bonnet, *Pharmaceuticals.* **2020**, *13*, 436.
8. J. Wahsner, E. M. Gale, A. Rodríguez-Rodríguez, P. Caravan, *Chem. Rev.* **2019**, *119*, 957–1057.
9. V. C. Pierre, S. M. Harris, S. L. Pailloux, *ACC Chem. Res.* **2018**, *51*, 342–351.
10. D. A. Hammoud, J. M. Hoffman, M. G. Pomper, *Radiology* **2007**, *245*, 21–42.
11. W. H. Li, S. E. Fraser, T. J. Meade, *J. Am. Chem. Soc.* **1999**, *121*, 1413–1414.
12. W. H. Li, G. Parigi, M. Fragai, C. Luchinat, T. J. Meade, *Inorg. Chem.* **2002**, *41*, 4018–4024.
13. K. W. MacRenaris, Z. Ma, R. L. Krueger, C. E. Carney, T. J. Meade, *Bioconjugate Chem.* **2016**, *27*, 465–473.
14. K. Dhingra, P. Fouskova, G. Angelovski, M. E. Maier, N. K. Logothetis, E. Toth, *J. Biol. Inorg. Chem.* **2008**, *13*, 35–46.
15. G. Angelovski, P. Fouskova, I. Mamedov, S. Canals, E. Toth, N. K. Logothetis, *ChemBioChem.* **2008**, *9*, 1729–1734.
16. A. Mishra, P. Fouskova, G. Angelovski, E. Balogh, A. K. Mishra, N. K. Logothetis, É. Tóth, *Inorg. Chem.* **2008**, *47*, 1370–1381.
17. T. Savić, G. Gambino, V. S. Bokharaie, H. R. Noori, N. K. Logothetis, G. Angelovski, *Proc. Natl. Acad. Sci.* **2019**, *116*, 20666–20671.
18. L. Connah, V. Truffault, C. Platas-Iglesias, G. Angelovski, *Dalton Trans.* **2019**, *48*, 13546–13554.
19. A. Mishra, N. K. Logothetis, D. Parker, *Chem. Eur. J.* **2011**, *17*, 1529–1537.
20. P. Kadjane, C. Platas-Iglesias, P. Boehm-Sturm, V. Truffault, G. E. Hagberg, M. Hoehn, N. K. Logothetis, G. Angelovski, *Chem. Eur. J.* **2014**, *20*, 7351–7362.
21. K. Dhingra, M. E. Maier, M. Beyerlein, G. Angelovski, N. K. Logothetis, *Chem. Commun.* **2008**, 3444–3446.
22. V. Kubicek, T. Vitha, J. Kotek, P. Hermann, L. Vander Elst, R. N. Muller, I. Lukeš, J. A. Peters, *CMMI* **2010**, *5*, 294–296.
23. I. Mamedov, S. Canals, J. Henig, M. Beyerlein, Y. Murayama, H. A. Mayer, N. K. Logothetis, G. Angelovski, *ACS Chem. Neuro.* **2010**, *1*, 819–828.
24. S. L. Sensi, P. Paoletti, A. I. Bush, I. Sekler, *Nat. Rev. Neurosci.* **2009**, *10*, 780–791.
25. K. Hanaoka, K. Kikuchi, Y. Urano, T. Nagano, J. Chem. Soc. Perkin *Trans.* **2001**, 1840–1843.
26. C. S. Bonnet, *Coord. Chem. Rev.* **2018**, *369*, 91–104.
27. A. C. Esqueda, J. A. Lopez, G. Andreu-de-Riquer, J. C. Alvarado-Monzón, J. Ratnakar, A. J. Lubag, A. D. Sherry, L. M. De León-Rodríguez, J. *Am. Chem. Soc.* **2009**, *131*, 11387–11391.
28. A. J. M. Lubag, L. M. De Leon-Rodriguez, S. C. Burgess, A. D. Sherry, *Proc. Natl. Acad. Sci.* **2011**, *108*, 18400–18405.
29. S. W. Chen, Z. J. Huang, H. Kidd, E. H. Ghazvini Zadeh, A. D. Sherry, P. E. Scherer, W. H. Li, *Front. Endoc.* **2021**, *12*, 613964.
30. M. Veronica Clavijo Jordan, S.-T. Lo, S. Chen, C. Preihs, S. Chirayil, S. Zhang, P. Kapur, W. H. Li, L. M. De Leon-Rodriguez, A. J. Lubag, N. M. Rofsky, *Proc. Natl. Acad. Sci.* **2016**, *113*, E5464–E5471.
31. J. Yu, A. F. Martins, C. Preihs, V. Clavijo Jordan, S. Chirayil, P. Zhao, Y. Wu, K. Nasr, G. E. Kiefer, A. D. Sherry, *J. Am. Chem. Soc.* **2015**, *137*, 14173–14179.
32. A. F. Martins, V. C. Jordan, F. Bochner, S. Chirayil, N. Paranawithana, S. Zhang, S. T. Lo, X. Wen, P. Zhao, M. Neeman, A. D. Sherry, *J. Am. Chem. Soc.* **2018**, *140*, 17456–17464.
33. J. L. Major, R. M. Boiteau, T. J. Meade, *Inorg. Chem.* **2008**, *47*, 10788–10795.

34. J. L. Major, G. Parigi, C. Luchinat, T. J. Meade, *Proc. Natl. Acad. Sci.* **2007**, *104*, 13881–13886.
35. L. M. Matosziuk, J. H. Leibowitz, M. C. Heffern, K. W. MacRenaris, M. A. Ratner, T. J. Meade, *Inorg. Chem.* **2013**, *52*, 12250–12261.
36. M. Regueiro-Figueroa, S. Guenduez, V. Patinec, N. K. Logothetis, D. Esteban-Gomez, R. Tripier, G. Angelovski, C. Platas-Iglesias, *Inorg. Chem.* **2015**, *54*, 10342–10350.
37. G. Wang, G. Angelovski, *Angew. Chem. Int. Ed.* **2021**, *60*, 5734–5738.
38. G. Wang, H. Martin, S. Amézqueta, C. Ràfols, C. S. Bonnet, G. Angelovski, *Inorg. Chem.* **2022**, *61*, 16256–16265.
39. Y.-Q. Xu, J. Luo, Z.-N. Chen, *Eur. J. Inorg. Chem.* **2014**, 3208–3215.
40. G. J. Stasiuk, F. Minuzzi, M. Sae-Heng, C. Rivas, H. P. Juretschke, L. Piemonti, P. R. Allegrini, D. Laurent, A. R. Duckworth, A. Beeby, G. A. Rutter, *Chem. Eur. J.* **2015**, *21*, 5023–5033.
41. C. S. Bonnet, F. Buron, F. Caillé, C. M. Shade, B. Drahoš, L. Pellegatti, J. Zhang, S. Villette, L. Helm, C. Pichon, F. Suzenet, *Chem. Eur. J.* **2012**, *18*, 1419–1431.
42. C. S. Bonnet, F. Caille, A. Pallier, J. F. Morfin, S. Petoud, F. Suzenet, É. Tóth, *Chem. Eur. J.* **2014**, *20*, 10959–10969.
43. K. P. Malikidogo, M. Isaac, A. Uguen, J. F. Morfin, G. Tircsó, É. Tóth, C. S. Bonnet, *Inorg. Chem.* **2023**, *62*, 17207–17218.
44. M. Bödenler, K. P. Malikidogo, J. F. Morfin, C. S. Aigner, É. Tóth, C. S. Bonnet, H. Scharfetter, *Chem. Eur. J.* **2019**, *25*, 8236–8239.
45. K. P. Malikidogo, M. Isaac, A. Uguen, S. Même, A. Pallier, R. Clémençon, J. F. Morfin, S. Lacerda, É. Tóth, C. S. Bonnet, *Chem. Commun.* **2023**, *59*, 12883–12886.
46. M. Isaac, A. Pallier, F. Szeremeta, P. A. Bayle, L. Barantin, C. S. Bonnet, O. Sénèque, *Chem. Commun.* **2018**, *54*, 7350–7353.
47. K. P. Malikidogo, A. Pallier, F. Szeremeta, C. S. Bonnet, O. Sénèque, *Dalton Trans.* **2023**, *52*, 6260–6266.
48. E. L. Que, C. J. Chang, J. Am. Chem. Soc. **2006**, *128*, 15942–15943.
49. E. L. Que, E. Gianolio, S. L. Baker, A. P. Wong, S. Aime, C. J. Chang, *J. Am. Chem. Soc.* **2009**, *131*, 8527–8536.
50. E. L. Que, E. J. New, C. J. Chang, *Chem. Sci.* **2012**, *3*, 1829–1834.
51. E. L. Que, E. Gianolio, S. L. Baker, S. Aime, C. J. Chang, *Dalton Trans.* **2010**, *39*, 469–476.
52. W.-S. Li, J. Luo, Z.-N. Chen, *Dalton Trans.* **2011**, *40*, 484–488.
53. X. Zhang, X. Jing, T. Liu, G. Han, H. Li, C. Duan, *Inorg. Chem.* **2012**, *51*, 2325–2331.
54. K. Bossak-Ahmad, T. Frączyk, W. Bal, S. C. Drew, *ChemBioChem.* **2020**, *21*, 331–334.
55. Y.-M. Xiao, G.-Y. Zhao, X. X. Fang, Y. X. Zhao, G. H. Wang, W. Yang, J. W. Xu, *RSC Adv.* **2014**, *4*, 34421–34427.
56. N. N. Paranawithana, A. F. Martins, V. Clavijo Jordan, P. Zhao, S. Chirayil, G. Meloni, A. D. Sherry, *J. Am. Chem. Soc.* **2019**, *141*, 11009–11018.
57. X. Zhang, Y. Lin, R. J. Gillies, *J. Nucl. Med.* **2010**, *51*, 1167–1170.
58. N. Nakanishi, M. Fukui, M. Tanaka, H. Toda, S. Imai, M. Yamazaki, G. Hasegawa, Y. Oda, N. Nakamura, *Kidney Blood Press. Res.* **2011**, *35*, 77–81.
59. R. G. Sawyer, M. D. Spengler, R. B. Adams, T. L. Pruett, *Ann. Surg.* **1991**, *213*, 253–260.
60. M. G. Joshua, D. P. Mark, Assessments of tumor metabolism with CEST MRI. *NMR Biomed.* **2018**, *32*, e3943.
61. C. S. Todd, W. Yunkou, A. Dean Sherry, *NMR Biomed.* **2012**, *26*, 829–38.
62. S. Bhuniya, K. S. Hong, *Investig. Magn. Reson. Imaging* **2019**, *23*, 17–25.

63. M. P. Lowe, D. Parker, O. Reany, S. Aime, M. Botta, G. Castellano, E. Gianolio, R. Pagliarin, *J. Am. Chem.* Soc. **2001**, *123*, 7601–7609.
64. L. Moriggi, M. A. Yaseen, L. Helm, P. Caravan, *Chem. Eur. J.* **2012**, *18*, 3675–3686.
65. M. Woods, G. E. Kiefer, S. Bott, A. Castillo-Muzquiz, C. Eshelbrenner, L. Michaudet, K. McMillan, S. D. Mudigunda, D. Ogrin, G. Tircsó, S. Zhang, *J. Am. Chem. Soc.* **2004**, *126*, 9248–9256.
66. G. B. Giovenzana, R. Negri, G. A. Rolla, L. Tei, Eur. *J. Inorg. Chem.* **2012**, *2012*, 2035–2039.
67. I. Mamedov, A. Mishra, G. Angelovski, H. A. Mayer, L. O. Pålsson, D. Parker, N. K. Logothetis, *Dalton Trans.* **2007**, , 5260–5267.
68. S. Zhang, K. Wu, A. D. Sherry, *Angew. Chem. Int. Ed.* **1999**, *38*, 3192–3194.
69. N. Raghunand, S. Zhang, A. D. Sherry, R. J. Gillies, *Acad. Radiol.* **2002**, *9* (Suppl 2), S481–S483.
70. N. Raghunand, C. Howison, A. Dean Sherry, S. Zhang, R. J. Gillies, *Magn. Reson. Med.* **2003**, *49*, 249–257.
71. M. L. Garcia-Martin, G. V. Martinez, N. Raghunand, A. Dean Sherry, S. Zhang, R. J. Gillies, *Magn. Reson. Med.* **2006**, *55*, 309–315.
72. M. Meser Ali, M. Woods, P. Caravan, A. C. Opina, M. Spiller, J. C. Fettinger, A. D. Sherry, *Chem. Eur. J.* **2008**, *14*, 7250–7258.
73. F. Mouffouk, H. Serrai, S. Bhaduri, R. Achten, M. Seyyedhamzeh, A. A. Husain, A. Alhendal, M. Zourob, *Molecules* **2020**, *25*, 1513.
74. T. Lee, L. X. Cai, V. S. Lelyveld, A. Hai, A. Jasanoff, *Science* **2014**, *344*, 533–535.
75. M. G. Shapiro, G. G. Westmeyer, P. A. Romero, J. O. Szablowski, B. Küster, A. Shah, C. R. Otey, R. Langer, F. H. Arnold, A. Jasanoff, *Nat. Biotech.* **2010**, *28*, 264–270.
76. V. Hsieh, S. Okada, H. Wei, I. García-Álvarez, A. Barandov, S. R. Alvarado, R. Ohlendorf, J. Fan, A. Ortega, A. Jasanoff, *J. Am. Chem. Soc.* **2019**, *141*, 15751–15754.
77. F. Oukhatar, S. Même, W. Meme, F. Szeremeta, N. K. Logothetis, G. Angelovski, E. Toth, *ACS Chem. Neuro.* **2015**, *6*, 219–225.
78. F. Oukhatar, H. Meudal, C. Landon, N. K. Logothetis, C. Platas-Iglesias, G. Angelovski, É. Tóth, *Chem. Eur. J.* **2015**, *21*, 11226–11237.
79. F. Oukhatar, S. V. Eliseeva, C. S. Bonnet, M. Placidi, N. K. Logothetis, S. Petoud, G. Angelovski, É. Tóth, *Inorg. Chem.* **2019**, *58*, 13619–13630.
80. D. Toljic, G. Angelovski, *ChemNanoMat.* **2019**, *5*, 1456–1460.
81. Y. F. Fan, Z. B. Guo, G. B. Ge, *Biosensors* **2023**, *13*, 476.
82. C. M. Ellis, J. Pellico, J. J. Davis, *Materials* **2019**, *12*, 4096.
83. D. V. Hingorani, B. Yoo, A. S. Bernstein, M. D. Pagel, *Chem. Eur. J.* **2014**, *20*, 9840–9850.
84. H. Kubo, Y. Murayama, S. Ogawa, T. Matsumoto, M. Yubakami, T. Ohashi, T. Kubota, K. Okamoto, M. Kamiya, Y. Urano, E. Otsuji, *Sci. Rep.* **2021**, *11*, 10664.
85. Q. Gao, B. Cheng, C. Chen, C. Lei, X. Lin, D. Nie, J. Li, L. Huang, X. Li, K. Wang, A. Huang, *Clin. Transl.* Med. **2022**, *12*, e995.
86. T. Chauvin, S. Torres, R. Rosseto, J. Kotek, B. Badet, P. Durand, É. Tóth, *Chem. Eur. J.* **2012**, *18*, 1408–1418.
87. R. A. Moats, S. E. Fraser, T. J. Meade, *Angew. Chem. Int. Ed.* **1997**, *36*, 726–728.
88. L. M. Urbanczyk-Pearson, F. J. Femia, J. Smith, G. Parigi, J. A. Duimstra, A. L. Eckermann, C. Luchinat, T. J. Meade, *Inorg. Chem.* **2008**, *47*, 56–68.
89. A. Y. Louie, M. M. Hüber, E. T. Ahrens, U. Rothbächer, R. Moats, R. E. Jacobs, S. E. Fraser, T. J. Meade, *Nat. Biotech.* **2000**, *18*, 321–325.
90. J. A. Duimstra, F. J. Femia, T. J. Meade, J. Am. Chem. Soc. 2005, *127*, 12847–12855.

91. J. He, C. S. Bonnet, S. V. Eliseeva, S. Lacerda, T. Chauvin, P. Retailleau, F. Szeremeta, B. Badet, S. Petoud, E. Toth, P. Durand, *J. Am. Chem. Soc.* **2016**, *138*, 2913–2916.
92. A. C. Lanoue, A. Dumitriu, R. H. Myers, J. J. Soghomonian, *Exp. Neurol.* **2010**, *226*, 207–217.
93. A. Daif, R. V. Lukas, N. P. Issa, A. Javed, S. VanHaerents, A. T. Reder, J. X. Tao, P. Warnke, S. Rose, V. L. Towle, S. Wu, *Epilepsy Behav.* **2018**, *80*, 331–336.
94. R. Napolitano, G. Pariani, F. Fedeli, Z. Baranyai, M. Aswendt, S. Aime, E. Gianolio, *J. Med. Chem.* **2013**, *56*, 2466–2477.
95. M. Aswendt, E. Gianolio, G. Pariani, R. Napolitano, F. Fedeli, U. Himmelreich, S. Aime, M. Hoehn, *NeuroImage* **2012**, *62*, 1685–1693.
96. I. Carballo-Carbajal, A. Laguna, J. Romero-Giménez, T. Cuadros, J. Bové, M. Martinez-Vicente, A. Parent, M. Gonzalez-Sepulveda, N. Peñuelas, A. Torra, B. Rodríguez-Galván, *Nat. Commun.* **2019**, *10*, 973.
97. H. Seo, H. A. Clark, *Analyst* **2020**, *145*, 1169–1173.
98. S. Laine, J.-F. Morfin, M. Galibert, V. Aucagne, C. S. Bonnet, É. Tóth, *Molecules* **2021**, *26*, 2176.
99. K. Hanaoka, K. Kikuchi, T. Terai, T. Komatsu, T. Nagano, *Chem. Eur. J.* **2008**, *14*, 987–995.
100. S.-H. Chen, Y.-T. Kuo, G. Singh, T. L. Cheng, Y. Z. Su, T. P. Wang, Y. Y. Chiu, J. J. Lai, C. C. Chang, T. S. Jaw, S. C. Tzou, *Inorg. Chem.* **2012**, *51*, 12426–12435.
101. A. L. Nivorozhkin, A. F. Kolodziej, P. Caravan, M. T. Greenfield, R. B. Lauffer, T. J. McMurry, *Angew. Chem. Int. Ed.* **2001**, *40*, 2903–2906.
102. A. Bogdanov, L. Matuszewski, C. Bremer, A. Petrovsky, R. Weissleder, *Mol. Imaging* **2002**, *1*, 15353500200200001.
103. J. W. Chen, W. Pham, R. Weissleder, A. Bogdanov Jr, *Magn. Reson. Med.* **2004**, *52*, 1021–1028.
104. E. Rodríguez, M. Nilges, R. Weissleder, J. W. Chen, *J. Am. Chem. Soc.* **2010**, *132*, 168–177.
105. R. Yan, Y. Hu, F. Liu, S. Wei, D. Fang, A. J. Shuhendler, H. Liu, H. Y. Chen, D. Ye, *J. Am. Chem. Soc.* **2019**, *141*, 10331–10341.
106. D. Trachootham, W. Lu, M. A. Ogasawara, N. R.-D. Valle, P. Huang, *Antioxid. Redox Signal.* **2008**, *10*, 1343–1374.
107. R. Banerjee, *J. Biol. Chem.* **2012**, *287*, 4397–4402.
108. T. C. Jorgenson, W. Zhong, T. D. Oberley, *Cancer Res.* **2013**, *73*, 6118–6123.
109. D. Gius, D. R. Spitz, *Antioxid. Redox Signal.* **2006**, 8 1249–1252.
110. E. Ortona, P. Margutti, P. Matarrese, F. Franconi, W. Malorni, *Autoimmun. Rev.* **2008**, 7579–584.
111. D. O. Perkins, C. D. Jeffries, K. Q. Do, *Biol. Psychiatry* **2020**, *88* 326–336.
112. Q. N. Do, J. S. Ratnakar, Z. Kovacs, A. Dean Sherry, *ChemMedChem.* **2014**, *9* 1116–1129.
113. S. M. Pinto, V. Tomé, M. J. F. Calvete, M. M. C. A. Castro, É. Tóth, C. F. G. C. Geraldes, *Coord. Chem. Rev.* **2019**, *390*, 1–31.
114. P. B. Tsitovich, P. J. Burns, A. M. McKay, J. R. Morrow, *J. Inorg. Biochem.* **2014**, *133*, 143–154.
115. D. V. Hingorani, A. S. Bernstein, M. D. Pagel, *Contrast Media Mol. Imaging* **2015**, *10* 245–265.
116. C. Tu, R. Nagao, A. Y Louie, *Angew. Chem. Int. Ed.* **2009**, *48* 6547–6551.
117. G. Berkovic, V. Krongauz, V. Weiss, *Chem. Rev.* **2000**, *100* 1741–1754.
118. P.-S. Chen, W.-T. Chiu, P. L. Hsu, S. C. Lin, I. C. Peng, C. Y. Wang, S. J. Tsai, *J. Biomed. Sci.* **2020**, *27* 63.

119. S. Iwaki, K. Hanaoka, W. Piao, T. Komatsu, T. Ueno, T. Terai, T. Nagano, *Bioorg. Med. Chem. Lett.* **2012**, *22* 2798–2802.
120. S. K. Biswas, I. Rahman, *Mol Aspects Med.* **2009**, *30* 60–76.
121. P. Ghezzi, *Int. J. General Med.* **2011**, *4*, 105–113.
122. J. S. Bains, C. A. Shaw, *Brain Res. Rev.* **1997**, *25* 335–358.
123. S. C. Lu, *Liver Res.* **2020**, *4* 64–73.
124. C. Carrera, G. Digilio, S. Baroni, D. Burgio, S. Consol, F. Fedeli, D. Longo, A. Mortillaro, S. Aime, *Dalton Trans.* **2007**, 4980–4987.
125. N. Raghunand, B. Jagadish, T. P. Trouard, J.-P. Galons, R. J. Gillies, E. A. Mash, *Magn. Reson. Med.* **2006**, *55*, 1272–1280.
126. N. Raghunand, G. P. Guntle, V. Gokhale, G. S. Nichol, E. A. Mash, B. Jagadish, *J. Med. Chem.* **2010**, *53*, 6747–6757.
127. T. H. Landowski, G. P. Guntle, D. Zhao, B. Jagadish, E. A. Mash, R. T. Dorr, N. Raghunand, *Transl. Oncol.* **2016**, *9*, 228–235.
128. G. P. Guntle, B. Jagadish, E. A. Mash, G. Powis, R. T. Dorr, N. Raghunand, *Transl. Oncol.* **2012**, *5*, 190–199.
129. J. Martinelli, M. Fekete, L. Tei, M. Botta, *Chem. Commun.* 2011, *47*, 3144–3146.
130. R. Barré, D. Mouchel dit Leguerrier, Q. Ruet, L. Fedele, D. Imbert, V. Martel-Frachet, P. H. Fries, J. K. Molloy, F. Thomas, *Chem. Asian J.* **2022**, *17*, e202200544.
131. L. Burai, E. Tóth, S. Seibig, R. Scopelliti, A. E. Merbach, *Chemistry A European Journal.* **2000**, *6*, 3761–3770.
132. N.-D. H. Gamage, Y. Mei, J. Garcia, M. J. Allen, *Angew Chem. Int. Ed.* **2010**, *49*, 8923–8925.
133. L. A. Ekanger, L. A. Polin, Y. Shen, E. Mark Haacke, P. D. Martin, M. J. Allen, *Angew. Chem. Int. Ed.* **2015**, *54*, 14398–14401.
134. S. Aime, F. Fedeli, A. Sanino, E. Terreno, *J. Am. Chem. Soc.* **2006**, *128*, 11326–11327.
135. L. Frullano, C. Catana, T. Benner, A. D. Sherry, P. Caravan, *Angew. Chem. Int. Ed.* **2010**, *49*, 2382–2384.
136. A. C. Pollard, J. L. Cerda, F. W. Schuler, T. R. Pollard, A. Kotrotsou, F. Pisaneschi, M. D. Pagel, *Biosensors* **2022**, *12*, 134.
137. M. Le Fur, N. J. Rotile, C. Correcher, V. Clavijo Jordan, A. W. Ross, C. Catana, P. Caravan, *Angew. Chem. Int. Ed.* **2020**, *59*, 1474–1478.
138. E. Gianolio, L. Maciocco, D. Imperio, G. B. Giovenzana, F. Simonelli, K. Abbas, G. Bisi, S. Aime, *Chem. Commun.* **2011**, *47*, 1539–1541.
139. K. P. Malikidogo, I. Da Silva, J. F. Morfin, S. Lacerda, L. Barantin, T. Sauvage, J. Sobilo, S. Lerondel, É. Tóth, C. S. Bonnet, *Chem. Commun.* **2018**, *54*, 7597–7600.
140. E. Gianolio, R. Napolitano, F. Fedeli, F. Arena, S. Aime, *Chem. Commun.* 2009, 6044–6046.
141. G. Gambino, T. Gambino, G. Angelovski, *Chem. Commun.* **2020**, *56*, 9433–9436.
142. G. Gambino, T. Gambino, R. Pohmann, G. Angelovski, *Chem. Commun.* **2020**, *56*, 3492–3495.

# 5 Mn-Based Small Complexes as MRI Contrast Agents

*Zoltán Garda, Sara Lacerda, and Éva Tóth*[*]
Centre de Biophysique Moléculaire,
CNRS UPR 4301, Université d'Orléans,
Rue Charles Sadron,
F-45071 Orléans, France
eva.jakabtoth@cnrs-orleans.fr

## CONTENTS



**Abstract**

The use of Gd-agents in magnetic resonance imaging (MRI) raises concerns related to their safety, as well as to their environmental impact, and this promotes intensive research to identify viable alternatives. $Mn^{2+}$ and $Mn^{3+}$ are paramagnetic in the high-spin state and can be good relaxation agents. Although manganese is biogenic which makes it more acceptable for *in vivo* use when injected at the elevated doses required for MRI, it also needs to be chelated in stable and inert complexes.

[*] Corresponding author.

DOI: 10.1201/9781003374688-5

For $Mn^{2+}$, most efforts have been devoted (1) to design novel chelators which satisfy these requirements, in particular kinetic inertness, (2) to improve relaxation efficiency and (3) to derive responsive and targeted $Mn^{2+}$ probes for the detection of specific biomarkers. For $Mn^{3+}$, porphyrin derivatives have met the most attention. Here, we survey progress from the last three years in the chemistry of $Mn^{2+}$ and $Mn^{3+}$ complexes dedicated to MRI probe development.

## KEYWORDS

Manganese(II); Manganese(III); MRI; Contrast Agents; Relaxivity; Molecular Imaging

## 1 INTRODUCTION

Complexes of paramagnetic metal ions, in particular $Gd^{3+}$, are commonly used for almost forty years now in medical magnetic resonance imaging (MRI). They enhance image contrast by decreasing the nuclear relaxation times of water protons in the surrounding tissues. The two major challenges today in the chemistry of MRI contrast agents involve (1) the design of probes which are safer and more biocompatible than the current, clinically applied $Gd^{3+}$ complexes, and (2) the development of molecular imaging approaches for the detection of specific biomarkers. While Gd-based agents were long considered as very safe, since 2006, it was demonstrated that Gd-injections can cause toxicity problems in patients with impaired kidney function [1], which soon led to the withdrawal of some less inert $Gd^{3+}$ complexes from commercial use. Later, long-term Gd-deposition has been found in various organs in patients [2], although no toxicity was observed. In parallel, environmental concerns are also seriously arising with respect to the use of large quantities of Gd in the clinics, which can be now detected not only in wastewater around hospitals, but also in natural waters [3,4]. In this context, there is an increasing interest in the development of $Mn^{2+}$ complexes, considered as the most viable alternatives to replace $Gd^{3+}$-based agents [5–7].

Manganese is an endogenous metal ion which obviously makes it more acceptable for *in vivo* use. With five unpaired electrons ($S = 5/2$) and a slow electron spin relaxation, $Mn^{2+}$ is strongly paramagnetic and acts as an efficient relaxation agent. One $Mn^{2+}$ chelate, MnDPDP (Scheme 1), was in clinical use, though its commercialization was later discontinued [8,9]. It is known that this chelate partially dissociates *in vivo* and its contrast enhancement in the body is largely related to the presence of free $Mn^{2+}$. Even if no toxicity issues were associated with this compound, it is evident that today no such unstable $Mn^{2+}$ complexes could be approved for human use. Contrast agents are injected in relatively elevated doses (0.05–0.30 mmol $kg^{-1}$ body weight), thus $Mn^{2+}$ complexes designed for MRI applications should have sufficiently high thermodynamic stability and kinetic inertness to avoid the release of $Mn^{2+}$, in spite of the endogenous character of this metal ion. Indeed, the body does not tolerate too high concentrations of free $Mn^{2+}$,

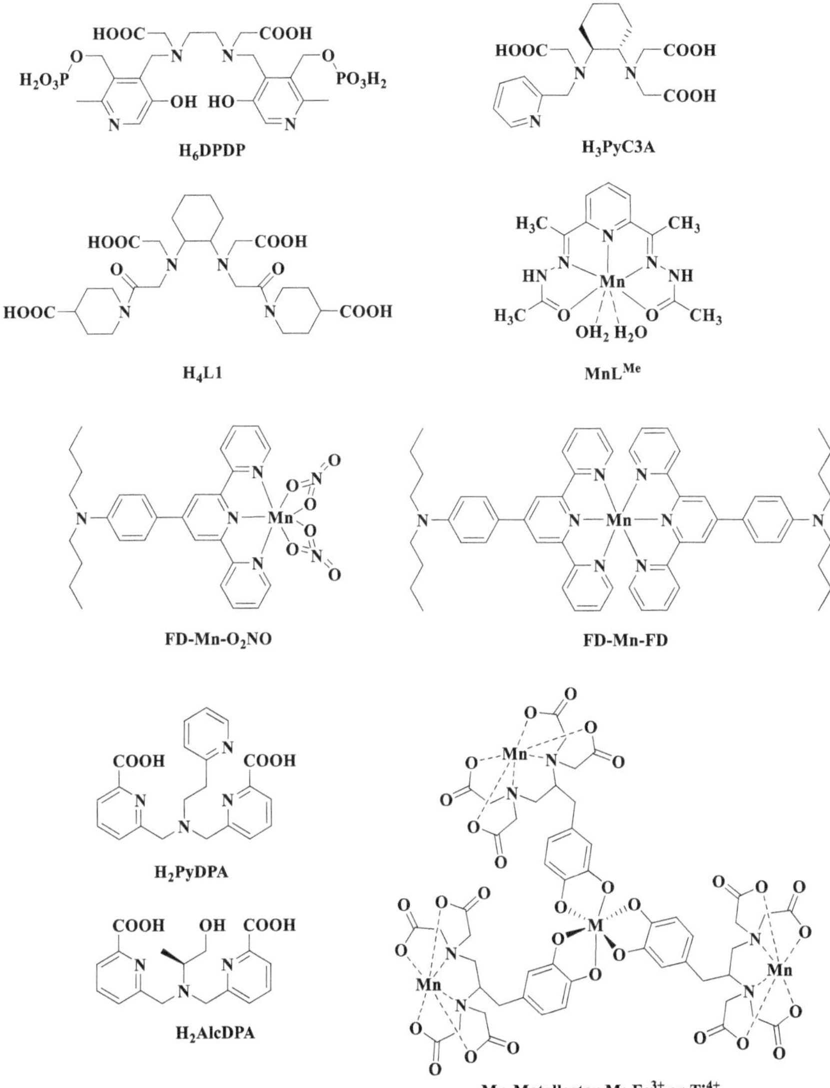


**SCHEME 1** Linear chelators investigated for $Mn^{2+}$ complexation.

which is associated with potential neurotoxicity problems (LD50 = 0.22 mmol $kg^{-1}$ for rats) [10].

However, it is by far not obvious to achieve high stability and inertness for $Mn^{2+}$. The relatively large ionic radius of $Mn^{2+}$ and the spherical distribution of its five *d* electrons typically lead to chelates which are intrinsically labile, i.e., have low resistance to dissociation and have thermodynamic stabilities lower than that of other first-raw transition metal ions, as it is predicted by the Irving–Williams series [11]. Moreover, for a good relaxation effect, the complex has to contain at least one inner-sphere water molecule in fast exchange with bulk water, which will transmit the paramagnetic effect to the surrounding solution/tissue. The typical coordination numbers of $Mn^{2+}$ are 6 or 7, implying that mostly penta- or hexadentate ligands will be the best suited for contrast agent applications.

Therefore, the determination of both the thermodynamic stability and the kinetic inertness is essential for the evaluation of a $Mn^{2+}$ complex for MRI. The thermodynamic stability is characterized by the formation or stability constant, defined as $K_{MnL} = [MnL]/[Mn^{2+}][L]$, where different terms in brackets correspond to the equilibrium concentrations of the MnL complex, the free metal ion $Mn^{2+}$ and the deprotonated ligand L. For a better and more direct comparison between ligands of different basicity, which obviously impacts the value of the thermodynamic stability constant, a conditional stability is often used. It is typically expressed by the pMn value, where $pMn = -\log[Mn^{2+}]$, conventionally calculated for $c_{Mn} = c_L = 10^{-5}$ M total concentrations and pH 7.4 [12]. Thermodynamic stability constants of $Mn^{2+}$ complexes are most commonly determined from pH-potentiometric titrations. The species distribution obtained in this way is often corroborated with pH-dependent relaxivity measurements which directly indicate the pH range where the formation of MnL, eventual protonated ($H_iMnL$), or hydrolyzed (MnL(OH)) complexes occurs. In some cases, complex formation might be slow, requiring the preparation of individual batch samples in the relevant pH region for the potentiometric measurements, with sufficiently long equilibration times.

The kinetic inertness of a $Mn^{2+}$ chelate is usually characterized by the rate of the dissociation reaction under well-defined conditions. Complex dissociation can be provoked either by competing $H^+$ at lower pH values where the complex is thermodynamically not stable, or by transmetalation, using an excess of an exchanging metal, typically $Zn^{2+}$ or $Cu^{2+}$ which are the most important in potential transmetalation in the body. A metal complex can dissociate *via* different pathways. The most important ones include (1) spontaneous dissociation which, for certain complexes, can be important at physiological pH, (2) proton-assisted dissociation implying the formation of a protonated complex, and (3) metal-assisted dissociation involving a transitionally formed dinuclear complex. Complexes with available protonation sites are more prone to faster proton-assisted dissociation, while complexes endowed with additional donor atoms can favor the metal-assisted pathway. A complete dissociation kinetic study involves the assessment of each of these pathways by determining the individual rate constants, and it requires measurements of the dissociation at varying pH and exchanging metal concentration [13]. A simpler protocol is commonly used at a single pH (typically pH 6) and $Zn^{2+}$ concentration (25 equiv.) [14], and the dissociation is monitored by relaxivity measurements (longitudinal or transverse relaxivity, the latter having the advantage of higher changes between the MnL complex and the free $Mn^{2+}$). The dissociation half-life, $t_{1/2}$, determined under these conditions can be then directly compared for various systems. While this is a straightforward approach, one has to be aware that the comparison does not necessarily hold for different conditions (for instance, for physiological pH), since the contribution of the three pathways might vary and their importance might drastically change with pH in particular.

In the last decade, there has been intensive research to identify $Mn^{2+}$ chelates which satisfy the criteria of high thermodynamic stability, kinetic inertness and good relaxation efficacy. High kinetic inertness has been particularly challenging to achieve, and in this respect, the rigidity and preorganization of the ligand proved to be very relevant structural features. Several rigidified ligand structures, mostly macrocyclic or bicyclic, have been now identified which ensure excellent resistance to dissociation for their $Mn^{2+}$ complexes. Bispidine ligands started to be explored and were shown to provide unprecedented coordination properties for $Mn^{2+}$, both in terms of inertness and $Mn^{2+}$ selectivity. In parallel to these coordination chemistry studies toward the optimization of the chelator structure, numerous $Mn^{2+}$ complexes were tested in preclinical MRI. Very importantly, two small molecular weight chelates successfully reached Phase I human trials as potential replacement for current clinical $Gd^{3+}$-agents [15,16]. One of them is MnPyC3A (Scheme 1), which is based on a rigidified open-chain chelator developed by the group of Caravan [14] and Reveal Pharmaceuticals Inc.; the other is a macrocyclic complex developed by GE Healthcare. Different strategies, which are well-known from the field of $Gd^{3+}$-agents, such as using nanosized systems, continued to be explored to enhance the proton relaxivity, thus the MRI efficiency of $Mn^{2+}$ complexes. An increasing number of $Mn^{2+}$ examples have also been designed for the detection of different biomarkers in molecular imaging applications, both as responsive and targeted probes.

In addition to $Mn^{2+}$, $Mn^{3+}$ is also paramagnetic in the high-spin state. $Mn^{3+}$ has a lower electron spin ($S = 4/2$), usually faster electron spin relaxation, and due to its smaller ionic radius and stronger propensity to hydrolysis, $Mn^{3+}$ complexes tend to have less inner-sphere water molecules, which together make them in general less efficient relaxation agents than the $Mn^{2+}$ counterparts. Nevertheless, some $Mn^{3+}$ complexes, in particular those formed with porphyrins, might be endowed with high relaxation potential [17] and have met a recently renewed interest as potential MRI contrast agents. Finally, the redox switch between the $Mn^{2+}$ and $Mn^{3+}$ states, mediated by biological redox species, can be exploited for the design of redox-responsive MRI probes [18]. Some recent examples demonstrate that the detection mode can be extended from classical $^{1}H$ to $^{19}F$ MRI as well, bringing about further advantages, but also novel chemical design requirements.

Several review articles [5–7,19–22] and book chapters [23,24] have been lately dedicated to this very rapidly evolving research area. Following our previous contribution to this book series [23], here, our objective is to overview exclusively the most recent developments in the field of $Mn^{2+}$- and $Mn^{3+}$-based MRI agents; thus, we restrict our selection to publications over the last three years. We do not intend to be exhaustive, but would like to specifically focus on molecular design, including open-chain, macrocyclic, porphyrin and bispidine ligands, by illustrating also molecular imaging applications *via* some representative examples.

## 2 MANGANESE(II) CHELATES

### 2.1 Open-Chain Manganese(II) Complexes

Most of the large body of previous literature data on linear $Mn^{2+}$ chelates refer to derivatives of EDTA, where either the ligand backbone was modified or the pending acetate functions have been partially replaced by other types of coordinating groups [23,24]. The *t*-CDTA ligand contains a cyclohexane backbone and its derivatives have been most widely investigated, as it was shown early that this rigid ligand skeleton contributes largely to an enhanced kinetic inertness of the $Mn^{2+}$ complexes [25].

In order to achieve higher relaxivities, Lalli et al. further modified this structure by synthesizing the bisamide derivative *t*-CDTA-bis(4-carboxypiperidinylamide) ligand, L1 (Scheme 1), and anchored its $Mn^{2+}$ complex to the surface of organo-modified silica nanoparticles (SiNPs) *via* amide bonds in *para* position of the two piperidines [26]. The nanoparticles prepared had homogeneous size distribution (80–90 nm) and could be loaded with a large amount of the complex. Detailed $^1$H NMRD (nuclear magnetic relaxation dispersion) and $^{17}$O NMR investigations were performed for the free MnL1 complex, as well as for the MnL1-SiNPs nanoparticles. As expected, the relaxivity increase for the MnL1-SiNPs nanoparticles is the most significant (250%) at 0.5 T ($r_1$ = 18.5 $mM^{-1}$ $s^{-1}$ *vs* 5.3 $mM^{-1}$ $s^{-1}$ for the free MnL1, 25°C), as it results from a slower rotational motion following immobilization of the complex, hence with the greatest effect at such medium magnetic fields. A relatively fast water exchange is maintained inside the nanoparticles which is thus not limiting relaxivity. The high relaxivity per Mn, combined with the large amount of MnL1, confined on the silica surface leads to a remarkable relaxivity per particle ($1.4 \times 10^5$ $mM^{-1}$ $s^{-1}$ at 1 T and 25°C). MRI images were acquired in mice at 1 T following intravenous administration of the nanoparticle contrast agent and indicated a significant signal enhancement in the liver 20 min post-injection.

Other $Mn^{2+}$ complexes were also immobilized within porous silica nanospheres, in a non-covalent fashion, in order to increase relaxivity and blood circulation time. Mukherjee and co-workers synthetized the ligand PyDPA (Scheme 1) and first investigated the equilibrium, kinetic and relaxometric properties of its $Mn^{2+}$ complex [27]. The relaxation behavior was also assessed for the complex entrapped in the silica nanosphere ([MnPyDPA]@$SiO_2$). MnPyDPA is thermodynamically stable, characterized with a thermodynamic stability constant $\log K_{MnL}$ = 14.80 and a pMn value of 8.97 (calculated for $c_L = c_{Mn} = 10^{-5}$ M total concentrations, pH 7.4). However, the complex is kinetically labile. It contains one inner-sphere water molecule and has a moderate relaxivity, $r_1$ = 2.88 $mM^{-1}$ $s^{-1}$ (pH~7.4, 1.41 T, 25°C). With a reverse microemulsion method, MnPyDPA was entrapped within porous (5.56 nm pore diameter) silica nanospheres of 14 ± 2 nm size. This led to a notable increase in its resistance to dissociation in the presence of anions, proton and $Zn^{2+}$. The relaxivity also increased 2.9-fold *vs* the free chelate to $r_1$ = 8.46 $mM^{-1}$ $s^{-1}$ (pH~7.4, 1.41 T, 25°C). The association of the nanoparticles with bovine serum albumin (BSA) further enhanced the relaxivity to $r_1$ = 24.76 $mM^{-1}$ $s^{-1}$.

The same group designed the AlcDPA ligand which forms a seven-coordinate, mono-hydrated $Mn^{2+}$ complex, characterized by a stability constant of log $K_{MnL}$= 15.06 and pMn = 9.56 (Scheme 1) [28]. This complex was also impregnated within porous silica nanospheres. Following further surface modification of the nanospheres with the $Zn^{2+}$-binding moiety 6-((bis(pyridin-2-ylmethyl)amino) methyl)picolinamide (Py2Pic), the nanoparticle was capable of selective interaction with BSA in the presence of $Zn^{2+}$ ions, in a similar way as previously reported for small molecular weight $Zn^{2+}$-sensitive agents. The relaxivity of MnAlcDPA increased from $r_1$ = 2.86 $mM^{-1}$ $s^{-1}$ in the free form to 13.26 $mM^{-1}$ $s^{-1}$ within the nanospheres (1.41 T, 37°C, pH ~ 7.4). In the presence of 0.6 mM BSA at pH 7.4, $r_1$ increased linearly with $Zn^{2+}$ concentration and reached 39.01 $mM^{-1}$ $s^{-1}$ in the presence of a 40-fold $Zn^{2+}$ excess.

A metallostar structure was built by using $Fe^{3+}$ or $Ti^{4+}$ as the core and three MnEDTA complexes attached to a catechol moiety (Scheme 1) [29]. Despite the general interest in metallostar structures which can provide high rigidity and consequently elevated relaxivities at medium frequencies, this structure is not optimal for physiological pH, since tris-coordination of the catechol around the central $Fe^{3+}$ occurs only above pH 7.8. Consequently, the longitudinal relaxivity measured at pH 7.4 is lower for the Fe-based metallostar, which exhibits an equilibrium between tris- and biscoordination, than for the fully formed $TiL_3Mn_3$ species (6.06 $mM^{-1}$ $s^{-1}$ *vs* 7.87 $mM^{-1}$ $s^{-1}$, respectively, 0.47 T, 32°C), though both of these values are higher than that of the free $Mn^{2+}$ complex (3.66 $mM^{-1}$ $s^{-1}$). Additionally, the Fe-based metallostar was tested for synergistic photothermal therapy and photodynamic therapy (PDT) effects as well upon 808 nm laser excitation, originating from ligand-to-metal charge transfer.

Martins, Stasiuk and collaborators reported an unusual, Schiff-base diacetylpyridyl carbohydrazide-derivative $Mn^{2+}$ complex ($MnL^{Me}$) obtained *via* an innovative single-pot template synthetic strategy, and studied the most important physicochemical properties *in vitro* [30]. This pentadentate ligand ($N_3O_2$) leaves enough room for two water molecules in the inner coordination sphere of the $Mn^{2+}$ center, which was experimentally proved by $^{17}O$ transverse relaxation rate measurements in aqueous solution yielding a hydration number of $q$ = 1.7. The solid-state crystal structure of the complex also shows a formal seven-coordination of the $Mn^{2+}$ ion in a pentagonal bipyramidal geometry. The relaxivities of the complex were determined at two different field strengths, $r_1$ = 4.88 $mM^{-1}$ $s^{-1}$ and 5.2 $mM^{-1}$ $s^{-1}$ at 400 MHz (9.4 T) and 64 MHz (1.5 T), respectively, (25°C). $MnL^{Me}$ displayed 21- and 50-fold higher kinetic inertness than the corresponding MnEDTA and MnDTPA complexes, respectively, in the presence of 25 molar equivalents of $Zn^{2+}$ in 50 mM MES buffer at pH 6.0, 25°C. The $^{17}O$ NMR measurements revealed relatively fast water exchange for $MnL^{Me}$ ($k_{ex}^{298} = 5.2 \times 10^6$ $s^{-1}$). The complex binds BSA, and the analysis of the relaxivity titration upon adding increasing amounts of BSA to the complex solution yielded an affinity constant of $K_A = 4.2 \times 10^3\ M^{-1}$ and a relaxivity of $r_1^b = 21.1\ mM^{-1}\ s^{-1}$ for the bound complex (32 MHz, 25°C). *In vivo* MRI studies in healthy mice showed longer blood circulation time for $MnL^{Me}$ as compared to the commercial $Gd^{3+}$-based contrast agent Magnevist® (GdDTPA), certainly related to albumin binding. The complex undergoes combined renal and hepatic clearance, and it was detected in an intact form in the bladder.

In a theranostic approach, Feng et al. synthetized two terpyridine-based $Mn^{2+}$ complexes (FD-Mn-$O_2$NO and FD-Mn-FD) for PDT guided by multiphoton excitation optical imaging and MRI [31]. FD-Mn-FD had higher longitudinal relaxivity ($r_1$ =2.6 $mM^{-1}\ s^{-1}$ at 1.5 T) and better stability relative to FD-Mn-$O_2$NO, which underwent rapid dissociation and oxidation. Under 808 nm irradiation, FD-Mn-FD showed prominent anticancer properties in PDT.

## 2.2 Macrocyclic Manganese(II) Complexes

The twelve-membered pyclen macrocycle containing one pyridine ring has been identified early on as a good base for ligand design for $Mn^{2+}$ [32,33]. More recently, Garda et al. studied two pyclen-diacetate ligands, 3,6-PC2A and 3,9-PC2A, where the two pendant acetate arms are in different positions related to the pyridine moiety (Scheme 2) [34]. The Mn(3,6-PC2A) and Mn(3,9-PC2A) complexes have high and slightly different thermodynamic stability, but the conditional stabilities at pH 7.4 are very similar (log $K_{MnL}$ = 15.53 and pMn = 8.09 for Mn(3,6-PC2A), and log $K_{MnL}$ = 17.09 and pMn = 8.64 for Mn(3,9-PC2A)). Relaxivities are also in the same range ($r_1$ = 2.72 $mM^{-1}\ s^{-1}$ and 2.91 $mM^{-1}\ s^{-1}$ for Mn(3,6-PC2A) and Mn(3,9-PC2A), respectively, 25°C and 0.49 T). Their inertness was investigated in detail at different pH levels against $Cu^{2+}$ as a scavenger metal ion. Both complexes are remarkably inert, with dissociation half-lives of 63 h and 21 h estimated for Mn(3,6-PC2A) and Mn(3,9-PC2A), respectively, at pH = 7.4, 0.01 mM $Cu^{2+}$ and 25°C. Superoxide dismutase (SOD) activity was also evidenced at pH = 7.8.

Devreux et al. also reported the relaxation properties for Mn(3,9-PC2A), as well as for several derivatives where an additional function was grafted in *ortho* position on the pyridine ring enabling the conjugation to targeting moieties ($L_2COO^-$, $L_3CH$ and $L_4NH_3^+$, Scheme 2) [35]. All these complexes contain one water molecule directly coordinated to the metal center and exhibit $r_1$ relaxivities between 2.6 and 3.2 $mM^{-1}\ s^{-1}$ at 20 MHz and 37°C. As expected, the addition of the extra side chain on the pyridine moiety to the 3,9-PC2A backbone did not significantly influence the inertness or any other properties of the $Mn^{2+}$ complexes.

In the objective of further increasing the inertness of such twelve-membered macrocyclic $Mn^{2+}$ complexes, Csupasz et al. synthetized a novel ligand (3,9-OPC2A) [36], in which the amine nitrogen opposite to the pyridine ring was substituted with an ether O-atom. The conditional stability of the $Mn^{2+}$ complex remained unaffected (pMn = 8.69) as compared to Mn(3,9-PC2A)

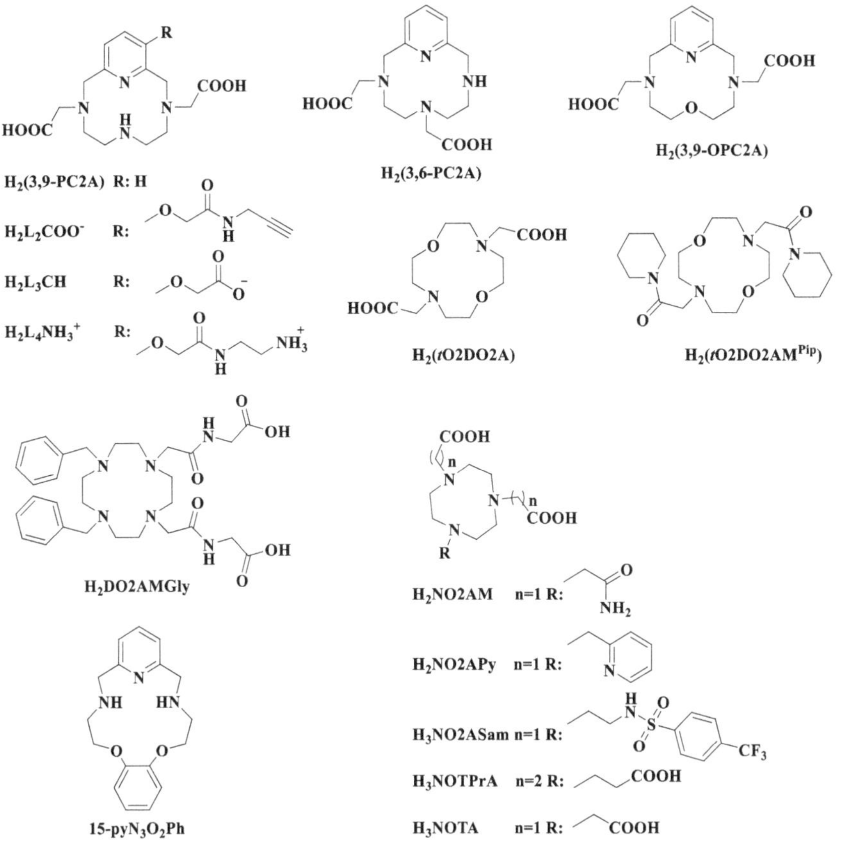

**SCHEME 2** Macrocyclic chelators investigated for $Mn^{2+}$ complexation.

(pMn = 8.64). The removal of the amine protonation site in the ligand results in two orders of magnitude decrease in the proton-assisted dissociation rate constant, $k_1$. Unfortunately, at the same time, the spontaneous dissociation pathways become more important, and altogether, a very similar overall dissociation half-life is obtained for the two complexes ($k_1$ = 2.81 $M^{-1}$ $s^{-1}$ *vs* 221 $M^{-1}$ $s^{-1}$ and $t_{1/2}$ = 21.9 h *vs* 21 h (pH 7.4), for Mn(3,9-OPC2A) and Mn(3,9-PC2A), respectively). The $^{17}O$ NMR measurements evidenced one inner-sphere water molecule for Mn(3,9-OPC2A), and a lower water exchange rate ($k_{ex}^{298}$ = 5.3 × $10^7$ $s^{-1}$) than that on Mn(3,9-PC2A) ($k_{ex}^{298}$ = 1.26 × $10^8$ $s^{-1}$). Its relaxivity is slightly more elevated ($r_1$ = 3.13 $mM^{-1}$ $s^{-1}$ at 20 MHz and 25°C). DFT calculations indicated that the geometry of the $Mn^{2+}$ complex does not substantially change upon the replacement of a nitrogen atom for an oxygen atom in the macrocycle. Also, the ligand could be successfully labelled with the positron-emitting isotope $^{52}Mn$, but only at more elevated temperatures than the pyclen-based 3,9-PC2A.

In an analogous way, replacement of two amine nitrogen atoms with ether oxygens has been also performed in the cyclen family. Kálmán et al. reported two macrocyclic ligands based on such 1,7-diaza-12-crown-4 platform, where the two N-atoms in the macrocyclic ring were substituted with acetates (*t*O2DO2A) or piperidineacetamides (*t*O2DO2AM$^{Pip}$) [37]. $Mn^{2+}$ complexes of these ligands have moderate conditional stability at pH = 7.4 (pMn = 6.67 for the acetate and 6.28 for the piperidineacetamide derivative), but comparable to those reported for the cyclen-based Mn*t*DO2A (pMn = 6.52). The solid-state X-ray structure of $[Mn(tO2DO2A)(H_2O)]{\cdot}2H_2O$ revealed that the metal ion is coordinated by six ligand donor atoms and one water molecule, and monohydration was also confirmed in solution by $^{17}O$ relaxation rate measurements on both $Mn^{2+}$ complexes. The presence of one inner-sphere water imparts relatively high relaxivities to the complexes. The $Mn^{2+}$ complex of *t*O2DO2A is remarkably inert with respect to dissociation, while the amide analog is surprisingly significantly more labile ($t_{1/2}$ = 39 h and 0.13 h, respectively, at pH = 7.4 and 0.01 mM $Cu^{2+}$ ion concentration). The limited inertness of the amide derivative can be attributed to the importance of the spontaneous dissociation pathway. This largely overbalances the reduced contribution from acid-assisted dissociation, which is typically observed for amide *vs* carboxylate derivatives.

Leone et al. proposed other modifications to the 1,4-DO2A ligand in order to promote albumin binding. A benzyl group was attached to each of the two unsubstituted macrocycle amine nitrogen atoms of DO2A and the two carboxylate functions were converted to amides (glycinate) to yield the DO2AMGly ligand [38]. They studied the interaction of the amphiphilic MnDO2AMGly complex with HSA, as well as the relaxometric properties of the free and the albumin-bound chelate. The relaxivity of MnDO2AMGly is 21% higher than that reported for bismethylamide derivative Mn(1,4-BnDO2AM) ($r_1$ = 4.6 $mM^{-1}$ $s^{-1}$ and 3.8 $mM^{-1}$ $s^{-1}$, respectively, at pH = 7.4, 25°C, 20 MHz). The presence of two benzyl groups in positions 7 and 10 allowed a good interaction with HSA ($K_A$ = 3.0 × $10^3$) and a 3-fold increase of the relaxivity in serum at 1 T. The complex did not show any cytotoxicity on J774 macrophage cells up to 0.1 mM concentration. An MRI study was realized at 1 T, following intravenous administration of MnDO2AMGly in a HER/2+ TS/A mouse model. MnDO2AMGly provided strong contrast enhancement of subcutaneous breast tumor lesions, which was persistent up to 3 h post-injection, and was comparable to that obtained with a clinical Gd-agent at the same dose (0.1 mmol $kg^{-1}$).

Among smaller macrocyclic ligands, several derivatives of the triazacyclononane (TACN) ring were previously described and shown to form six-coordinated complexes with $Mn^{2+}$ [23]. Thus, complexes formed with hexadentate ligands do not contain inner-sphere water and have correspondingly low relaxation effect. Recently, Uzal-Varela et al. investigated the $Mn^{2+}$ complexes of five TACN derivatives bearing different pendant arms (NOTA, NO2ASAm, NO2AM, NO2APy and NOTPrA) [39]. By combining $^1H$ NMR relaxometry and theoretical calculations, their objective was a better understanding of electron spin relaxation. As expected, the relaxivity of these $Mn^{2+}$ complexes is low and reflects the lack of water molecules coordinated to the metal ion (1.1–1.3 $mM^{-1}$ $s^{-1}$ at 20 MHz and 25°C). At low magnetic field (< 1 MHz proton Larmor frequency), where the effect of electron spin relaxation is important, the relaxivities are significantly different for different systems. The zero field splitting (ZFS) parameters were calculated by using DFT and CASSCF methods and indicated little influence of the nature of the donor atoms. On the other hand, the twist angle of the two tripodal faces that delineate the coordination polyhedron, defined by the N atoms of the TACN unit (lower face) and the donor atoms of the pendant arms (upper face), seems to be important for electronic relaxation. As another important finding of this study, the authors concluded that the classical Solomon–Bloembergen–Morgan description of the electron relaxation theory is reasonably satisfactory to describe the low-field relaxivities for small $Mn^{2+}$ complexes.

Pota et al. investigated a larger macrocyclic ligand (15-py$N_3O_2$Ph), which incorporates both a pyridine and an *ortho*-phenylene building block within the cycle in order to decrease ligand flexibility [40]. Equilibrium studies proved that the introduction of the phenylene moiety into the ligand structure seriously decreased the thermodynamic stability of the $Mn^{2+}$ complex ($\log K_{ML}$ = 5.62), implying that complex formation is not complete below pH 8. Solid-state X-ray diffraction results proved a seven-coordinate metal center with two water molecules and a $N_3O_2$ coordinating ligand donor atom set. In solution, the two coordinated water molecules yield a high longitudinal relaxivity of 5.16 $mM^{-1}$ $s^{-1}$ for Mn(15-py$N_3O_2$Ph) (0.49 T and 25°C). The chelate is very labile and shows catalytic activity in superoxide dismutation.

## 2.3 Bispidine Manganese(ii) Complexes

Ligands based on the 3,7-diazabicyclo[3.3.1]nonane (bispidine) platform have been widely explored in coordination chemistry, including applications in imaging [41–43]. This bicyclic amine skeleton can be appended with a variety of coordinating functions in different positions as a function of the coordination requirements of very different metal ions. Depending on the position and the steric demand of the substituents, the bicyclic core can take different conformations. Among these, the chair–chair conformation is the best suited for the coordination of metal ions (Scheme 3a). Bispidine chelators are typically endowed with a highly preorganized and rigid structure, two properties which make them particularly attractive to achieve selective metal coordination and high kinetic inertness. $Mn^{2+}$ complexes of bispidines have been explored in the last 3 years and proved to provide several unprecedented and very interesting features with respect to potential MRI applications.

The $Mn^{2+}$ complex of the 2,4-pyridyl-disubstituted bispidol derivative Bispl (Scheme 3) has a relatively modest stability, as assessed by pH-potentiometry ($\log K_{MnL}$ = 12.21, pMn = 6.65 calculated for $c_{Mn} = c_L = 10^{-5}$ M; pH 7.4) [44]. Interestingly, the formation of the complex is slow, preventing direct potentiometric titrations and requiring batch samples. The kinetic inertness of MnBispl was assessed in transmetalation experiments in the presence of 10 or 50 equivalents

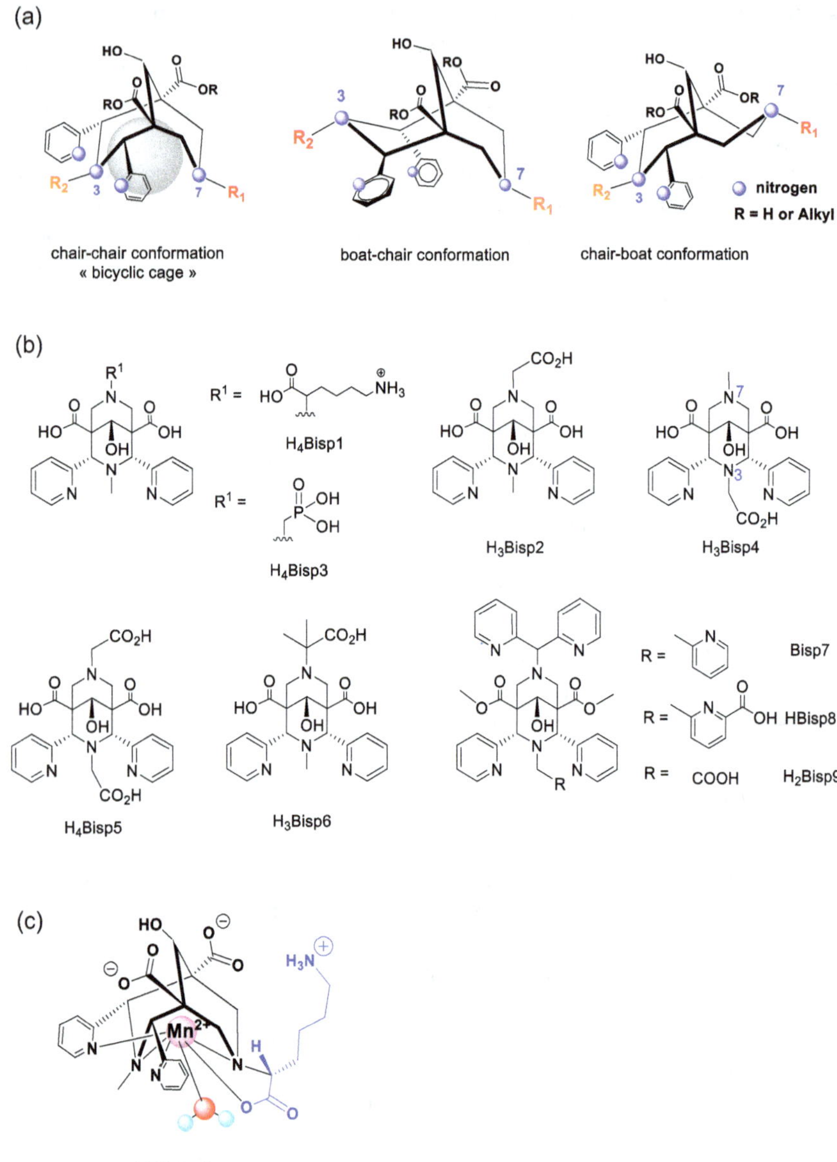


**SCHEME 3** (a) Bispidine conformations, (b) bispidine ligands for $Mn^{2+}$ complexation and (c) the putative structure of MnHBisp1.

of $Zn^{2+}$, at pH 6 (50 mM MES buffer), 37°C, which revealed an unprecedented resistance to dissociation. Indeed, there is no relaxivity change and hence no dissociation could be observed in these samples for at least 140 days. This result is to be compared to dissociation half-lives, $t_{1/2}$, of 0.285 h and 54.4 h reported under similar conditions for MnPyC3A [14] and MnPC2A-EA [33], respectively, two reference chelates considered as potential MRI probes with good kinetic inertness. The dissociation was further investigated under more acidic conditions where the proton-assisted dissociation pathway could be assessed. The rate constant characterizing this mechanism, $k_1 = (1.6 \pm 0.1) \times 10^{-3}\ s^{-1}\ M^{-1}$, is two orders of magnitude lower than $k_1$ for MnPC2A-EA, again demonstrating the high inertness of this bispidine chelate.

The presence of one inner-sphere water in MnBisp1 was evidenced by $^{17}O$ transverse relaxation (see Scheme 3c for a putative structure). Its water exchange is relatively slow for a $Mn^{2+}$ complex (the rate of the exchange is $k_{ex}^{298} = (5.1 \pm 0.7) \times 10^7\ s^{-1}$) and follows an associatively activated mechanism, as expected for a six-coordinate $Mn^{2+}$ chelate. Indeed, typical coordination numbers of $Mn^{2+}$ in solution are six and seven, thus water exchange tends to be associatively activated for six-coordinate and dissociatively activated for seven-coordinate complexes. The proton relaxivity of MnBisp 1 is remarkably high ($r_1 = 4.28\ mM^{-1}\ s^{-1}$ and $3.37\ mM^{-1}\ s^{-1}$ at 25°C and 37°C, respectively, 20 MHz, pH 7 in water). This high relaxivity can be attributed to its slightly higher molecular weight as compared to typical, small molecular weight chelates, as well as to the contribution of a second sphere relaxation mechanism to the relaxivity, promoted by the presence of the two non-coordinating, charged carboxylates. Bisp1 was successfully radiolabelled by the PET radioisotope $^{52}$Mn, and the radiocomplex showed good stability under biologically relevant conditions. Finally, the MRI contrast agent potential of MnBisp1 was validated in a preclinical imaging experiment. Healthy mice have been injected with MnBisp1 at 0.06 mmol $kg^{-1}$ dose and the MR signal intensity was monitored over time in different organs. Good signal enhancement was detected in the kidney which is the main organ of MnBisp1 excretion, while liver uptake was limited. Fast elimination of the compound from the mice was confirmed by *ex vivo* ICP-OES measurements.

Following these very promising data, several other ligands in this bispidine family have been studied for $Mn^{2+}$ complexation. The lysine substituent was replaced by an acetate (Bisp2) or a methylenephosphonate (Bisp3) group (Scheme 3) [45]. The MnBisp2 and MnBisp3 complexes retain similarly good characteristics as MnBisp1. Their conditional stabilities are slightly higher than those of MnBisp1. $Zn^{2+}$ transmetalation experiments between pH 3 and 6 and a dissociation study at higher acidity without $Zn^{2+}$ evidenced extremely high kinetic inertness. At pH 6, in the presence of 50 equiv. of $Zn^{2+}$, even after 150 days, dissociation is limited to ~20% and ~40% for MnBisp2 and MnBisp3, respectively (37°C). Interestingly and in contrast to literature examples, the phosphonate derivative retains excellent inertness, in spite of the protonation sites on the phosphonate group which are generally known to accelerate the proton-catalyzed dissociation process. This high resistance to dissociation was further supported by the excellent stability of the $^{52}$Mn-labeled radiocomplex. The relaxivity of the phosphonate analogue, 4.31 $mM^{-1}\ s^{-1}$, is remarkably higher than that of MnBisp2, 3.64 $mM^{-1}\ s^{-1}$ (20 MHz, 25°C), which was attributed to an enhanced second sphere relaxivity contribution. It is also interesting to note that phosphonate group induces a ~50 times slower water exchange with respect to the acetate derivative ($k_{ex}^{298}$ is 0.12 × $10^7$ $s^{-1}$ for MnBisp3 *vs* 5.1 × $10^7$ $s^{-1}$ for MnBisp2). This can be related to the associative character of the water exchange mechanism where the rate-determining step, i.e., the entering of the second water molecule into the metal coordination sphere, is disfavored for the phosphonate complex with a higher negative charge (–2 for MnBisp3 *vs* –1 for MnBisp2).

A recent study involving MnBisp2, MnBisp4 and MnBisp5 complexes (Scheme 3) addressed how positional isomerism of bispidine ligands affects the conformation, the stability, the inertness and the relaxivity of their $Mn^{2+}$ complexes [46]. ROESY NMR and pH-potentiometric data confirmed that depending on the substituent of N3 amine nitrogen (methyl or acetate), the conformation of the ligand at pH 7 switches from chair–chair (Bisp1, Bisp2) to boat–chair (Bisp4, Bisp5), and these conformations are conserved in the $Zn^{2+}$ and $Mn^{2+}$ complexes. This difference has important consequences on the physical-chemical properties. The chelates in boat–chair conformation are characterized by several orders of magnitude faster formation and dissociation. Their hydration number is less than one, while the complexes in chair–chair conformation are all monohydrated, as it is indicated by $^{17}O$ relaxation rates. This translates to lower relaxivities for MnBisp4 and MnBisp5 (2.59 $mM^{-1}$ $s^{-1}$ and 2.79 $mM^{-1}$ $s^{-1}$, respectively) in comparison to MnBisp2 (3.17 $mM^{-1}$ $s^{-1}$, 25°C, 60 MHz). Overall, it could be concluded that the N3 substituent is determinant for the ligand conformation, and exclusively, the complexes in chair–chair conformation are endowed with the extraordinary kinetic inertness.

The influence of steric crowding on the α carbon of the N7 substituent was also studied, and similar to previous observations on cyclen-type lanthanide complexes, it was shown to generate increased resistance to dissociation [46]. Indeed, the dissociation was found to be twice as slow for MnBisp6 as for MnBisp2 (at pH 4, in the presence of 50 equiv. of $Zn^{2+}$).

Higher denticity bispidine ligands have been also investigated for $Mn^{2+}$ complexation and proved to provide, for the first time, selectivity for $Mn^{2+}$ over $Zn^{2+}$ [47,48]. Thanks to the spherical distribution of the five *d* electrons and to the relatively large ionic radius of $Mn^{2+}$, $Mn^{2+}$ complexes are not only intrinsically labile, but for any particular ligand, the MnL thermodynamic stability is usually below that of other first row transition metal chelates. Taking advantage of the preorganized and greatly rigid structure of bispidines, it becomes possible to tailor the coordination sphere and discriminate $Mn^{2+}$ *vs* its major biological competitor $Zn^{2+}$. This was demonstrated by ligands Bisp7, Bisp8 and Bisp9 bearing a dipyridyl pendant arm in the N7 position. These hepta- or octadentate ligands form unusual, eight-coordinate complexes with $Mn^{2+}$ in the solid state, supported by X-ray crystal structures, while the $Zn^{2+}$ analogues are six-coordinate, with one or two non-coordinating donor atoms, as the $Zn^{2+}$ ion is too small to accommodate them all in its coordination sphere (Figure 1). DFT calculations led to identical conclusions. In the solution state, $^{17}O$ NMR indicated that the heptadentate ligands Bisp8 and Bisp9 form monohydrated $Mn^{2+}$ chelates with correspondingly high relaxivities (5.04 $mM^{-1}$ $s^{-1}$ and 4.44 $mM^{-1}$ $s^{-1}$ at 20 MHz, 25°C). As expected

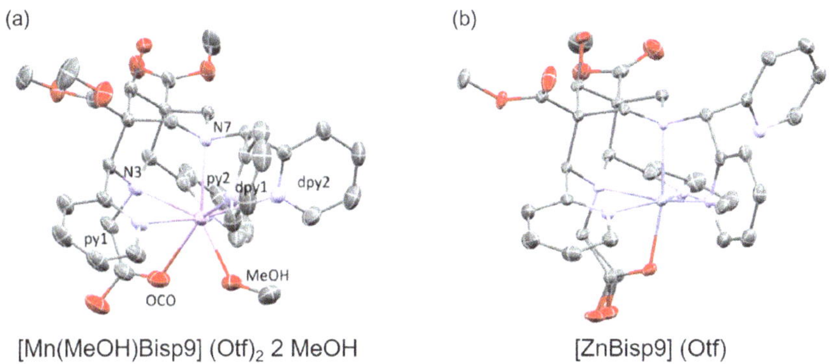


**FIGURE 1** Plots of the solid state structures of (a) [MnBisp9(MeOH)](OTf)$_2$ and (b) [ZnBisp9](OTf) showing different coordination modes for $Mn^{2+}$ (octacoordinate) and $Zn^{2+}$ (hexacoordinate). Atoms shown at 50% probability, hydrogen atoms, solvent molecules and anions have been omitted for clarity. (Adapted with permission from Ref. [48].)

for an eight-coordinate chelate, the water exchange on MnBisp9 occurs *via* a dissociative mechanism, as shown by the negative activation entropy.

Most importantly, the thermodynamic stability is 10 orders of magnitude higher for the $Mn^{2+}$ complexes than for the $Zn^{2+}$ analogues of Bisp7 and Bisp8, both of which contain a large bulky substituent (pyridine or picolinate) on the N3 position. Interestingly, for Bisp9 appended with a smaller and more flexible acetate function in the N3 position, this difference is less, 5 orders of magnitude, but still remains in favor of $Mn^{2+}$. This again demonstrates the importance of rigidity of the ligand coordination cage and its perfect size match with the metal ion size to achieve such impressive selectivity.

Although the kinetic inertness is lower for these complexes than it is for MnBispl, it remains excellent with a dissociation half-life of 14.3 days for MnBisp9 at pH 6, 37°C and in the presence of 50 equiv. of $Zn^{2+}$. A more detailed dissociation kinetic study on MnBisp9 revealed another unusual behavior, which is the complete lack of pH dependence of the dissociation in the investigated pH range 3–6. This implies that the proton-assisted dissociation is negligible in this pH range, in contrast to what is generally observed for transition metal or lanthanide chelates of poly(amino carboxylates). In the complex, the metal ion is locked within the coordination cage, and the orientation of the lone pairs of the tertiary amines toward the center of the coordination cavity and the rigidity of the adamantane-derived backbone prevent any competing protonation. $OH^-$ attack, which could lead to dissociation, remains also unimportant, since MnBisp9 was intact for several days at pH 10.

MnBisp9 was injected into healthy mice at a low dose of 0.02 mmol kg$^{-1}$. Good contrast enhancement was registered in the kidneys at short post-injection times (maximum at 2 min), indicating that the good relaxation efficiency is retained *in vivo*, and no significant liver uptake was observed (Figure 2). An *ex vivo*

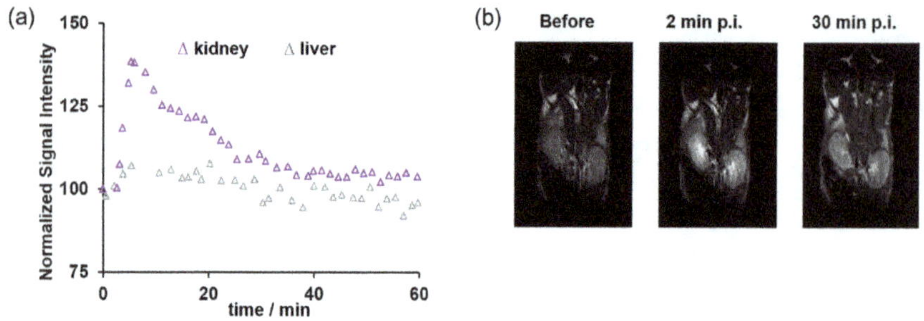


**FIGURE 2** (a) Normalized signal intensity in the kidney and the liver plotted as a function of time. Measurements were performed for 3 mice. Standard deviations are not presented for better readability. (b) Coronal $T_1$-weighted MR images prior to injection and at 2 and 30 min post-intravenous injection of 0.02 mmol kg$^{-1}$ of MnBisp9. (Adapted with permission from Ref. [48].)

biodistribution study confirmed the complete elimination of the complex after 24 h post-injection.

## 3 MANGANESE(III) COMPLEXES

The paramagnetic $Mn^{3+}$ ($3d^4$ configuration) ion can also generate high-spin complexes ($S = 2$), but their nuclear relaxation efficiency is often limited by short longitudinal electronic relaxation times [49]. Interestingly, $Mn^{3+}$ relaxivities might show a very different dependency on the magnetic field as compared to $Mn^{2+}$ or $Gd^{3+}$ chelates leading to potentially elevated relaxivity values at high magnetic field [17]. This feature can be particularly interesting for preclinical applications, which are often performed at ≥7 T.

Porphyrins represent the most widely investigated scaffold for $Mn^{3+}$ complexation, extensively reviewed recently [19,50]. The porphyrin structure yields bishydrated $Mn^{3+}$ complexes with correspondingly enhanced relaxivities, where the metal is hexacoordinate, involving the complexation of four N-atoms of the planar ligand and two water molecules in apical position below and above the porphyrin plane. Because Mn-porphyrin complexes are thermodynamically stable and kinetically inert, and naturally target tumor cells, their features generated high interest in different application areas.

In the context of MRI, $Mn^{3+}$ttps (Scheme 4) is the most extensively studied $Mn^{3+}$ complex, it is water soluble, stable in air and biological media, including in human plasma, and presents high relaxivity at all fields [17]. Other porphyrin structures involving different substitution and functionalization have been explored to improve water solubility and *in vivo* properties [19,50,51]. Recent additions to this field include work by Liu et al. [52, 53].

In a quest of better balance between blood retention and body clearance, Liu et al. have proposed an improved second-generation version of previously reported dimers. It consists of two tpps units connected *via* a twisted *mesa*-diphenyl link

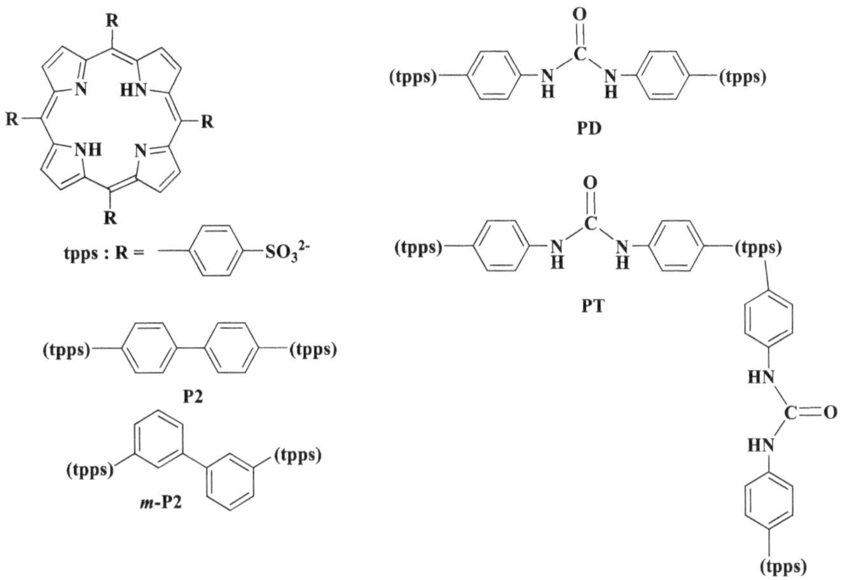

**SCHEME 4** Porphyrin chelators for $Mn^{3+}$ complexation.

(*m*-MnP2) (Scheme 4) [52]. This novel regioisomer presents a slightly lower relaxivity than the first-generation compound, $r_1$ = 15.4 $mM^{-1}$ $s^{-1}$ for *m*-MnP2 *vs* 20.2 $mM^{-1}$ $s^{-1}$ for MnP2, at 1.6 T, 25°C (in HEPES, pH 7.2). This difference is most probably related to the lower hydrodynamic size of *m*-MnP2 compared to MnP2. These complexes are able to bind serum albumin. Regardless of the lower affinity of *m*-MnP2 which will probably contribute to better clearance, the ternary, HSA-bound complexes for both MnP2 and *m*-MnP2 present similar relaxivities at magnetic fields above 1 T (~15.8 $mM^{-1}$ $s^{-1}$ at 1 T), which is relevant for *in vivo* imaging.

The dimer MnPD and the trimer MnPT were synthesized using a synthetic strategy to construct porphyrin oligomers in a three-step process under mild reaction conditions [53]. As expected, their $Mn^{3+}$ complexes exhibit high $r_1$ values of 16.76 and 19.98 $mM^{-1}$ $s^{-1}$ for MnPD and MnPT, respectively (1 T, 25°C), which are higher than $r_1$ determined for Mnttps4 (12.8 $mM^{-1}$ $s^{-1}$) at the same field and temperature. Phantom MRI studies at 3 T (25°C) revealed that the novel complexes are detectable at concentrations as low as 0.01 mM. Their toxicity was evaluated in three different cell lines (HepG2, HUVEC and A549; MTT studies) indicating cell viability higher than 86% even at 400 μM, thus revealing no toxicity for both oligomers. In *in vivo* studies conducted on healthy mice injected with a low dose of 0.025 mmol Mn $kg^{-1}$ (4-fold lower than the typical dose used for Gd-based complexes), the two complexes showed different distribution profiles, with the trimer displaying liver accumulation and kidney retention.

More recently, other $Mn^{3+}$ porphyrins have been proposed as targeted or responsive probes and are described below.

## 4 MOLECULAR IMAGING BASED ON MANGANESE COMPLEXES

Molecular imaging probes are in extensive development for more than two decades now, with an increasing attention paid to Mn-based agents. The classical concepts, widely explored for $Gd^{3+}$ complexes, also apply for $Mn^{2+}$-based agents: the binding of a targeted probe to a specific biomarker [54] or the modulation of the relaxivity of a responsive agent *via* interactions with a biomarker [55]. In addition, the redox properties of Mn offer further opportunities for the design of redox-sensitive probes [18].

Among various molecular imaging approaches, the detection of pH, redox changes, zinc concentrations and the presence of specific proteins have been the most investigated in the last three years, which will be overviewed here.

### 4.1 pH Detection

Botár et al. proposed a pyclen-based $Mn^{2+}$ complex as pH-responsive probe [33]. The PC2A-EA chelator (Scheme 5) contains two acetate groups in positions 3 and 9, while a pH-sensitive ethylamine moiety is appended on the third amine nitrogen. The protonation of the ethylamine occurs in the physiological pH region, with a log $K^{H} = 6.88$, inducing a pH-dependent, reversible coordination/decoordination of the pendant arm with a concomitant change in the hydration number of the metal between $q = 0$ at higher and $q = 1$ at lower pH. Accordingly, the relaxivity $r_1 = 3.54$ $mM^{-1}$ $s^{-1}$ measured in the pH range of 3.7–5.8 decreases to $r_1 = 2.04$ $mM^{-1}$ $s^{-1}$ at pH 8.4, at 0.49 T and 25°C. In the presence of HSA, a 5 times more important variation is observed in the transverse relaxivity between pH 6.7 and 7.5 (at 3 T), resulting in an excellent pH discrimination. It is worth noting that this complex exhibits a very high thermodynamic and kinetic inertness, making it a promising $Mn^{2+}$ based pH-responsive probe candidate for further *in vivo* tests.

Uzal-Varela et al. developed two pH-responsive probes, MnDPASAm and MnNO2ASAm, bearing *N*-ethyl-4-(trifluoromethyl)benzenesulfonamide groups

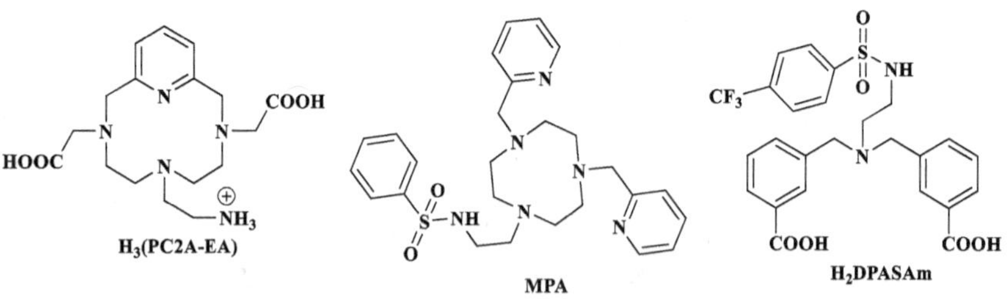

**SCHEME 5** Chelators used for pH-sensitive $Mn^{2+}$ probes.

(Scheme 5) [56]. The electron-withdrawing $CF_3$ group was incorporated in order to shift the log $K^H$ of the sulfonamide to the biologically relevant window, in view of the lower charge of $Mn^{2+}$, as compared to $Gd^{3+}$ analogues. The open-chain ligand DPASAm leads to the formation of a seven-coordinate $Mn^{2+}$ complex, while $Mn^{2+}$ is six-coordinated in the macrocyclic complex MnNO2ASAm. Sulfonamide protonation should trigger the replacement of the sulfonamide N atom by one water molecule in the coordination sphere of the metal, thus anticipating an increase in relaxivity and MRI signal intensity. MnDPASAm is stable in the pH range of 4–10. The complex undergoes protonation close to physiological pH (log $K^H$ = 6.44) provoking an increase in the $r_1$ relaxivity from 3.8 $mM^{-1}$ $s^{-1}$ at pH 9.0 to 8.9 $mM^{-1}$ $s^{-1}$ at pH 4.0 (20 MHz, 25°C). On the other hand, MnNO2ASAm protonates at a lower pH (log $K^H$ = 5.5), making it less suitable for *in vivo* applications. The solubility of the complexes studied was limited, and their kinetic inertness should be also improved for further *in vivo* validation. Nevertheless, this study demonstrated that the reversible binding of sulfonamide groups might be adequate for the design of pH-responsive probes.

The same pH-responsive aryl sulfonamide moiety, but without $CF_3$, was used by Shen et al. to prepare a triazacyclononane macrocycle bearing two additional picolyl pendant arms (MPA, Scheme 5) [57]. The longitudinal relaxivity of MnMPA is 2.3 $mM^{-1}$ $s^{-1}$ at 0.5 T and 35°C, indicating the formation of a monohydrated $Mn^{2+}$ complex. This relaxivity is constant between pH 5.5 and 10.0. At lower pH, a 3-fold increase is observed, with $r_1$ = 6.9 $mM^{-1}$ $s^{-1}$ obtained at pH 3.5. The kinetic inertness was assessed in the presence of 25 equiv. of $Zn^{2+}$ at pH 6.0, revealing a dissociation half-life of $t_{1/2}$ = 19.6 h. A phantom study at 1.0 T showed a difference in MRI signal intensity between the tubes containing MnMPA at different pH values. The authors prepared a series of complexes with varying substituents to demonstrate the feasibility of tuning the pH window *via* the substituents on the aromatic ring or by modifying the picolyl pendants. The good thermodynamic stability and kinetic inertness, the effective relaxivity response to pH stimuli, and the easy modulation of the structure, thus the log $K_a$, together make MnMPA interesting for further development.

## 4.2 Zinc Detection

Sherry and co-workers have extensively reported on zinc-responsive MRI probes using $Gd^{3+}$. Recently, a $Mn^{2+}$ agent was proposed as an alternative to the Gd-based ones [58]. The monohydrated MnPyC3A–BPEN complex (Scheme 6) shows superior kinetic inertness to GdDTPA. The design of this probe is based on the very promising MnPyC3A complex which is currently in clinical trials [14,15], to which the authors conjugated a BPEN unit, known for high-affinity $Zn^{2+}$ binding. MnPyC3A-BPEN has a modest increase in $r_1$ relaxivity in the presence of $Zn^{2+}$ or of HSA alone ($r_1$ = 3.7 $mM^{-1}$ $s^{-1}$, 4.1 $mM^{-1}$ $s^{-1}$, and 10.8 $mM^{-1}$ $s^{-1}$ in the absence and presence of 1 eq. $Zn^{2+}$ or 0.6 mM HSA, respectively, 0.5 T, 37°C). When both $Zn^{2+}$ and HSA are present, a ternary complex MnPyC3A–BPEN–$Zn^{2+}$–HSA forms and $r_1$ is significantly increased ($r_1$ = 17.4 $mM^{-1}$ $s^{-1}$).

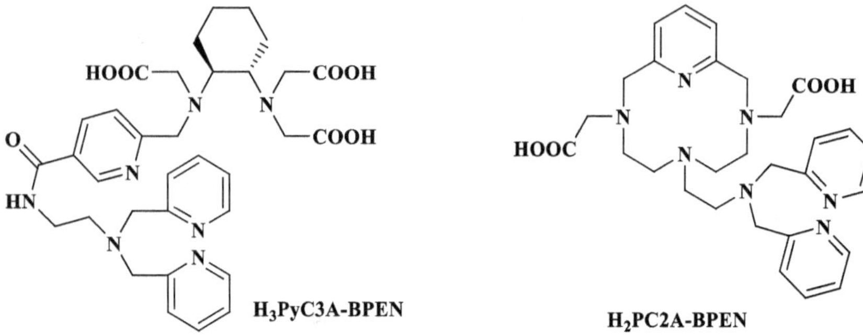

$H_3PyC3A$-BPEN $H_2PC2A$-BPEN

**SCHEME 6** Chelators for $Zn^{2+}$-sensitive probes.

The ability of this probe to detect $Zn^{2+}$ secretion in tissues was tested *in vivo*. MnPyC3A–BPEN was injected (0.07 mmol $kg^{-1}$) in healthy fasted male mice, and the pancreas and prostate were imaged before and after injection of a glucose bolus allowing the visualization of zinc secretion, both at 4.7 T and 9.4 T. More detailed biodistribution studies revealed an excretion pathway about 50% biliary and 50% renal. These results show that MnPyC3A-BPEN has translational value for prostate cancer early detection and for monitoring $\beta$-cell function during the development of type 2 diabetes.

A macrocyclic $Mn^{2+}$ complex responsive to zinc was studied by Botár et al. [59]. It consists of a pyclen-3,9-diacetate (3,9-PC2A) functionalized with the same $Zn^{2+}$ recognition moiety as above. This MnPC2A–BPEN complex (Scheme 6) has a high thermodynamic stability and possesses an outstanding inertness toward $Zn^{2+}$ transmetalation, with a dissociation half-life of $t_{1/2} = 64.5$ h (pH = 6). As for the previously described example, the $r_1$ relaxivity of the complex (3.24 $mM^{-1}$ $s^{-1}$ at 1.41 T and 37°C) shows no significant increase in the presence of $Zn^{2+}$ alone, but in the presence of both $Zn^{2+}$ and HSA, a 3-fold increase in $r_1$ is observed ($r_1$ = 12.14 $mM^{-1}$ $s^{-1}$; 1.41 T and 37°C). The *in vivo* profile of this complex was evaluated in healthy mice, at a 0.1 mmol $kg^{-1}$ dose, before and after a glucose bolus injection. Zinc secretion could be visualized on both $T_1$- and $T_2$-weighted images of the prostate.

## 4.3 Redox Detection

Goldsmith and co-workers have previously reported on redox sensors which are based on the redox properties of quinol-containing ligands, instead of the redox modulation of manganese. Indeed, they showed that $H_2O_2$ primarily oxidizes the ligand rather than the metal ion. In 2021, aiming at higher thermodynamic stability and kinetic inertness of both reduced and oxidized forms of the complexes, they proposed a new chelator that consists of two quinols covalently tethered to a cyclam macrocycle, $H_4$qp4 [60]. The $Mn^{2+}$ complex responds to $H_2O_2$ with a relaxivity increase from $r_1$ = 3.16 $mM^{-1}$ $s^{-1}$ to 5.09 $mM^{-1}$ $s^{-1}$ after 1 h and reaching

7.35 $mM^{-1}$ $s^{-1}$ after 2 h reaction (3 T, 25°C). This relaxivity change, attributed to an increase in the hydration number, is 4 times higher compared to that observed for a previously reported linear complex Mnqp2 [61], despite a similar set of donor atoms involving two quinolates and four neutral amine nitrogen atoms. In biologically relevant conditions, low but rapidly replenishing concentrations of $H_2O_2$ are expected. The authors showed that, at such low $H_2O_2$ levels, the intramolecular oxidation of the quinols to *p*-quinones occurs in the complex. This process appears to be slow relative to analogous reactions seen with manganese complexes of linear quinol-containing ligands. With a large excess of $H_2O_2$, there is a noticeable induction period before quinol oxidation and $r_1$ enhancement occurs, and this could be explained by the catalase activity of Mnqp4, which initially outcompetes ligand oxidation.

A tyrosine derivative ethylenediaminetetraacetate ligand, TyrEDTA, was proposed to design a peroxidase-activatable $Mn^{2+}$ complex capable of detecting myeloperoxidase (MPO) activity (Scheme 6) [62]. MPO is a biomarker of innate immune cell function and activity, and it contributes to tissue oxidation in inflammation. The complex MnTyrEDTA displays relaxivities of $r_1$ = 3.3 $mM^{-1}$ $s^{-1}$, $r_2$ = 4.6 $mM^{-1}$ $s^{-1}$ and $r_1$ = 8.0 $mM^{-1}$ $s^{-1}$, $r_2$ = 10.7 $mM^{-1}$ $s^{-1}$ in PBS and 4.5% BSA, respectively (0.47 T, 32°C). Transmetalation studies in the presence of 25 equiv. of $Zn^{2+}$ (50 mM MES buffer, pH 6.0, 32°C), monitored *via* changes in the transverse $T_2$ relaxation time, revealed that similar kinetic inertness for MnTyrEDTA and GdDTPA.

In the absence of peroxidase, MnTyrEDTA is stable with no $R_2$ changes observed in air or in $H_2O_2$ solution. Incubation of MnTyrEDTA with 3.5 equiv. of $H_2O_2$ and 500 U horseradish peroxidase yielded a relaxivity increase to $r_1$ = 8.2 $mM^{-1}$ $s^{-1}$, $r_2$ = 9.9 $mM^{-1}$ $s^{-1}$ in PBS and $r_1$ = 13.8 $mM^{-1}$ $s^{-1}$, $r_2$ = 17.8 $mM^{-1}$ $s^{-1}$ in 4.5% BSA solution. Mass spectrometry studies indicated that this complex acts as the substrate of peroxidase/$H_2O_2$ in the 'peroxidase cycle' leading to the formation of oligomers with enhanced relaxivity, thereby amplifying the MR signal.

*In vivo* studies in healthy mice and in a murine model of MSU crystal-induced acute gouty arthritis (left foot of the mouse) showed fast clearance of the probe *via* mixed hepatobiliary and renal pathways. Contrast-to-noise ratio of the inflammation sites in the left foot was 1.5- to 2.5-fold higher than with GdDTPA and MnBnOTyrEDTA, injected in the same conditions; however, no significant differences were observed in the non-inflamed right feet of the animals. These promising results suggest that this peroxidase-activatable $Mn^{2+}$ MRI probe could potentially be used for non-invasive detection of MPO activity *in vivo*.

## 4.4 Redox Detection in $^{19}$F MRI

$^{19}$F MRI is a promising complementary technique to classical $^1$H MRI. Due to the absence of fluorine in the human body (except for teeth and bones in the form of solid salts), $^{19}$F MRI minimizes the problems associated with the background $^1$H signal of bulk water. Complexes combining a paramagnetic metal ion and fluorine are explored to accelerate the typically slow $^{19}$F relaxation [63–65]. Here,

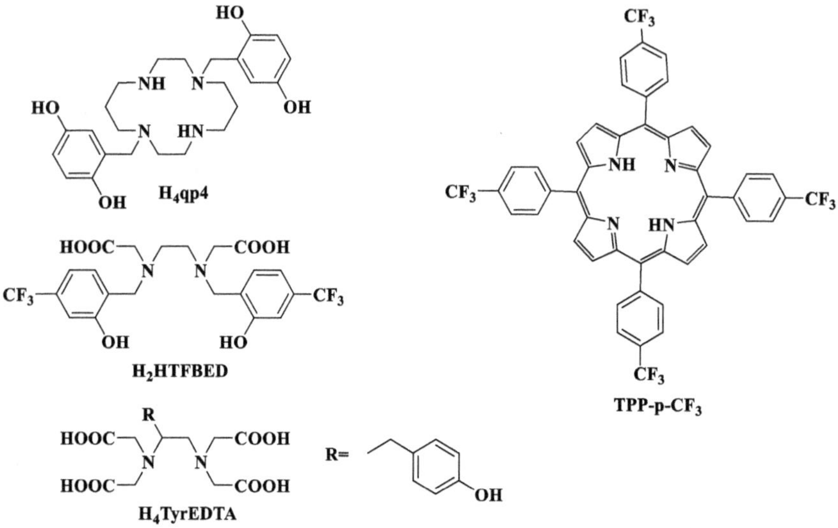

**SCHEME 7** Chelators for redox-sensitive agents.

we focus on the examples involving manganese complexes, published in the last three years.

In 2020, Chen et al. reported the $Mn^{2+/3+}$ complex of the HTFBED ligand containing two $CF_3$ moieties as water-soluble molecular probes with a reversible redox response in both $^1H$ and $^{19}F$ MRI (Scheme 7) [66]. Both paramagnetic $Mn^{3+}$ and $Mn^{2+}$ ions cause a huge decrease in the relaxation times ($T_1$ and $T_2$) of the fluorine nucleus as compared to the free ligand; nevertheless, this relaxation effect is very different for the two redox states ($T_1$ = 6.4, 1.6 and 955 ms; $T_2$ = 4.0, 0.6 and 629 ms, respectively, at 9.4 T). The complex responds to various biological redox stimuli, like reactive oxygen species (ROS) such as $H_2O_2$, ascorbic acid (AA), or glutathione (GSH), resulting in a significant variation in $^{19}F$ as well as in $^1H$ MRI signal intensity. ROS could be also successfully detected in HepG2 cells which were treated with MnTFBED and pyocyanin.

Another example of dual redox-sensitive $^1H/^{19}F$ probes is the Mn-TPP-*p*-$CF_3$ chelate where four *para*-$CF_3$-benzene moieties were appended on the porphyrin ligand (Scheme 7) [67]. Despite its poor water solubility, this complex is an interesting model compound, and it was subject of a detailed physical-chemical investigation in both redox states, in DMSO-$d_6$/$H_2O$ mixed solvent. The $Mn^{3+}$-F distance was estimated to be $d_{Mn\text{-}F}$ = 9.7–10 Å from DFT calculations. Mn-TPP-*p*-$CF_3$ has a reversible $Mn^{2+}/Mn^{3+}$ redox potential of 0.574 V *vs* NHE. In order to prove the reversibility of the system, the authors used ascorbic acid to promote the reduction of the $Mn^{3+}$ complex and air to re-oxidize the $Mn^{2+}$ form. The redox switch could be monitored by UV–Vis spectroscopy, $^{19}F$ NMR and $^{19}F$ MRI. Both $T_1$ and $T_2$ $^{19}F$ relaxation times are highly reduced in the $Mn^{3+}$/$Mn^{2+}$ complexes as compared to the free porphyrin ($T_1$ = 11.5, 6.5 and 968 ms;

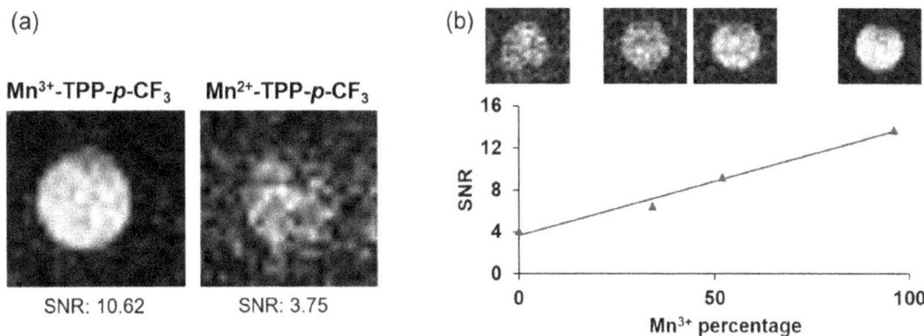


**FIGURE 3** $^{19}$F MRI phantom images acquired at 7 T: (a) $Mn^{3+/2+}$-TPP-*p*-CF3 ($c_{MnL}$ = 1.1 mM in DMSO); (b) monitoring reoxidation of $Mn^{2+}$-TPP-*p*-CF3 ($c_{MnL}$ = 1.6 mM) by air oxygen: signal-to-noise ratio (with respect to the $Mn^{3+}$-TPP-*p*-$CF_3$ peak) as a function of the oxidized $Mn^{3+}$TPP-*p*-$CF_3$ content (%); the signal of the $Mn^{2+}$ complex was saturated with a selective pulse. B = 7 T; TR = 60 ms; TE = 1.3 ms; FA = 90°; NA = 256; acquisition time 8 min 11 s. (Reproduced with permission from Ref. [67].)

$T_2$ = 7.7, 1.9 and 325 ms, respectively, at 9.4 T, 25°C in DMSO). Consequently, on the $^{19}$F MRI phantom images at 7 T, signal intensity is considerably weaker for the reduced $Mn^{2+}$ than for the oxidized $Mn^{3+}$ complex, with a threefold decrease in the signal-to-noise ratio (SNR) (Figure 3). Additionally, $^{1}$H NMRD curves were recorded for both $Mn^{2+}$ and $Mn^{3+}$ complexes of Mn-TPP-*p*-$CF_3$ in mixed DMSO/$H_2O$ (20/80) solvent. They reflect the typical shape of relaxivity profiles of $Mn^{3+}$ and $Mn^{2+}$ porphyrins. However, the absolute relaxivity values are very low and concentration-dependent, indicative of important aggregation in these solutions.

## 4.5 Protein Targeting

Several targeted probes, based on both $Mn^{2+}$ and $Mn^{3+}$ complexes, have been designed. Liver fibrogenesis was recently visualized using $Mn^{2+}$-based contrast agents targeted to the aldehyde containing amino acid, called allysine ($Lys^{Ald}$). In fact, overexpression of lysyl oxidase is a characteristic of fibrogenesis. This enzyme catalyzes the oxidation of lysine $\varepsilon$-amino groups of extracellular matrix proteins, forming allysine, which is then a good biomarker of fibrogenesis. Ning et al. derivatized Mn(1,4-DO2A) on the non-substituted amine nitrogen atoms with hydrazine-bearing arms with the objective of imaging liver fibrogenesis [68]. The $Mn^{2+}$ complexes of the ligands 2Hyd, 2CHyd and 1CHyd have been studied for their redox stability, longitudinal relaxivity ($r_1$), and inertness. Mn-2CHyd was found to be least susceptible to oxidation. The number of inner-sphere water molecules ($q$), determined by $^{17}$O NMR, was lower than 1 for all three $\alpha$-substituted derivatives, indicating that the hydrazine substituents generate a steric crowding which limits water access to the metal, as compared to the parent Mn(1,4-DO2A). This results in limited $r_1$ relaxivities (1.6–1.8 $mM^{-1}$ $s^{-1}$, in PBS, pH 7.4, 2 h incubation, 37°C, 60 MHz). On the other hand, the $\alpha$-substituted complexes are 3 times more inert to $Mn^{2+}$ release than Mn(1,4-DO2A) in the presence of 25 equiv.

of $Zn^{2+}$ (pH 6, 37°C). The piperazino-hydrazine moiety showed higher reactivity and affinity for $Lys^{Ald}$ compared to their hydrazide analogues. Moreover, $Mn^{2+}$ is less susceptible to oxidation in Mn2CHyd, and the latter displays 4-fold relaxivity turn-on response upon $Lys^{Ald}$ binding, namely when bound to $Lys^{Ald}$ bearing BSA ($r_{1(BSAAld)}$ = 7.7 mM$^{-1}$ s$^{-1}$, at 1.41 T, pH 7.4 and 37°C). Taking advantage of the $^{52}$Mn isotope, with suitable properties for PET imaging, *in vivo* application of Mn2CHyd was assessed in healthy and $CCl_4$-induced fibrogenesis mice, by simultaneous PET-MRI studies. Liver fibrogenesis was detected by significantly enhanced liver MR signal in fibrogenic mice compared to healthy mice. The bimodal PET-MRI approach used offers further *in vivo* quantification of such probes.

It is known that for the clinically used, liver-targeted agent GdEOB-DTPA (Primovist®), the ethoxybenzyl group provides an amphiphilic character to the molecule and this mediates hepatocyte uptake through organic anion transporting polypeptides (OATPs), highly expressed on hepatocyte membranes. Recently, several $Mn^{2+}$ and $Mn^{3+}$ complexes targeting OATPs have been proposed for liver imaging.

First, a $Mn^{3+}$ porphyrin, $Mn^{3+}$TriCP-PhOEt, was explored for *in vivo* imaging of OATP1, as a Mn-based alternative to the clinically approved GdEOB-DTPA [69]. $Mn^{3+}$TriCPPhOEt displays a significantly higher relaxivity than GdEOB-DTPA at relevant clinical fields: 8.32 *vs* 7.32 mM$^{-1}$ s$^{-1}$ at 1.5 T (37°C) and 11.25 *vs* 8.34 mM$^{-1}$ s$^{-1}$ at 3 T (18°C). This novel probe was tested in nude mice. The animals were injected intraperitoneally at a low dose 0.025 mmol kg$^{-1}$ (in 30% DMSO/saline), which yielded high liver, gallbladder and kidney uptakes up to 60 min post-injection (Figure 4). The enhancement observed was validated by *in vivo* $R_1$ mapping and *ex vivo* relaxivity measurements of the harvested organs. Although this $Mn^{3+}$ complex performs better than its $Gd^{3+}$ analogue, its limited water solubility compromises intravenous administration. Further functionalization to overcome the

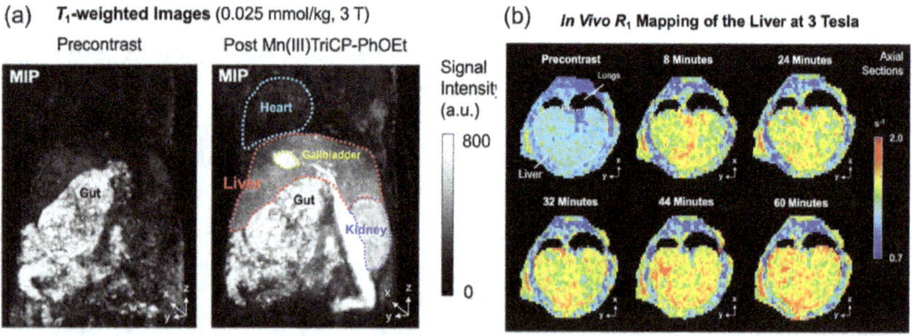


**FIGURE 4** (a) $T_1$-weighted images of a representative mouse before and after intraperitoneal injection of 0.025 mmol kg$^{-1}$ $Mn^{3+}$TriCP-PhOEt. (b) Representative *in vivo* $R_1$ map (in s$^{-1}$) of axial cross-section of the mouse before and following injection of 0.025 mmol kg$^{-1}$ $Mn^{3+}$TriCP-PhOEt. (Adapted with permission from Ref. [69]; copyright 2022, American Chemical Society.)

solubility issues is required in order to envisage application of this probe endowed with advantageous prolonged blood circulation.

Following their previous efforts [70], McRae et al. developed a simplified synthesis pathway to derive a series of five anionic, amphiphilic MnPyC3A derivatives, Mn-1, Mn-2, Mn-3, Mn-4 and Mn-5 (Scheme 8), with varied lipophilic side chains [71]. The $r_1$ relaxivities range between 2.3 and 3.0 $mM^{-1}$ $s^{-1}$ (3 T). The pharmacokinetics was assessed in mice and uptake by human OATPs was investigated *in vitro*. The authors demonstrated that despite the lower relaxivity of the complexes, the appropriate choice of the lipophilic targeting group can promote significant liver uptake in mice. Overall, the contrast enhancement capability of these probes *in vivo* is comparable to that of $Gd^{3+}$-based agents at equal dose.

Chen et al. functionalized the EDTA scaffold that was with a lipophilic group to yield the complex MnBnO-TyrEDTA [72]. MnBnO-TyrEDTA displays a relaxivity of 4.34 $mM^{-1}$ $s^{-1}$ (0.47 T, in 50 mM Tris buffer, pH = 7.4, 32°C) which increases to 15.81 $mM^{-1}$ $s^{-1}$ in the presence of BSA. BSA interaction correlates

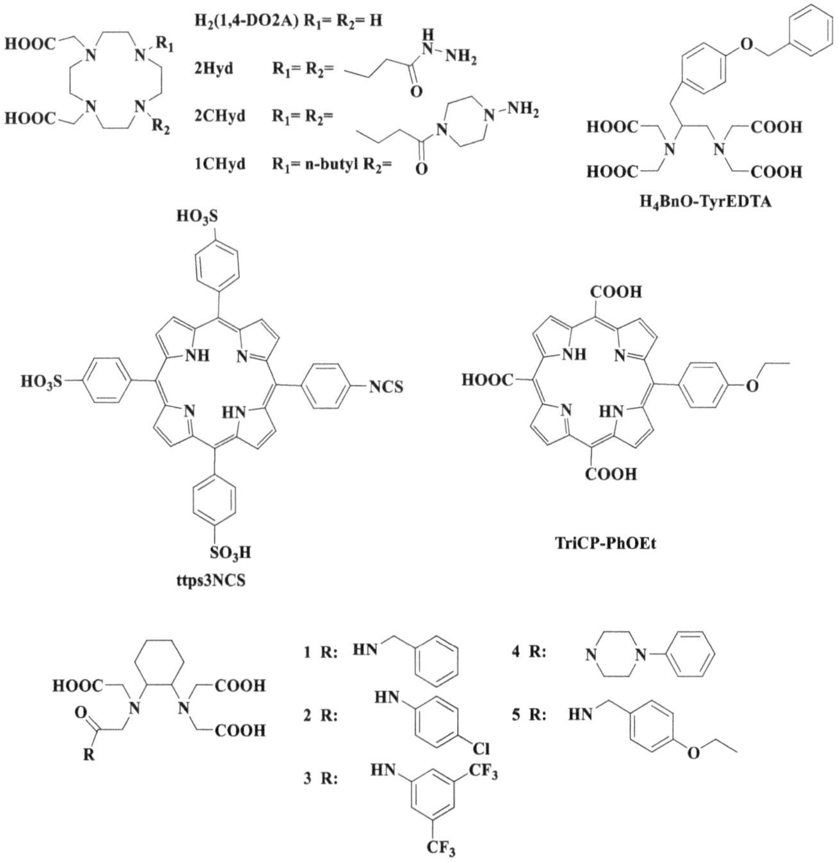

**SCHEME 8** Chelators for targeted agents.

with the improved lipophilic behavior of MnBnO-TyrEDTA (log $P$ = 0.18 *vs* log $P_{MnEDTA}$ = − 2.27), which anticipates good liver uptake. Healthy mice were imaged at 3 T upon intravenous injection (0.1 mmol $kg^{-1}$), showing the overwhelmingly highest signal-to-noise ratio in the liver 20 min post-injection and the strongest liver-to-muscle contrast-to-noise ratio 5 min post-injection when compared to MnEDTA and GdEOB-DTPA. MnBnO-TyrEDTA undergoes combined renal and hepatobiliary clearance. The hepatic OATP-mediated uptake can be inhibited by bromosulfophthalein (BSP); thus, mice administered intravenously with 0.1 mM of BSP before the injection of MnBnO-TyrEDTA were also imaged. In these animals, a less pronounced liver enhancement and lower SNR were observed, indicating that the liver uptake is specifically mediated by OATPs.

Islam et al. studied a naphthalene-bearing NOTA-derivative MnNOTA-NP [73]. This complex exhibits an $r_1$ relaxivity of 3.57 in water and 9.01 $mM^{-1}$ $s^{-1}$ in the presence of HSA at 3 T. Its log $P$ (−0.93) contributes certainly to the observed HSA binding which was quantified ($K_a$ = 36.6 $M^{-1}$). Docking studies revealed that OATP1B1 is a crucial transporter for MnNOTANP uptake. *In vivo* administration of 0.05 mmol $kg^{-1}$ MnNOTA-NP in healthy mice proved a dual pathway for excretion, which is highly similar to the clinically used GdDTPAEOB. In human hepatocellular carcinoma (HCC) tumor-bearing mice, MnNOTANP injection at the same dose enabled elevated tumor-to-normal liver contrast ratio, up to 2 h post-injection, allowing for high-sensitivity early diagnosis of HCC.

In order to propose a generalized protein conjugation platform with the aim of providing conjugation versatility for porphyrin complexes and facile one-step conjugation to proteins, the electrophilic $Mn^{3+}$ttps3NCS, obtained from $Mn^{3+}$ttps3$NH_2$, was investigated [74]. As a proof of principle, $Mn^{3+}$ttps3NCS was conjugated to albumin (as model protein) and collagen hydrogels (scaffold material in tissue engineering applications). After covalent conjugation to these proteins, the relaxation rates $T_1$ and $T_2$ decreased 3- to 5-fold in both cases (at 1.5 T) proving efficient protein labeling. $(Mn(ttps3))_n$-HSA ($n$ = 3.5 or 6) were injected intravenously in rats and MR imaging at 1.5 T revealed large $T_1$ decrease in the intravascular space and stable $T_1$ values in the blood for at least one hour. The Mnttps3NCS platform offers efficient protein labeling with potential interest for MRI tracking of biomaterials in the context of regenerative medicine.

## 5 GENERAL CONCLUSIONS

In the last few years, progress has been remarkable in the development of manganese-based MRI agents. Novel ligands have been designed for $Mn^{2+}$ complexation which ensure very high kinetic inertness combined with good relaxation properties. $Mn^{2+}$ complexes of PC2A and bispidine derivatives are particularly interesting, with unprecedented resistance to dissociation or first-ever selectivity for $Mn^{2+}$ with respect to the major endogenous competitor, $Zn^{2+}$.

Various molecular imaging agents have been derived to detect important biological biomarkers. $Mn^{3+}$ porphyrins are increasingly investigated and the $Mn^{2+}/Mn^{3+}$ redox couple continues to be explored for redox detection, including in $^{19}F$ MRI. In parallel, the first two $Mn^{2+}$ complexes entered Phase I human studies toward the replacement of Gd-based MRI agents in the clinics.

## ACKNOWLEDGMENTS

We acknowledge funding from the French Academy of Science via the Prix Mariano Gago 2022. Zoltán Garda has received funding from the European Union's Horizon 2021 research and innovation programme under the Marie Skłodowska-Curie Grant agreement no. 101065389.

## ABBREVIATIONS AND DEFINITIONS

| | |
|---|---|
| **BPEN** | N′,N′-bis(pyridine-2-ylmethyl)ethane-1,2-diamine |
| **BSA** | bovine serum albumin |
| **CDTA** | 1,2-cyclohexylenedinitrilotetraacetic acid |
| **DMSO** | dimethyl sulfoxide |
| **DOTA** | 1,4,7,10-tetraazacyclododecane-1,4,7,10-tetraacetic acid |
| **DPA** | dipicolinic acid |
| **DPDP** | N,N′-dipyridoxylethylenediamine-N,N′-diacetate-5,5′-bis(phosphate) |
| **DTPA** | diethylenetriamine pentaacetic acid |
| **EDTA** | ethylenediamine tetraacetic acid |
| **EOB** | ethoxybenzyl group |
| $H_2O_2$ | hydrogen peroxide |
| **HEPES** | 2-[4-(2-hydroxyethyl)piperazin-1-yl]ethanesulfonic acid |
| **HSA** | human serum albumin |
| **MES** | 2-(N-morpholino)ethanesulfonic acid |
| **MPO** | myeloperoxidase |
| **MRI** | magnetic resonance imaging |
| **NMRD** | nuclear magnetic relaxation dispersion |
| **NOTA** | 1,4,7-triazacyclononane-1,4,7-triacetic acid |
| **OATP** | organic anion transporting polypeptides |
| **OBn** | benzoyl group |
| **PBS** | phosphate-buffered saline |
| **PET** | positron emission tomography |
| **Pyclen** | 3,6,9,15-tetrazabicyclo[9.3.1]pentadeca-1(14),11-dien-13-one |
| $R_1 / R_2$ | longitudinal/transverse relaxation rate |
| $r_1 / r_2$ | longitudinal/transverse relaxivity |
| **SiNPs** | silica nanoparticles |
| **SOD** | sodium dismutase |
| $T_1 / T_2$ | longitudinal/transverse relaxation time |
| **TACN** | 1,4,7-triazacyclononane |
| ***t*-CDTA** | *trans*-1,2-diaminocyclohexane-N,N,N′,N′-tetraacetate |

## REFERENCES


1. T. Grobner, *Nephrol. Dial. Transplant.* **2006**, *21*, 1104–1108.
2. E. Di Gregorio, G. Ferrauto, C. Furlan, S. Lanzardo, R. Nuzzi, E. Gianolio, S. Aime, *Invest. Radiol.* **2018**, *53*, 167–172.
3. S. Le Goff, J. A. Barrat, L. Chauvaud, Y. M. Paulet, B. Gueguen, D. Ben Salem, *Sci. Rep.* **2019**, *9*, 8015.
4. K. Inoue, M. Fukushi, A. Furukawa, S. K. Sahoo, N. Veerasamy, K. Ichimura, S. Kasahara, M. Ichihara, M. Tsukada, M. Torii, M. Mizoguchi, Y. Taguchi, S. Nakazawa, *Mar. Pollut. Bull.* **2020**, *154*, article number 111148.
5. M. Botta, F. Carniato, D. Esteban-Gomez, C. Platas-Iglesias, L. Tei, *Future Med. Chem.* **2019**, *11*, 1461–1483.
6. A. Gupta, P. Caravan, W. S. Price, C. Platas-Iglesias, E. M. Gale, *Inorg. Chem.* **2020**, *59*, 6648–6678.
7. B. Drahos, I. Lukes, E. Toth, *Eur. J. Inorg. Chem.* **2012**, 1975–1986.
8. S. M. Rocklage, W. P. Cacheris, S. C. Quay, F. E. Hahn, K. N. Raymond, *Inorg. Chem.* **1989**, *28*, 477–485.
9. L. Josephson, M. Rudin, *J. Nucl. Med.* **2013**, *54*, 329–332.
10. M. Aschner, K. M. Erikson, D. C. Dorman, *Crit. Rev. Toxicol.* **2005**, *35*, 1–32.
11. H. Irving, R. J. P. Williams, *Nature* **1948**, *162*, 746–747.
12. B. Drahos, J. Kotek, P. Hermann, I. Lukes, E. Toth, *Inorg. Chem.* **2010**, *49*, 3224–3238.
13. B. Drahos, V. Kubicek, C. S. Bonnet, P. Hermann, I. Lukes, E. Toth, *Dalton Trans.* **2011**, *40*, 1945–1951.
14. E. M. Gale, I. P. Atanasova, F. Blasi, I. Ay, P. Caravan, *J. Am. Chem. Soc.* **2015**, *137*, 15548–15557.
15. https://clinicaltrials.gov/ct2/show/NCT05413668.
16. https://www.gehealthcare.com/about/newsroom/press-releases/ge-healthcare-announces-completion-of-phase-i-subject-recruitment-in-early-clinical-development-program-for-a-first-of-its-kind-macrocyclic-manganese-based-mri-contrast-agent.
17. S. H. Koenig, R. D. Brown III, M. Spiller, *Magn. Reson. Med.* **1987**, *4*, 252–260.
18. S. M. Pinto, V. Tomé, M. J. F. Calvete, M. M. C. A. Castro, É. Tóth, C. F. G. C. Geraldes, *Coord. Chem. Rev.* **2019**, *390*, 1–31.
19. F. Wan, L. Wu, X. Chen, Y. Zhang, L. Jiang, *Polyhedron* **2023**, *242*, 116489.
20. S. Daksh, A. Kaul, S. Deep, A. Datta, *J. Inorg. Biochem.* **2022**, *237*, 112018.
21. C. Henoumont, M. Devreux, S. Laurent, *Molecules* **2023**, *28*, 7275.
22. P. Caravan, *Invest. Radiol.* **2024**, *59*, 187–196.
23. S. Lacerda, D. Ndiaye, É. Tóth, in *Metal Ions in Bio-Imaging Techniques, Vol. 22*, Eds.: A. Sigel, E. Freisinger, R. K. O. Sigel,, De Gruyter, Berlin, Boston, MA, 2021, pp. 71–100.
24. S. Lacerda, D. Ndiaye, É. Tóth, in *Advances in Inorganic Chemistry, Vol. 78*, Eds.: C. D. Hubbard, R. van Eldik, Academic Press, N.Y., 2021, pp. 109–142.
25. F. K. Kalman, G. Tircso, *Inorg. Chem.* **2012**, *51*, 10065–10067.
26. D. Lalli, G. Ferrauto, E. Terreno, F. Carniato, M. Botta, *J. Mat. Chem. B* **2021**, *9*, 8994–9004.
27. R. Mallik, M. Saha, C. Mukherjee, *ACS Appl. BioMat.* **2021**, *4*, 8356–8367.
28. R. Mallik, M. Saha, V. Singh, H. Mohan, S. S. Kumaran, C. Mukherjee, *J. Mater. Chem. B* **2023**, *11*, 8251–8261.
29. H. Wu, Z. Li, Y. Liu, X. Shi, Y. Xue, Z. Zeng, F. Mi, H. Wang, J. Zhu, *Mater. Adv.* **2023**, *4*, 6682–6693.
30. S. Anbu, S. H. L. Hoffmann, F. Carniato, L. Kenning, T. W. Price, T. J. Prior, M. Botta, A. F. Martins, G. J. Stasiuk, *Angew. Chem. Int. Ed.* **2021**, *60*, 10736–10744.

31. Z. Feng, T. Zhu, L. Wang, T. Yuan, Y. Jiang, X. Tian, Y. Tian, Q. Zhang, *Inorg. Chem.* **2022**, *61*, 12652–12661.
32. B. Drahos, J. Kotek, I. Cisarova, P. Hermann, L. Helm, I. Lukes, E. Toth, *Inorg. Chem.* **2011**, *50*, 12785–12801.
33. R. Botár, E. Molnár, G. Trencsényi, J. Kiss, F. K. Kálmán, G. Tircsó, *J. Am. Chem. Soc.* **2020**, *142*, 1662–1666.
34. Z. Garda, E. Molnár, N. Hamon, J. L. Barriada, D. Esteban-Gómez, B. Váradi, V. Nagy, K. Pota, F. K. Kálmán, I. Tóth, N. Lihi, C. Platas-Iglesias, É. Tóth, R. Tripier, G. Tircsó, *Inorg. Chem.* **2021**, *60*, 1133–1148.
35. M. Devreux, C. Henoumont, F. Dioury, S. Boutry, O. Vacher, L. V. Elst, M. Port, R. N. Muller, O. Sandre, S. Laurent, *Inorg. Chem.* **2021**, *60*, 3604–3619.
36. T. Csupász, D. Szücs, F. K. Kálmán, O. Hollóczki, A. Fekete, D. Szikra, É. Tóth, I. Tóth, G. Tircsó, *Molecules* **2022**, *27*, 371.
37. F. K. Kálmán, V. Nagy, R. Uzal-Varela, P. Pérez-Lourido, D. Esteban-Gómez, Z. Garda, K. Pota, R. Mezei, A. Pallier, É. Tóth, C. Platas-Iglesias, G. Tircsó, *Molecules* **2021**, *26*, 1524.
38. L. Leone, A. Anemone, A. Carella, E. Botto, D. L. Longo, L. Tei, *ChemMedChem* **2022**, *17*, e202200508.
39. R. Uzal-Varela, L. Valencia, D. Lalli, M. Maneiro, D. Esteban-Gomez, C. Platas-Iglesias, M. Botta, A. Rodriguez-Rodriguez, *Inorg. Chem.* **2021**, *60*, 15055–15068.
40. K. Pota, E. Molnár, F. K. Kálmán, D. M. Freire, G. Tircsó, K. N. Green, *Inorg. Chem.* **2020**, *59*, 11366–11376.
41. A. M. Nonat, A. Roux, M. Sy, L. J. Charbonniere, *Dalton Trans.* **2019**, *48*, 16476–16492.
42. P. Comba, M. Kerscher, W. Schiek, in *Progress in Inorganic Chemistry, Vol. 55*, Ed.: K. D. Karlin, John Wiley & Sons, Inc., 2007, pp. 613–704.
43. P. Comba, M. Kerscher, K. Rück, M. Starke, *Dalton Trans.* **2018**, *47*, 9202–9220.
44. D. Ndiaye, M. Sy, A. Pallier, S. Même, I. de Silva, S. Lacerda, A. M. Nonat, L. J. Charbonnière, É. Tóth, *Angew. Chem. Int. Ed.* **2020**, *59*, 11958–11963.
45. M. Sy, D. Ndiaye, I. da Silva, S. Lacerda, L. J. Charbonniere, E. Toth, A. M. Nonat, *Inorg. Chem.* **2022**, *61*, 13421–13432.
46. D. Ndiaye, M. Sy, W. Thor, L. J. Charbonniere, A. M. Nonat, E. Toth, *Chem. Eur. J.* **2023**, *29*, e202301880 (1 of 13).
47. P. Cieslik, P. Comba, B. Dittmar, D. Ndiaye, É. Tóth, G. Velmurugan, H. Wadepohl, *Angew. Chem. Int. Ed.* **2022**, *61*, e202115580.
48. D. Ndiaye, P. Cieslik, H. Wadepohl, A. Pallier, S. Même, P. Comba, É. Tóth, *J. Am. Chem. Soc.* **2022**, *144*, 22212–22220.
49. I. Bertini, C. Luchinat, G. Parigi, E. Ravera, *NMR of Paramagnetic Molecules-Applications to Metallobiomolecules and Models*, Elsevier, Amsterdam, 2017.
50. C. F. G. C. Geraldes, M. M. C. A. Castro, J. A. Peters, *Coord. Chem. Rev.* 2021, *445*, 214069.
51. P. Fathi, D. Pan, *Nanomedicine* **2020**, *15*, 2493–2515.
52. H. Liu, W. Cheng, S. Dong, D. F. Xu, K. Tang, X.-a. Zhang, *Pharmaceuticals* **2020**, *13*, 282.
53. H. S. Lu, M. Y. Wang, F. P. Ying, Y. Y. Lv, *Bioorg. Med. Chem.* **2021**, *35*, 116090.
54. S. Lacerda, *Inorganics* **2018**, *6*, 129.
55. C. S. Bonnet, É. Tóth, *Curr. Opin. Chem. Biol.* **2021**, *61*, 154–169.
56. R. Uzal-Varela, A. Rodríguez-Rodríguez, M. Martínez-Calvo, F. Carniato, D. Lalli, D. Esteban-Gómez, I. Brandariz, P. Pérez-Lourido, M. Botta, C. Platas-Iglesias, *Inorg. Chem.* **2020**, *59*, 14306–14317.

57. X. Shen, Y. Pan, G. Liang, *Eur. J. Inorg. Chem.* **2023**, *26*, e202200786.
58. S. Chirayil, V. C. Jordan, A. F. Martins, N. Paranawithana, S. J. Ratnakar, A. D. Sherry, *Inorg. Chem.* **2021**, *60*, 2168–2177.
59. R. Botár, E. Molnár, Z. Garda, E. Madarasi, G. Trencsényi, J. Kiss, F. K. Kálmán, G. Tircsó, *Inorg. Chem. Front.* **2022**, *9*, 577–583.
60. S. Karbalaei, E. Knecht, A. Franke, A. Zahl, A. C. Saunders, P. R. Pokkuluri, R. J. Beyers, I. Ivanovic-Burmazovic, C. R. Goldsmith, *Inorg. Chem.* **2021**, *60*, 8368–8379.
61. M. Yu, M. B. Ward, A. Franke, S. L. Ambrose, Z. L. Whaley, T. M. Bradford, J. D. Gorden, R. J. Beyers, R. C. Cattley, I. Ivanović-Burmazović, D. D. Schwartz, C. R. Goldsmith, *Inorg. Chem.* **2017**, *56*, 2812–2826.
62. Y. Li, Q. Xia, C. Zhu, W. Cao, Z. Xia, X. Liu, B. Xiao, K. Chen, Y. Liu, L. Zhong, B. Tan, J. Lei, J. Zhu, *J. Inorg. Biochem.* **2022**, *236*, 111979.
63. M. Zalewski, D. Janasik, A. Wierzbicka, T. Krawczyk, Inorg. Chem. **2022**, *61*, 19524–19542.
64. E. Hequet, C. Henoumont, R. N. Muller, S. Laurent, *Future Med. Chem.* **2019**, *11*, 1157–1175.
65. K. L. Peterson, K. Srivastava, V. C. Pierre, *Front. Chem.* **2018**, *6*, 160.
66. H. Chen, X. Tang, X. Gong, D. Chen, A. Li, C. Sun, H. Lin, J. Gao, *Chem. Comm.* **2020**, *56*, 4106–4109.
67. S. M. A. Pinto, A. R. R. Ferreira, D. S. S. Teixeira, S. C. C. Nunes, A. L. M. Batista de Carvalho, J. M. S. Almeida, Z. Garda, A. Pallier, A. A. C. C. Pais, C. M. A. Brett, É. Tóth, M. P. M. Marques, M. M. Pereira, C. F. G. C. Geraldes, *Chem. Eur. J.* **2023**, *29*, e202301442.
68. Y. Ning, I. Y. Zhou, N. J. Rotile, P. Pantazopoulos, H. Wang, S. C. Barrett, M. Sojoodi, K. K. Tanabe, P. Caravan, *J. Am. Chem. Soc.* **2022**, *144*, 16553–16558.
69. N. N. Nyström, H. Liu, F. M. Martinez, X.-a. Zhang, T. J. Scholl, J. A. Ronald, *J. Med. Chem.* **2022**, *65*, 9846–9857.
70. J. Wang, H. Wang, I. A. Ramsay, D. J. Erstad, B. C. Fuchs, K. K. Tanabe, P. Caravan, E. M. Gale, *J. Med. Chem.* **2018**, *61*, 8811–8824.
71. S. W. McRae, M. Cleary, D. DeRoche, F. M. Martinez, Y. Xia, P. Caravan, E. M. Gale, J. A. Ronald, T. J. Scholl, *J. Med. Chem.* **2023**, *66*, 6567–6576.
72. K. Chen, P. Li, C. Zhu, Z. Xia, Q. Xia, L. Zhong, B. Xiao, T. Cheng, C. Wu, C. Shen, X. Zhang, J. Zhu, *J. Med. Chem.* **2021**, *64*, 9182–9192.
73. M. K. Islam, A.-R. Baek, B.-W. Yang, S. Kim, D. W. Hwang, S.-W. Nam, G.-H. Lee, Y. Chang, *Pharmaceuticals* **2023**, *16*, 602.
74. K. D. W. Vollett, D. A. Szulc, H.-L. M. Cheng, *Int. J. Mol. Sci.* **2023**, *24*, 9532.

# 6 Other First-Row Transition Metal-Based Complexes as MRI Contrast Agents

*Janet R. Morrow and Jaclyn J. Raymond*
Department of Chemistry, Natural Sciences Complex, University at Buffalo, the State University of New York, Amherst, NY 14260, USA
jmorrow@buffalo.edu, jaclynra@buffalo.edu

**CONTENTS**



DOI: 10.1201/9781003374688-6

**ABSTRACT**

First-row transition metal ions form paramagnetic complexes that have applications as magnetic resonance imaging (MRI) probes. Relaxivity probes ($T_1$ agents) that may be alternatives to Gd(III) agents are discussed with a focus on high-spin Fe(III) complexes. Challenges include control of spin, oxidation state, and solution chemistry including inner-sphere and second-sphere interactions with water protons. Strategies to increase relaxivity of the Fe(III) agents through the development of multimeric complexes are presented. Another class of probe that is discussed is paramagnetic chemical exchange saturation transfer (paraCEST) agents. TM paraCEST agents have diverse coordination chemistry, as represented by high-spin Co(II), Fe(II), and Ni(II) complexes with amide, hydroxypropyl, amine, or heterocyclic groups with exchangeable NH or OH protons. ParaCEST agents are best suited for development as responsive agents to register pH, temperature, and biological redox status. An important challenge is to develop these agents for *in vivo* MRI.

## KEYWORDS

MRI Contrast Agents; Iron Coordination Complexes; Cobalt Coordination Complexes; Shift Agents; Chemical Exchange Saturation Transfer

## 1 INTRODUCTION

Since the 1980s, clinical magnetic resonance imaging (MRI) methods have incorporated paramagnetic metal complexes as contrast agents (CAs) [1]. Coordination complexes of Gd(III) are the most commonly used CAs [2]. These Gd(III) complexes give enhanced contrast as $T_1$ probes by causing a shortening of the water proton relaxivity, as described in earlier chapters. Recent interest in alternatives to Gd(III)-based MRI CAs has piqued interest in Mn(II) and Fe(III) complexes, which also effectively relax water protons [3–6]. Mn(II) and Fe(III) have symmetrically half-filled valence subshells ($3d^5$) and relatively slow electronic relaxation times, making them suitable for development as $T_1$ MRI CAs. However, there are hurdles to the development of transition metal (TM) contrast agents. For Fe(III) probes, a focus of this chapter, water interactions including control of aqueous chemistry, spin, and oxidation state are major challenges [4]. The first part of this chapter will center on Fe(III) complexes and the chemistry relevant to MRI probe development that is unique to this highly Lewis acidic metal ion. There are limited studies on other first-row TM ion complexes as relaxivity agents that will be briefly mentioned, including Cu(II), Ni(II), Co(II), and V(IV) as given in Table 1. Note that Mn(II) probes are discussed in an earlier chapter.

The second part will focus on TM ions complexes under development as paramagnetic chemical exchange saturation transfer (paraCEST) agents. These metal ions have electronic relaxation times that are shorter by several orders of magnitude compared to Mn(II) or Fe(III), and correspondingly exhibit proton resonances which are shifted and relatively sharp. The paramagnetic metal center of a paraCEST agent should produce paramagnetically-shifted ligand proton

**TABLE 1**
**Common Oxidation and Spin States of Transition Metal-Based MRI Probes and Their Most Common Application**

| | | | |
|---|---|---|---|
| | **V(IV)** ($T_1$ agent) | | |
| | **Mn(III) high-spin** ($T_1$ agent) | **Mn(II) high-spin** ($T_1$ agent) | |
| **Fe(II) high-spin** (paraCEST) | **Fe(III) high-spin** ($T_1$ agent) | **Fe(III) low-spin** (paraCEST) | **Fe(II)-Fe(III) magnetically coupled** (paraCEST) |
| | **Co(II) high-spin** (paraCEST) | **Co(II)-Co(II) high-spin** (paraCEST) | |
| | **Ni(II) high-spin** (paraCEST) | | |
| | **Cu(II)** ($T_1$ agent) | **Cu(II)-Cu(II) magnetically coupled** (paraCEST) | |

resonances without a broadening effect and are known as "shift agents". The variation in the chemical shift of the proton or other nucleus of interest in the paramagnetic complex compared to its diamagnetic analog is known as the hyperfine shift. For TM paraCEST agents, ligand protons are shifted far from the bulk water with shifts that are often greater than 100 ppm. Highly hyperfine-shifted NH or OH protons on ligands that exchange with water are key to CEST imaging applications and are described further below.

The two types of imaging agents, $T_1$ and paraCEST agents, will be discussed here for several first-row TM ions. Importantly, first-row TM complexes commonly have several accessible oxidation and spin states, and the development of TM MRI probes requires control over these different oxidation and spin states. These metal ion spin and oxidation states are shown in Table 1 along with their most common application.

## 2 TRANSITION METAL RELAXIVITY AGENTS

### 2.1 Mechanisms of Proton Relaxation by Paramagnetic Metal Ions

A growing interest in alternatives to Gd(III) MRI probes has stimulated research into the development of $T_1$ MRI probes based on TM ion complexes [7]. This section will focus on Fe(III) MRI probes as one promising alternative. There are several recent reviews that summarize challenges in coordination chemistry, as

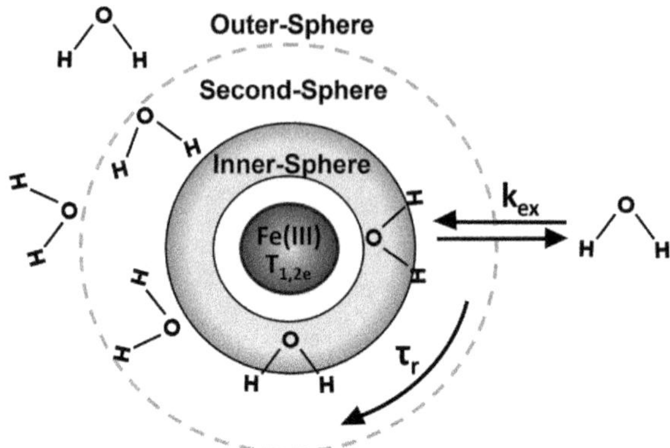


**FIGURE 1** Important parameters for water proton relaxivity in high-spin Fe(III) complexes.

well as challenges in understanding the theoretical basis for Fe(III) probes and mechanisms of proton relaxation [4,8,9]. These recent efforts pick up the thread of research begun several decades ago on the relaxivity of iron salts and polyaminocarboxylate complexes as introduced by Bertini and Koenig [10,11]. An introduction to the theory of paramagnetic complexes and their effect on proton relaxation is given in earlier chapters, so only a few basic principles will be summarized here.

Paramagnetic complexes promote proton relaxation by interaction with inner-sphere, second-sphere, and outer-sphere water (Equation 1 and Figure 1) [12,13]. The longitudinal relaxivity, $r_1$, is the sum of all three contributions. The contribution that involves inner-sphere water exchange is thought to be the most important, and optimizing the rate of water exchange in paramagnetic complexes has been studied. As shown in Equation 2, the inner-sphere water contribution involves the number of bound water molecules ($q$), their lifetime ($\tau_m$), and the relaxation time of the bound water ($T_{1m}$). Transverse relaxation ($r_2$) by paramagnetic complexes also involves inner-sphere, second-sphere, and outer-sphere water contributions. For $T_1$ MRI contrast agents containing Gd(III), $r_2/r_1$ ratios are close to unity [1]. The $r_2/r_1$ ratio may be a little higher for Mn(II) or Fe(III), but it is generally less than two or three. Given the focus on $T_1$ probes in this chapter, equations describing transverse relaxivity ($r_2$) will not be discussed further here.

$$r_1 = r_1^{\text{IS}} + r_1^{\text{SS}} + r_1^{\text{OS}} \tag{1}$$

$$r_1^{\text{IS}} = \frac{q/[\text{H}_2\text{O}]}{T_{1\text{m}} + \tau_\text{m}} \tag{2}$$

Longitudinal relaxation by paramagnetic complexes is dominated by dipolar interactions and is described by Solomon–Bloembergen–Morgan (SBM) theory [2]. As shown in Equation 3, $1/T_{1m}$ is proportional to the number of unpaired electrons as expressed by electronic spin state ($S$), showing that larger values should lead to higher relaxivity. Thus, Fe(III) with an electronic spin state with $S=5/2$ is at a disadvantage compared to Gd(III) with an $S=7/2$. However, the metal proton distance ($r_{MH}$) is substantially shorter for Fe(III) due to the smaller size of this metal ion, which is an advantage given that $T_{1m} \propto r_{MH}^6$. The expression also contains the Bohr magneton ($\mu_0$), the gyromagnetic ratio of proton ($\gamma_H$), the electron $g$ factor ($g_e$), the proton Larmor frequency ($\omega_H$), and the correlation time ($\tau_c$) as the time constant for the fluctuating magnetic dipole that mediates proton relaxation.

The fastest process will be the dominating factor in the correlation time, whether it is electron spin relaxation ($T_{1e}$), the lifetime of bound water ($\tau_m$), or the rotational motion of the complex ($\tau_R$). However, for a low-molecular-weight complex at field strengths used in biomedical scanners, the lifetime of an inner-sphere water may be longer than the rotational correlation time, making rotational movement the limiting factor under these conditions. As for second-sphere water lifetimes, these are typically much shorter than inner-sphere water exchange times [14]. Second-sphere water relaxation models have been proposed, but are limited by the difficulty of estimating the number of second-sphere waters, their strength of interaction, and rate of exchange [15]. Finally, electron spin relaxation of Fe(III) may be limiting at low field strengths as it is more rapid than Gd(III) or Mn(II), but is less of a limiting factor at high field strengths (>3 T) [8,9].

$$\frac{1}{T_{1m}} = \frac{2}{15}\left(\frac{\mu_0}{4\pi}\right)\frac{\gamma_H^2 g_e^2 \mu_B^2 S(S+1)}{r_{MH}^6}\left[\frac{3\tau_c}{1+\omega_H^2\tau_c^2}\right] \tag{3}$$

$$\frac{1}{\tau_c} = \frac{1}{\tau_R} + \frac{1}{T_{1e}} + \frac{1}{\tau_m} \tag{4}$$

## 2.2 Considerations Specific to Fe(III) MRI Probes

There are several additional considerations for the development of Fe(III) complexes as MRI probes [4,9]. First, similar to other TM complexes such as Mn(II), the oxidation and spin states of the metal must be controlled through coordination chemistry. Second, a unique property of Fe(III) complexes is the very high Lewis acidity of the metal center. The high Lewis acidity results in slow water exchange rates for six-coordinate Fe(III) complexes [16,17] and results in the deprotonation of the water ligands to give hydroxide or bridging oxide ligands in both six- and seven-coordinate complexes [9,18]. Finally, in common with other metal complexes used as MRI contrast agents, the thermodynamic stability and kinetic inertness of Fe(III) probes under biological conditions is an important consideration.

Fe(II) and Fe(III) are the two most common oxidation states for iron coordination complexes under physiological conditions [19]. The biological environment tends to be reducing, especially in the intracellular space by biological redox buffers such as glutathione, cysteine, and the thiols of proteins [20]. However, most MRI contrast agents are extracellular and the redox potential of the extracellular environment is neither as reducing nor as tightly controlled [21]. Nonetheless, it is generally thought that the redox potential of the Fe(III)/Fe(II) center should be negative *versus* NHE to prevent reduction to Fe(II). A recent analysis of electrode potentials of biologically important molecules that are present in buffering concentrations and involved in redox suggests that the Fe(II)/Fe(III) couple should be less than 0.1 V to stabilize Fe(III) or greater than 0.9 V to stabilize Fe(II) in biological environments [22]. Being outside of this window inhibits redox cycling through oxidation of Fe(II) by peroxide and then reduction by ascorbate. For example, Fe(III) complexes of EDTA derivatives have redox potentials ranging from 90 to 290 mV (NHE) [9]. Fe(EDTA) shows significant aromatic hydroxylation in the presence of ascorbate and peroxide, whereas complexes with negative electrode potentials do not produce hydroxylation [16].

The spin state of Fe(III) depends on the type of donor groups and geometry of the complex. Many high-spin Fe(III) complexes have a mix of nitrogen and oxygen donor ligands. For example, catechols are good ligands for high-spin Fe(III), as are polyaminocarboxylates, or azamacrocycles with hydroxypropyl or phenol pendants [4,5]. Coordination environments that feature all nitrogen donor groups in a six-coordinate complex may favor the low-spin form of Fe(III) [23,24].

It is important to point out that donor groups may deprotonate in aqueous solution due to the highly polarizing nature of the Fe(III) center. Of course, this is true for all metal coordination complexes, but ligand ionization at near neutral pH is especially common for Fe(III) complexes. Stabilization of the Fe(III) center increases upon deprotonation of the donor group on the ligand. In other words, deprotonation of donor groups produces more negative redox potentials for the Fe(III)/Fe(II) couple. For polyaminocarboxylate complexes of Fe(III), the redox potential shifts more negative upon deprotonation of the bound water molecule [25]. For the macrocyclic complexes with hydroxypropyl groups, the redox potential becomes more negative by 600 mV on going from pH 3 to pH 7.2 [16]. Thus, protons on ligand donor groups that may ionize when the group is bound to the highly polarizing Fe(III) center should be considered in the design of Fe(III) MRI probes.

In this regard, ionization of the inner-sphere water ligand of Fe(III) complexes is also quite common [18,25–27]. Production of a hydroxide ligand from a water ligand through deprotonation is undesirable in that hydroxide is bound strongly and is unlikely to undergo ligand exchange. Production of a $\mu$-hydroxide or $\mu$-oxo bridged dimer through combination of two Fe(III) centers is even more of a disadvantage as these dimers are often not very water soluble and feature antiferromagnetic coupling between iron centers, which results in a reduction of relaxivity [18]. More on this process will be covered as each class of Fe(III) probe is considered.

Finally, the stability and inertness of the iron complex must be considered. One type of study involves incubation of the complex in serum or with carbonate, phosphate, and chloride anions at concentrations that are prevalent in the blood [16]. More demanding assays include incubation in concentrated acid or base, transmetalation assays with Zn(II) or Cu(II), and studies with transferrin, the iron transport protein in humans [28]. In these assays, kinetic inertness to dissociation is of prime importance. For example, some of the Fe(III) macrocyclic complexes discussed here do not dissociate in strong acid or in the presence of transferrin, yet have small formation constants and low thermodynamic stability, including the Fe(III) macrocyclic complexes with three hydroxypropyl pendants [29]. However, some of the Fe(III) complexes discussed here have very high stability as well as kinetic inertness to dissociation [28,29].

## 2.3 Classes of Fe(III) Complexes

Fe(III) MRI probes can be categorized into a few major classes. Among the earliest complexes studied were based on polyaminocarboxylate ligands, including EDTA (L1) or DTPA (DTPA=diethylene triamine pentaacetate) complexes of Fe(III) (Scheme 1) [12,30]. More recently, the relaxivity of several Fe(III) complexes

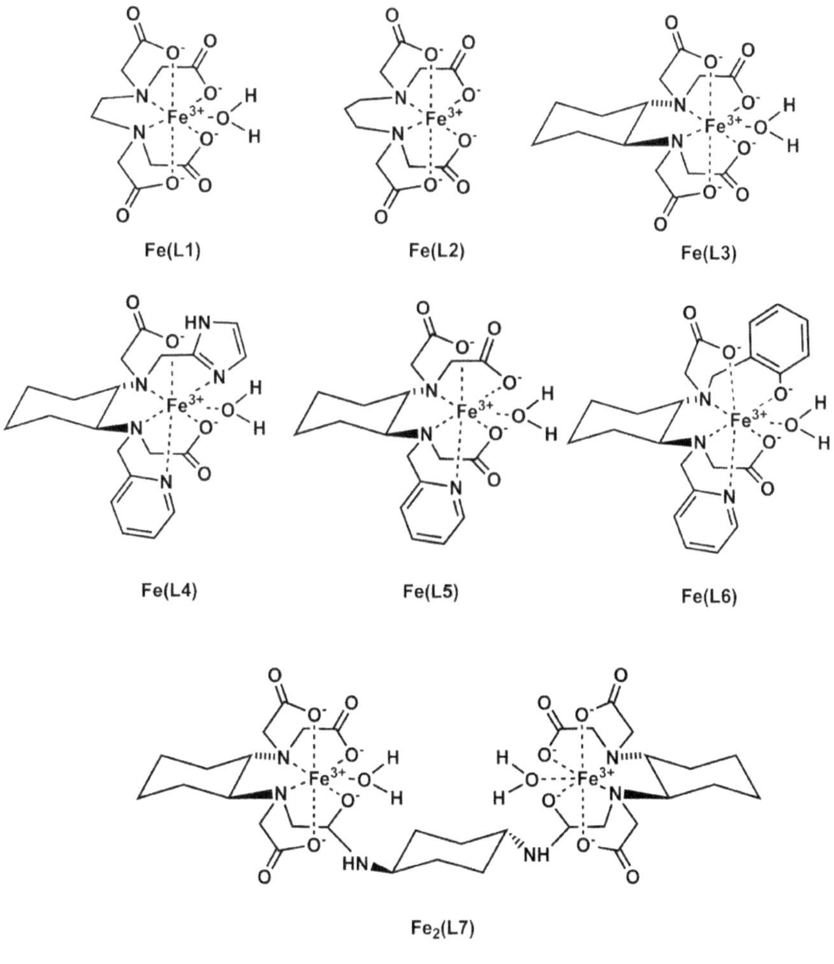

**SCHEME 1** Polyaminocarboxylate Fe(III) complexes.

of EDTA derivatives was studied for magnetic field and pH dependence [9]. These studies showed that Fe(III) complexes with an exchangeable water ligand (Fe(L3)) had higher relaxivities than those that lacked an inner-sphere water, such as Fe(L2), as shown in Table 1. However, all members of this class that had a bound water showed an ionization of the water to form hydroxide with $pK_a$ values ranging from 7.5 to 8.6 [9]. The formation of hydroxide or dimerization to form $\mu$-oxo complexes resulted in a large decrease in relaxivity at pH values approaching the $pK_a$ values. All of the Fe(III) complexes of EDTA-analogs had positive redox potentials ranging from 97 to 443 mV *versus* NHE. Notably, while the Fe(L3) analogs are robust, some of the analogs with a less rigid linker or less basic amine groups dissociated at basic pH. Another relatively recent report featured Fe(L1) and Fe(L3) as MRI probes in murine tumor models at two different magnetic field strengths. These studies showed that contrast enhancement in murine tumors with a dose of 0.50 mmol $kg^{-1}$ of Fe(L1) or 0.20 mmol $kg^{-1}$ of Fe(L3) was similar to that observed for 0.10 mmol $kg^{-1}$ Gd(DTPA). Shown in Figure 2 are dynamic contrast-enhanced (DCE) MR images of mouse tumor xenografts upon dosing with Fe(DTPA), Fe(t-CDTA) (Fe(L3)), or Gd(DTPA). Enhanced contrast is observed in $T_1$-weighted imaging procedures for all three complexes, but at different doses that correspond to their relaxivity in phantoms [31].

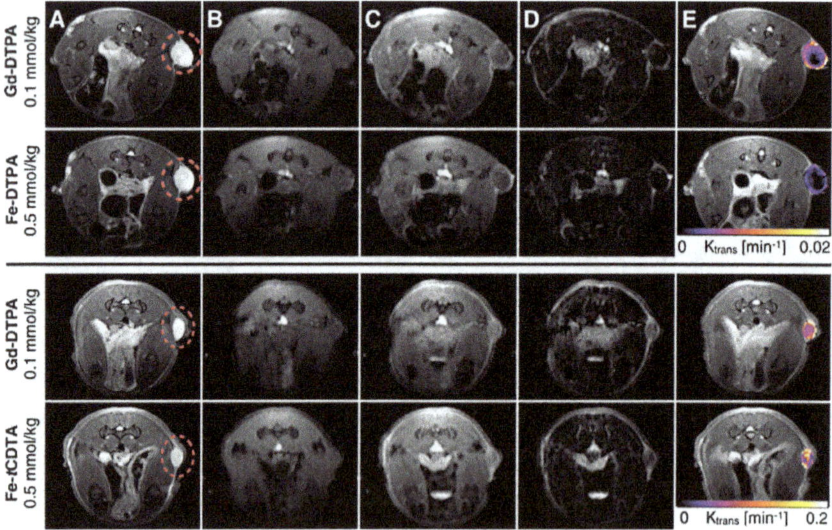


**FIGURE 2** MR images showing *in vivo* $T_1$ contrast effects and DCE MR imaging parameter mapping with Fe(DTPA) and Fe(tCDTA) (Fe(L3)), and Gd(DTPA) for comparison, at 7 T. (A) $T_2$-weighted preinjection images with tumors indicated by dashed circles; (B) $T_1$-weighted preinjection images in same position; (C) $T_1$-weighted images several minutes after injection in same position; (D) subtraction images of (B) and (C); and (E) color-coded $K^{trans}$ parametric maps of tumor overlaid on $T_2$-weighted images. (From Ref. [31]; Licensed under CC BY 4.0.)

A related class of compounds has the same *trans*-1,2-diaminocyclohexane scaffold as L3 but has pyridyl or imidazole pendants, including L4 [18], L5 [28], and L6 [26]. This type of Fe(III) complex has an exchangeable inner-sphere water. Fe(L5) is isolated in the Fe(II) form and is oxidized to Fe(III) as an inflammation-sensitive probe [28]. *In vivo* studies show that the complexes can be used to detect pancreatitis in mice models. In other reports, related Fe(III) complexes dimerize in solution as a function of pH and concentration, which has been promoted as a way to form pH-responsive and enzyme-responsive probes [18,26]. On the other hand, dinuclear Fe(III) probes that lack $\mu$-oxo bridges are under investigation to increase relaxivity. For example, the dinuclear analog $Fe_2$(L7) had $r_1$ values that are slightly higher than the mononuclear analog, Fe(L3). The lower than expected relaxivity values are attributed to the water ligands being deprotonated at neutral pH [32].

A recently studied class of probes features macrocyclic complexes of Fe(III) (Scheme 2) [4]. Macrocyclic complexes are generally more inert toward dissociation than those of linear chelates. Most macrocyclic Fe(III) $T_1$ MRI probes reported to date contain 1,4,7-triazacyclononane (TACN) with either hydroxypropyl or phosphonate pendants [16,17,29,33]. The Fe(III) macrocyclic complexes of TACN have inner-sphere water ligands for macrocycles with two pendant groups or lack an inner-sphere water for ligands with three pendants. The TACN macrocyclic framework stabilizes the small Fe(III) cation and provides additional sites for water and for pendant groups, whereas chiral hydroxypropyl groups have

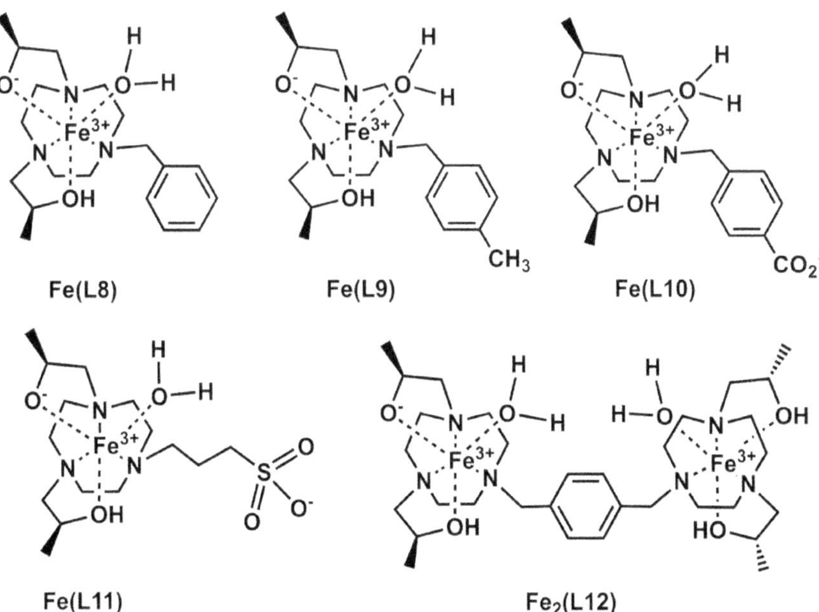


**SCHEME 2** Macrocyclic complexes of Fe(III) with inner-sphere water.

additional protons to provide relaxation through proton exchange. Moreover, the chiral hydroxypropyl pendant group tends to produce fewer isomers in a less dynamic and more rigid complex. Solid-state X-ray crystal structures for two of the complexes in this class show that six-coordination is typical [16,17].

The solution chemistry of the Fe(III) macrocyclic complexes with an inner-sphere water (Scheme 2) was studied in detail. These studies suggest that there is a single deprotonated hydroxypropyl group at pH 7.4 and that all complexes are inert toward dissociation in strongly acidic solutions and in the presence of serum and carbonate/phosphate [16,17,33]. Variable temperature $^{17}O$ NMR spectroscopy studies are consistent with water exchange being too slow to substantially contribute to the relaxivity of the complex. However, the complexes have higher relaxivity values than would be expected for complexes that lack an exchangeable water ligand. For example, the $r_1$ of Fe(L8) is similar to that of Fe(L3), which has an exchangeable inner-sphere water ligand (Table 2). To further increase relaxivity, a series of dinuclear complexes with analogous macrocyclic ligands were studied in solution and *in vivo*. $Fe_2$(L12) has an $r_1$ of $5.6\,mM^{-1}\,s^{-1}$ at 4.7 T, which is slightly more than double that of the mononuclear complex [33].

The complexes in Scheme 2 apparently do not form complexes with bridged $\mu$-hydroxide or $\mu$-oxo dinuclear centers under conditions of millimolar concentrations of complex and neutral pH. The deprotonation event at near-neutral pH values is attributed to the pendant hydroxypropyl groups based on comparison of similar complexes, although deprotonation of the bound water is another option [16,17,33,34]. Curiously, the Fe(L8) complex acts as a strong Lewis acid by binding to anionic bidentate oxygen ligands in ternary interactions [35] and to the anionic surfaces of yeast cells [36]. This property has been used to load nanoparticles formed from yeast cell walls with Fe(L8) as a $T_1$ MRI probe [35].

Another important class of Fe(III) MRI probes lack an inner-sphere water, as shown in Scheme 3. The macrocyclic complexes in this group typically have lowered relaxivity of less than $1.0\,mM^{-1}\,s^{-1}$. Exceptions to this include Fe(L13) and Fe(L14) [29] (Table 2). The higher values are attributed to strong second-sphere water interactions through hydroxy or phosphonate groups. For TACN complexes, the hydroxypropyl group produces higher relaxivity than does the phosphonate group, despite expectation that phosphonate groups [29] would be superior for increasing second-sphere water contributions as they do for Gd(III) complexes [14]. Curiously, the complex Fe(L15) [37] has a relaxivity that is less than half that of Fe(L14) under similar conditions. This lowered relaxivity is attributed to disruption of the hydrophilic coordination sphere by the benzyl groups or possibly a change in electronic relaxation from breaking of the symmetry of the complex. Adding to this class, 12- and 14-membered macrocyclic complexes of Fe(III) [38,39] have been reported, including Fe(L16) [39] (Scheme 3).

The Fe(III) bis-complex of deferasirox shows effective relaxivity for a closed coordination complex [40]. Studies in mice show that it can be used for tumor imaging and has a similar pharmacokinetic profile to Gd(DOTA). Another recent study shows that the Fe(HBEDP) complex has improved relaxivity compared to the well-known Fe(HBED) complexes that were studied decades ago [5,41]. The

**TABLE 2**
**Table of $r_1$ ($mM^{-1}s^{-1}$) Values for Selected Iron Complexes at 37°C, Along with the Values Normalized to Fe, and Whether *In Vivo* Studies Were Performed**

| Complex | $r_1$ 1.4 T ($mM^{-1}s^{-1}$) | $r_1$ 4.7 T ($mM^{-1}s^{-1}$) | $r_1$ 4.7 T with HSA ($mM^{-1}s^{-1}$) | $r_1$* normalized to Fe at 4.7 T ($mM^{-1}s^{-1}$) | *In vivo* studies | Ref |
|---|---|---|---|---|---|---|
| **Fe(L3)** | 1.9±0.1 (0.94T) | | | 1.9 | ✓ | [31,32] |
| **Fe(L4)** | 1.8±0.1 | | | 1.8 | ✓ | [28] |
| **$Fe_2$(L7)** | 2.0±0.1 (0.94T) | | | 1.0 | | [32] |
| **Fe(L8)** | | 2.2±0.30 | 2.5±0.10 | 2.2 | ✓ | [16] |
| **Fe(L9)** | 1.5±0.1 | 1.8±0.3 | 2.7±0.2 | 1.8 | | [33] |
| **Fe(L10)** | | 1.7±0.1 | 2.2±0.1 | 1.7 | ✓ | [16] |
| **Fe(L11)** | | 2.0±0.19 | 2.5±0.6 | 2.0 | ✓ | [17] |
| **$Fe_2$(L12)** | 3.5±0.3 | 5.3±0.1 | 6.7±0.04 | 2.6 | ✓ | [33] |
| **Fe(L13)** | 1.0±0.1 | 0.86±0.03 | 1.3±0.1 | 0.86 | ✓ | [29] |
| **Fe(L14)** | 1.5±0.2 | 0.97±0.12 | 1.2±0.2 | 0.97 | ✓ | [17, 29] |
| **Fe(L15)** | 0.66 | | | 0.66 | | [37] |
| **Fe$(L17)_2$** | 1.4±0.22 (1 T) | | | 1.4 | ✓ | [40] |
| **Fe(L18)** | 1.5 | | | 1.5 | ✓ | [41] |
| **$Fe_4A_6$** | 8.3±0.3 | 8.7±0.2 | 21±0.3 | 2.2 | ✓ | [88] |

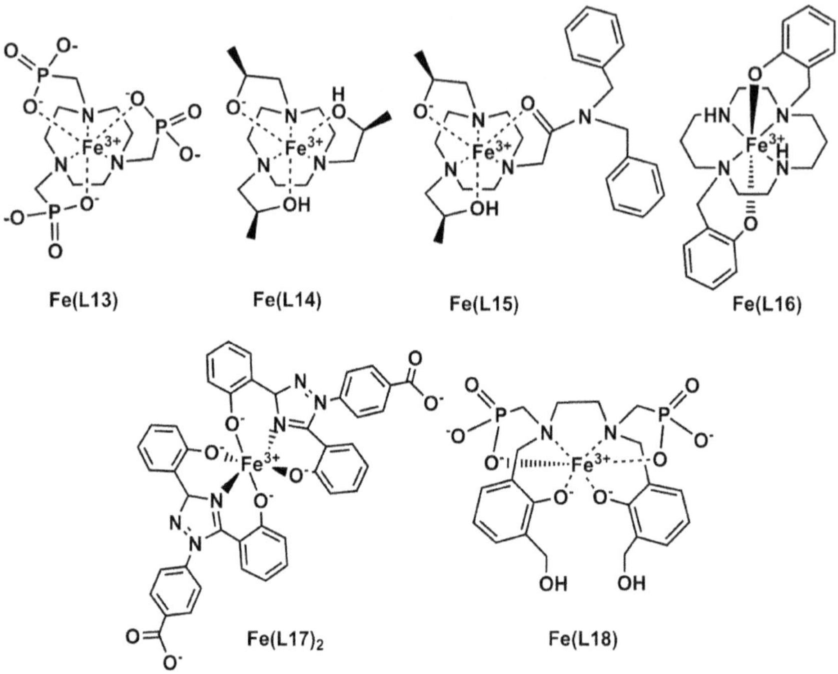

Fe(L13) Fe(L14) Fe(L15) Fe(L16)

$Fe(L17)_2$ Fe(L18)

**SCHEME 3** Fe(III) complexes lacking an inner-sphere water.

original parent complexes lacked a bound water and had relaxivities of less than $1.0\,mM^{-1}\,s^{-1}$. The more recently reported one shown in Scheme 3 has phosphonate groups and hydroxyl groups to enhance second-sphere interactions. This complex shows an improved relaxivity of $1.5\,mM^{-1}\,s^{-1}$ at 1.5 T and 37 °C and enhanced contrast in murine tumor models [41]. Other complexes studied *in vivo* that lack a bound water include ferrioxamine, which was studied in human clinical trials [42].

Several of the Fe(III) complexes of TACN have been studied in mice. The highly cationic complexes such as Fe(L8), $Fe_2$(L12), or Fe(L10) clear rather slowly through a mostly renal route and show excellent contrast enhancement of kidneys [16,17,33]. Addition of an anionic group to decrease the overall charge (Fe(L10) or Fe(L11)) and increase the hydrophilicity of the complex as shown by the log P measurements enhances the rate of clearance of the complex [16,17].

The final class of Fe(III) MRI probes to be discussed are those that contain three catechol donors per iron. There are several examples of tris-catechol complexes that have been studied in solution and in animals [44,45]. Unfortunately, these complexes are neither thermodynamically nor kinetically stable and form the bis-complex at millimolar concentrations of catechol and iron(III) [46]. There are many examples of nanoparticles of Fe(III) with catechol donor groups to form particles with good relaxivity [47]. One solution to overcome the lack of stability or inertness to dissociation is to prepare a complex that uses catechol donors and

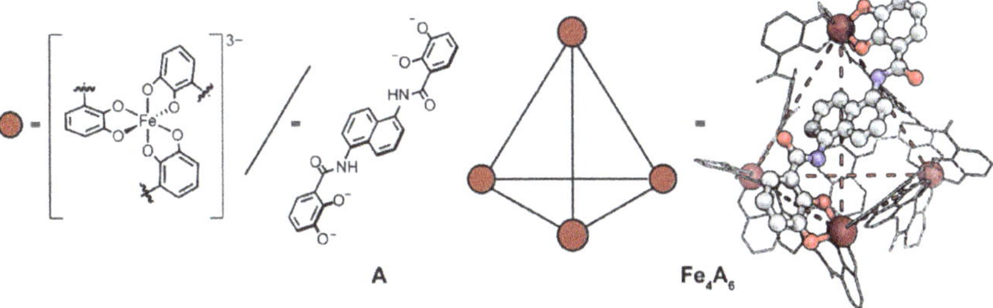


**FIGURE 3** Catechol complexes and a tetrahedral cage of Fe(III). (Reproduced by permission from Ref. [43]; copyright 2022, American Chemical Society.)

links iron centers together in a self-assembled complex [4]. This tetrahedral metal organic polyhedron (MOP) or cage, shown in Figure 3, shows good relaxivity in buffered solution and in serum or with serum albumin. Notably, the Fe(III) centers lack any bound water, although it is possible that the catechol groups interact with water through second-sphere interactions. The strong binding of this complex to serum proteins produces a blood-pool agent that clears slowly with a half-life of 2.5 h from mice. The cage also accumulates in tumors over several hours. Approaches that utilize multiple Fe(III) centers are thus promising for the development of new high-relaxivity iron probes.

## 2.4 Additional First-Row Transition Metal Probes

Many first-row TM complexes have oxidation and spin states that have unpaired electrons. Paramagnetic TM ion complexes are anticipated to enhance relaxation of water protons, although not as effectively as Fe(III) or Mn(II) do with their half-filled d-shells. The electronic relaxation times also affect paramagnetic proton relaxation [48,49]. For example, as discussed in the next section, Fe(II), Ni(II), and Co(II) have electronic relaxation times that are orders of magnitude more rapid than the proton Larmor frequency and are thus good shift agents with relatively low water proton relaxivity [48,50]. These complexes will be discussed as paraCEST agents. Despite these considerations, Cu(II) [51], high-spin Fe(II) [52], and V(IV) [53] have been studied as relaxivity probes. Ni(II) and Co(II) have electronic relaxation times that vary substantially with coordination geometry and thus may have potential for developments as MRI probes [48].

There are a few examples of Cu(II) agents that have been studied for MRI probe development. One report measured water proton relaxation in a study focused on redox-activated copper probes. Cu(II) complexes showed modest relaxivity values of 0.11–0.27 $mM^{-1} s^{-1}$ at 9.4 T [51]. A more recent report featured a Cu(II) center in a designed coiled-coil protein scaffold that showed promising relaxivity [54]. A recent report on V(IV) complexes showed remarkably good $r_1$ relaxivity for a complex with a single unpaired electron, especially at low magnetic field strengths [53].

Both Co(II) and Ni(II) complexes have been reported in applications that employ relaxivity agents, often as part of oxidation or spin state switches. For example, enhanced MRI contrast was used to track the accumulation of the Co(II) probe in hypoxic area of cell spheroids [55]. Six-coordinate Ni(II) complexes with a photoactive pendant were shown to have a relaxivity value of $0.03\,mM^{-1}s^{-1}$ as part of a photo-switchable contrast agent [56]. Several examples of Fe(II) macrocyclic complexes with heterocyclic pendants were studied as pH or enzyme-activated $T_1$ MRI probes [52,57,58]. Activation involved the removal of one pendant group to give an inner-sphere water and a complex with increased relaxivity. The original Fe(II) complex was low spin and diamagnetic, but the high-spin complex that produces signal is most likely Fe(III) and not high-spin Fe(II), given the low relaxivity of most high-spin Fe(II) complexes [48,59].

## 3 TRANSITION METAL ParaCEST AGENTS

### 3.1 Basic Theory for ParaCEST Agents

The spin and oxidation states of first-row TM complexes that are typically used as paraCEST agents are shown in Table 1. ParaCEST agents have ligand proton resonances that are not only observable in $^1H$ NMR but may be relatively sharp, although proton-coupling is obscured. Resonances are typically shifted by hundreds of parts per million (ppm). An early description of shift agents posits that the protons experience an average electronic spin polarization of the rapidly relaxing metal center and this interaction produces a shift of the proton resonances [60,61]. Such paramagnetic complexes that produce large hyperfine shifts of protons or other nuclei are known as shift agents. Shift agents are used as MRI probes in two main ways. First, the hyperfine-shifted proton or other nucleus itself can be used in magnetic spectroscopy imaging. In this case, the complex itself is directly imaged by tracking the resonances of the complex, typically either by $^1H$ or $^{19}F$ imaging [62,63]. These agents are referred to as paraSHIFT agents [62]. Second, the complex may be used as a paraCEST agent. In this application, an exchangeable NH or OH proton is shifted from bulk water and is used in a saturation transfer experiment to modulate the intensity of the water signal, as described further below [64].

First-row TM ions have been used as shift agents for several decades and have been especially useful as paramagnetic probes in metalloproteins [48], as well as in paraCEST [65] and paraSHIFT [66,67] probes. Trivalent lanthanide complexes that contain unpaired *f*-electrons (apart from Gd(III)) are used as shift agents in MRI probes [62] or for applications in the elucidation of protein structure [68]. Chapter 7 in this book focuses on lanthanide paraCEST agents and Chapter 2 focuses on paraSHIFT agents of lanthanide and TM complexes, so these topics are not covered further here.

The hyperfine shift is the difference in the chemical shift of the proton (or other nucleus) in the paramagnetic metal ion complex *versus* the shift in an analogous diamagnetic complex [69,70]. The hyperfine shift is a result of the electron-nuclear dipolar interaction that gives rise to the pseudocontact shift, in addition to the through-bond or Fermi or scalar contact contribution [49]. The scalar contact shift involves excess electron spin density at the observed nucleus and involves two mechanisms: through direct-spin delocalization or spin-polarization. TM complexes have substantial covalent bond character and more substantial contact contributions to nuclei several bonds away from the metal center than do the Ln(III) complexes [69]. Pseudocontact shifts arise from the through-space interaction of the unpaired electron with the nucleus. There is an angular and distance dependence that depends on the position of the nucleus with respect to the magnetic axis [71,72]. Pseudocontact shifts are the main contributor to Ln(III) shift agents, especially for protons that are not directly bound [49].

The defining feature of paraCEST agents is ligand protons that exchange with bulk water protons, typically an NH or OH group on a ligand such as an amide (NH) or a hydroxyalkyl (OH) or a water ligand [73]. The rate of exchange of the ligand proton with water is important and must meet certain requirements to produce the CEST effect. A primary requirement is that the ligand proton exchanges sufficiently slowly to produce a distinct resonance, but not so rapidly that the resonances merge due to fast exchange on the NMR time scale. The frequency difference between the exchangeable proton resonance and bulk water proton resonance is labeled as "$\Delta\omega$" (Figure 4). The ligand to water exchange rate constant ($k_{ex}$) must be less than this difference ($k_{ex} \leq \Delta\omega$). The CEST experiment involves selective saturation of the exchangeable proton with a radiofrequency (RF) pulse. Exchange of the magnetically-saturated ligand proton with bulk water protons produces a reduction in the water proton magnetization ($M_z$), which is used to generate contrast. The resulting "Z-spectra" show the CEST effect and are created through a series of experiments, each carried out at a distinct RF of the presaturation pulse. Z-spectra feature the $M_z$ produced upon saturation transfer divided by the original water proton signal ($M_0$). As the RF presaturation pulse matches that of the exchangeable proton, the water proton resonance decreases, appearing as a dip in the plot of $M_z/M_0$. The presaturation RF pulse is

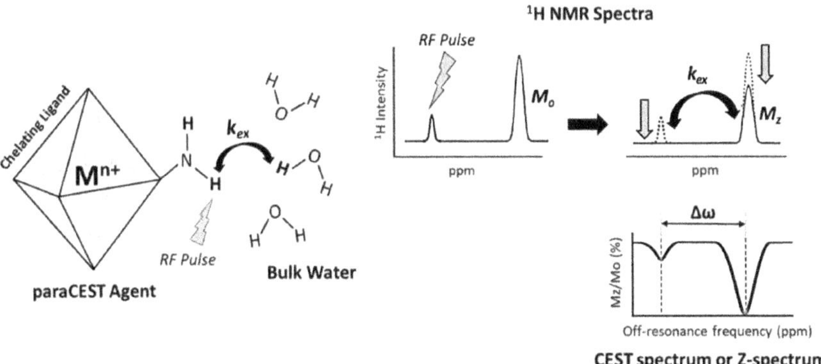


**FIGURE 4** Schematic of the CEST effect and corresponding Z-spectra.

varied to irradiate symmetrically both upfield and downfield of the bulk water $^1H$ resonances. The duration of the presaturation pulse may be varied. Commonly, the presaturation pulse is maintained for 2–4 s to ensure magnetic saturation of the proton. Another type of plot shows the magnetization transfer asymmetry ($MTR_{asym}$), which represents the difference in magnetization upon irradiation at $\Delta\omega$ and the frequency $-\Delta\omega$ (the symmetric mirror frequency relative to the bulk water frequency). The $MTR_{asym}$ removes the effect of magnetization transfer and other effects not due to CEST [74]. Under steady-state conditions, the $MTR_{asym}$ is proportional to the number of magnetically equivalent exchangeable protons ($n$) and concentration of paraCEST agent ($c$), their exchange rate constant ($k_{ex}$), and $R_1$ is the longitudinal relaxation rate constant for bulk water protons in absence of exchange (Equation 5) [73].

$$MTR_{asym} = \frac{(c \times n/111)k_{ex}}{R_1 + (c \times n/111)k_{ex}} \tag{5}$$

The choice of RF pulse power for the presaturation pulse ($B_1$) depends on the magnitude of $k_{ex}$ [75]. More rapidly exchanging protons typically show higher saturation transfer effects at higher power. The relationship between rate constant and pulse power is $k_{ex} = \omega_1 \times \gamma \times B_1$, where $B_1$ is the pulse power in Tesla and $\gamma$ is the gyromagnetic ratio. However, there are power limitations for *in vivo* studies. In humans, clinical parameters are $B_1 = 5\,\mu T$ at $t = 0.5$ s, while in animals, the pulse power may be studied at higher levels of $B_1 = 10\,\mu T$ and 5 s [73,76]. Thus, $k_{ex}$ should then be on the order of 2700 $s^{-1}$ at 10 μT in preclinical studies or about 1300 $s^{-1}$ at 5 μT for studies in humans. For more details on the experiment, see reviews of the mechanism of paraCEST agents [49,64,73,77].

## 3.2 Other Considerations for ParaCEST Agents

TM ions that are commonly used as paraCEST agents are given in Table 1. It is important to note that TM ions have several accessible oxidation states and corresponding spin states, and control over these different spin and oxidation states is a requisite for the development of paraCEST agents. While nearly any paramagnetic TM shift agent may be developed into a paraCEST agent, most are found in the first row of the series. Thus, mononuclear complexes containing high-spin Fe(II), low-spin Fe(III), high-spin Co(II), Ni(II), and dinuclear complexes that feature coupling of two metal ions to modulate the magnetic properties (Cu(II) and Fe(II)/Fe(III)) have been reported as paraCEST agents [65].

Many of the metal ions used for paraCEST agents are in the divalent state and rather labile to ligand exchange. For example, Co(II) and Fe(II) typically undergo rapid ligand exchange with monodentate ligands. Thus, all TM paraCEST agents reported to date contain either macrocyclic frameworks or multidentate linear chelates to prevent dissociation under physiological conditions [65]. The design of robust paraCEST agents is facilitated by total encapsulation of the metal ion given that an inner-sphere water is not required. Rather, the protons on the chelating ligand itself produce the CEST signal. It is, however, feasible to use an inner-sphere water to produce a CEST effect. Ln(III) paraCEST agents often have an exchangeable water [75,78], whereas only a few examples of bound water CEST are known for TM paraCEST agents [65].

Many reported paraCEST agents contain either high-spin Co(II) or Fe(II) [79]. The divalent state is stabilized by neutral oxygen or nitrogen donors, such as amines or amides or hydroxyalkyl groups. High spin *versus* low spin can be tuned by changes in the coordination sphere, such as variation in bond lengths. For example, Fe(II) or Co(II) complexes that have seven or eight coordination tend to be stabilized in the divalent state in comparison to analogous six-coordinate complexes [80,81].

## 3.3 Classes of ParaCEST Agents

ParaCEST agents can be categorized by metal ion, type of ligand including macrocyclic or linear framework, and donor groups with exchangeable protons [79]. The TM complexes most commonly used for paraCEST agents contain high-spin Co(II), Ni(II), or high-spin Fe(II) [65]. Macrocyclic ligands are commonly used to confer kinetic inertness to the complexes. Macrocyclic complexes may contain triaza, tetraaza, or incorporate mixed aza-oxa macrocyclic rings (Schemes 4 and 5). Other examples contain multidentate conformationally rigid ligands to prevent dissociation for either mononuclear or dinuclear agents (Scheme 6). The most common donor group for TM paraCEST agents are amide pendants, which give CEST peaks from NH exchange. Rate constants are in a favorable range of 600–3000 $s^{-1}$ for CEST at pH 7.4 and pulse powers of 10 μT [73,76]. Hydroxyl OH pendant groups are also common, but the CEST effect is strongest at acidic pH and exchange is typically too fast at neutral pH to observe strong CEST peaks [82]. Other classes of pendants include heterocyclic groups, such as amino-pyridines [83], imidazoles [23,84], pyrazoles [85], or triazoles [86]. These pendant groups produce CEST effects through NH exchange. However, in most cases, exchange rate constants are too rapid for the five-membered heterocycles to be practical as CEST agents.

Triazacyclononane (TACN) is among the most common macrocyclic ligand used for TM paraCEST agents as it forms kinetically inert, six-coordinate complexes with TM ions. A variety of different pendant groups have been used with TACN, including heterocyclic ones with exchangeable NH (L21, L22, L25) [23,59,83,85], as well as the amide groups NH (L19) [59,87], or hydroxypropyl pendants OH (L20) [82] (Scheme 7). Co(L23) and Co(L24) were prepared to study inner-sphere water in CEST [88], but this water exchanged too rapidly to observe a CEST effect. One drawback of the TACN macrocycle with three pendants is that the complexes tend to be dynamic in solution, giving broadened proton resonances and correspondingly broad CEST peaks. Bulky groups such as the 2-amino-pyridines in L22 [59], the chiral hydroxypropyl groups in L20 [82], and the attached fluorophore in L40 [89] (Scheme 4) tend to produce more rigid complexes that are less dynamic in solution. Another unique feature of TACN

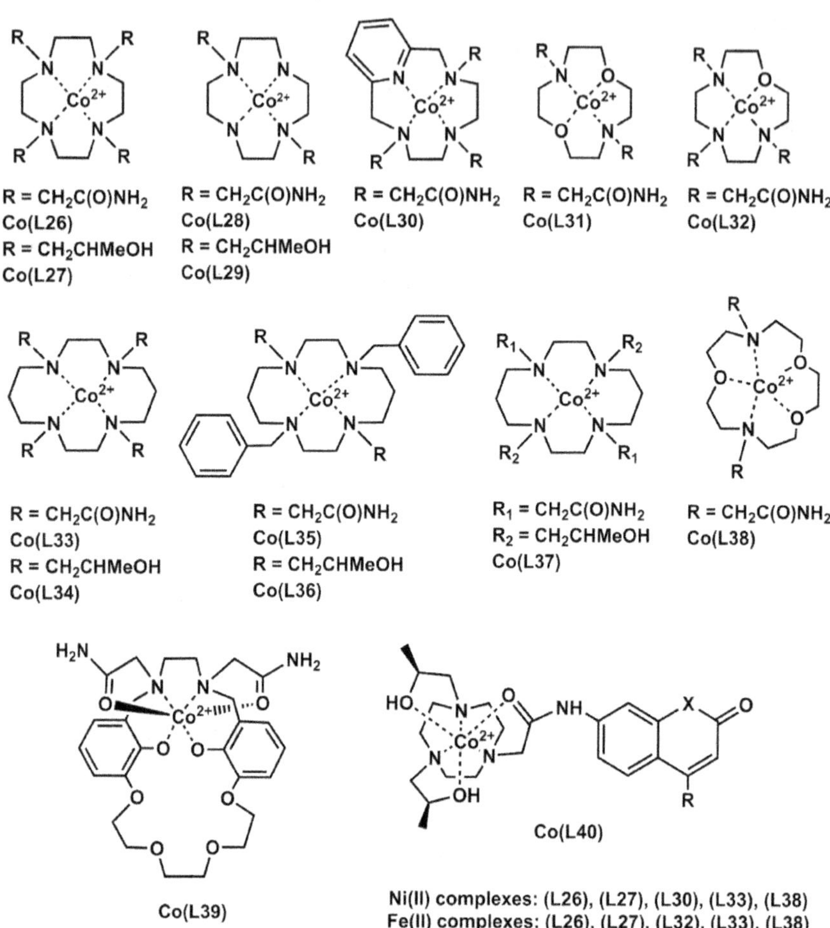


**SCHEME 4** Examples of Co(II) paraCEST agents featuring 12- or 14-membered tetraaza (CYCLEN and CYCLAM) or mixed oxa-aza macrocycles.

macrocycles is that, coupled to a heterocyclic pendant group, they stabilize the trivalent form of the metal complex (Co(III) or Fe(III)) in low-spin form [85]. Thus, Fe(L25) is a rare example of a low-spin iron-based paraCEST agent [23]. However, the small shift of the imidazole protons and their propensity to deprotonate highlights the challenges of low-spin Fe(III) paraCEST agents.

Larger macrocyclic ring sizes such as 12-membered or 14-membered tetraaza-macrocycles (CYCLEN L26, L27 and CYCLAM L33, L35, L37) or ones with mixed oxa-aza macrocycles (L31, L32, L38, and L39) are common in TM paraCEST agents (Scheme 4). CYCLEN-based macrocycles with four pendants are interesting as they form 7- or 8-coordinate complexes with Co(II) or Fe(II) [81,87,88,90]. In this coordination environment, the high-spin divalent state is stabilized with Fe(II)/Fe(III) redox potentials of approximately 1 V *versus* NHE for

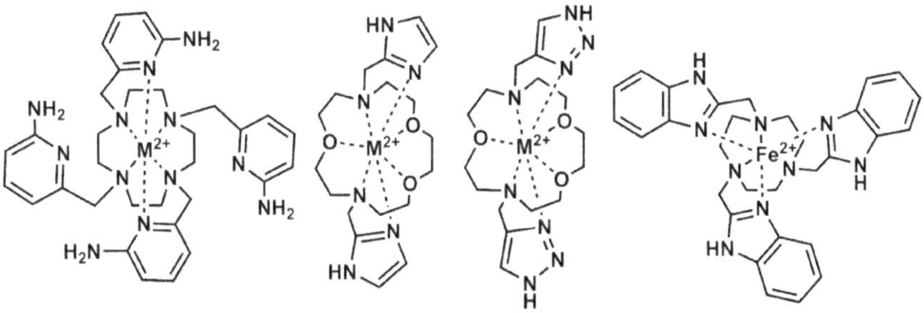

M(L41) M = Co, Fe M(L42) M = Co, Ni, Fe M(L43) M = Co, Fe Fe(L44)

**SCHEME 5** Examples of paraCEST agents containing heterocyclic pendants.

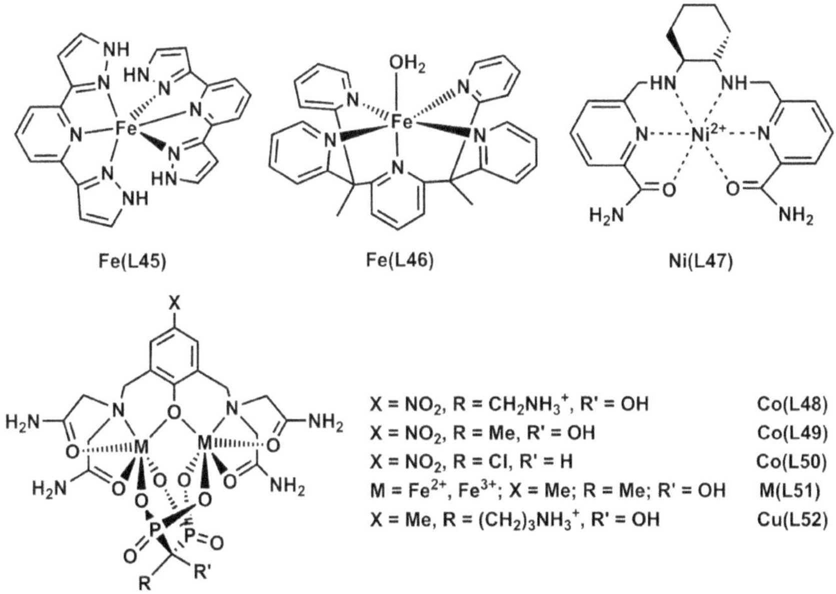

Fe(L45) Fe(L46) Ni(L47)

| | |
|---|---|
| X = $NO_2$, R = $CH_2NH_3^+$, R' = OH | Co(L48) |
| X = $NO_2$, R = Me, R' = OH | Co(L49) |
| X = $NO_2$, R = Cl, R' = H | Co(L50) |
| M = $Fe^{2+}$, $Fe^{3+}$; X = Me; R = Me; R' = OH | M(L51) |
| X = Me, R = $(CH_2)_3NH_3^+$, R' = OH | Cu(L52) |

**SCHEME 6** paraCEST agents that contain linear multidentate chelates.

Fe(L26) [80]. However, CYCLEN-based complexes with four pendants also form Co(II) or Fe(II) complexes that are dynamic, especially with amide pendants. Several examples of paraCEST agents containing Co(II) or Fe(II) complexes of amide-appended CYCLEN are reported [87,88,90–92] The 12-membered macrocycles with oxa-aza backbones are less fluxional in solution, but are also less kinetically inert to loss of metal ion [88]. The complexes of ligand L38 are relatively rigid and give sharp resonances for all three metal ions Fe(II), Co(II), and Ni(II), but complexes are not as kinetically inert as they are for all azamacrocyclic complexes [93]. The CYCLAM macrocyclic complexes are 6-coordinate and

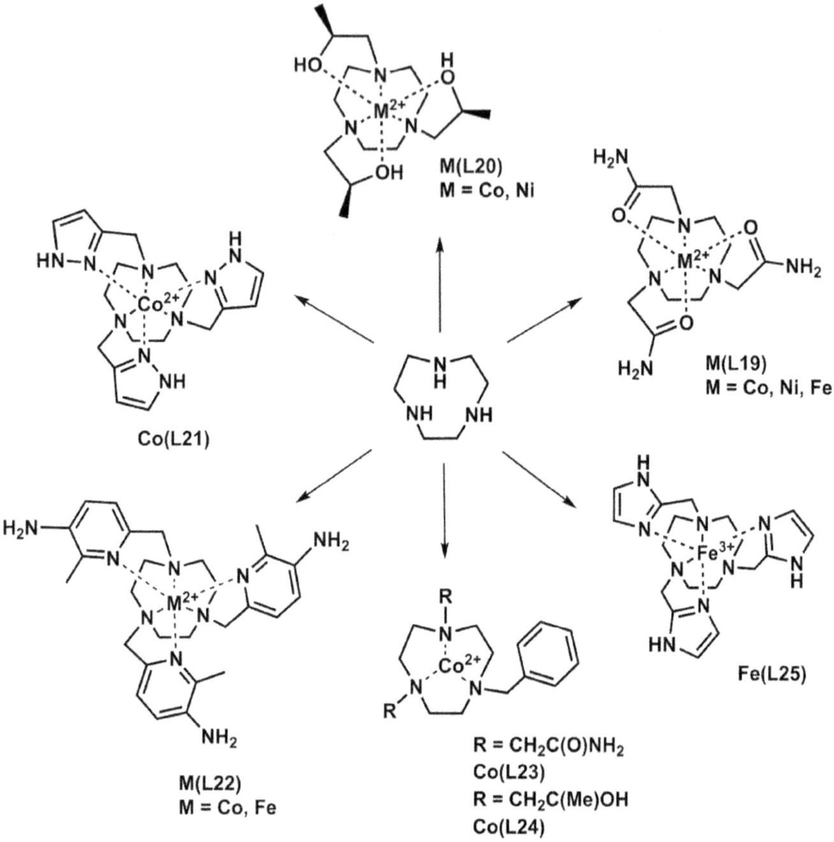

**SCHEME 7** TACN-based complexes with various pendants for paraCEST applications.

there are three different isomers that commonly form. The 1,4- and 1,8-isomers have *trans*-coordinated pendants [91,94], whereas the 1,8-*cis* conformation is favored in L35 and L36 due to the bulky benzyl groups [95,96]. These complexes show some of the most highly-shifted proton resonances, as discussed below. The Co(II) complex of L39 has a pocket for binding of Ca(II) to give a shift of the CEST peaks as a responsive agent [97].

ParaCEST agents with amide pendants are of particular interest because each amide pendant has two NH protons that are magnetically and chemically inequivalent due to slow rotation about the CN bond. Thus, there are multiple CEST peaks in TM amide complexes as shown in Figure 5 for Co(II) complexes of L26 and L35. These amide proton resonances often have very different chemical shifts. For example, Co(L37) has two NH resonances that give rise to CEST peaks at 102 and 99 ppm, and a second resonance that is very close to that of bulk water and cannot be observed as a CEST peak except at very low pulse power. This approximately 100 ppm chemical shift separation for amide protons that is observed for TM complexes is attributed largely to contact shift contributions.

For example, the distinct NH proton resonances of a macrocyclic Fe(II) amide complex were shown by theoretical calculations to arise from spin delocalization on one NH [98]. In contrast, Ln(III) macrocyclic complexes with amide pendants do not show this large difference in amide NH proton resonances, as contact shift contributions are not large over several bonds [99,100].

Examples of complexes with heterocyclic pendants with exchangeable NH protons are shown in Scheme 5. The amino-pyridine pendants produce six-coordinate Co(II) and Fe(II) complexes even with a CYCLEN backbone, presumably due to the bulky pyridine pendants. The upfield-shifted CEST peak for Fe(L41) is attributed to the NH group on the pendant [83]. Imidazole pendants deprotonate at neutral pH to stabilize the Fe(III) state, which is high-spin in Fe(L42) [84] but low-spin in Fe(L22) [24]. The Co(II) or Ni(II) complexes of L42 show a CEST peak at neutral pH, but the Fe(II) complex is not stable toward oxidation under those conditions [84]. Triazoles, however, deprotonate under acidic conditions for Co(II) and Fe(II) complexes, so they are not useful for paraCEST, at least in this scaffold [86]. Complexes with several benzimidazole pendants are not sufficiently water soluble for development as paraCEST agents [66].

Some paraCEST agents that contain linear multidentate chelates are shown in Scheme 7. The Fe(II) complex of L35 has exchangeable pyrazole NH protons for CEST [101]. It has interesting spin-cross over properties that produce a large increase in the CEST effect with increasing temperature. The Fe(II) complex Fe(L46) is one of the few examples of a paraCEST agent that has a CEST peak

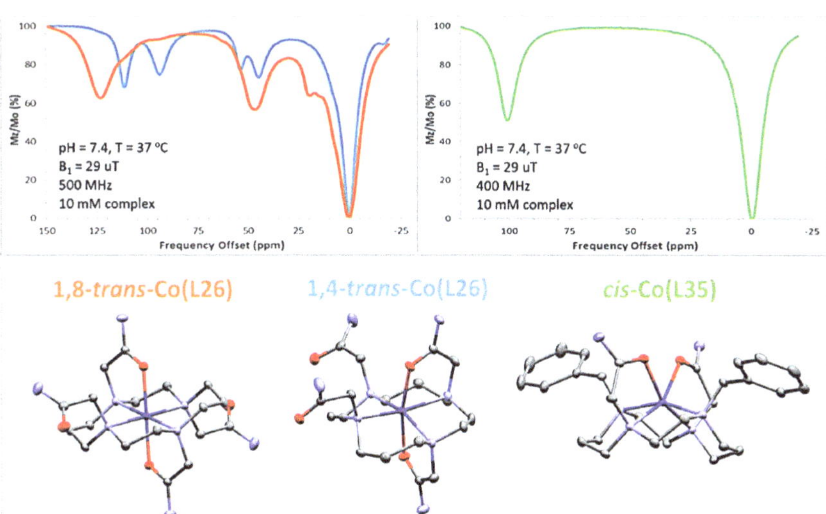


**FIGURE 5** Z-spectra for the isomers 1,4- and 1,8-Co(L33) (left) and Co(L35) (right) with corresponding crystal structures of the complex cations shown below. (Reproduced by permission from Ref. [94] and Ref. [96]; copyright 2020 and 2023, Royal Society of Chemistry.)

due to an exchangeable water ligand [101]. Ni(L47) has amide NH protons on aromatic groups that produce a CEST effect and undergo rapid NH exchange [102].

The dinuclear complexes of Co(II), Fe(II)/Fe(III), and Cu(II) in Scheme 6 show interesting properties as paraCEST agents with the NH groups of amide pendants, and in some cases OH or NH groups of the phosphonate ligand, giving rise to CEST peaks. The dicobalt complexes produce two CEST peaks from amide NH groups that can be used for ratiometric pH imaging, as detailed further below [103–105]. Moreover, different groups in the phosphonate ligands with either OH or NH protons give rise to an additional CEST signal for ratiometric pH analysis. The dinuclear copper complex is of interest as the magnetic coupling exchange between the two Cu(II) centers produces a paraCEST agent [106]. The amide protons produce a moderately-shifted CEST peak at 29 ppm. The dinuclear iron complex features two Fe(II) centers that give rise to a CEST amide NH peak. Upon oxidation to Fe(II)/Fe(III), the magnetic coupling of the two iron centers gives rise to new CEST peaks [107] as an unusual example of a redox-activated paraCEST sensor, as described further below.

## 3.4 PARACEST AGENTS RESPONSIVE TO PH, TEMPERATURE, OR REDOX

TM paraCEST agents are uniquely suited for development as responsive MRI probes. Responsive probes give distinct signals as a function of biologically important environment attributed to changes in pH, temperature, or redox status [108]. Responsive MRI probes are of interest for development in molecular imaging applications that require a probe to report on biological processes that may signify disease states [109]. The development of $T_1$ agents as responsive MRI probes is an active field of research [110–112]. However, paraCEST probes have certain advantages, such as the production of multiple peaks that register different probes states that can be used for ratiometric imaging [113]. In other cases, the CEST peak position may shift in response to environment, making it feasible to track probe response in a concentration independent way [83,89,114]. By comparison for $T_1$ MRI probes, it may be difficult to distinguish the change in signal attributed to the probe response from the difference in tissue probe concentration [108].

### 3.4.1 ParaCEST Agents as pH-Responsive Probes

CEST peak intensity is pH responsive, as proton exchange is often acid or base catalyzed. For example, amide NH groups undergo base-catalyzed proton exchange [99]. Thus, the CEST effect is larger due to faster exchange as pH increases from 6 to 8, as shown by an increase in rate constant ($k_{ex}$). At higher pH values, the rate constant may become too large, resulting in exchange broadening and a decreased CEST effect. Complexes that have more than one CEST peak can be used for ratiometric CEST imaging, as each NH proton typically has a distinct $k_{ex}$ and corresponding base-catalyzed rate constant [87]. As shown in Figure 6, the two different isomers of the Co(II) complex of L33 have multiple CEST peaks that are sufficiently far shifted to avoid the large magnetization transfer effect due to

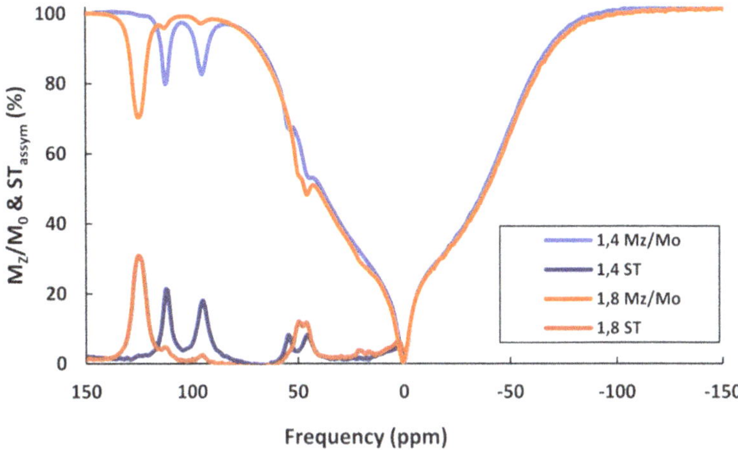


**FIGURE 6** Overlaid Z-spectra and corresponding saturation transfer spectra of the 1,4- and 1,8-Co(L33) complexes incubated in chicken thigh homogenate at 37°C, $B_1 = 12.4\,\mu T$ at 11.4 T and % CEST = (% $ST_{neg}$ – % $St_{pos}$). Samples had 14 mM Co(II) complex with a pH of 7.4. (Reproduced by permission from Ref. [94]; © 2020 Royal Society of Chemistry.)

tissue, as modeled by chicken thigh homogenate [94]. Each of the CEST peaks has a different response to pH, and a plot of two different CEST peak intensities as a function of pH gives a straight line that can be used to calibrate pH in the chicken thigh homogenate (Figure 7). In other examples, dinuclear Co(II) complexes with amide pendants have been used as pH-responsive agents. These complexes have multiple CEST peaks from amide NH and additional NH or OH peaks from the bridging phosphonate ligands [103–105].

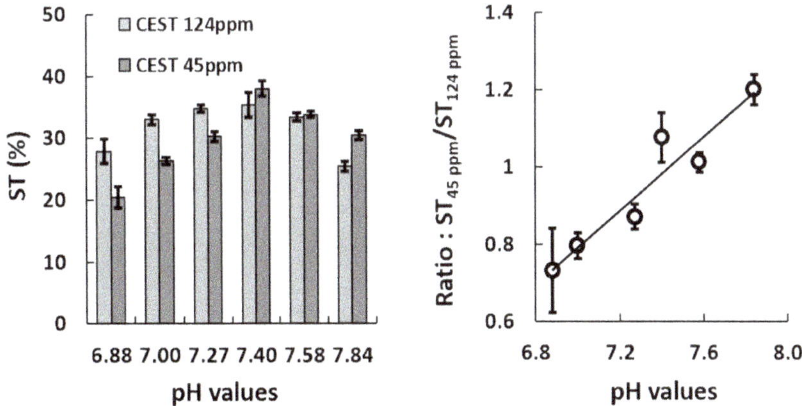


**FIGURE 7** The saturation transfer ($ST = (1 - M_z/M_o) \times 100$) is graphed at each pH (left), and the ratio of the CEST peaks (45 ppm/124 ppm) at a given pH value is plotted (right), and it fits to a line with a slope of 0.48 and $R^2 = 0.91$ at 11.7 T. (Reproduced by permission from Ref. [94]; copyright 2020, Royal Society of Chemistry.)

Another approach to pH-responsive TM paraCEST agents relies on a shift of the CEST signal in response to change in pH. For example, the amino-pyridine pendants in Fe(L41) produce a pH-dependent CEST peak shift between pH 5.2 and 7.7 [83]. Other paraCEST agents show dramatic changes in CEST peaks that are attributed to deprotonation of a ligand group with corresponding change in the coordination environment, such as those observed for Co(L40) and other Co(II) complexes with appended amide fluorophores [89].

### 3.4.2 Temperature-Responsive ParaCEST Probes

ParaCEST probes show pronounced temperature dependence arising from the effect of temperature on the hyperfine shift [115]. The shift of the exchangeable proton with temperature produces a shift in the CEST peak position, as shown by Ln(III)-based paraCEST agents that were studied as temperature probes [116,117]. A study on Co(II) paraCEST agents showed a pronounced temperature dependence of the far-shifted CEST peaks (Figure 8) with a change of –0.66 ppm/°C [91]. Improved temperature-dependent paraCEST probes such as Fe(L46) undergo spin-crossover. In this study, an increase in temperature produces a change from low-spin to high-spin Fe(II) with corresponding increase in temperature response to 1.02 ppm/°C [101].

### 3.4.3 Redox-Responsive ParaCEST Agents

The connection between disease, inflammation, and redox status has piqued interest in redox-responsive MRI probes [118–120]. For TM complexes, it is relatively straightforward to tune the redox potential of the probe through coordination chemistry changes. The responsive probe undergoes a metal oxidation state change, which induces a change in CEST effect [118,119]. For paraCEST agents, the most studied redox couples are Fe(II)/Fe(III) and Co(II)/Co(III). An early example featured Co(L21), which is an active paraCEST agent in the Co(II) state, but upon oxidation with $O_2$ produces diamagnetic and paraCEST silent Co(III) [85]. The complex has a Co(III)/Co(II) redox potential of –107 mV *versus* NHE.

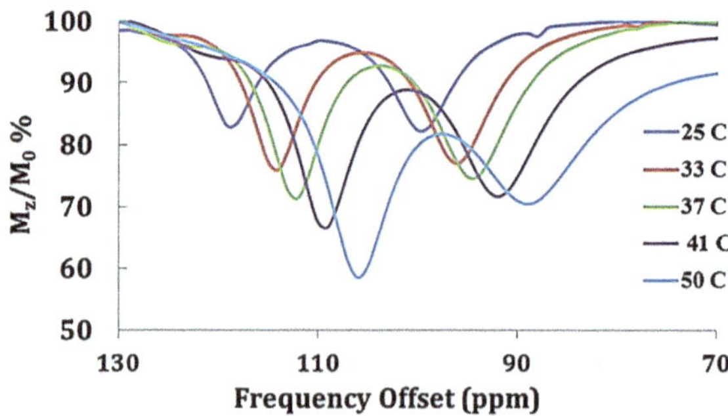


**FIGURE 8** Z-spectra showing temperature dependence of the CEST peak shift and intensity for Co(L33). (Reproduced by permission from Ref. [91]; copyright 2015, John Wiley and Sons.)

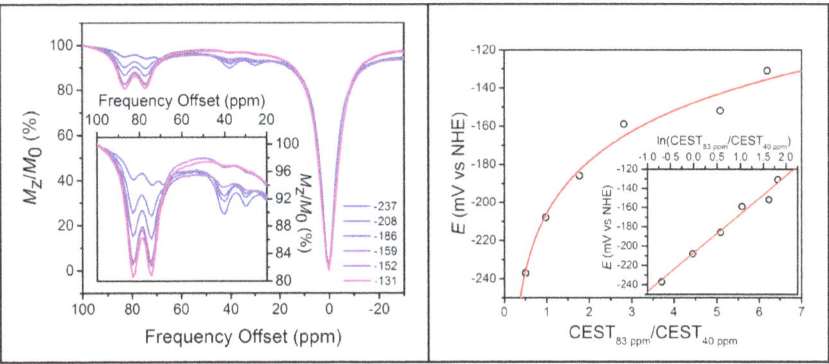


**FIGURE 9** Left: CEST spectra with a mix of diferrous (1) or mixed valence (Fe(II)/Fe(III)) $Fe_2$(L51) at 4.9 mM at 37°C and pH 7.4 with ratios of 1:2 from 9:1 (blue) to 1:9 (red). The legend contains the obtained OCP of each sample (mV *versus* NHE). Right: The OCP of aqueous $Fe_2$ samples *versus* ratios of CEST peak intensities corresponding to 83 and 40 ppm. The inset contains the semilog form of the plot. (From Ref. [107]; Licensed under CC BY 3.0 Unported.)

However, kinetic studies of the oxidation of the Co(II) complex showed that the reactivity of the complex with $O_2$ was too low to register $O_2$ changes in animal hypoxia. A half-life of 2.6 h for the Co(II) complex in arterial blood conditions or 26 h under hypoxic conditions in tumors was predicted from studies at different $O_2$ partial pressures. Another study featured a dinuclear iron complex, $Fe_2$(L51), as the first example of a ratiometric paraCEST probe to study redox environment [107]. Complexes containing $Fe(II)_2$, Fe(II)/Fe(III), and $Fe(III)_2$ were isolated and all iron centers were high spin. Significant shortening of the electronic relaxation time of Fe(III) due to magnetic coupling with Fe(II) and fast electron transfer between two Fe centers led to sharp carboxamide resonances and CEST peaks. For example, CEST peaks at 74 and 83 ppm for the Fe(II)/Fe(III) complex and at 29, 40, and 68 ppm for the Fe(II)/Fe(II) complex were found. A calibration curve from the ratio of the CEST effect at 83 over 40 ppm from Fe(II)/Fe(II) and Fe(II)/Fe(III) complexes was determined and plotted *versus* the open circuit potential (OCP). The ratio of the CEST effect at 83 and 40 ppm *versus* the OCP was used as a measure of the reducing or oxidizing nature of a solution registered by the paraCEST agent (Figure 9) [107].

## 3.5 Toward *In Vivo* Applications of ParaCEST Agents

All of the studies on TM paraCEST agents covered in this chapter were carried out in the laboratory on high-field NMR spectrometers or on MRI scanners in phantoms, but not in animals. In fact, there are only a few *in vivo* studies of Ln(III) paraCEST agents published to date [113,114,121,122] and none with TM paraCEST agents, to the best of our knowledge. A recent report on paraCEST agents suggests reasons for the lack of *in vivo* studies [78]. Reasons include the

10- to 50-fold higher concentration of paraCEST agents that are required in order to observe a signal in comparison to $T_1$ probes. Also, the extreme sensitivity of the agents to pH, temperature, and environment may make it challenging to locate the CEST signal. Moreover, the CEST experiment is more technically challenging than are proton relaxation measurements or $T_1$-weighted imaging. There are more variables such as pulse power and duration, and the long duration of time to collect Z-spectra in phantoms is not necessarily amenable to studies in mice where MRI probes are used for dynamic imaging procedures. Yet, the field of diamagnetic CEST probes is thriving [123], and it is reasonable to propose that paraCEST agents, with additional effort to optimize probes, may also be successful *in vivo*.

In this vein, a recent analysis of paraCEST agents is useful [73]. Part of this study analyzed the relationship between $k_{ex}$, pulse power, and the CEST effect in Ln(III) and TM paraCEST agents. This study reported on a comparison [76] between different paraCEST agents at pulse powers amenable to *in vivo* studies and showed that the paraCEST agents with $k_{ex}$ of 1300–2600 $s^{-1}$ were promising to produce a CEST effect at pulse powers of 5 or 10 μT. In comparison, agents with rapidly exchanging water ligands such as those of the most commonly studied Ln(III) paraCEST agents did not produce as large a signal at the lower pulse powers. These studies point toward the development of paraCEST agents with moderate $k_{ex}$ and multiple equivalent protons as the way toward agents for *in vivo* studies.

One way to increase the number of magnetically-equivalent exchangeable protons is to use nanoparticles such as liposomes, micelles, cells, or silica particles [124–127]. These topics are under study for Ln(III)-based paraCEST agents, but there are few examples with TM agents [89], so it will not be covered here. However, increasing the number of exchangeable protons is one solution to the low sensitivity of paraCEST agents.

## 4 GENERAL CONCLUSIONS

The development of TM MRI probes has experienced a renaissance over the past decade [3,4,7,8,119]. This is due in part to the desire to find alternatives to Gd(III) or other Ln(III)-based probes, but also to the desire to exploit the diverse chemistry of paramagnetic TM complexes to address applications in molecular imaging. For $T_1$ agents, Fe(III) MRI probes show promise as alternatives to Gd(III)-based probes [4,9]. Such probe development is important given the unique role of iron as the most abundant TM ion in the body. Such probes may capitalize on the body's ability to store and recycle iron. Yet there are challenges that are associated with incorporation of the strongly Lewis acidic Fe(III) center. Designing complexes that have an exchangeable inner-sphere water that does not deprotonate or form bridging hydroxy-species is one challenge [9]. Another challenge is to stabilize the trivalent state and to form complexes of high aqueous solubility. Linking multiple Fe(III) centers together is one way to produce probes with effective relaxivity [33,43] to compete with Gd(III) agents. TM paraCEST agents provide premier examples of responsive probes for pH, temperature, and redox status monitoring [65]. TM paraCEST agents are diverse, but there is a focus on Co(II) and Fe(II) as the most promising shift agents. The paraCEST agents have their own set of challenges, including the need to study *in vivo* applications rather than a focus on probe studies in phantoms [73,78]. This challenge will require focused efforts from coordination chemists and their collaborators in the field of imaging.

## ACKNOWLEDGMENTS

JRM thanks the NSF (CHE-2004135) and (STTR-1951127) for support of our research on TM CEST and $T_1$ probes.

## NOTES

JRM is the cofounder of Ferric Contrast, a company that develops iron-based MRI contrast agents.

## ABBREVIATIONS AND DEFINITIONS

| | |
|---|---|
| $B_1$ | presaturation pulse power |
| $c$ | concentration of paraCEST agent |
| **CAs** | contrast agents |
| **CEST** | chemical exchange saturation transfer |
| **DCE** | dynamic contrast enhanced |
| **DTPA** | diethylene triamine pentaacetate |
| $g_e$ | the electron g factor |
| **HS** | high-spin |
| $k_{ex}$ | water exchange rate constant |
| $K^{trans}$ | relative volume transfer constant |
| **LS** | low-spin |
| $MTR_{asym}$ | magnetization transfer asymmetry |
| $M_0$ | the original water proton signal |
| **MOP** | metal organic polyhedron |
| **MRI** | magnetic resonance imaging |
| **MT** | magnetization transfer |
| $M_z$ | water proton magnetization |
| $n$ | number of magnetically equivalent exchangeable protons |
| **paraCEST** | paramagnetic chemical exchange saturation transfer |
| **paraSHIFT** | paramagnetic chemical shift |
| **PICS** | paramagnetic induced chemical shift |
| **ppm** | parts per million |
| $q$ | number of bound water molecules |
| $r_2$ | transverse relaxivity |
| $r_{MH}$ | metal proton distance |
| **RF** | radiofrequency |

| | |
|---|---|
| *S* | electronic spin state |
| **SBM** | Solomon–Bloembergen–Morgan |
| $T_{1e}$ | electron spin relaxation |
| $T_{1m}$ | relaxation time of the bound water |
| **TACN** | triazacyclononane |
| **TM** | transition metal |
| $\Delta_{\omega}$ | frequency difference between the exchangeable proton resonance and bulk water proton resonance |
| $\gamma_H$ | the gyromagnetic ratio of proton |
| $\mu_0$ | Bohr magneton |
| $\omega_H$ | proton Larmor frequency |
| $\tau_c$ | correlation time |
| $\tau_m$ | lifetime of bound water molecules |
| $\tau_R$ | rotational motion of the complex |

## REFERENCES

1. P. Caravan, J. J. Ellison, T. J. McMurry, R. B. Lauffer, *Chem. Rev.* **1999**, *99*, 2293–2352.
2. J. Wahsner, E. M. Gale, A. Rodriguez-Rodriguez, P. Caravan, *Chem. Rev.* **2019**, *119*, 957–1057.
3. M. Botta, F. Carniato, D. Esteban-Gómez, C. Platas-Iglesias, L. Tei, *Future Med. Chem.* **2019**, *11*, 1461–1483.
4. E. A. Kras, E. M. Snyder, G. E. Sokolow, J. R. Morrow, *Acc. Chem. Res.* **2022**, 55, 1435–1444.
5. N. Kuźnik, M. Wyskocka, *Eur. J. Inorg. Chem.* **2016**, 445–458.
6. B. Drahoš, I. Lukeš, É. Tóth, *Eur. J. Inorg. Chem.* **2012**, 1975–1986.
7. A. Gupta, P. Caravan, W. S. Price, C. Platas-Iglesias, E. M. Gale, *Inorg. Chem.* **2020**, *59*, 6648–6678.
8. Z. Baranyai, F. Carniato, A. Nucera, D. Horvath, L. Tei, C. Platas-Iglesias, M. Botta, *Chem. Sci.* **2021**, *12*, 11138–11145.
9. R. Uzal-Varela, F. Lucio-Martinez, A. Nucera, M. Botta, D. Esteban-Gomez, L. Valencia, A. Rodriguez-Rodriguez, C. Platas-Iglesias, *Inorg. Chem. Front.* **2023**, *10*, 1633–1649.
10. I. Bertini, F. Capozzi, C. Luchinat, Z. C. Xia, *J. Phys. Chem.* **1993**, *97*, 1134–1137.
11. S. H. Koenig, C. M. Baglin, R. D. Brown, *Magn. Reson. Med.* **1985**, *2*, 283–288.
12. R. B. Lauffer, *Chem. Rev.* **1987**, *87*, 901–927.
13. I. Bertini, C. Luchinat, G. Parigi, E. Ravera, *Coord. Chem. Rev.* **1996**, *150*, 77–110.
14. M. Botta, *Eur. J. Inorg. Chem.* **2000**, 399–407.
15. E. Toth, L. Helm, A. E. Merbach, in *The Chemistry of Contrast Agents in Medical Magnetic Resonance Imaging*, Eds.: A. Merbach, L. Helm, E. Toth, Wiley, West Sussex, UK, 2013, pp 25–82.
16. E. M. Snyder, D. Asik, S. M. Abozeid, A. Burgio, G. Bateman, S. G. Turowski, J. A. Spernyak and J. R. Morrow, *Angew. Chem. Int. Ed.* **2020**, *132*, 2435–2440.
17. D. Asik, R. Smolinski, S. M. Abozeid, T. B. Mitchell, S. G. Turowski, J. A. Spernyak, J. R. Morrow, *Molecules* **2020**, *25*, 2291.
18. H. Wang, A. Wong, L. C. Lewis, G. R. Nemeth, V. C. Jordan, J. W. Bacon, P. Caravan, H. S. Shafaat and E. M. Gale, *Inorg. Chem.* **2020**, *59*, 17712–17721.

19. D. J. Kosman, *J. Biol. Chem.* **2010**, *285*, 26729–26735.
20. D. P. Jones, H. Sies, *Antioxid. Redox. Sign.* **2015**, *23*, 734–746.
21. R. Banerjee, *J. Biol. Chem.* **2012**, *287*, 4397–4402.
22. W. H. Koppenol, R. H. Hider, *Free Radical Bio. Med.* **2019**, *133*, 3–10.
23. P. B. Tsitovich, F. Gendron, A. Y. Nazarenko, B. N. Livesay, A. P. Lopez, M. P. Shores, J. Autschbach, J. R. Morrow, *Inorg. Chem.* **2018**, *57*, 8364–8374.
24. P. B. Tsitovich, A. M. Kosswattaarachchi, M. R. Crawley, T. Y. Tittiris, T. R. Cook, J. R. Morrow, *Chem. Eur. J.* **2017**, *23*, 15327–15331.
25. A. Brausam, J. Maigut, R. Meier, P. A. Szilágyi, H.-J. Buschmann, W. Massa, Z. Homonnay, R. van Eldik, *Inorg. Chem.* **2009**, *48*, 7864–7884.
26. H. Wang, M. B. Cleary, L. C. Lewis, J. W. Bacon, P. Caravan, H. S. Shafaat, E. M. Gale, *Angew. Chem. Int. Ed.* **2022**, *61*, e202114019.
27. T. Schneppensieper, S. Seibig, A. Zahl, P. Tregloan, R. van Eldik, *Inorg. Chem.* **2001**, *40*, 3670–3676.
28. H. Wang, V. C. Jordan, I. A. Ramsay, M. Sojoodi, B. C. Fuchs, K. K. Tanabe, P. Caravan, E. M. Gale, *J. Am. Chem. Soc.* **2019**, *141*, 5916–5925.
29. E. A. Kras, S. M. Abozeid, W. Eduardo, J. A. Spernyak, J. R. Morrow, *J. Inorg. Biochem.* **2021**, *225*, 111594.
30. M. Tweedle, G. Gaughan, J. Hagan, P. Wedeking, P. Sibley, L. J. Wilson, D. W. Lee, *Nucl. Med. Biol.* **1988**, *15*, 31–36.
31. P. Boehm-Sturm, A. Haeckel, R. Hauptmann, S. Mueller, C. K. Kuhl, E. A. Schellenberger, *Radiology* **2018**, *286*, 537–546.
32. J. Xie, A. Haeckel, R. Hauptmann, I. P. Ray, C. Limberg, N. Kulak, B. Hamm, E. Schellenberger, *Magn. Reson. Med.* **2021**, *85*, 3370–3382.
33. D. Asik, S. M. Abozeid, S. G. Turowski, J. A. Spernyak, J. R. Morrow, *Inorg. Chem.* **2021**, *60*, 8651–8664.
34. T. J. Swift, R. E. Connick, *J. Chem. Phys.* **1962**, *37*, 307–315.
35. A. Patel, D. Asik, E. M. Snyder, A. E. Delillo, P. J. Cullen, J. R. Morrow, *ChemMedChem* **2020**, *15*, 1050–1057.
36. A. Patel, D. Asik, E. M. Snyder, J. A. Spernyak, P. J. Cullen, J. R. Morrow, *Magnetochemistry* **2020**, *6*, 41.
37. M. S. I. Chowdhury, E. Kras, S. G. Turowski, J. A. Spernyak, J. R. Morrow, *Biomater. Sci.* **2023**, *11*, 5942–5954.
38. R. T. Kadakia, R. T. Ryan, D. J. Cooke, E. L. Que, *Chem. Sci.* **2023**, *14*, 5099–5105.
39. S. Karbalaei, A. Franke, A. Jordan, C. Rose, P. R. Pokkuluri, R. J. Beyers, A. Zahl, I. Ivanovic-Burmazovic, C. R. Goldsmith, *Chem. Eur. J.* **2022**, *28*, e202201179.
40. L. Palagi, E. Di Gregorio, D. Costanzo, R. Stefania, C. Cavallotti, M. Capozza, S. Aime, E. Gianolio, *J. Am. Chem. Soc.* **2021**, *143*, 14178–14188.
41. B. C. Bales, B. Grimmond, B. F. Johnson, M. T. Luttrell, D. E. Meyer, T. Polyanskaya, M. J. Rishel, J. Roberts, *Contrast Media. Mol. Imaging* **2019**, *2019*, 8356931.
42. D. Worah, A. E. Berger, K. R. Burnett, H. H. Cockrill, E. Kanal, C. Kendall, P. T. Leese, K. P. Lyons, E. Ross, G. L. Wolf, S. C. Quay, *Invest. Radiol.* **1988**, *23*, S281–S285.
43. G. E. Sokolow, M. R. Crawley, D. R. Morphet, D. Asik, J. A. Spernyak, A. J. R. McGray, T. R. Cook, J. R. Morrow, *Inorg. Chem.* **2022**, *61*, 2603–2611.
44. D. Maheshwaran, T. Nagendraraj, T. S. Balaji, G. Kumaresan, S. S. Kumaran, R. Mayilmurugan, *Dalton Trans.* **2020**, *49*, 14680–14689.
45. D. D. Schwert, N. Richardson, G. Ji, B. Radüchel, W. Ebert, P. E. Heffner, R. Keck, J. A. Davies, *J. Med. Chem.* **2005**, *48*, 7482–7485.

46. J. T. Weisser, M. J. Nilges, M. J. Sever, J. J. Wilker, *Inorg. Chem.* **2006**, *45*, 7736–7747.
47. M. Botta, C. F. G. C. Geraldes, L. Tei, *Wires Nanomed. Nanobi.* **2023**, *15*, e1858.
48. I. Bertini, C. Luchinat, G. Parigi, E. Ravera, *NMR of Paramagnetic Molecules: Applications to Metallobiomolecules and Models*, 2nd edn., Elsevier, Amsterdam, **2017**, pp. 1–24.
49. S. Viswanathan, Z. Kovacs, K. N. Green, S. J. Ratnakar, A. D. Sherry, *Chem. Rev.* **2010**, *110*, 2960–3018.
50. I. Bertini, P. Turano, A. J. Vila, *Chem. Rev.* **1993**, *93*, 2833–2932.
51. L. Dunbar, R. J. Sowden, K. D. Trotter, M. K. Taylor, D. Smith, A. R. Kennedy, J. Reglinski, C. M. Spickett, *Biometals* **2015**, *28*, 903–912.
52. F. Touti, A. K. Singh, P. Maurin, L. Canaple, O. Beuf, J. Samarut, J. Hasserodt, *J. Med. Chem.* **2011**, *54*, 4274–4278.
53. V. Lagostina, F. Carniato, D. Esteban-Gomez, C. Platas-Iglesias, M. Chiesa, M. Botta, *Inorg. Chem. Front.* **2023**, *10*, 1999–2013.
54. A. Shah, M. J. Taylor, G. Molinaro, S. Anbu, M. Verdu, L. Jennings, I. Mikulska, S. Diaz-Moreno, H. El Mkami, G. M. Smith, M. M. Britton, J. E. Lovett, A. F. A. Peacock, *Proc. Natl. Acad. Sci. USA* **2023**, *120*, e2219036120.
55. E. S. O'Neill, A. Kaur, D. P. Bishop, D. Shishmarev, P. W. Kuchel, S. M. Grieve, G. A. Figtree, A. K. Renfrew, P. D. Bonnitcha, E. J. New, *Inorg. Chem.* **2017**, *56*, 9860–9868.
56. M. Dommaschk, J. Grobner, V. Wellm, J. B. Hovener, C. Riedel, R. Herges, *Phys. Chem. Chem. Phys.* **2019**, *21*, 24296–24299.
57. F. Touti, P. Maurin, L. Canaple, O. Beuf, J. Hasserodt, *Inorg. Chem.* **2012**, *51*, 31–33.
58. J. Wang, C. Gondrand, F. Touti, J. Hasserodt, *Dalton Trans.* **2015**, *44*, 15391–15395.
59. S. J. Dorazio, P. B. Tsitovich, K. E. Siters, J. A. Spernyak, J. R. Morrow, *J. Am. Chem. Soc.* **2011**, *133*, 14154–14156.
60. R. S. Drago, J. I. Zink, R. M. Richman, W. D. Perry, *J. Chem. Educ.* **1974**, *51*, 464–467.
61. R. S. Drago, J. I. Zink, R. M. Richman, W. D. Perry, *J. Chem. Educ.* **1974**, *51*, 371–376.
62. A. C. Harnden, D. Parker, N. J. Rogers, *Coord. Chem. Rev.* **2019**, *383*, 30–42.
63. D. Xie, M. Yu, R. T. Kadakia, E. L. Que, *Acc. Chem. Res.* **2020**, 53, 2–10.
64. S. R. Zhang, M. Merritt, D. E. Woessner, R. E. Lenkinski, A. D. Sherry, *Acc. Chem. Res.* **2003**, *36*, 783–790.
65. J. R. Morrow, S. M. Abozeid, A. Patel, J. J. Raymond, *in Encyclopedia of Inorganic and Bioinorganic Chemistry*, Eds.: R. A. Scott, T. Storr, John Wiley & Sons, Ltd., West Sussex, UK, **2020**. https://doi.org/10.1002/9781119951438.eibc2749.
66. P. B. Tsitovich, J. R. Morrow, *Inorg. Chim. Acta* **2012**, *393*, 3–11.
67. P. B. Tsitovich, J. M. Cox, J. B. Benedict, J. R. Morrow, *Inorg. Chem.* **2016**, *55*, 700–716.
68. G. Pintacuda, M. John, X. C. Su, G. Otting, *Acc. Chem. Res.* **2007**, *40*, 206–212.
69. C. F. G. C. Geraldes, C. Luchinat, in *Lanthanides and Their Interrelations with Biosystems (Metal Ions in Biological Systems), Vol. 40*, Ed.: A. Sigel, H. Sigel, Marcel Dekker, New York, **2003**, pp. 513–588.
70. I. Bertini, C. Luchinat, G. Parigi, E. Ravera, *Coord. Chem. Rev.* **1996**, *150*, 29–75.
71. L. Banci, I. Bertini, C. Luchinat, M. S. Viezzoli, *Inorg. Chem.* **1990**, *29*, 1438–1440.
72. S. Viswanathan, Z. Kovacs, K. N. Green, S. J. Ratnakar, A. D. Sherry, *Chem. Rev.*, **2010**, *110*, 2960.
73. A. Rodriguez-Rodriguez, M. Zaiss, D. Esteban-Gomez, G. Angelovski, C. Platas-Iglesias, *Int. Rev. Phys. Chem.* **2021**, *40*, 51–79.
74. M. T. McMahon, A. A. Gilad, J. Y. Zhou, P. Z. Sun, J. W. M. Bulte, P. C. M. van Zijl, *Magn. Reson. Med.* **2006**, *55*, 836–847.

75. A. D. Sherry, Y. K. Wu, *Curr. Opin. Chem. Biol.* **2013**, *17*, 167–174.
76. T. Gambino, V. Laura, P. Perez-Lourido, D. Esteban-Gomez, M. Zaiss, C. Platas-Iglesias, G. Angelovski, *Inorg. Chem. Front.* **2020**, *7*, 2274–2286.
77. E. Vinogradov, A. D. Sherry, R. E. Lenkinski, *J. Magn. Reson.* **2013**, *229*, 155–172.
78. A. D. Sherry, D. D. Castelli, S. Aime, *NMR Biomed.* **2022**, e4698. https://doi.org/10.1002/nbm.4698.
79. S. J. Dorazio, A. O. Olatunde, P. B. Tsitovich, J. R. Morrow, *J. Biol. Inorg. Chem.* **2014**, *19*, 191–205.
80. S. J. Dorazio, P. B. Tsitovich, S. A. Gardina, J. R. Morrow, *J. Inorg. Biochem.* **2012**, *117*, 212–219.
81. S. J. Dorazio, J. R. Morrow, *Inorg. Chem.* **2012**, *51*, 7448–7450.
82. S. M. Abozeid, E. M. Snyder, T. Y. Tittiris, C. M. Steuerwald, A. Y. Nazarenko, J. R. Morrow, *Inorg. Chem.* **2018**, *57*, 2085–2095.
83. P. B. Tsitovich, J. M. Cox, J. A. Spernyak, J. R. Morrow, *Inorg. Chem.* **2016**, *55*, 12001–12010.
84. P. J. Burns, J. M. Cox, J. R. Morrow, *Inorg. Chem.* **2017**, *56*, 4545–4554.
85. P. B. Tsitovich, J. A. Spernyak, J. R. Morrow, *Angew. Chem. Int. Ed.* **2013**, *52*, 13997–14000.
86. E. M. Snyder, S. I. Chowdhury, J. R. Morrow, *Inorg. Chim. Acta* **2020**, 509, 119649.
87. S. J. Dorazio, A. O. Olatunde, J. A. Spernyak, J. R. Morrow, *Chem. Commun.* **2013**, *49*, 10025–10027.
88. C. J. Bond, G. E. Sokolow, M. R. Crawley, P. J. Burns, J. M. Cox, R. Mayilmurugan, J. R. Morrow, *Inorg. Chem.* **2019**, *58*, 8710–8719.
89. A. Patel, S. M. Abozeid, P. J. Cullen, J. R. Morrow, *Inorg. Chem.* **2020**, *59*, 16531–16544.
90. K. Srivastava, G. Ferrauto, V. G. Young, S. Aime, V. C. Pierre, *Inorg. Chem.* **2017**, *56*, 12206–12213.
91. A. O. Olatunde, C. J. Bond, S. J. Dorazio, J. M. Cox, J. B. Benedict, M. D. Daddario, J. A. Spernyak, J. R. Morrow, *Chem. Eur. J.* **2015**, *21*, 18290–18300.
92. M. Yu, B. S. Bouley, D. Xie, E. L. Que, *Dalton Trans.* **2019**, 48, 9337–9341.
93. A. O. Olatunde, J. M. Cox, M. D. Daddario, J. A. Spernyak, J. B. Benedict, J. R. Morrow, *Inorg. Chem.* **2014**, *53*, 8311–8321.
94. C. J. Bond, R. Cineus, A. Y. Nazarenko, J. A. Spernyak, J. R. Morrow, *Dalton Trans.* **2020**, *49*, 279–284.
95. S. M. Abozeid, M. S. I. Chowdhury, D. Asik, J. A. Spernyak, J. R. Morrow, *ACS Appl. Bio. Mater.* **2021**, *4*, 7951–7960.
96. J. J. Raymond, S. M. Abozeid, G. E. Sokolow, C. J. Bond, C. E. Yap, A. Y. Nazarenko, J. R. Morrow, *Dalton Trans.* **2023**, *52*, 9831–9839. https://doi.org/10.1039/D3DT01572F.
97. K. Du, A. E. Thorarinsdottir, T. D. Harris, *J. Am. Chem. Soc.* **2019**, *141*, 7163–7172.
98. B. Martin, J. Autschbach, *Phys. Chem. Chem. Phys.* **2016**, *18*, 21051–21068.
99. S. Aime, A. Barge, D. Delli Castelli, F. Fedeli, A. Mortillaro, F. U. Nielsen, E. Terreno, *Magn. Reson. Med.* **2002**, 47, 639–648.
100. J. A. Peters, K. Djanashvili, C. F. G. C. Geraldes, C. Platas-Iglesias, in *The Chemistry of Contrast Agents in Medical Magnetic Resonance Imaging*, Eds.: A. Merbach, L. Helm, E. Toth, Wiley, West Sussex, UK, **2013**, pp. 209–276.
101. I. R. Jeon, A. I. Gaudette, J. G. Park, C. R. Haney, T. D. Harris, *Chem. Sci.* **2014**, 5, 2461–2465.
102. L. Caneda-Martinez, L. Valencia, I. Fernandez-Perez, M. Regueiro-Figueroa, G. Angelovski, I. Brandariz, D. Esteban-Gomez, C. Platas-Iglesias, *Dalton Trans.* **2017**, *46*, 15095–15106.

103. A. E. Thorarinsdottir, K. Du, J. H. P. Collins, T. D. Harris, *J. Am. Chem. Soc.* **2017**, *139*, 15836–15847.
104. A. E. Thorarinsdottir, S. M. Tatro, T. D. Harris, *Inorg. Chem.* **2018**, *57*, 11252–11263.
105. A. E. Thorarinsdottir, T. D. Harris, *Chem. Commun.* **2019**, *55*, 794–797.
106. K. Du, T. D. Harris, *J. Am. Chem. Soc.* **2016**, *138*, 7804–7807.
107. K. Du, E. A. Waters, T. D. Harris, *Chem. Sci.* **2017**, *8*, 4424–4430.
108. L. M. De Leon-Rodriguez, A. J. M. Lubag, C. R. Malloy, G. V. Martinez, R. J. Gillies, A. D. Sherry, *Acc. Chem. Res.* **2009**, *42*, 948–957.
109. S. Shuvaev, E. Akam, P. Caravan, *Invest. Radiol.* **2021**, *56*, 20–34.
110. C. Q. Tu, E. A. Osborne, A. Y. Louie, *Ann. Biomed. Eng.* **2011**, *39*, 1335–1348.
111. A. Louie, *J. Magn. Reson. Imaging* **2013**, *38*, 530–539.
112. J. L. Major, T. J. Meade, *Acc. Chem. Res.* **2009**, *42*, 893–903.
113. G. Rancan, D. D. Castelli, S. Aime, *Magn. Reson. Med.* **2016**, *75*, 329–336.
114. Y. K. Wu, S. R. Zhang, T. C. Soesbe, J. Yu, E. Vinogradov, R. E. Lenkinski, A. D. Sherry, *Magn. Reson. Med.* **2016**, *75*, 2432–2441.
115. S. Aime, M. Botta, M. Fasano, E. Terreno, P. Kinchesh, L. Calabi, L. Paleari, *Magn. Reson. Med.* **1996**, *35*, 648–651.
116. S. R. Zhang, C. R. Malloy, A. D. Sherry, *J. Am. Chem. Soc.* **2005**, *127*, 17572–17573.
117. A. X. Li, F. Wojciechowski, M. Suchy, C. K. Jones, R. H. E. Hudson, R. S. Merton, R. Bartha, *Magn. Reson. Med.* **2008**, *59*, 374–381.
118. S. M. Pinto, V. Tome, M. J. F. Calvete, M. M. C. A. Castro, E. Toth, C. F. G. C. Geraldes, *Coord. Chem. Rev.* **2019**, *390*, 1–31.
119. J. R. Morrow, J. J. Raymond, M. S. I. Chowdhury, P. R. Sahoo, *Inorg. Chem.* **2022**, *61*, 14487–14499.
120. S. Karbalaei, C. R. Goldsmith, *J. Inorg. Biochem.* **2022**, *230*, 111763.
121. G. Ferrauto, E. Di Gregorio, V. Auboiroux, M. Petit, F. Berger, S. Aime, H. Lahrech, *NMR Biomed.* **2018**, *31*, e4005.
122. N. McVicar, A. X. Li, M. Suchy, R. H. E. Hudson, R. S. Menon, R. Bartha, *Magn. Reson. Med.* **2013**, *70*, 1016–1025.
123. N. N. Yadav, J. D. Xu, H. Y. Heo, P. C. M. van Zijl, *NMR Biomed.* **2023**, *36*, e4960.
124. D. D. Castelli, E. Terreno, D. Longo, S. Aime, *NMR Biomed.* **2013**, *26*, 839–849.
125. G. Ferrauto, E. Terreno, *NMR Biomed.* **2023**, 36, e4791.
126. F. Carniato, G. Ferrauto, M. Munoz-Ubeda, L. Tei, *Magnetochemistry* **2020**, *6*, 38. https://doi.org/10.3390/magnetochemistry6030038.
127. G. Ferrauto, D. Delli Castelli, E. Di Gregorio, E. Terreno, S. Aime, *Nanomed. Nanobiotechnol.* **2016**, *8*, 602–618.

# 7 Lanthanide-Based Paramagnetic Chemical Exchange Saturation Transfer (paraCEST) Agents for MRI

*A. Dean Sherry*
Advanced Imaging Research Center,
University of Texas Southwestern Medical Center,
Dallas, TX 75390, USA
Department of Chemistry & Biochemistry,
University of Texas at Dallas,
Richardson, TX 75080, USA
sherry@utdallas, dean.sherry@utsouthwestern.edu

*Mark Woods*
Department of Chemistry, Portland State
University, Portland, OR 97201, USA
Advanced Imaging Research Center, Oregon Health &
Science University,
Portland, OR 97239, USA
markw@pdx.edu

## CONTENTS



DOI: 10.1201/9781003374688-7

**Abstract**

The discovery of paramagnetic lanthanide ion complexes having either slowly exchanging inner-sphere water molecules or rapidly exchanging ligand protons has catalyzed widespread interest in using them as contrast agents for magnetic resonance imaging. Unlike paramagnetic complexes that alter water proton relaxation ($T_1$ or $T_2$ agents), paraCEST agents are unique in that image contrast can be turned "on and off" by including a frequency selective saturation pulse in any imaging sequence without altering the relaxation characteristics of tissue water. One clear advantage of paraCEST over diaCEST agents is that the chemical shift of an exchanging water molecule or ligand proton is typically much further away from bulk water protons (large $\Delta\omega$) so that a wider variety of faster exchanging species can be considered while still meeting the basic CEST limitation, $\Delta\omega \geq k_{ex}$. This chapter briefly summarizes the history of paraCEST agents and describes in some detail the chemical design features that allow one to optimize water/proton exchange rates for optimal CEST sensitivity. While significant chemical progress has been made, only a few lanthanide-based paraCEST agents have been successfully used *in vivo* to date. The reasons why paraCEST agents seem to be less sensitive *in vivo* than typical $T_1/T_2$ agents is addressed, and future potential applications are described.

## KEYWORDS

Chemical Exchange Saturation Transfer (CEST) Agents; Paramagnetic Lanthanide Complexes as paraCEST Agents; Biologically Responsive Imaging Agents

## 1 INTRODUCTION

The field of chemical exchange saturation transfer (CEST) applied to biological imaging has grown dramatically since first introduced by Ward, Aletras, and Balaban in 2000 [1]. The basic concept of "saturating" one pool of spins and monitoring chemical exchange of those saturated spins into other chemically distinct pool or spins was described by Forsén and Hoffman nearly 40 years earlier [2] and widely applied in living tissues to measure ATP-$P_i$ exchange rates by $^{31}P$

NMR since the late 1970s [3]. However, Ward et al. were first to apply saturation transfer principles to $^1H$ imaging by saturating a pool of exchangeable protons on a biomolecule (–OH, –NH, and $NH_2$) to indirectly detect those molecules through the water signal. This seminal paper catalyzed a flurry of activity in applying CEST principles to detect a variety of endogenous metabolites and biomolecules or to use exogenous molecules designed for CEST as biological sensors. The basic principles of CEST theory have been described in many articles and several recent reviews [4] so will not be repeated here.

Before one can apply a saturating pulse at the frequency of any exchangeable proton, two conditions must be met: (1) the proton must have a chemical shift sufficiently different from water ($\Delta\omega$) to allow for saturation without altering the signal intensity itself and (2) the proton must exchange with water protons at a rate ($k_{ex}$) that meets the basic CEST limitation, $\Delta\omega \geq k_{ex}$. This simple relationship serves as a valuable guide for the design of new CEST probes. The range of proton chemical shifts for most common biomolecules is quite small, typically within ± 5 ppm relative to bulk water protons, so proton exchange rates that meet this slow-to-intermediate exchange requirement is also quite small, i.e., $k_{ex}$ must be relatively slow. For example, consider an exchanging proton with chemical shift of 3.5 ppm downfield of water and a CEST experiment is being performed at 3 T (in this case, $\Delta\omega$ = 2815 rad/s) so, depending upon linewidths, protons in exchange with water protons at rates up to about 2815 s–1 should in principle be detectable by CEST. This general principle is satisfied by many endogenous molecules containing exchangeable –NH, $–NH_2$, or –OH protons species (Table 1) so the opportunities to detect individual metabolites such as these using a 3 T clinical scanner are almost limitless. Performing experiments

**TABLE 1**
**Proton Exchange Rates for a Few Biological Molecules (diaCEST)**

| Exchanging Species | Reported $k_{ex}$ |
|---|---|
| Lysine ε-$NH_2$ | 4000 $s^{-1}$ |
| Arginine ε-NH | 1200 $s^{-1}$ |
| Arginine η-$NH_2$ | 700 $s^{-1}$ |
| Serine –OH | 900 $s^{-1}$ |
| Threonine –OH | 700 $s^{-1}$ |
| Glucose –OH | 500–10,000 $s^{-1}$ |
| Histidine –NH | 1700 $s^{-1}$ |
| Cysteine –SH | 1400 $s^{-1}$ |
| Protein amide | 10–100 $s^{-1}$ |
| Poly-L-lysine | 400 $s^{-1}$ |

*Source:* Data collected from C. M. van Zijl, N. N. Yadav, *Magn. Reson. Med.* **2011**, *65*, 927–948.

at even higher magnetic fields (7 T and above) brings even more molecules into play for detection by CEST.

Although a few novel chemical approaches for expanding $\Delta\omega$ beyond the usual ±5 ppm diamagnetic window have been reported [5], the primary motivation for expanding $\Delta\omega$ has been to allow for more selective CEST activation of the probe with less interference from endogenous exchanging protons. There has been less emphasis on optimizing proton exchange rates in new diaCEST probes even though, as pointed out in Ward et al. [1], $k_{ex}$ should be "as large as possible to maximize the CEST effect". The number of currently reported diaCEST agents that have proton exchange rates exceeding the slow-to-intermediate exchange limitation, $\Delta\omega \geq k_{ex}$, is rare so maximizing $k_{ex}$ remains a worthwhile goal. One can easily increase $\Delta\omega$ by performing experiments at higher magnetic fields or by identifying compounds with unusually large proton chemical shifts but optimizing $k_{ex}$ in diamagnetic CEST agents is less obvious. As we shall see, this is not true for paraCEST agents where $k_{ex}$ can be quite easily modified.

Weakly paramagnetic metal ion complexes acting as paraCEST agents offer several advantages over diaCEST agents. Not only can one expand the frequency range of exchanging protons (200 ppm or more) but one also has more choices in terms of the types of proton exchange species and exchange rates that meet the basic CEST limitation, $\Delta\omega \geq k_{ex}$. As we shall see in the remaining sections of this review, the concept of paraCEST has catalyzed chemistry thought leaders worldwide to reimagine new designs of "responsive" agents for detecting a variety of physiological indices including pH, temperature, ROS, biological ions and select metabolites, enzyme activities, and other functional measures. However, *in vivo* applications of these agents to address clinically relevant questions have been disappointingly slow.

## 1.1 The Discovery of paraCEST Agents

Given the 40+ year history of gadolinium-based MRI contrast agent development, it is not surprising that many of the first paraCEST agents were also derived from paramagnetic lanthanide ion complexes. The first agent, EuDOTA-(gly-OEt)$_4^{3+}$, is a tetra-amide analog of EuDOTA. This 9-coordinate complex has eight ligand donor atoms and a single inner-sphere water molecule that give rise to a distinct, proton resonance at 50 ppm (downfield of water) in the $^1$H NMR spectrum of this complex in pure water [6]. This was a very surprising result because water exchange in lanthanide complexes at that time was assumed to be too fast to detect an inner-sphere water molecule by $^1$H NMR. It wasn't until André Merbach and colleagues [7] developed $^{17}$O NMR methods for measuring water exchange rates that we became aware that chelated lanthanide ions, $LnL^{n\pm}$, display much slower water exchange rates than anticipated. It was shown that many common lanthanide chelates such as $LnDTPA^{2-}$ and $LnDOTA^-$ have a single inner-sphere water molecule that exchanges with bulk solvent on the order of $10^5$–$10^6$ s$^{-1}$, about 3-fold slower than the unchelated aqua ions. Even more surprising was the water exchange rate in the first paraCEST agent, EuDOTA-(gly-OEt)$_4^{3+}$, was even slower, ~$4.5 \times 10^3$ $s^{-1}$ at 25°C [8]. Given that the chemical shift of the exchanging water molecule in this complex is 50 ppm downfield of water, this system clearly meets the slow-to-intermediate exchange limitation, $\Delta\omega >> k_{ex}$, at common imaging fields. At 37°C, the rate is considerably faster, $10^4$ $s^{-1}$, yet the CEST signal from the exchanging water molecule is easily detected at 9.4 T in water samples. This early result was encouraging, and it didn't take long to learn that water exchange rates in $Eu^{3+}$ complexes such as this could be easily modified using simple coordination chemistry principles.

Another advantage of lanthanide ion-based paraCEST agents is the rich history and knowledge of lanthanide induced shifts (LIS) for the entire lanthanide series. If one detects an NMR signal of an exchanging water molecule or exchanging –NH or –OH proton in any given lanthanide chelate, then the chemical shift of that same resonance can be reasonably predicted for the entire series of lanthanide complexes, unless of course there is a dramatic change in structure along the lanthanide series. In 1972, Bleaney [9] published LIS values ($C_J$) for a series of isostructural lanthanide complexes based upon comparisons of magnetic anisotropy of the lanthanide ions. These "pseudocontact shift" values were widely used for comparisons of $^1H$ NMR spectra of chelated lanthanide ions for over 50 years and were particularly helpful in identifying structural changes along any series of lanthanide complexes previously thought to be isostructural [10]. More recently, Parker and colleagues [11] have pointed out some quantitative limitations of Bleaney's theory and suggested alternative shift values. Nonetheless, Bleaney's $C_J$ values remain useful as a guide for predicting the chemical shifts of exchanging protons or water molecules in CEST experiments. As the data in Table 2 illustrate, the predicted chemical shifts of the Ln-$H_2O$ resonances in all other LnDOTA-$(gly\text{-}OEt)_4^{3+}$ complexes based comparisons of Bleaney's $C_J$ values and observed water resonance in EuDOTA-$(gly\text{-}OEt)_4^{3+}$ (+50 ppm) are not perfect but nonetheless qualitatively useful for locating exchanging species that may be too broad to observe directly.

## 1.2 Impact of Water Proton $T_1$ on CEST Intensity

In 2005, Woessner et al. [12] reported one of the first quantitative descriptions of CEST based upon the NMR Bloch equations modified for exchange. One simple prediction was that the water intensity ($M_Z/M_0$) after a long saturation pulse on an exchanging pool (pool B) (for metal-based systems, this is often referred to as pool M) is sensitive to both the $T_1$ of the bulk water proton pool (pool A) and the lifetime ($\tau_a$) a proton remains in that pool (Equation 1).

$$\left(\frac{Mz}{M_O}\right) = \frac{\tau_a}{(T_{1a} + \tau_a)} \tag{1}$$

This simple relationship indicates that a short water proton lifetime ($\tau_a$) and long water proton $T_{1a}$ will have the largest impact on reducing the water signal, $M_Z/M_0$. The data in Table 2 show that the paramagnetic ions $Dy^{3+}$, $Tb^{3+}$, $Ho^{3+}$, and $Er^{3+}$

**TABLE 2**
**Bleaney's Theoretical $C_J$ Values, $\delta_{obs}$ of the Water Resonances in the LnDOTA-(gly-OEt)$_4$$^{3+}$ Complexes, Paramagnetic Contribution to Water Relaxation ($1/T_{1P}$) for Each $Ln^{3+}$ Complex and an Index ($|\delta| / T_{1P}^{-1}$) for Predicting Which $Ln^{3+}$ Complex Is Most Efficient for paraCEST**

| $Ln^{3+}$ | $C_J$ Value | $\delta(^1H)$ ppm | $1/T_{1P}$ $s^{-1}$ | $\|\delta\|/T_{1P}^{-1}$ |
|---|---|---|---|---|
| $Pr^{3+}$ | −11.0 | −60 | 0.18 | 333 |
| $Nd^{3+}$ | −4.2 | −32 | 0.29 | 110 |
| $Sm^{3+}$ | −0.7 | −4 | 0.08 | 50 |
| $Eu^{3+}$ | +4.0 | +50 | 0.08 | 625 |
| $Tb^{3+}$ | −86 | −600 | 5.60 | 107 |
| $Dy^{3+}$ | −100 | −720 | 9.33 | 77 |
| $Ho^{3+}$ | −39 | −360 | 5.56 | 65 |
| $Er^{3+}$ | +33 | +200 | 5.56 | 36 |
| $Tm^{3+}$ | +53 | +500 | 4.72 | 106 |
| $Yb^{3+}$ | +22 | +200 | 0.65 | 308 |

*Source:* Data adapted from M. Milne, Y. Wu, A. D. Sherry, Chapter 11 in *ParaCEST Agents: Design, Discovery, and Implementation*, Eds.: McMahon, Gilad, Bulte, & van Zijl, Pan Stanford Publishing Ltd., 2017.

Bleaney's $C_J$ values predict pseudocontact shifts only and some water protons have an added contact shift contribution to their $\delta_{obs}$. T1P was measured on 10 mM samples of each $Ln^{3+}$ in pure water.

have the largest impact on water proton $T_1$ relaxation, while $Eu^{3+}$, $Sm^{3+}$, $Pr^{3+}$, $Nd^{3+}$, and $Yb^{3+}$ have the least. Given that a large chemical shift, $\delta$, is also an advantage for CEST, the ratio $|\delta| / T_{1P}$ predicts which $Ln^{3+}$ ions are best choices for CEST. These ratios predict that the order $Eu^{3+} >> Pr^{3+} \cong Yb^{3+}$ is best for CEST and explains why $Eu^{3+}$ and $Yb^{3+}$ complexes most often appear in the paraCEST literature. $Pr^{3+}$ complexes are also attractive as paraCEST agents but, as we shall see, the smaller chemical shift of water and other protons in $Pr^{3+}$ complexes limits possible applications *in vivo* due to overlap with the bulk water signal.

The second factor to consider in Equation 1 is $\tau_a$. In this case, if the lifetime of a proton in the water pool is to be minimized for optimal CEST, it follows that the lifetime a water molecule or a proton residing on a paramagnetic complex ($\tau_M$ in this case) should be maximized. Hence, $Ln^{3+}$ complexes with long Ln-$H_2O$, Ln-NH, or Ln-OH proton lifetimes should give a maximum reduction in $M_Z/M_0$. This seems at odds with the advice of Ward et al. that $k_{ex}$ be "as fast as possible", but one should remember that $k_{ex}$ can be neither too slow nor too fast, i.e., it must be in the slow-to-intermediate exchange regime for CEST. If exchange is too fast, a separate resonance will not be observed for the exchanging species so it cannot be saturated, and if exchange is too slow, prolonged

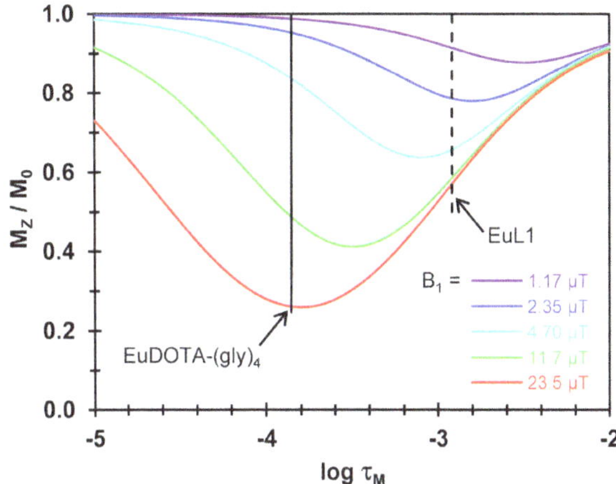


**FIGURE 1** Simulated curves showing changes in $M_z/M_0$ as a function of water exchange rate for a paraCEST agent at 200 MHz. The Bloch simulation assumed 20 mM agent, a chemical shift of 50 ppm for the exchanging water molecule, and $T_1$'s of 2.5 s for bulk water and 0.2 s for the exchanging Eu-$H_2O$ molecule. (Figure modified and adapted from Refs. [12,13].)

saturation of the exchanging species will not result in a significant reduction in the water signal intensity. To be more specific, the advice offered by Ward et al. could be modified to, "as fast as possible while also meeting the slow-to-intermediate exchange limitation".

An example of this classic "Goldilocks" effect is shown by a plot of $M_Z/M_0$ *versus* Eu-$H_2O$ residence lifetime ($\tau_M$) for a paramagnetic complex predicted by the Bloch equations (Figure 1). This simulation shows that one should be able to saturate a significant fraction of the bulk water signal by presaturating the Eu-$H_2O$ resonance for a prolonged period, typically 2–3 s, prior to collection of an NMR spectrum, but only if the rate of water exchange ($k_{ex} = \tau_M^{-1}$) is optimal. In this example, the optimal rate lies between $10^3$ and $10^4$ $s^{-1}$ depending upon the applied power ($B_1$) used for saturation. As seen by different curves for $B_1$, faster exchanging systems (shorter lifetimes as shown in Figure 1) require higher power to achieve maximum CEST; however, if one is $B_1$ limited due to equipment limitations or sample heating, then even slower water exchange paraCEST systems are required to achieve maximal CEST. One example of such a slow water exchange complex was given by a EuDOTA-tetraamide complex having both charged polar groups (glutamyl) and phosphonate ester side chains [13]. This combination of side-chain groups resulted in a complex (labeled EuL1 in Figure 1) with such slow water exchange ($k_{ex} = 1.36 \times 10^3$ $s^{-1}$ or $\tau_M = 0.74 \times 10^{-3}$ s at 25°C) that it appears on the slow side (right side) of the maximum shown in Figure 1. At 37°C, the rate of water exchange in this complex increased to $3.68 \times 10^3$ $s^{-1}$, and as predicted by Bloch theory, the water exchange CEST peak increased in intensity at the higher temperature [13]. This example illustrates the influence of side-chain polarity and bulkiness on water exchange rates in EuDOTA-tetraamide complexes of this type, but there are other chemical features of these complexes that influence water exchange rates as well.

## 2 FACTORS AFFECTING Ln-$H_2O$ EXCHANGE RATES

Probably, the most widely studied class of compounds in the context of paraCEST are the DOTA-tetraamide chelates of the rare earths. There are several reasons for this, not least of which is the need to produce chelates that are safe for *in vivo* applications. The macrocyclic backbone of the DOTA framework imparts considerable robustness to the chelate so that the chelate can be excreted intact more quickly than the metal ion can be released. A typical first-order rate constant for acid-catalyzed release of a lanthanide ion from a DOTA-tetraamide macrocycle is about $k_1 \sim 10^{-4}–10^{-5}$ $s^{-1}$ [14], a rate consistent with a robust and safe chelate. However, this kinetic definition of robustness is often confused with thermodynamic considerations. In general terms, the chelates of DOTA-tetraamides are found to have comparatively low thermodynamic stability constants in the range $K_{ML} = 10^{12}–10^{14}$ $M^{-1}$ [14]. This reflects the overall basicity of the ligand and does not directly correlate with the likelihood of metal ion release *in vivo*. The reduction in basicity arises in part from the presence of the tetra-amide ligating groups of the pendant arms. Amide oxygen donors are not strongly basic, and this has implications for other aspects of a chelate's properties, specifically the kinetics of water exchange.

### 2.1 Nature of the Ligating Group

It is generally accepted that in chelates based around the DOTA framework, the coordinated water molecule (if present, see below) exchanges with the bulk through a dissociative mechanism. Breaking the bond between metal and water is therefore the rate-determining step for exchange. When the metal ion is a rare earth, the nature of this bond is primarily electrostatic and the strength of the bond is governed by the Lewis acidity of the metal. This means that the electron donating capacity of the ligand will play a substantial role in determining the exchange kinetics of the coordinated water. Stronger electron donating ligating groups place more electron density on the metal ion which translates to a weaker M-$OH_2$ bond and faster water exchange (shorter $\tau_M$) [15]. Conversely, less basic ligating groups place less electron density on the metal ion which translates to a stronger M-$OH_2$ bond and slower (longer $\tau_M$) water exchange. Thus, a considerable amount of control of the water exchange rate can be exerted simply by changing the ligating groups on the pendant arms of the DOTA framework. Although $Gd^{3+}$ is unsuitable for paraCEST (it does not induce hyperfine shifts and substantially shortens $T_1$ and $T_2$), it is often convenient to compare the exchange kinetics of Gd complexes. The methodology for measuring exchange in $Gd^{3+}$ chelates is robust [16] and comparing across the same metal ion avoids the potentially confounding effects of change ionic radius (discussed later). As water exchange in GdDOTA is fast ($\tau_M$ = 360 ns) [17], in order to be useful for paraCEST, $\Delta\omega$ would need to be of the order of 2–3 MHz, far beyond the known shifting abilities of any rare earth or transition metal ion. From the preceding, it is clear that incremental substitution of acetate pendant arms with acetamide pendant arms will progressively slow water exchange (Figure 2(a)). The ranges shown are approximate and based on the range of published values [17,18]. It is also important to note that exchange lifetimes in LnDOTA-tetraamide complexes can be significantly impacted by factors such as solvent and counter ion identity [19]. Once all four pendant arms have been substituted with acetamides, water exchange is slowed into the regime required for paraCEST.

A logical extension of altering ligating group basicity to modulate water exchange would be to use even less basic groups, such as ketones [20]. In general, ketones are poor ligating groups and preparing stable metal ion chelates is challenging, although structures that promote enolate formation represent better ligands. Nonetheless, incorporating ketone side-chain donating groups onto a DOTA framework slows water exchange rates even further, although this may result in chelate that is exchanging water more slowly than is optimal for CEST. Using a strategy of selecting ligating groups of different basicity, it is possible to fine tune water exchange rates over a wide range from much faster than permissible for CEST to much slower. Although this strategy is effective at tuning water exchange rates that lie within the permissible range, it is less effective when it comes to fine-tuning exchange within that permissible range. Fortunately, other aspects of coordination chemistry can be manipulated to further fine-tune water exchange rates.

## 2.2 Coordination Geometry in LnDOTA-Type Chelates

An appreciation of the coordination chemistry of LnDOTA-type complexes begins with understanding the stereochemistry of the ligand itself. Although DOTA contains no stereocenters, once a lanthanide ion is introduced, it possesses several elements of chirality. Four chelate rings are formed by the interaction of the metal ion with macrocyclic ring, each of which form a helix which can be designated $\delta$ or $\lambda$ for a clockwise or anticlockwise rotation, respectively. Similarly, the pendant arms wrap around the metal ion in a helix denoted $\Delta$ or $\Lambda$ for a clockwise or anticlockwise rotation, respectively. Accordingly, four stereoisomers are possible: $\Delta(\lambda\lambda\lambda\lambda)$, $\Lambda(\delta\delta\delta\delta)$, $\Delta(\delta\delta\delta\delta)$, and $\Lambda(\lambda\lambda\lambda\lambda)$ as shown in Figure 2(b). It is evident that the two stereoisomers in which the helicity of the arms and the ring are the same are enantiomers, as are the two isomers in which the helicity of the arms and the ring are opposed. Thus, there are two diastereoisomers defined as a square antiprism (SAP) and a twisted square antiprism (TSAP) [21]. The four stereoisomers can interconvert, either through rotation of the pendant arms or through inversion of the macrocyclic ring. This means each isomer of a DOTA-type chelate may exist in solution; i.e., both SAP and TSAP isomers may be present in proportion to the relative energies of the structures [22]. This is significant because these two structures exhibit vastly different water exchange kinetics [23]. The SAP isomer exchanges water 50–100 times slower than the TSAP isomer and so SAP isomers are generally preferred for paraCEST applications. Furthermore,

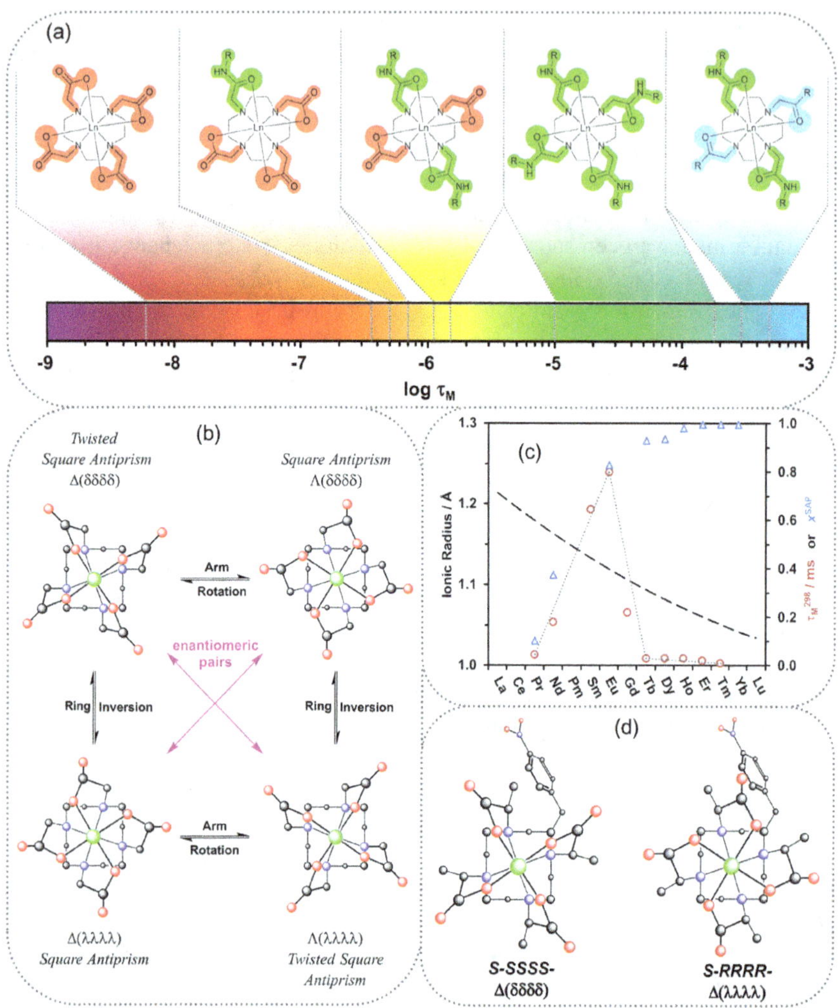


**FIGURE 2** (a) A schematic illustrating the relationship between water exchange lifetime for a series of LnDOTA-(amide)$_x$ complexes *versus* the electron donating capacity of the ligating groups on the pendant arms. (b) A schematic illustrating the structure and dynamics of the diastereoisomers present in LnDOTA-like complexes. The symbols Λ and Δ refer to the helicity of the acetate arms, while (*λλλλ*) and (*δδδδ*) refer to the helicity of the 5-membered Ln- ethylenediamine chelate rings of the macrocycle. (c) The water exchange lifetime for the series of LnDOTA-(gly-OEt)$_4$ complexes (open circles) increase sharply from $Pr^{3+}$ to $Eu^{3+}$, decrease sharply from $Eu^{3+}$ to $Tb^{3+}$, then gradually decrease for the smaller ionic radii $Ln^{3+}$ complexes. This illustrates that the water exchange rates do not correlate directly to changes in ionic radii (dashed line). The mole fraction of SAP isomer ($x^{SAP}$) for LnDOTAM chelates is also shown (blue triangles) – pendant arm = $CH_2CONH_2$. (d) Schematic showing how LnDOTA chelates can be frozen into a single configuration by adding a nitro-benzyl group onto a single backbone ethylenediamine carbon of the macrocycle and methyl groups to the α-position of each pendant arm. (Figure adapted and modified from Refs. [21,25,27].)

the SAP isomer generally produces larger hyperfine chemical shifts in all ligand and water resonances thereby further increasing its attractiveness for paraCEST applications. Certain structural modifications can be made to the DOTA framework that will either increase the proportion of SAP isomer present in solution or select it exclusively.

## 2.3 Cation Size

One of the more significant factors in determining the relative energies of the SAP and TSAP isomers is the ionic radius (*r*) of the metal cation. The TSAP conformation of a DOTA-type ligand has a taller coordination cage owing to the smaller N–C–C–O torsion angle. This means that the early lanthanides (larger metal cations) are better accommodated within a TSAP conformation. For DOTA-type chelates of $Ce^{3+}$ ($r$ = 1.2 Å) [24], the TSAP isomer is almost always observed [17,22c], but as the ionic radius decreases across the $Ln^{3+}$ series, the proportion of SAP isomer increases. Typically, before the ionic radius decreases to ~1.0 Å (near $Gd^{3+}$) [24], the SAP isomer has already become the lower energy isomer. The precise ionic radius at which this change occurs depends upon the ligand. Although there are some exceptions [17], the SAP isomer will usually become the only isomer present in solution (Figure 2(c)) [25]. As the ionic radius of the $Ln^{3+}$ continues to decrease along the series, either the SAP isomer will remain the exclusive isomer in solution or the trend will reverse and a TSAP isomer, likely missing an inner-sphere water molecule, appears. This reversal in the isomeric trend is related to the coordination of the axial water molecule. The lighter $Ln^{3+}$ ions (early in the series) are larger in ionic radius so can easily accommodate a capping water molecule. In these cases, the metal ion sits high in the coordination cage so the water molecule experiences little repulsion from the four oxygen donors in DOTA-type chelates. However, with decreasing ionic radius, the $Ln^{3+}$ ion moves further into the ligand coordination cage and this results in greater repulsion with the four carboxylate oxygen atoms and in a longer Ln-$OH_2$ bond distance. The average position of such "inner-sphere" water molecules might be located at a distance that exceeds what might be considered "bonding" but, nonetheless, such water molecules do play a role in the overall energetics of the coordination complex [26].

The ionic radius of the metal cation itself also plays a role in determining water exchange kinetics. As discussed earlier, the rate of water exchange depends upon the strength of metal ion–water interaction, with the strength of this interaction governed by several electrostatic interactions, the most obvious of which is between the water oxygen atom and the metal cation. It follows that as the metal cation increases in mass and decreases in size, this electrostatic interaction tends to increase in strength. However, as the size of the metal ion decreases and the metal moves further into the coordination cage [26], this has the effect of pushing the pendant arms toward the coordination site occupied by water. This in turn increases the repulsive electrostatic interaction between the coordinating oxygens of the pendant arms and the water oxygen. Thus, for any given ligand

and metal ion, the strength of the metal–water interaction (hence the rate of water exchange) is governed by competing factors that depend upon the precise ligand structure. The impact on water exchange as a function of cation size cannot be easily predicted. The degree to which the water coordination site is crowded may also impact the mechanism of exchange [19a]. This shift in mechanism can, at most, be a subtle shift from a purely dissociative mechanism to a dissociative mechanism with small amount of interchange character ($I_d$ with a late or early transition state). A shift in the exact mechanism of exchange has been proposed to explain changes in exchange rate constants for chelates of the same ligand across the lanthanide series [7,19a]. In summary, the only way to truly understand the impact of metal cation size on water exchange is to measure it. For example, a study of water exchange kinetics of different LnDOTA-$(glyOEt)_4$) [27] shows an unusual dependence on ionic radii that cannot be simply explained in terms of any one of these factors (Figure 2(c)).

## 2.4 Coordination Isomer Selection through Stereochemical Effects

Introducing a substituent onto any single carbon in the DOTA framework generates a stereocenter. Synthetically, it is easiest to substitute the $\alpha$-carbon of all four pendant arms of the DOTA ligand with the same substituent and stereochemistry. Substituting the arms in this way renders four stereoisomers shown in Figure 2 diastereoisomeric. However, the $^1$H NMR spectra of these tetra-$\alpha$-substituted DOTA-type chelates unambiguously contain only two coordination isomers [28]. This is because two of the coordination isomers are effectively inaccessible – when the configuration of the ethylene diamine backbone carbons opposes the helicity of the arms (i.e., R-Δ or S-$\Lambda$), the energy of the stereoisomer is too high to access at temperatures possible in aqueous solution. This means that only the RRRR-$\Lambda(\delta\delta\delta\delta)$ (SAP) and RRRR-$\Lambda(\lambda\lambda\lambda\lambda)$ (TSAP) isomers (or their corresponding enantiomers) are accessible – the stereochemistry of the configuration at carbon has determined the helicity of the pendant arms. It is notable that substituting the $\alpha$-position of the pendant arms is generally found to have the effect of increasing the proportion of TSAP present in solution [23a,29], which is the opposite configuration needed for CEST.

It follows that the same logic can be applied to determining the helicity of the macrocyclic ring. If substituents are introduced onto the ethylene bridges of the macrocycle, they will adopt an equatorial position which in turn defines the helicity of the ring: an S- configuration affords a $\delta\delta\delta\delta$ conformation and an R- configuration $\lambda\lambda\lambda\lambda$ (Figure 2(d)). It is worth noting at this point that the size and number of substituents are important considerations for control of helicity at both the arms and the ring. In general, a single substituent is sufficient to achieve control over helicity as long as this substituent has sufficient steric bulk [30]. The lower limit of required steric bulk is unclear, but it is known that a single methyl group is insufficient to exact complete control [31]. This means that substituting both the macrocyclic ring and the pendant arms will freeze out the exchange processes by which the four coordination isomers of DOTA-type chelates interconvert.

By considering the configuration at each stereocenter, it is possible to select the most favorable coordination isomer for the application at hand. If the configuration at the ring and the arms are the same, a TSAP isomer will be favored (an advantage for $T_1$ agent design), whereas if the configurations are opposed, a SAP isomer will be favored (an advantage for paraCEST) [23b,32].

## 2.5 Impact of Amide Substituents

Another significant factor in the coordination geometry observed for DOTA-tetraamide derivatives is the number and nature of the amide nitrogen substituents. It is convenient to consider the effect of amide substituents on the $Eu^{3+}$ chelates of DOTA-type chelates, owing to its size and favorable NMR properties. When amide substituents are small, both SAP and TSAP isomers are observed in $Eu^{3+}$ chelates [33]. When no substituents are present (i.e., primary amides), the SAP/TSAP ratio is comparable to that in EuDOTA, about 4:1 in favor of the SAP isomer. The proportion of SAP isomer decreases slightly when one methyl substituent is added (secondary amide, 3.2:1), but when two methyl substituents are added (tertiary amide), a more dramatic change occurs with the TSAP isomer which now favored 1:2 [33]. The preference of chelates incorporating tertiary amides for the TSAP isomer is widely observed [34]. Presumably, this effect arises from steric interactions between the rest of the chelate and substituents in the *cis*-position of the amide, which are reduced in the taller TSAP conformation of the chelate. It is also worth noting that in chelates incorporating tertiary amides, the solvent was also found to have an effect on the SAP/TSAP ratio [34]. Although for *in vivo* applications the medium will always be aqueous, the structure and solvating properties of water are affected by its constituent parts [35] so it may not be safe to assume that the SAP/TSAP ratio remains constant in all aqueous media.

The nature, principally the size, of the amide substituent in chelates based on secondary amides also affects the SAP/TSAP ratio. It is generally observed that larger substituents favor the SAP isomer in an $Eu^{3+}$ chelate. Other factors, such as charge, hydrophobicity/hydrophilicity, and steric demand, do not seem to change the general observation that substituents with a mass greater than about 50–60 D preferentially adopt the SAP coordination geometry [36]. This means that, in general, selecting for a SAP geometry for a EuDOTA-tetraamide is relatively easy – include larger substituents. It is not clear why this should be the case. There is a general observation from crystallographic data [37] that more bulky substituents tend to have a preference for a specific orientation within the chelate framework. This orientation and how it can affect water exchange and CEST are discussed in the next section.

## 2.6 Impact of Amide Substituent Orientation

With careful consideration of the preceding factors, a chelate can be prepared that incorporates an $Ln^{3+}$ ion with an adequate $\Delta\omega$ and a water exchange rate appropriate CEST. This chelate will most likely be a DOTA-tetraamide derivative that adopts the SAP isomer exclusively. However, it is likely that the kinetics of water exchange may require further refinements that can be made by considering conformational aspects of the amide substituents. In the chelates of DOTA-tetraamide derivatives, the space surrounding the coordination site of water on the metal ion is crucial to determining water exchange kinetics. Parker and co-workers found that increasing the hydrophobicity of the groups that surround this space increases the water exchange lifetime [19a,19c]. In addition to adding hydrophobic amide substituents, the orientation of these substituents should also be considered. The crystallographic record suggests bulky substituents may be oriented in one of two ways: either pseudo-axial to the chelate (the volume shaded blue in Figure 3(a)) or pseudo-equatorial to the chelate (the volume shaded in green, Figure 3(a)). Intuitively, if a hydrophobic substituent is oriented in the pseudo-axial position, its impact on the water exchange lifetime is likely to be substantial, much greater than if oriented pseudo-equatorial. However, the crystallographic record suggests that for most bulky substituents, the pseudo-equatorial position is the lowest in energy. One notable exception is phenyl substituents which are found in the pseudo-axial orientation [38], although the reason for this difference remains unclear.

Introducing a stereocenter into the $\delta$-position (onto the amide substituent) may permit some control over the position of substituents. The crystallographic record appears to show that the lowest energy conformation for any DOTA-tetraamide chelate with a $\delta$-stereocenter is for the proton to be oriented inwards towards the metal ion (Figure 3(b, c)). This leaves two positions for the bulky substituent, either pseudo-axial or pseudo-equatorial. It is observed that the pseudo-equatorial position seems to be the preferred (though not exclusive) position for the bulkiest substituent of the stereocenter and this creates a preference for the helicity of pendant arm rotation. This situation is reflected in panel (b). The influence of a $\delta$-stereocenter is weaker than that of an $\alpha$-stereocenter, and although it exerts a preference on the helicity of arm rotation, it will not be determinative. This means that if the pendant arms are induced to rotate against the influence of the $\delta$-stereocenter, then the bulky substituent will be pushed upwards and into the pseudo-axial position (panel (c)). In many cases, this conformation ($R$-$\Delta$/$S$-$\Lambda$) is thermally accessible [36] which led to the hypothesis that the pseudo-axial orientation will slow water exchange [39]. This hypothesis has not been fully verified.

One such example is provided by the $Eu^{3+}$ chelate of the ligand shown in Figure 3(d) [39a]. This ligand has a nitrobenzyl substituent on a single macrocyclic ring carbon in an $S$- configuration which has the effect of "freezing" the macrocycle into a $\delta\delta\delta\delta$ configuration. The pendant amide arms are based on $R$- and $S$-alanine where the configuration at the $\delta$-carbon could be varied. This amide substituent is relatively bulky which tends to favor a SAP conformation. Indeed, when the $\delta$-configuration was $R$-alanine, the pendant arms favored $\Lambda$ helicity and the SAP isomer was obtained exclusively. However, when the $S$-alanine was introduced at the $\delta$-position, then the stereochemistry favors $\Delta$ rotation of the pendant arms

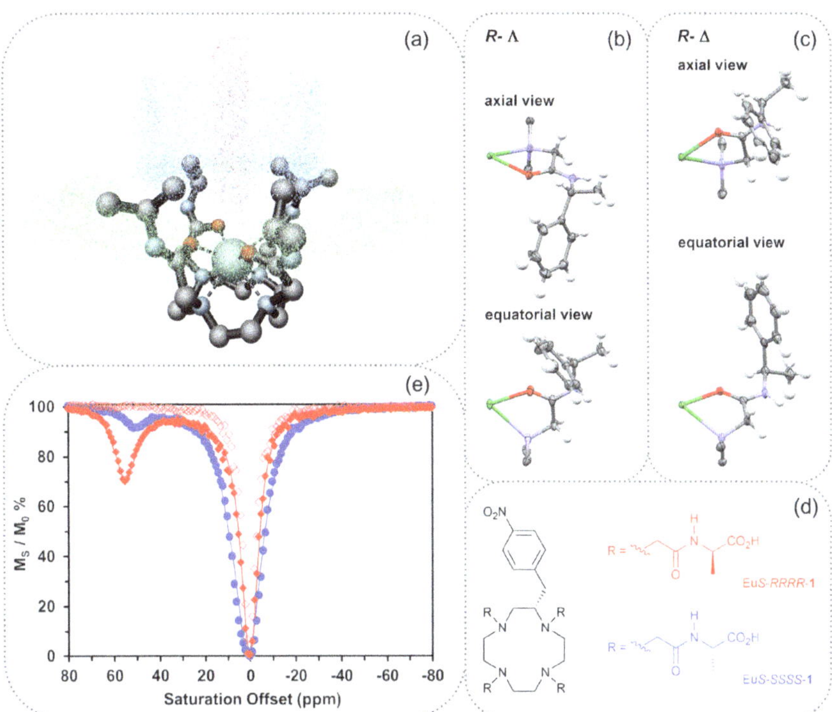


**FIGURE 3** (a) The structure of a DOTA-tetraamide chelate showing the possible orientations of bulky amide substituents. Three volumes represent the space occupied by an exchanging coordinating water molecule (red); the space occupied by substituents in a pseudo-equatorial orientation (green); and the space occupied by substituents in a pseudo-axial orientation (blue). (b) A single pendant arm and chelate ring (shown in axial and equatorial views: the primary axis of the chelate is defined by the Ln-$OH_2$ bond) taken from the crystal structure of an LnDOTA-type chelate containing a -CHMePh substituent with an R- configuration. This configuration favors a Λ helicity at the pendant arms. In this orientation, the bulky phenyl group is angled in the equatorial plane of the chelate. (c) The effect of reversing the helicity of the pendant arms: Δ helicity while retaining the R-configuration at stereocenter. (d) Connectivity structures of chelates in panel (e). (e) The CEST spectra of EuS-RRRR (274 μM, closed red diamonds), EuS-RRRR (13.7 μM, open red diamonds), and EuS-SSSS ([EuL] = 274 μM, [SAP] = 13.7 μM, closed blue circles). The spectra were recorded in $CD_3CN$ (10% $H_2O$) at 400 MHz; 298 K; irr. time = 10 s; $B_1$ = 30 μT. (Figure adapted from Refs. [30,39].)

and a TSAP isomer. In this case, the effects of substituent bulkiness and stereochemistry are opposed, rendering both SAP and TSAP isomers accessible. Indeed, it was found that 5% of Eu*S*-*SSSS*-**1** adopted the SAP configuration, even though the bulkier carboxylate group must adopt a pseudo axial position. In Eu*S*-*RRRR*-**1**, however, the carboxyl group is forced to be in a pseudo-equatorial position, and consequently, this isomer adopts the SAP configuration exclusively. The CEST spectra of both chelates are shown in Figure 3(e) [39a]. From these data, it is evident that Eu*S*-*RRRR*-**1** isomer (100% SAP) displays the largest CEST signal at ~50 ppm. As expected, the CEST signal of Eu*S*-*SSSS*-**1** (5% SAP) is smaller, but surprisingly large considering that the SAP isomer was only ~5% abundant. When Eu*S*-*RRRR*-**1** was diluted to the same concentration as present in the Eu*S*-*SSSS*-1 sample, essentially no CEST was observed. This indicates that Eu*S*-*SSSS*-**1** actually produces a much stronger CEST signal than does Eu*S*-*RRRR*-**1** when compared at equivalent SAP concentrations. Clearly, conformational considerations are acutely important in determining the effectiveness of these types of paraCEST agents.

## 3 RESPONSIVE paraCEST AGENTS AND RATIOMETRIC METHODS

This topic has been recently reviewed [40] so the goal in this chapter is not to summarize once again the entire literature on this topic but to select a few papers that we think will be most instructive in terms of future designs. The basic coordination chemistry principles described above demonstrate the exquisite sensitivity of water exchange rates to relatively small changes in chemical structure and bonding for all chelated forms of the lanthanide ions. For the EuDOTA-tetraamide complexes, even small inductive effects from a single side chain donor atom can have a dramatic impact on the water exchange rate and consequently a CEST signal. This is best exemplified by the work of Ratnakar et al. [41] who demonstrated that the water exchange lifetimes varied from a low of 144 μs (p-F derivative) to a high of 352 μs (p-$CO_2{}^tBu$ derivative) for a series of p-substituted aniline amide complexes and intensity of the water exchange CEST signal varied about 5–6 fold (Figure 4(a)). These early results led to other agent designs whereby the inductive effects of substituents were used to create both redox- and pH-responsive paraCEST agents.

### 3.1 Redox Sensor

Imaging the redox state of tissue by MRI has been an important goal of the molecular imaging community for many years. An example of how this could potentially be done using a simple Eu-based paraCEST is illustrated again by the work of Ratnakar et al. [42]. The reduction potential of the disubstituted N-methylquinolinium amide complex shown in Figure 4(b) is comparable to that of $NAD^+$ so addition of NADH resulted in rapid reduction of the N-methylquinolinium cation to dihydroquinoline. The impact of adding more electrons into the quinoline ring resulted in less electron density on the $Eu^{3+}$ ion, a slowing of water exchange and a subsequently increase in CEST and, unexpectedly, an approximate 7 ppm downfield shift in frequency of CEST water exchange peak. Although it would be difficult to measure tissue redox by simple readout of the intensity of this single CEST peak, this example does illustrate the potential of imaging tissue redox using such an agent provided that another CEST

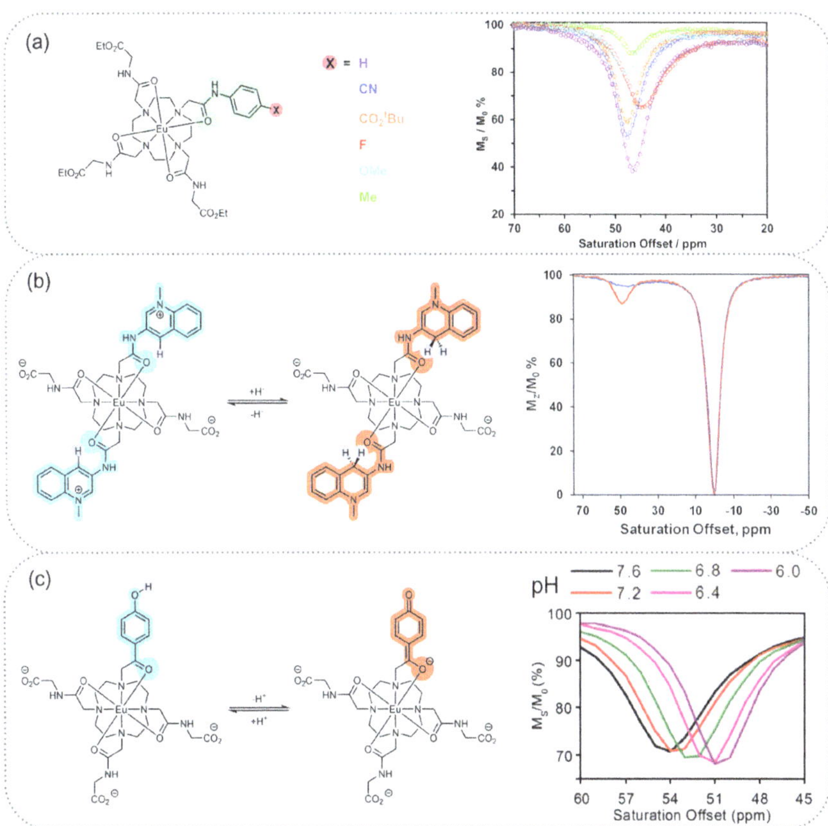


**FIGURE 4** Three examples of how electronic inductive effects can alter the water exchange rate and hence CEST intensity in various EuDOTA-amide complexes. (a) The impact a p-substituent of a single side chain amide on the CEST intensity of the exchanging $Eu^{3+}$-$OH_2$ molecule. (From Ref. [41].) (b) a EuDOTA-tetraamide complex with two N-methylquinolinium amide side chains is readily reduced by NADH. The reduction results in a 10-fold increase in CEST near 50 ppm. (Adapted from Ref. [42].) (c) A $Eu^{3+}$-based paraCEST agent designed to induce changes in chemical shift rather than CEST signal intensity. This allows a direct readout of pH by measuring $M_s/M_0$ at two different chemical shifts. (Adapted from Ref. [43].)

signal, perhaps arising from one or more of the amide –NH groups on the ligand side-chains could be used as an internal concentration marker. This is yet to be demonstrated.

## 3.2 pH Sensor

One of the lessons learned from the redox sensor study was that chemical modifications on one or more of a side-chain amide groups might not only result in electronic inductive effects that alter water exchange rates, but such modifications might also result in detectable differences in chemical shift of the exchanging water molecule. An example of this concept is given by the work of Wu et al. [43] who used it to develop a responsive paraCEST agent for imaging tissue pH *in vivo* (Figure 4(c)). Deprotonation of the phenolic –OH group in this complex ($pK_a = 6.7$) resulted in greater negative charge being placed on the carbonyl oxygen that serves as a direct ligand donor to the central $Eu^{3+}$. Surprisingly, this had little impact on the rate of water exchange over the pH range of interest but had a much larger impact on the chemical shift of the exchanging water molecule due to the increased negative charge placed on the ligand oxygen donor atom that occurs upon deprotonation. As shown in Figure 4, the chemical shift of the water CEST peak shifted downfield ~4 ppm as the pH increased from 5 to 8 in both water and plasma [43]. This was an important observation because it offered the opportunity to measure CEST ($M_s/M_o$) at two different chemical shift offsets and use the ratio of those two CEST signals as a concentration independent index of tissue pH [44].

## 3.3 Glucose and $Zn^{2+}$ Sensors

It has also been shown that one can also modify water exchange rates by simple interference with the Eu-$OH_2$ binding pocket. Two examples of this effect are illustrated in Figure 5. In the first example, two phenylboronate groups were added in trans-positions to a EuDOTA-tetraamide complex to create a paraCEST-based glucose sensor [45]. It was shown that glucose binds with both boronate groups to form a bridge over the "top" of the molecule and this resulted in a 2-fold decrease in the rate of water exchange. This nicely illustrates the possible use of paraCEST agents of this type for imaging important metabolites *in vivo* [45].

A second example is given by the zinc ion sensor also shown in Figure 5. In this case, formation of a $Zn^{2+}$ complex again resulted in the formation of a "bridge" above the Eu-$OH_2$ exchange site, but unlike the glucose sensor, the $Zn^{2+}$ ion acted as a catalyst to promote proton exchange between the $Eu^{3+}$-bound water molecule and bulk water. Consequently, the CEST signal arising from Eu-$OH_2$ exchange disappeared in the $Zn^{2+}$ adduct because water proton exchange became too rapid for CEST [46]. These two examples illustrate the exquisite sensitivity of paraCEST agents such as these to a variety of molecular interactions. In fact, it could be argued that agents like these are so sensitive to their physical environment that when placed in a complex milieu such as tissue, their CEST characteristics could differ considerably from that measured *in vitro*. Nonetheless, the glucose sensor actually worked quite well in perfused livers for monitoring changes in extracellular glucose [45]. The zinc agent could also potentially work *in vivo*, but the affinity of the agent shown here was simply too high for monitoring μM concentrations of extracellular $Zn^{2+}$. More recently, Zhao et al. [47] have shown that it is possible to monitor $Zn^{2+}$ at this concentration using the more sensitive hyper-CEST technique and a genetically altered $Zn^{2+}$ binding protein with hyperpolarized $^{129}Xe$ as the readout signal [47].

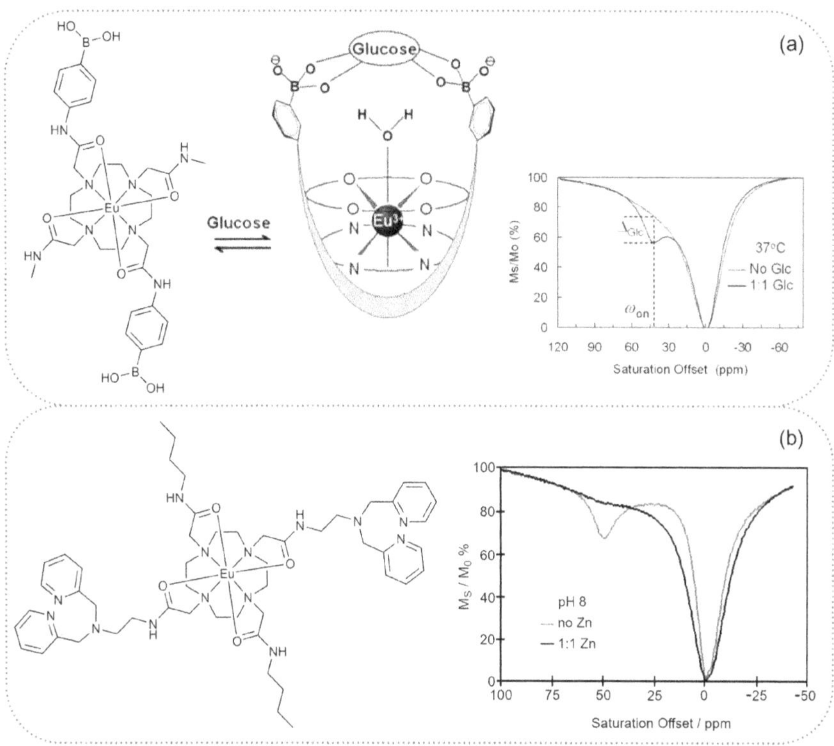


**FIGURE 5** Two examples of paraCEST designs that respond to changes in either water exchange or proton exchange from an inner-sphere Eu-$OH_2$ water molecule. Upon adding glucose to the bis-phenylboronate complex shown in (a), water exchange slows and the CEST signal is turned "on". (Adapted from Ref. [45].) The $Zn^{2+}$ sensor has the opposite effect (b); in this case, when $Zn^{2+}$ binds to all 4 pyridine ligands above the Eu-$OH_2$ binding site, the $Zn^{2+}$ ion catalyzes proton exchange from the Eu-bound water molecule (much like $Zn^{2+}$ in carbonic anhydrase) and this appears to make water exchange faster thereby turning "off" CEST. (Adapted from Ref. [46].)

## 3.4 Ratiometric CEST

It is clear that the intensity of any single CEST signal alone cannot be used to measure pH, glucose, metal ion concentrations, or tissue redox without a separate measure of agent concentration. One of the more common ways to quantify metabolic indices such as these is to use two or more different CEST signals from the same agent that have differing sensitivities to the biological parameter of interest. Aime et al. [48] were first to recognize that the intensity of a water exchange CEST signal is independent of pH, except at the extremes of either very high or very low pH, while the CEST signal arising from an exchanging –NH proton(s) on the ligand is quite sensitive to pH (base catalyzed). This is quite useful because it means that the water CEST signal can be used as a reference and the ratio of these two CEST signals is in principle independent of agent concentration and thereby provides a direct readout of pH. This was the first published example of using the ratio of two different CEST signals to quantify pH. Since this first report, ratiometric CEST has become a standard approach for imaging numerous other biological parameters including enzyme activity, tissue redox, and hypoxia.

This early example illustrates another important point about the differences between water exchange CEST in comparison to amide exchange CEST. In the LnDOTA-$(gly)_4^-$ complexes reported by Aime et al. [48] (where Ln = Pr, Nd, and Eu), the water exchange CEST signal was 2–4 fold more intense than the amide CEST signal at all pH values, even though one exchanging water molecule has only two protons compared to four exchanging –NH protons on each ligand. This advantage reflects the fact that water is exchanging at a much faster rate (~4–10 × $10^4$ $s^{-1}$) compared to the amide protons (~125–130 $s^{-1}$ at pH 6). This ~160-fold advantage for water exchange CEST over proton exchange CEST nicely illustrates the importance of maximizing $k_{ex}$ in the design of any CEST probe. Identifying unique –NH or –OH groups in molecules with much faster proton exchange rates, perhaps as rapid as 4–10 × $10^4$ $s^{-1}$ or more, while maintaining the intermediate-to-slow intermediate exchange limitation for CEST, should be an important goal.

Lui et al. [49] were first to report the use of a YbDO3A-monoamide derivative containing two different types of exchangeable protons, an amide –NH and arylamine –$NH_3^+$, in the same molecule (Figure 6(a)). $Yb^{3+}$ was chosen in this case to induce the largest paramagnetic shift possible in two different exchangeable proton resonances. Interestingly, CEST spectra of the complex identified the amide resonance upfield of water (−11 ppm) and the arylamine resonance downfield of water (+8 ppm), each having a somewhat different pH dependency (Figure 6). The arylamine showed largest CEST signal at all pH values below ~7.5 due to a combination of more exchanging protons (3 for the amine *versus* 1 for the amide) plus a slightly more favorable exchange rate. The fact that the two exchangeable proton resonances straddled the water resonance precluded an asymmetry analysis of the spectra so the CEST magnitude from each exchanging site was determined by fitting the spectra to a three-pool model using Lorentzian line shapes. Although the agent needed to be directly injected into a tumor to reach sufficient concentration for a pH measurement, the agent did report an extracellular pH of 6.5 for a large MCF-7 mammary tumor grown in the flank of a rat. To demonstrate that the agent itself does not modify tissue pH after direct injection, a later study demonstrated that the pH of resting muscle after direct injection of the agent reported a value of 7.26, as expected [50].

## 3.5 paraCEST Agents for Detection of Enzyme Activity

Yoo and Pagel [51] were first to report a paraCEST agent designed to detect enzyme activity. Their design was based on the attachment of caspase-sensitive peptide sequence (DEVD) attached to an unique amino carboxylate side-chain of TmDOTA (Figure 6(b)) which upon cleavage by caspase-3 converted the

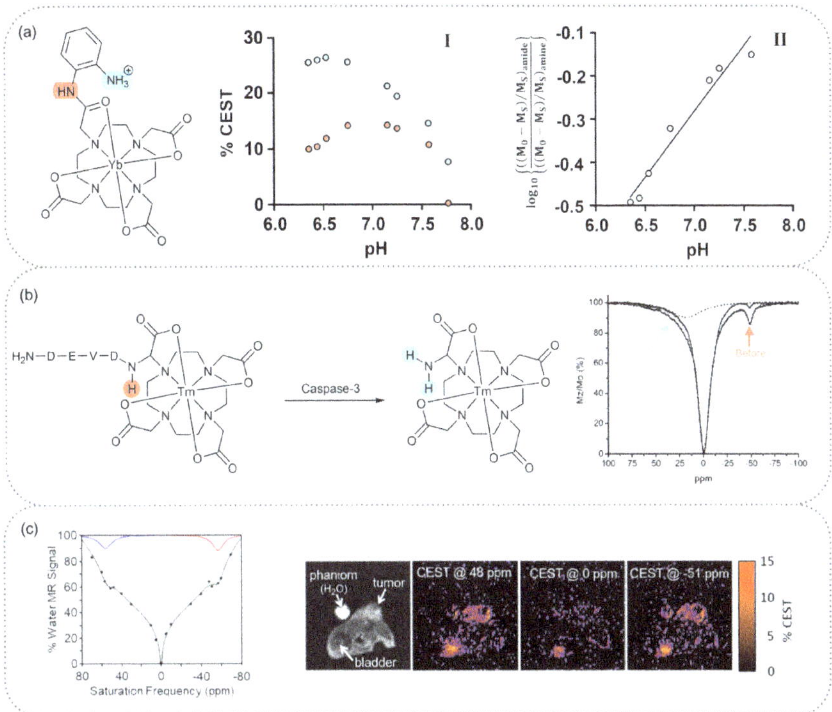


**FIGURE 6** (a) A Yb-based paraCEST agent having two different types of exchanging protons was successfully used to measure extracellular $pH_e$ of tumors and muscle using ratiometric CEST. (Adapted from Ref. [49].) (b) A DEVD-TmDOTA paraCEST agent for monitoring caspase-3 activity. Cleavage of the peptide-like bond on the carboxyl side of aspartate converts the initial CEST-active amide (–NH) into a less CEST-active amine ($-NH_2$) which was monitored by collection of CEST spectra over time. (Adapted from Ref. [51].) (c) The first *in vivo* demonstration of catalyCEST. A cocktail of two paraCEST agents were injected simultaneously and subsequent Lorentzian line shape fitting of each CEST spectra collected over time provided evidence for enzymatic cleavage of ZGGR-amino-TmDOTA (–51 ppm signal). (Adapted from Ref. [52].)

CEST-active amide group (slower proton exchange) to a less CEST-active amino group (faster proton exchange). Hence, a decrease in intensity of the upfield-shifted amide CEST peak (–51 ppm) and the simultaneous appearance of the new broad peak near +8 ppm (assigned to the TmDOTA-amine group) provided a direct measure of caspase-3 activity after deconvolution of several CEST spectra collected over 1 h. These initial experiments were performed *in vitro* where the total agent concentration (substrate + product) was constant so it was of interest to test whether this method might work *in vivo* where the agent concentration in tissue would likely be changing over time.

The first *in vivo* measurement of enzyme activity using a responsive paraCEST agent was reported by Yoo et al. [52] using a similar ZGGR-amino-TmDOTA derivative sensitive to the activity of urokinase plasminogen activator (uPA), a protease biomarker of pancreatic tumor invasion (Figure 6(c)). Cleavage of the peptide from the agent yielded the same TmDOTA-amine product as described above in the caspase study. Tumors expressing uPA were grown in the flank of immune compromised mice and CEST spectra were collected every 5.1 s using fast the CEST-FISP method [53]. In addition to the enzyme responsive Tm agent, the non-enzyme responsive EuDOTA-$(gly)_4$ was injected simultaneously and used as a concentration marker. A comparison of CEST intensities of the two paraCEST agents appearing in the bladder over time indicated that renal clearance of the two agents had similar time constants, while the intensity of the unreacted ZGGR-amino-TmDOTA (-NH CEST signal at −51 ppm) decreased faster than the water CEST signal of EuDOTA-$(gly)_4$ at +48 ppm. This was ascribed to the extracellular activity of uPA in the tumor. This example nicely illustrates that paraCEST agents can potentially be used to detect enzyme activity *in vivo.*

Other designs of paraCEST agents for detection of enzyme activity have also been proposed. Chauvin et al. [54] reported a general approach to target a large variety of enzymes by coupling an enzyme-specific substrate to a YbDOTA-amine through a self-immolative spacer. This platform design is attractive because virtually any enzyme substrate can be attached to the self-immolative spacer and the same product, YbDOTA-amine, is formed in every assay. Unfortunately, this versatile paraCEST design has not, to our knowledge, been applied *in vivo.*

## 3.6 paraCEST Agents for Imaging Extracellular pH ($pH_E$) *In vivo*

Given that all proton exchanging CEST agents (–NH, –OH) are base catalyzed and hence sensitive to changes in solution pH, it is not surprising that many different paraCEST designs have been reported. Many different transition metal ion complexes having exchanging –NH or –OH protons are described elsewhere in this book (Chapter 6). Unfortunately, only a few lanthanide-based paraCEST agents have been applied *in vivo.* Arguably, the most successful agent to date has been YbHPDO3A (Figure 7(a)) [55]. The macrocylic ligand, HP-DO3A, was selected as a paraCEST platform for a couple of reasons. First, this ligand is the same as that found in GdHPDO3A, a widely used clinically approved contrast agent (ProHance®) so stability, safety, and toxicity data had already been established for this system. Second, the –OH proton on the hydroxypropyl side chain bonded directly to the paramagnetic $Yb^{3+}$ ion exchanges with water protons at an appropriate rate required for CEST. Furthermore, this macrocyclic complex exists in aqueous solution as a mixture of SAP and TSAP isomers with differing chemical shifts and, fortuitously, the –OH protons in the two isomers display slightly different proton exchange rates with water. This provided the basis of using this single agent to measure $pH_e$ in two different murine tumor models *in vivo* using ratiometric CEST imaging methods [56].

A second class of paraCEST agent for imaging pH is those that change CEST frequency with changes in pH. The prototype of this class of paraCEST agent, described earlier in Section 3.1, is shown again in Figure 7(b), along with calibration curves and an *in vivo* CEST spectrum collected from a mouse kidney 6 min

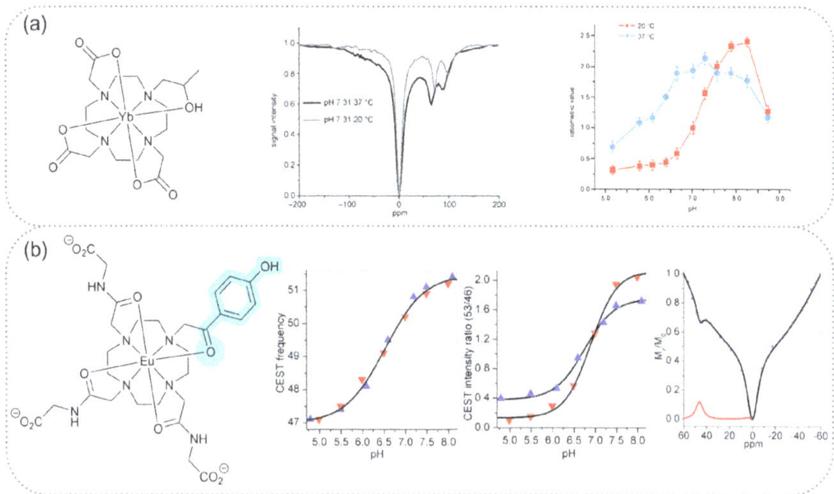


**FIGURE 7** (a) The structure of YbHPDO3A, a prototypical paraCEST agent for imaging tissue pH using ratiometric CEST imaging methods. The two –OH exchange CEST peaks reflect SAP and TSAP isomers, the ratio of which is sensitive to pH over the physiological range of interest. (Adapted from Ref. [55].) (b) The chemical structure of a prototype frequency-dependent paraCEST agent. The two center plots show *in vitro* calibration curves for this agent in two different media, in aqueous solution (red) and in plasma (blue). A measure of CEST frequency *versus* pH is independent of media composition, whereas ratiometric CEST *versus* pH differs in the two media. The plot on the right shows an *in vivo* CEST spectrum of a mouse kidney 6 min after injection of the agent. (Adapted from Ref. [44].)

after IV injection of 0.4 mmol $kg^{-1}$ agent [44]. Two *in vitro* calibration curves were collected, one with the agent dissolved in aqueous biological media (red triangles) and another with the agent dissolved in plasma (blue triangles). The plot of CEST frequency *versus* pH showed identical curves for the two different media, while the ratiometric CEST *versus* pH calibration curve showed quite different results for the two media. This highlights one of the potential problems in using ratiometric CEST measurements in that the calibration curves used to inform the *in vivo* pH values can be quite sensitive to the media in which they are collected. Nonetheless, the frequency-dependent agent described here also had some limitations; first, the exchanging water CEST peak collected during renal excretion of the agent was quite broad so the measured chemical shift had larger errors than expected, and second, the agent was excreted from the mouse kidney quite quickly so the amount of time needed to collect CEST spectra with sufficient resolution for an accurate chemical shift determination was also limited. Some progress has been made toward identifying a frequency-dependent agent with much sharper exchange features (based on –OH exchange rather than water exchange) [57], but further improvements and discoveries will likely appear.

A different non-ratiometric approach for imaging tissue pH *in vivo* was reported by McVicar et al. [58] who used the linewidth of a –NH CEST exchange peak as a direct readout of pH. In that study, they demonstrated that the linewidth of a single –NH amide CEST peak in TmDOTAM-gly-lys was quite sensitive to changes in pH but was independent of temperature over the narrow range anticipated for *in vivo* measurements and, surprisingly, agent concentration as well. Although they demonstrated *in vivo* pH measurements in skeletal muscle after injection of TmDOTAM-gly-lys, the extracellular concentration of agent needed to make this measurement (not reported) was likely quite high, perhaps as much as 3–5 mM. Like the other two methods reported above, this method also requires that sufficient agent be injected to reach a tissue concentration that allows an accurate measure of the CEST –NH exchange peak linewidth. Thus, paraCEST agent sensitivity again appears to be a limiting factor in imaging tissue pH *in vivo*.

Another related approach championed by Coman et al. is to use $^1H$ spectroscopic data collected on paramagnetic complexes to measure both tissue temperature and pH using a method called biosensor imaging of redundant deviation in shifts (BIRDS) [59]. Given CEST signals are also sensitive to these same parameters, the combination of BIRDS [60] and CEST data [61] using a single Tm-based paraCEST agent provides additional redundant data to allow accurate measurement of both temperature and pH in the rodent brain [62]. This powerful combination of $^1H$ spectroscopy and CEST data holds considerable promise for future human imaging applications.

## 3.7 Metabolite Specific CEST Detection of D- and L-Lactate

Aime et al. [63] were first to demonstrate the use of a 7-coordinate paramagnetic complex, $YbMB\text{-}DO3AM^{3+}$, for quantitative detection of lactate by CEST. This complex has two exchanging water binding sites which allow lactate to form a weak bidentate complex with the agent. The CEST spectrum of $YbMB\text{-}DO3AM^{3+}$ alone displays a single, broad CEST exchange peak near –29 ppm (upfield of water), reflecting the six exchanging amide protons, but upon formation of a binary complex with lactate, the amide CEST signal of $YbMB\text{-}DO3AM^{3+}$• lactate shifts downfield to −15.5 ppm. Although the CEST signals of lactate bound and lactate free complexes were not well resolved, a simple titration of L-lactate into a sample of $YbMB\text{-}DO3AM^{3+}$ yielded a binding constant of $8 \times 10^3$ $M^{-1}$ and an amide proton exchange rate, $k_{ex}$, of 1180 $s^{-1}$. This early study demonstrated the possibility of designing paraCEST complexes for metabolite-specific imaging.

Metabolite-specific imaging of lactate was later investigated by direct CEST detection of the exchanging –OH proton in the complex, EuDO3A•lactate [64]. As in the Yb complex, EuDO3A is a 7-coordinate complex that provides two open coordination positions for lactate to bind as a bidentate ligand. The –OH proton of free lactate in solution does have a CEST signal, but it is very near water itself (~0.5 ppm downfield) so it is difficult to detect selectively *in vivo*. The EuDO3A•lactate complex however displays a unique CEST signature near

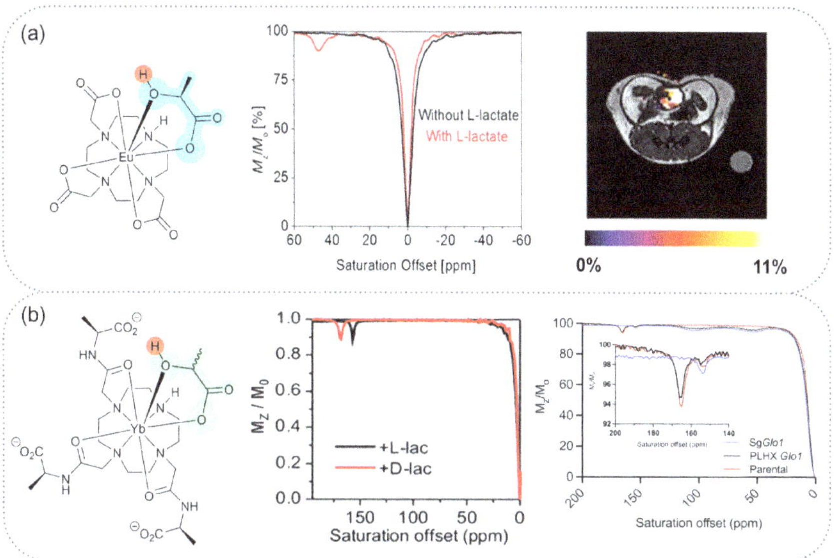


**FIGURE 8** (a) The 7-coordinate complex, EuDO3A, forms a CEST-active binary complex with lactate. The CEST peak near 47 ppm reflects the EuDO3A•lactate complex (exchanging–OH proton). The coronal image of a mouse shows that the complex forms *in vivo* and can be detected by CEST. (Adapted from Ref. [64].) (b) YbDO3A-tris(S-alanylamide) acts as a chiral shift reagent for distinguishing D-lactate *versus* L-lactate by CEST. The chiral SR has been used to show that some cancer cells produce both D- and L-lactate (right). (Adapted from Refs. [65,66].)

47 ppm so it is relatively easy to detect without interference from other endogenous metabolites. Figure 8(a) shows that the binary adduct remains CEST active *in vivo.*

A series of YbDO3A complexes with chiral side chain arms were also prepared to examine whether D- *versus* L-lactate could be distinguished by CEST using a chiral shift reagent [65]. Interestingly, well-separated –OH CEST peaks were observed for the two enantiomeric complexes formed with lactate (16 ppm different), and this formed the basis of using YbDO3A-tris(S-alanylamide) to specifically detect D-lactate production in cancer cells (Figure 8(b)) [66]. These examples demonstrate that metabolite-specific shift reagents can be prepared for selective imaging of important metabolites *in vivo.*

## 4 SUMMARY AND CONCLUSIONS

This short review of paraCEST agents illustrates the early excitement involving this novel class of magnetic resonance imaging agents. Unlike more traditional $T_1$- or $T_2$-based agents that alter image contrast by changes in water relaxation rates, paraCEST agents in general have much less impact on water relaxation rates but rather provide image contrast only after selective saturation of either a highly shifted water molecule or a ligand proton that exchanges into the water pool. Selective detection of the agent itself or some endogenous metabolite that forms a complex with a paraCEST complex does require collection of at least two images with presaturation pulses ON *versus* OFF but this small difference in gathering CEST images is relatively easy using today's modern scanners. The larger concern is the lower sensitivity for detection of paraCEST agents compared to traditional relaxation based contrast agents. This remains the biggest barrier for routine clinical use of these agents so an important question is what are the major factors that contribute to their lower sensitivity *in vivo*?

Given that the number of reported *in vivo* applications of paraCEST agents has been disappointingly small, a worthwhile question is why is this true? Even after optimizing all parameters involved, are paraCEST agents simply too insensitive for *in vivo* applications or are we missing something? Referring once again to Bloch theory, Equation (1) can be modified to include a concentration term using the relationship, $\tau_a = \tau_M$ / ([agent]/110M water protons), where $\tau_b$ in Equation 1 is given by $\tau_M$ (the lifetime of a proton on a metal-based paraCEST agent) [67]. This substitution results in the following equation:

$$\frac{M_Z}{Mo} = \frac{1}{1 + \left([\text{agent}] * T_{1a} / 110M * \tau_M\right)} \quad (2)$$

If one accepts a 5% decrease in the water intensity sufficient for detecting a paraCEST agent *in vivo*, then the term in the denominator, $[\text{agent}]*T_{1a}/110M*\tau_M$, should be about 0.05. If one assumes $T_{1a} = 2$ s for tissue water and a paraCEST agent with a single exchanging proton near the optimal rate shown in Figure 1 ($k_{ex} = 5000\ s^{-1}$ or $\tau_M$ of $2 \times 10^{-4}$), then an agent concentration of 2.2 mM should be detectable by CEST. If another paraCEST agent has 4 exchanging protons per molecule all exchanging at this same rate, then the amount of agent needed to produce a 5% reduction in water intensity would be even more favorable, 0.55 mM. This value is not unlike the typical extracellular tissue concentration of a $Gd^{3+}$-based MRI contrast agent.

Why then has this not been observed *in vivo*? One likely confounding factor is that other fast exchanging protons in tissues, including protons that produce the broad MT signal in many tissues, impact the sensitivity of a paraCEST agent. Woessner et al. [12] predicted such an effect several years ago by pointing out that Bloch theory predicts that other proton exchanging species in tissues can substantially reduce the sensitivity of a paraCEST agent (see Figure 4 in Woessner et al. [12]). This effect has also been demonstrated experimentally. Evbuomwan et al. [68] first demonstrated this effect by mixing a Eu-based TSAP complex having very fast water exchange (no CEST signal) with another Eu-based SAP complex having slow water exchange and a large CEST signal. The net result was that as the TSAP complex was added to a sample of the SAP complex, the CEST signal of the SAP complex was partially quenched. Milne et al. [38] also compared amide CEST sensitivities for a series of TmDOTA-tetraamide-aryl derivatives and found that those lacking an exchanging inner-sphere water molecule showed a larger amide CEST signal compared to the amide CEST signal of TmDOTA-(gly-lys)$_4$ which does have an exchanging inner-sphere water molecule. This also demonstrates that any source of $T_2$ water relaxation, whether it arises from multiple short $T_2$ endogenous species or from a paramagnetic agent, will impact the sensitivity of CEST.

So, after 20+ years of proof-of-concept studies of paraCEST agents as biological reporters, what have we learned? As a CEST community, we have learned how to manipulate water exchange rates in paramagnetic lanthanide complexes over a wide range using coordination chemistry principles, we have learned how to modify –NH and –OH proton exchange rates in metal ion binding ligands using hydrogen bonding, inductive effects, and other acid–base principles, we have learned how to optimize water and proton exchange rates to achieve the maximum CEST signal, we have learned best-practice methods for selective saturation of exchanging proton resonances and analysis of CEST imaging data, and we have learned that paraCEST agents are exquisitely sensitive to other exchanging protons in tissues. It would be disappointing to conclude that paraCEST agents are simply not sensitive enough for most *in vivo* applications. Nonetheless, given the amount of new knowledge gained during development of paraCEST agents plus the advances being made in deep learning and AI techniques, we have reasons to remain optimistic that these interesting agents will one day be applied *in vivo* as biological reporters. Considering that paraCEST agents appear to be quite sensitive to the presence of other exchanging protons in tissues, one potential example to consider is to use a paraCEST agent as MRI reporters of proton pump activity [69] or other biological processes governed proton exchange.

## ACKNOWLEDGMENTS

ADS acknowledges financial support received from multiple NIH grants (R01-CA115531, P41-RR02584, R01-EB004582, P41-EB015908) and the Robert A. Welch Foundation (AT-584) over the years in support of development of paraCEST agents.

## REFERENCES

1. K. M. Ward, A. H. Aletras, R. S. Balaban, *J. Magn. Reson.* **2000**, *143*, 79–87.
2. S. Forsén, R. A. Hoffman, *J. Chem. Phys.* **1963**, *39*, 2892–2901.
3. a). J. Alger, J. Den Hollander, R. Shulman, *Biochemistry* **1982**, *21*, 2957–2963; b). T. R. Brown, K. Ugurbil, R. G. Shulman, *Proc. Natl. Acad. Sci.* **1977**, *74*, 5551–5553; c). D. G. Gadian, G. K. Radda, T. R. Brown, E. M. Chance, M. J. Dawson, D. R. Wilkie, *Biochem. J.* **1981**, *194*, 215–228; d). P. M. Matthews, J. L. Bland, D. G. Gadian, G. K. Radda, *Bioche. Biophys. Res. Commun.* **1981**, *103*, 1052–1059.
4. a). P. C. M. van Zijl, N. N. Yadav, *Magn. Reson. Med.* **2011**, *65*, 927–948; b). M. Zaiss, T. Jin, S.-G. Kim, D. F. Gochberg, *NMR Biomed.* **2022**, *35*, e4789; c). A. Rodríguez-Rodríguez, M. Zaiss, D. Esteban-Gómez, G. Angelovski, C. Platas-Iglesias, *Int. Rev. Phys. Chem.* **2021**, *40*, 51–79.

5. X. Yang, X. Song, Y. Li, G. Liu, S. R. Banerjee, M. G. Pomper, M. T. McMahon, *Angew. Chem. Int. Ed. Engl.* **2013**, *52*, 8116–8119.
6. S. Zhang, P. Winter, K. Wu, A. D. Sherry, *J. Am. Chem. Soc.* **2001**, *123*, 1517–1518.
7. D. Pubanz, G. Gonzalez, D. H. Powell, A. E. Merbach, *Inorg. Chem.* **1995**, *34*, 4447–4453.
8. W. T. Dixon, J. Ren, A. J. M. Lubag, J. Ratnakar, E. Vinogradov, I. Hancu, R. E. Lenkinski, A. D. Sherry, *Magn. Reson. Med.* **2010**, *63*, 625–632.
9. B. Bleaney, *J. Magn. Reson. (1969)* **1972**, *8*, 91–100.
10. J. A. Peters, K. Djanashvili, C. F. G. C. Geraldes, C. Platas-Iglesias, *Coord. Chem. Rev.* **2020**, *406*, 213146.
11. a). A. M. Funk, K. N. A. Finney, P. Harvey, A. M. Kenwright, E. R. Neil, N. J. Rogers, P. Kanthi Senanayake, D. Parker, *Chem. Sci.* **2015**, *6*, 1655–1662; b). E. A. Suturina, K. Mason, C. F. G. C. Geraldes, I. Kuprov, D. Parker, *Angew. Chem. Int. Ed.* **2017**, *56*, 12215–12218.
12. D. E. Woessner, S. Zhang, M. E. Merritt, A. D. Sherry, *Magn. Reson. Med.* **2005**, *53*, 790–799.
13. W. S. Fernando, A. F. Martins, P. Zhao, Y. Wu, G. E. Kiefer, C. Platas-Iglesias, A. D. Sherry, *Inorg. Chem.* **2016**, *55*, 3007–3014.
14. A. Pasha, G. Tircsó, E. T. Benyó, E. Brücher, A. D. Sherry, *Eur. J. Inorg. Chem.* **2007**, *2007*, 4340–4349.
15. A. D. Sherry, Y. Wu, *Curr. Opin. Chem. Biol.* **2013**, *17*, 167–174.
16. D. H. Powell, O. M. N. Dhubhghaill, D. Pubanz, L. Helm, Y. S. Lebedev, W. Schlaepfer, A. E. Merbach, *J. Am. Chem. Soc.* **1996**, *118*, 9333–9346.
17. S. Aime, M. Botta, Z. Garda, B. E. Kucera, G. Tircso, V. G. Young, M. Woods, *Inorg. Chem.* **2011**, *50*, 7955–7965.
18. a). R. Pujales-Paradela, T. Savić, D. Esteban-Gómez, G. Angelovski, F. Carniato, M. Botta, C. Platas-Iglesias, *Chem. Eur. J.* **2019**, *25*, 4782–4792; b). B. C. Webber, K. M. Payne, L. N. Rust, C. Cassino, F. Carniato, T. McCormick, M. Botta, M. Woods, Inorg. Chem. 2020, *59*, 9037–9046; c). J. Yu, A. F. Martins, C. Preihs, V. Clavijo Jordan, S. Chirayil, P. Zhao, Y. Wu, K. Nasr, G. E. Kiefer, A. D. Sherry, *J. Am. Chem. Soc.* **2015**, *137*, 14173–14179.
19. a). S. Aime, A. Barge, A. S. Batsanov, M. Botta, D. D. Castelli, F. Fedeli, A. Mortillaro, D. Parker, H. Puschmann, *Chem. Commun.* **2002**, 1120–1121; b). A. Barge, M. Botta, D. Parker, H. Puschmann, *Chem. Commun.* **2003**, 1386–1387; c). A. L. Thompson, D. Parker, D. A. Fulton, J. A. K. Howard, S. U. Pandya, H. Puschmann, K. Senanayake, P. A. Stenson, A. Badari, M. Botta, S. Avedano, S. Aime, *Dalton Trans.* **2006**, 5605–5616.
20. K. N. Green, S. Viswanathan, F. A. Rojas-Quijano, Z. Kovacs, A. D. Sherry, *Inorg. Chem.* **2011**, *50*, 1648–1655.
21. S. Aime, M. Botta, M. Fasano, M. P. M. Marques, C. F. G. C. Geraldes, D. Pubanz, A. E. Merbach, *Inorg. Chem.* **1997**, *36*, 2059–2068.
22. a). S. Aime, A. Barge, M. Botta, M. Fasano, J. Danilo Ayala, G. Bombieri, *Inorg. Chim. Acta* **1996**, *246*, 423–429; b). S. Aime, M. Botta, G. Ermondi, F. Fedeli, F. Uggeri, *Inorg. Chem.* **1992**, *31*, 1100–1103; c). S. Aime, M. Botta, G. Ermondi, E. Terreno, P. L. Anelli, F. Fedeli, F. Uggeri, *Inorg. Chem.* **1996**, *35*, 2726–2736; d). M. P. M. Marques, C. F. G. C. Geraldes, A. D. Sherry, A. E. Merbach, H. Powell, D. Pubanz, S. Aime, M. Botta, *J. Alloys Compd.* **1995**, *225*, 303–307.
23. a). M. Woods, S. Aime, M. Botta, J. A. K. Howard, J. M. Moloney, M. Navet, D. Parker, M. Port, O. Rousseaux, *J. Am. Chem. Soc.* **2000**, *122*, 9781–9792; b). M. Woods, Z. Kovacs, S. Zhang, A. D. Sherry, *Angew. Chem. Int. Ed.* **2003**, *42*, 5889–5892.
24. R. D. Shannon, *Acta Crystallogr. Sect.* A **1976**, *32*, 751–767.

25. J. Vipond, M. Woods, P. Zhao, G. Tircsó, J. Ren, S. G. Bott, D. Ogrin, G. E. Kiefer, Z. Kovacs, A. D. Sherry, *Inorg. Chem.* **2007**, *46*, 2584–2595.
26. M. Woods, K. M. Payne, E. J. Valente, B. E. Kucera, V. G. Young Jr., *Chem. Eur. J.* **2019**, *25*, 9997–10005.
27. S. Zhang, K. Wu, A. D. Sherry, *J. Am. Chem. Soc.* **2002**, *124*, 4226–4227.
28. a). H. G. Brittain, J. F. Desreux, *Inorg. Chem.* **1984**, *23*, 4459–4466; b). J. A. Howard, A. M. Kenwright, J. M. Moloney, D. Parker, M. Woods, M. Port, M. Navet, O. Rousseau, *Chem. Commun.* **1998**, 1381–1382.
29. E. C. Wiener, M.-C. Abadjian, R. Sengar, L. Vander Elst, C. Van Niekerk, D. B. Grotjahn, P. Y. Leung, C. Schulte, C. E. Moore, A. L. Rheingold, *Inorg. Chem.* **2014**, *53*, 6554–6568.
30. M. Woods, Z. Kovacs, R. Kiraly, E. Brücher, S. Zhang, A. D. Sherry, *Inorg. Chem.* **2004**, *43*, 2845–2851.
31. R. S. Ranganathan, N. Raju, H. Fan, X. Zhang, M. F. Tweedle, J. F. Desreux, V. Jacques, *Inorg. Chem.* **2002**, *41*, 6856–6866.
32. M. Woods, M. Botta, S. Avedano, J. Wang, A. D. Sherry, *Dalton Trans.* **2005**, 3829–3837.
33. S. Aime, A. Barge, J. I. Bruce, M. Botta, J. A. K. Howard, J. M. Moloney, D. Parker, A. S. de Sousa, M. Woods, *J. Am. Chem. Soc.* **1999**, *121*, 5762–5771.
34. K. J. Miller, A. A. Saherwala, B. C. Webber, Y. Wu, A. D. Sherry, M. Woods, *Inorg. Chem.* **2010**, *49*, 8662–8664.
35. K. M. Payne, J. M. Wilds, F. Carniato, M. Botta, M. Woods, Isr. *J. Ch*em. **2017**, *57*, 880–886.
36. T. Mani, G. Tircsó, P. Zhao, A. D. Sherry, M. Woods, *Inorg. Chem.* **2009**, *48*, 10338–10345.
37. a). R. S. Dickins, S. Aime, A. S. Batsanov, A. Beeby, M. Botta, J. I. Bruce, J. A. K. Howard, C. S. Love, D. Parker, R. D. Peacock, H. Puschmann, *J. Am. Chem. Soc.* **2002**, *124*, 12697–12705; b). R. S. Dickins, A. S. Batsanov, J. A. K. Howard, D. Parker, H. Puschmann, S. Salamano, *Dalton Trans.* **2004**, 70–80.
38. M. Milne, M. Lewis, N. McVicar, M. Suchy, R. Bartha, R. H. Hudson, *RSC Adv.* **2014**, *4*, 1666–1674.
39. a). C. E. Carney, A. D. Tran, J. Wang, M. C. Schabel, A. D. Sherry, M. Woods, *Chem. Eur. J.* **2011**, *17*, 10372–10378; b). T. Mani, A. C. L. Opina, P. Zhao, O. M. Evbuomwan, N. Milburn, G. Tircso, C. Kumas, A. D. Sherry, J. Biol. *Inorg. Chem.* **2014**, *19*, 161–171; c). J. R. Slack, M. Woods, J. Biol. *Inorg. Chem.* **2014**, *19*, 173–189.
40. a). A. D. Sherry, D. D. Castelli, S. Aime, *NMR Biomed.* **2023**, *36*, e4698; b). É. Tóth, C. S. Bonnet, *Inorganics* **2019**, *7*, 68.
41. S. J. Ratnakar, M. Woods, A. J. M. Lubag, Z. Kovács, A. D. Sherry, *J. Am. Chem. Soc.* **2008**, *130*, 6–7.
42. S. J. Ratnakar, S. Viswanathan, Z. Kovacs, A. K. Jindal, K. N. Green, A. D. Sherry, *J. Am. Chem.* Soc. **2012**, *134*, 5798–5800.
43. Y. Wu, T. C. Soesbe, G. E. Kiefer, P. Zhao, A. D. Sherry, *J. Am. Chem. Soc.* **2010**, *132*, 14002–14003.
44. Y. Wu, S. Zhang, T. C. Soesbe, J. Yu, E. Vinogradov, R. E. Lenkinski, A. D. Sherry, *Magn. Reson. Med.* **2016**, *75*, 2432–2441.
45. J. Ren, R. Trokowski, S. Zhang, C. R. Malloy, A. D. Sherry, *Magn. Reson. Med.* **2008**, *60*, 1047–1055.
46. R. Trokowski, J. Ren, F. K. Kálmán, A. D. Sherry, Angew. Chem. *Int. Ed. Engl.* **2005**, *44*, 6920–6923.
47. Z. Zhao, M. Zhou, S. D. Zemerov, R. Marmorstein, I. J. Dmochowski, *Chem. Sci.* **2023**, *14*, 3809–3815.

48. a). S. Aime, A. Barge, D. Delli Castelli, F. Fedeli, A. Mortillaro, F. U. Nielsen, E. Terreno, *Magn. Reson. Med.* **2002**, *47*, 639–648; b). S. Aime, D. Delli Castelli, E. Terreno, *Angew. Chem. Int. Ed.* **2002**, *41*, 4334–4336.
49. G. Liu, Y. Li, V. R. Sheth, M. D. Pagel, *Mol. Imaging* **2012**, *11*, 47–57.
50. V. R. Sheth, Y. Li, L. Q. Chen, C. M. Howison, C. A. Flask, M. D. Pagel, *Magn. Reson. Med.* **2012**, *67*, 760–768.
51. B. Yoo, M. D. Pagel, *J. Am. Chem. Soc.* **2006**, *128*, 14032–14033.
52. B. Yoo, V. R. Sheth, C. M. Howison, M. J. Douglas, C. T. Pineda, E. A. Maine, A. F. Baker, M. D. Pagel, *Magn. Reson. Med.* **2014**, *71*, 1221–1230.
53. T. Shah, L. Lu, K. M. Dell, M. D. Pagel, M. A. Griswold, C. A. Flask, *Magn. Reson. Med.* **2011**, *65*, 432–437.
54. T. Chauvin, P. Durand, M. Bernier, H. Meudal, B. T. Doan, F. Noury, B. Badet, J. C. Beloeil, E. Tóth, Angew. *Chem. Int. Ed. Engl.* **2008**, *47*, 4370–4372.
55. D. Delli Castelli, E. Terreno, S. Aime, Angew. *Chem. Int. Ed.* **2011**, *50*, 1798–1800.
56. a). D. Delli Castelli, G. Ferrauto, J. C. Cutrin, E. Terreno, S. Aime, *Magn. Reson. Med.* **2014**, *71*, 326–332; b). G. Ferrauto, E. Di Gregorio, V. Auboiroux, M. Petit, F. Berger, S. Aime, H. Lahrech, *NMR Biomed.* **2018**, *31*, e4005.
57. S. J. Ratnakar, S. Chirayil, A. M. Funk, S. Zhang, J. F. Queiró, C. F. G. C. Geraldes, Z. Kovacs, A. D. Sherry, *Angew. Chem. Int. Ed.* **2020**, *59*, 21671–21676.
58. N. McVicar, A. X. Li, M. Suchý, R. H. E. Hudson, R. S. Menon, R. Bartha, *Magn. Reson. Med.* **2013**, *70*, 1016–1025.
59. D. Coman, H. K. Trubel, R. E. Rycyna, F. Hyder, *NMR Biomed.* **2009**, *22*, 229–239.
60. D. Coman, H. K. Trubel, F. Hyder, *NMR Biomed.* **2010**, *23*, 277–285.
61. D. Coman, G. E. Kiefer, D. L. Rothman, A. D. Sherry, F. Hyder, *NMR Biomed.* **2011**, *24*, 1216–1225.
62. A. B. M. Zakaria, Y. Huang, D. Coman, S. K. Mishra, J. M. Mihailovic, S. Maritim, F. A. Rojas-Quijano, P. Jurek, G. E. Kiefer, F. Hyder, *NMR Biomed.* **2022**, *35*, e4687.
63. S. Aime, D. Delli Castelli, F. Fedeli, E. Terreno, *J. Am. Chem. Soc.* **2002**, *124*, 9364–9365.
64. L. Zhang, A. F. Martins, Y. Mai, P. Zhao, A. M. Funk, M. V. Clavijo Jordan, S. Zhang, W. Chen, Y. Wu, A. D. Sherry, *Chemistry* **2017**, *23*, 1752–1756.
65. L. Zhang, A. F. Martins, P. Zhao, M. Tieu, D. Esteban-Gómez, G. T. McCandless, C. Platas-Iglesias, A. D. Sherry, *J. Am. Chem. Soc.* **2017**, *139*, 17431–17437.
66. E. H. Suh, C. F. G. C. Geraldes, S. Chirayil, B. Faubert, R. Ayala, R. J. DeBerardinis, A. D. Sherry, *Cancer Metab.* **2021**, *9*, 38.
67. S. Zhang, M. Merritt, D. E. Woessner, R. E. Lenkinski, A. D. Sherry, *Acc. Chem. Res.* **2003**, *36*, 783–790.
68. O. M. Evbuomwan, J. Lee, M. Woods, A. D. Sherry, *Inorg. Chem.* **2014**, *53*, 10012–10014.
69. a). M. Minoshima, J. Kikuta, Y. Omori, S. Seno, R. Suehara, H. Maeda, H. Matsuda, M. Ishii, K. Kikuchi, *ACS Cent. Sci.* **2019**, *5*, 1059–1066; b). P. Irrera, L. Consolino, M. Roberto, M. Capozza, C. Dhakan, A. Carella, A. Corrado, D. Villano, A. Anemone, V. Navarro-Tableros, M. Bracesco, W. Dastrù, S. Aime, D. L. Longo, *Cancers* **2022**, *14*, 4916.

# 8 LipoCEST and Similar Systems

*Silvio Aime*
Department of Molecular Biotechnology and Health Sciences, University of Turin,
Via Nizza, 52
I-10126, Turin, Italy
IRCCS SDN SynLAB
Via Gianturco 113, Naples, Italy
silvio.aime@unito.it

*Enzo Terreno*
Department of Molecular Biotechnology and Health Sciences, University of Turin,
Via Nizza, 52
I-10126, Turin, Italy
enzo.terreno@unito.it

## CONTENTS



**Abstract**

This chapter illustrates how the amplification of the properties of paramagnetic lanthanide(III) complexes, entrapped in compartmentalized nano- and micro-sized systems, may offer innovative solutions for the development of MRI CEST-based procedures. The encapsulation of lanthanide shift reagents in the aqueous cavity

DOI: 10.1201/9781003374688-8 

and/or their incorporation into the membrane of liposomes has led to the development of a new class of agents, known as LipoCEST, which have represented a significant advancement in the sensitivity of CEST-based MRI acquisitions. They leverage the compartmental exchange of water molecules, thereby exploiting the vast pool of intraliposomal water proton spins to generate the CEST effect. The chemical shift of the inner water resonance is determined either by the effects associated with the coordination to the paramagnetic metal centre or to the overall shape of the compartment that affects the contribution arising from the changes in the bulk susceptibility term. The outstanding results observed with LipoCEST agents prompted the use of cells as compartments for the lanthanide shift reagents (CellCEST). The peculiar properties of LipoCEST and CellCEST can be jointly exploited to report on their relative biodistribution *in vivo*.

## KEYWORDS

CEST-MRI Agents; Lanthanide Shift Reagents; Liposomes; Cell Labelling

## 1 INTRODUCTION

The peculiar magnetic properties of lanthanide(III) ions (that is by far the most common, and stable, oxidation state) are at the basis of the important role these metal ions have in the study of biological systems. Already in the first decades of the development of NMR, they were often employed to mimic calcium sites on proteins (e.g., in Calmodulin) and to extract detailed information on the prosthetic sites thanks to the marked effect their magnetic properties induce on the shifts and relaxation rates of the protein proton resonances. The advent of high magnetic fields and the fancy 2D sequences had, to some extent, limited their use as structural probes in the definition of 3D-molecular structure of macromolecular systems. However, there are very important legacies from those times, for instance, the use of Gd(III) complexes as MRI-contrast agents that are currently used in about 40% of the MRI exams carried out at clinical settings.

Since 2001, much attention has been devoted to a new procedure to generate contrast in MR images via the transfer of saturated magnetization to the bulk water signal [1]. This occurs because a second radiofrequency field is applied at the resonance frequency of protons exchanging at slow/intermediate regime with the water. The sensitivity of such contrast media, dubbed CEST (Chemical Exchange Saturation Transfer) agents, is in the mM range in terms of exchanging protons, i.e., definitively lower than the one shown by relaxation enhancers, which act at $\mu$M concentration [2]. Although the efficiency of a CEST agent can be maximized by increasing the exchange rate of the saturable spins, $k_{ex}$, this approach is limited by two conditions: (1) the necessity to meet the slow-to-intermediate exchange regime ($\Delta\omega > k_{ex}$), and (2) large $k_{ex}$ values require high energy saturation pulses that may exceed the limits imposed by specific absorption rate (SAR) constraints. On this basis, the search for improving the sensitivity of CEST agents considered the possibility of using systems containing a high number of magnetically equivalent (or pseudo-equivalent) exchanging protons. This led to propose macromolecular and supramolecular systems that showed a marked improvement of the detection threshold down to the $\mu$M range.

The breakthrough in terms of attainable sensitivity came from considering as the pool of exchanging protons the whole ensemble of water molecules contained in a given compartment. This view relies on two basic points: (1) the condition of slow/intermediate exchange regime with the water molecules outside the compartment is determined by the permeability of its barrier, and (2) the $^1$H-chemical shift of the water molecules inside the compartment must be different from the one of the "bulk" waters in the outer space. Thus, for lanthanide shift reagents (LSRs) entrapped inside a nanovesicle such as a liposome, saturation transfer takes place in the form of compartmental exchange. Although other routes (*vide infra*) are possible to generate different chemical shift values for the water molecules in the inner and in the outer compartment, the most efficient approach relies on the interaction with suitable paramagnetic ions in the inner compartment that induce a marked effect on the shift of the metal-coordinated water molecule. Of course, the metal-coordinated water molecules must be in fast exchange with all the free water molecules present in the same compartment to endow all of them with a shift of their resonance frequency. Paramagnetic lanthanide (III) ions (all but Gd(III), at least for spherical liposomes) are the ions of choice for this application as they were already widely investigated in the 1970s of the previous century as LSRs. When specific interactions occur between the paramagnetic ion and the coordinated molecule, a number of useful information can be easily obtained. The lanthanide ion acts as shift reagent when its outer electrons have an asymmetric distribution in the f orbitals, thus generating a magnetic anisotropy and short electronic relaxation time. In turn, the coordination to a paramagnetic Ln(III) yields a marked effect on the chemical shift of the coordinated water protons.

## 2 NMR/MRI AND COMPARTMENTS

In biological systems, the occurrence of compartmentalized spaces (often crucial for the existence of life) is provided by dedicated membranes that are complex natural barriers that isolate the content of the cells or the organelle from the outer environments while controlling exchanges with the external milieu. The membrane and its constituents are involved in essential biochemical processes such as molecular transport, signalling, catalysis, cell–cell interactions, and fusion. In the context of the molecular transport, very important is the control of water cycling as it reflects key aspects of the ongoing cellular metabolism.

The passage of water molecules across the cell membrane occurs either through passive mechanisms (largely dependent on the presence of unsaturated phospholipids and the level of cholesterol) or active mechanisms involving dedicated transmembrane proteins (i.e., aquaporins and ions/nutrients/metabolites transporters).

Biological membranes are lipid bilayers, ~4 nm thick, in a liquid crystalline phase. They are hydrated systems, with a primary hydration shell of a lipid composed of about 20 water molecules [3]. Several membrane mimetics have been developed and are generally prepared using lipids or detergents. The composition of these mimetics strongly influences physical properties such as shape, curvature, thickness, lateral pressure, dielectric constant, and hydration, all factors expected to act as determinants in the water cycling across the membrane. Cholesterol is particularly important as it rigidifies the lipid acyl chains and tunes the membrane fluidity.

Spontaneously, phospholipids in water form large multilamellar vesicles (MLVs) that are inhomogeneous in composition, with a size of the order of 1 μm diameter and up to a dozen bilayers. Although MLV would have the advantage to mimic cells because of their vesicle size and curvature, they may differ markedly as far as the exchange of water is considered as the multilamellarity that obviously generates multiple obstacles to its passage across it. The advantage of the size is recovered with giant unilamellar vesicles (GUVs) whose diameter can reach up to 10 μm. A key point for the intended applications deals with the bilayer fluidity, i.e., the property that controls the physical exchange of water across the membrane and the transfer of saturated magnetization protons from the inside to the extraliposomal bulk water. Higher temperatures result in an increased bilayer fluidity because lipids exist in an ordered gel phase, but as they approach their gel-to-liquid transition point ($T_m$), they assume an intermediate pretransition state (also termed the ripple phase), and finally at temperatures above $T_m$, they enter a disordered fluid state associated with trans-gauche isomerization. Thus, the selection of the components of the bilayer is crucial for the control of the intercompartmental water exchange.

Magnetic interactions are often orientation dependent. Much work on oriented phospholipid containing systems has been done in recent years with bicelles as tools to investigate the structure of membrane proteins. Bicelles are known to spontaneously align in the magnetic field and have a thickness of ~4 nm (like a natural bilayer) as well as a planar surface. The addition of paramagnetic lanthanides with their large magnetic susceptibility flips the orientation of the bicelle normal from perpendicular to parallel to the static magnetic field direction. This effect is achieved at very small lanthanides/lipids molar ratios.

As far as concerns water molecules in the inner compartment of liposomes, the magnetic interactions are isotropic in the case of spherical systems, but become highly anisotropic in the case of non-spherical ones.

## 3 LipoCEST: LIPOSOMES ENCAPSULATING LANTHANIDE SHIFT REAGENTS

LipoCEST agents are systems that act as highly sensitive CEST agents. In principle, the exchanging pool may correspond to the intraliposomal water molecules whose chemical shift is made different from the one of the extravesicular "bulk" pool thanks to the interaction with paramagnetic metal centers suitably encapsulated in the inner cavity of the liposomes (Scheme 1) [4].

Liposomes are nanovesicles made of a phospholipid bilayer that forms spontaneously through self-assembling in water. Being constituted by biocompatible constituents, they have been under intense scrutiny as nanoparticulate drug carriers, and several liposome formulations have been approved for the clinical use.

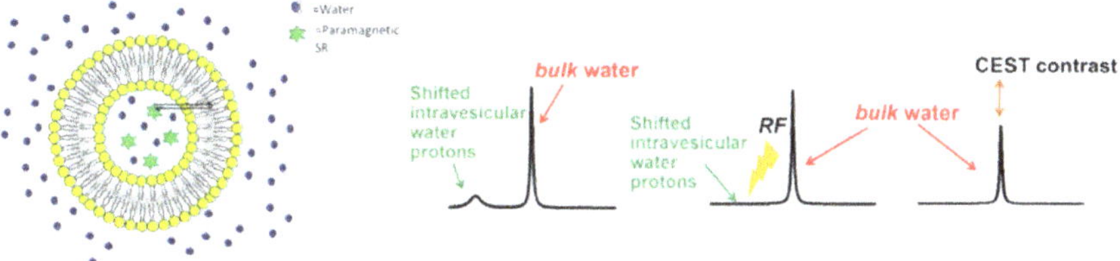


**SCHEME 1** General scheme describing the mechanism of action of a LipoCEST agent. (Adapted from Ref. [4].)

**TABLE 1**
**Water Permeability Values (at 25°C) for a Series of Liposomes with Different Membrane Composition**

| Membrane Formulation | $P_w$ (cm $s^{-1}$)$\times 10^{-5}$ |
|---|---|
| DPPC/DSPE-PEG2000 | 2.6±0.6 |
| POPC/CHOL/DSPE-PEG2000 | 720.0±210.0 |
| POPC/CHOL/DSPE-PEG2000 | 253.4±60.4 |

DPPC: dipalmitoylphosphatidylcholine; POPC: palmitoyl-oleylphosphatidylcholine; DSPE-PEG2000:distearoylphosphoethanolamine-polyethyleneglycol 2000

They can be loaded either with hydrophilic compounds (in the aqueous core) or with hydrophobic chemicals that are embedded in the bilayer or with amphiphilic compounds that intercalate in the liposomal membrane. The presence of cholesterol into the phospholipid bilayers affects the mechanical properties of the membrane and provides an additional stability. Often, long polyethylene glycol (PEG) chains are incorporated on the membrane to increase the physico-chemical stability of the colloidal system, prevent the interaction with macromolecules, and hamper the *in vivo* phagocytosis by the reticuloendothelial system. The water permeability ($P_w$) of the liposome membrane can be modulated by acting on the phospholipid composition (saturated *vs.* unsaturated lipids) and the cholesterol level. Table 1 reports the $P_w$ values of liposomes with different membrane composition [5].

The first report on LipoCEST dated 2005 [6]. In this study, the paramagnetic shift reagent [Tm-DOTMA]$^-$ was used to induce a downfield chemical shift effect on the water protons in the aqueous inner cavities of 3.1 ppm (Figure 1). The intra-liposomal water is continuously exchanging with water molecules in the bulk space, thus yielding a detectable CEST response by compartmental exchange. This approach produced extremely sensitive results and yielded an *in vitro* detection limit of 90 pM (liposome concentration).

The applied procedure did not allow to entrap more LSR in the liposomes, due to the necessity to keep an osmotic equilibrium between the inner and the bulk compartment. Thus, the maximum achievable shift for the intravesicular water resonance was limited to ca. 4 ppm (at higher or lower frequencies according to the encapsulated Ln(III) complex).

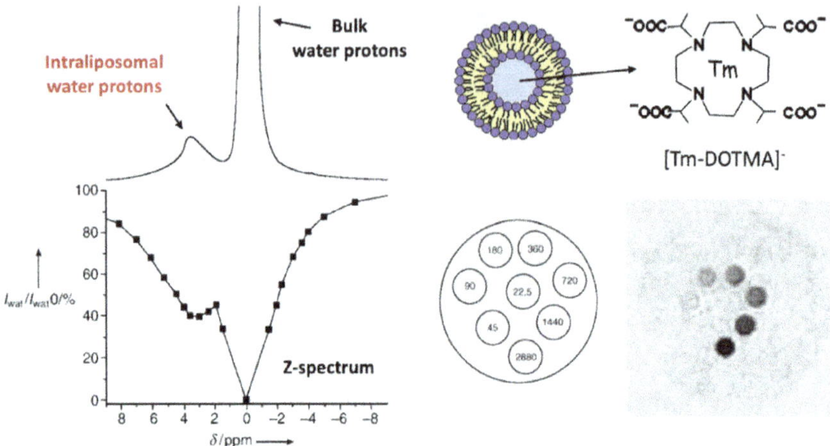


**FIGURE 1** Top left: [1]H-NMR spectrum of a suspension of liposomes encapsulating [Tm-DOTMA]-150 mM. Bottom left: corresponding Z-spectrum. Bottom right: CEST image of a phantom containing different concentration (expressed in pM) of liposomes. (Adapted from Ref. [6].)

A possible strategy to improve the sensitivity of LipoCEST agents is the modulation of the water exchange rate across the membrane. In addition to modify the liposome formulation, as already discussed above, it is important to consider that also the size of the liposomes may affect the $k_{ex}$ value. In principle, size (affecting the curvature) does not influence the water permeability of the membrane as much as the bilayer composition, but it affects the water exchange as the consequence of the change in the volume of the inner cavity of the liposome. In fact, the smaller the volume, the shorter the average residence lifetime of the intraliposomal water molecules and the faster their exchange rate. Zhao et al. showed that the LipoCEST contrast of liposomes loaded with [Tm-DOTA]$^-$ increased by almost sixfold due to the markedly increased exchange rate (from 0.059 $s^{-1}$ to 1.02 $s^{-1}$, ~17 folds) consequent to a size reduction of the nanovesicles from 536 nm to 99 nm [7].

In addition to the detection sensitivity, even the chemical shift offset of the intraliposomal water protons is a relevant factor to be considered for the design of efficient LipoCEST agents.

Although it has been demonstrated that such offset can be increased by encapsulating a LSR with two water molecules coordinated with the metal centre [8], a remarkable improvement was achieved by working with non-spherical liposomes, i.e., taking advantage of the resulting anisotropic magnetic interactions that significantly affect the chemical shifts of the intraliposomal water molecules (Figure 2) [9]. The route to generate non-spherical liposomes relies on the osmometric properties of these vesicles. In fact, when the thin lipidic film is hydrated with a hypotonic solution of the LSR and the liposomes are next suspended in an isotonic medium, the vesicles shrink and lose their spherical shape. This process

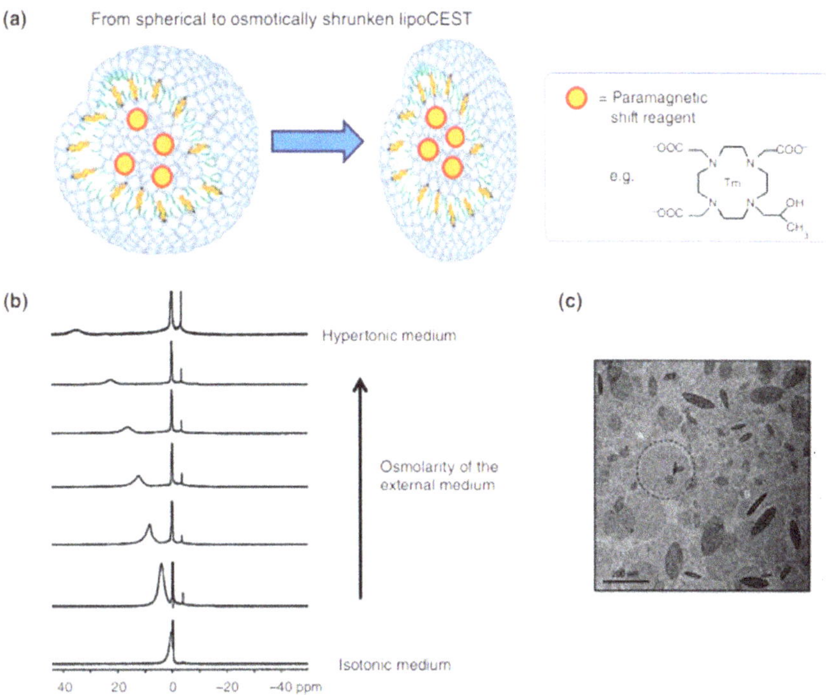


**FIGURE 2** (a) Osmotic shrinkage of a LipoCEST agent; (b) NMR spectra of LipoCEST agents suspended in media with increasing osmolarity; and (c) a representative cryo-TEM image of osmotically shrunken LipoCEST. (Adapted from Ref. [31].)

results in the formation of discoidal or cigar-like liposomes. In such asymmetrical environment, the chemical shift of the intraliposomal water ($\delta^{IL}$) receives a strong contribution from the bulk magnetization susceptibility (BMS) contribution, which adds to the one arising from the contact (usually negligible for water protons coordinated to Ln(III) ions) and dipolar terms:

$$\delta^{IL} = \delta^{IL}_{DIP} + \delta^{IL}_{CON} + \delta^{IL}_{BMS}$$

It has been shown that a further strong increase of the chemical shift of the intraliposomal water protons can be gained by (1) incorporating amphiphilic paramagnetic complexes in the liposomal membrane [10,11] and (2) encapsulating electroneutral multimers of LSRs in the inner aqueous cavity (to increase the actual concentration of paramagnetic shift centres without affecting osmolarity) [12].

Thus, the osmotic shrinkage procedure coupled with the use of properly designed amphiphilic complexes and multimeric derivatives has allowed the development of improved generations of LipoCEST agents, each characterized by its own chemical shift of the intraliposomal water resonance.

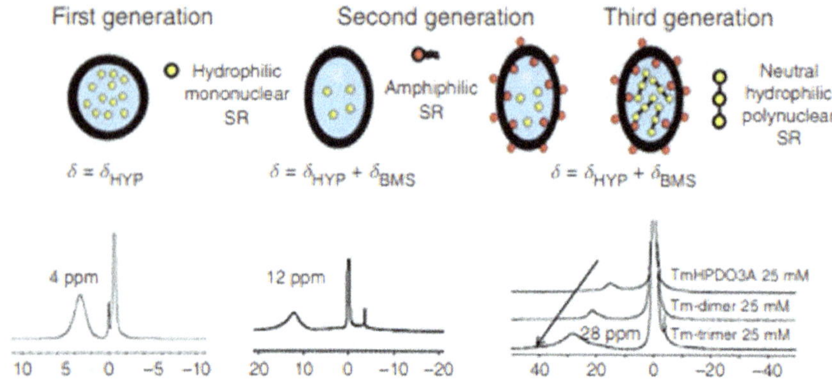


**FIGURE 3** Three generations of LipoCEST agents: (i) spherical, (ii) aspherical with Ln(III) complexes in the inner aqueous cavity and in the membrane, and (iii) as in (ii) with multimeric Ln(III) complexes in the inner cavity. (Adapted from Ref. [13].)

As a result, the CEST offsets of LipoCEST signal were extended from several ppm for spherical liposomes to a much large range (up to 30–45 ppm) for non-spherical systems (Figure 3) [13].

Very interestingly, the occurrence of BMS effects enables to affect the chemical shift of the intraliposomal water protons in aspherical liposomes entrapping a Gd(III) complex (for which $\delta^{DIP}$ and $\delta^{CON}$ terms are null), thus allowing, in principle, to use such systems as triple-modal ($T_1$, $T_2$, and CEST) MRI agents [14].

In addition to lanthanides, also paramagnetic d-metals with electronic configuration suitable for acting as SR have been considered for designing LipoCEST agents. An example is represented by the work from the Morrow's lab [15], where liposomes encapsulating hydrophilic and/or incorporating amphiphilic Co(II) complexes displayed intravesicular chemical shift offset comparable to the systems loaded with Ln(III) ions.

A very insightful experiment showing how the direction of the chemical shift of aspherical liposomes is governed by the alignment of the particle in the external magnetic field was reported by Burdinski et al. [16]. They prepared lens-shaped systems containing LSRs on the liposome membrane (incorporated through an aliphatic chain brought at the ligand's surface) and encapsulating [Ln-HPDO3A] complexes selecting lanthanide ions with different Bleaney's constants ($C_j$) (Dy(III) $C_j<0$ and Tm(III) $C_j>0$) in the inner aqueous cavity as SRs for the water molecules.

The experiment relies on the immobilization of the LipoCEST at the surface of a capillary properly functionalized with cyclodextrins for which the nanoparticle exhibited high affinity thanks to the adamantane moieties present on its surface. The enforced orientation may, however, be different from the magnetic alignment, which is dictated by the magnetic anisotropy of the liposomes (Figure 4). Consequently, the chemical shift offset of the bound LipoCEST may differ from that of the unbound agent, and the authors showed that the alignment can be

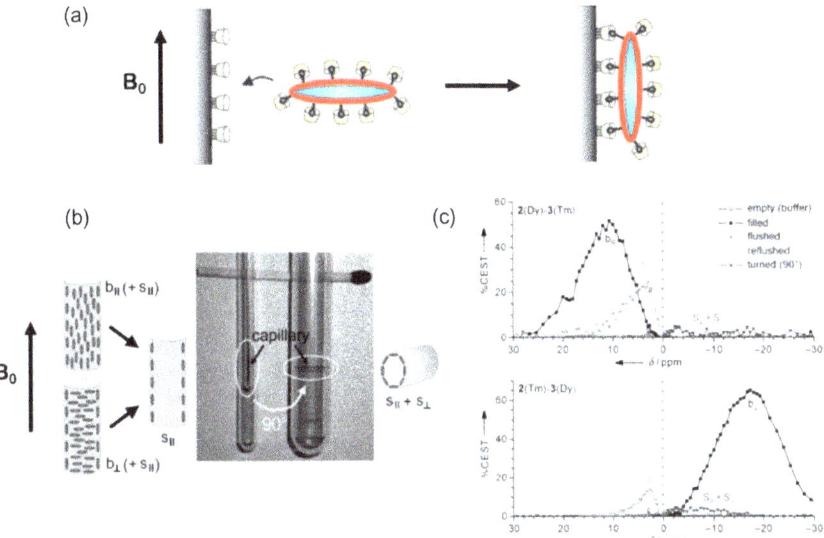


**FIGURE 4** (a) Reorientation with respect to an external magnetic field $B_0$ of aspherical (e.g. oblate) liposomes upon binding to a target surface. The orientation is in turn dictated by the sign of the magnetic anisotropy ($\Delta\chi$) of the incorporated amphiphilic lanthanide complex (Tm or Dy). (b) Parallel and perpendicular alignment of aspherical LipoCEST in bulk solution (b) and at the capillary surface (s). Bulk liposomes were removed by flushing with buffer solution. Photo: capillaries in NMR tubes with fixed orientations with respect to $B_0$. (c) CEST asymmetry spectra of capillaries loaded with two liposome formulations differing in the $\Delta\chi$ value of the Ln(III) metals. (Adapted from Ref. [16].)

nicely controlled upon the proper choice of the magnetic characteristic of the lanthanide complex on the outer surface of the vesicles.

The potential of these observations, i.e., that the orientation induced by the binding of aspherical liposomes to a target surface could be determined by using routine CEST MR methods, is huge. In principle, as outlined by the authors, the use of the chemical shift of the intraliposomal water for target-bound multivalent agents can provide a direct measure for the orientation of the target surface, such as membranes, vessel walls, or nerve fibers, thus adding an anatomical information that would further enhance and complement the molecular imaging approach.

Overall, the access to aspherical LipoCEST has allowed the generation of a broad range of systems that, in principle, can be visualized simultaneously in the same anatomical region simply upon matching their respective absorption frequency (Figure 5).

The *in vivo* potential of LipoCEST in molecular imaging was nicely demonstrated by Flament et al., who formulated a Tm(III)-based LipoCEST decorated with the RGD tripeptide for targeting $\alpha_v\beta_3$ integrin receptors as neo-angiogenesis marker [17], validating the probe in an orthotopic mouse model of a human glioblastoma (Figure 6).

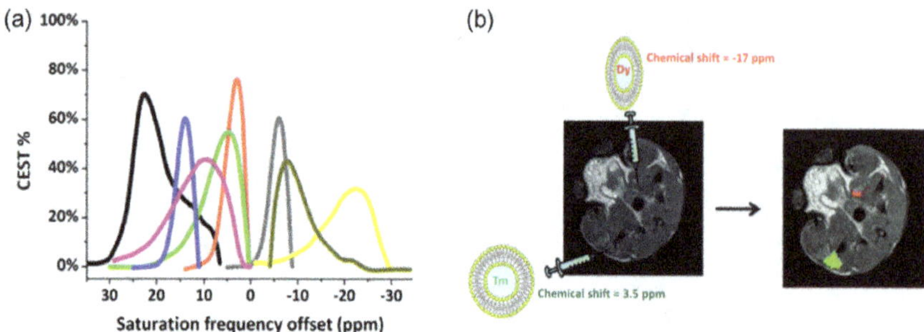


**FIGURE 5** (a) Representative ST-spectra of a mixture of different LipoCEST agents. (b) *In vivo* detection of two LipoCEST agents in the same slice with different chemical shift offsets. (Adapted from Ref. [32].)

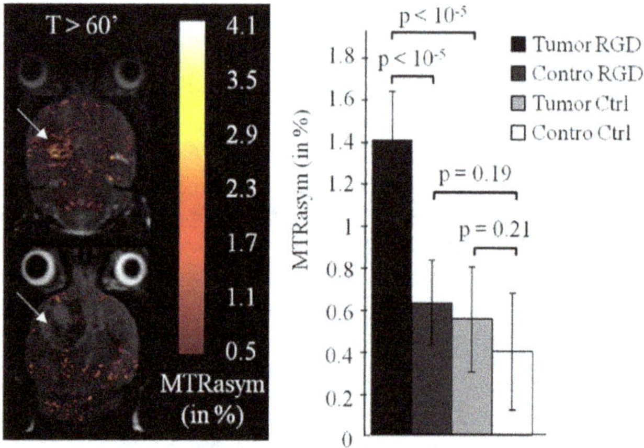


**FIGURE 6** Left: CEST MR maps of mice bearing an orthotopic human glioblastoma (white arrows) 1 h after the injection of LIPOCEST-RGD (top) or LIPOCEST-ctrl (bottom). Right: corresponding CEST contrast. (Adapted from Ref. [17].)

LipoCEST agents can be very useful not only to monitor the delivery of a drug to a given target tissue but also to report about its release. An early example nicely depicts that this property was reported by Langereis et al. [18], who showed that the temperature-sensitive liposomes co-loaded with the LSR [Tm-HPDO3A] and $NH_4PF_6$ have a switchable $^1$H-CEST and $^{19}$F-MRI contrast. In fact, below the $T_m$ point of the liposomes, the proposed agent can be visualized by $^1$H-CEST contrast only (at 11 ppm, spherical vesicles and −17 ppm, aspherical vesicles) because the $^{19}$F of the $PF_6^-$ ion signal is "quenched" by the strong spectral broadening effect caused by the high concentration of paramagnetic shift agents encapsulated in the liposomes. Therefore, when the temperature reaches the $T_m$ value, the $^1$H-CEST signal was no longer detected (as a result of releasing the LSR and losing its effect on the intravesicular water), whereas $^{19}$F-MRI was turned on because the release of both LSR and $PF_6^-$ enabled the removal of the "quenching" effect induced by the LSR. This provided a way to detect not only the location of liposomes but also the subsequent drug release.

The potential of LipoCEST to report about drug delivery and release has also been emphasized by Delli Castelli et al. [19], who applied the multispectral properties of different generations of LipoCEST agents to validate a combined triggered release/imaging strategy for the selective stimulation of the release of material from a mixture of liposomal carriers.

Additionally, the same team effectively demonstrated that the multicontrast capabilities of lanthanide-loaded liposomes, serving as $T_1$, $T_2$, and CEST agents, can provide valuable insights into the *in vivo* intratumoral fate of liposomes [20].

Given that the LipoCEST approach relies on nanovesicular carriers, its applicability can be extended to other liposome-like nanosystems. As an illustrative example, Grüll et al. utilized polymer-based nanovesicles (polymersomes) loaded with hydrophilic and amphiphilic Tm(III) complexes. They demonstrated the similarity of these polymersomes to liposomes in terms of CEST contrast and their response to osmotic forces, forming highly shifted aspherical nanoparticles [21].

## 4 AN ANALYTICAL ASSAY BASED ON LipoCEST AGENTS

The marked sensitivity shown by LipoCEST agents and their exquisite ability to report on the overall shape of the particles prompted the design of an analytical assay based on the changes of the magnetic anisotropy [22]. First, it was showed that the intraliposomal water resonance of a spherical LipoCEST acts as a sensitive reporter of the distribution of streptavidin proteins anchored at the liposome surface by biotinylated phospholipids, i.e. the experimental design relies on the transformation of the spherical system into an asymmetrical one taking advantage of the presence of the biotinyl residues on the outer surface of the liposome membrane (Figure 7).

The chemical shift of intraliposomal water protons increased upon the addition of streptavidin, reaching a maximum value of 12.2 ppm when approximately half of the biotin moieties exposed on the liposome surface were bound to streptavidin. With a further increase in the amount of added streptavidin, the trend reversed, indicating a shift towards a more symmetrical shape and reaching a value of 6 ppm.

On this basis, a MMP-2 responsive LipoCEST agent was designed by inserting a properly tailored enzyme-cleavable peptide linking the biotin/streptavidin moieties to the liposome surface (Figure 8).

As a proof of concept, this approach was applied for assessing the amount of MMP-2 in the serum of breast tumor bearing mice. The dependence of the intraliposomal LipoCEST chemical shift intraliposomal was measured and compared with the one of control healthy animals. By using the calibration curve measured in serum, the amount of MMP-2 was assessed to correspond to ca. 0.83 μM in tumor bearing mice.

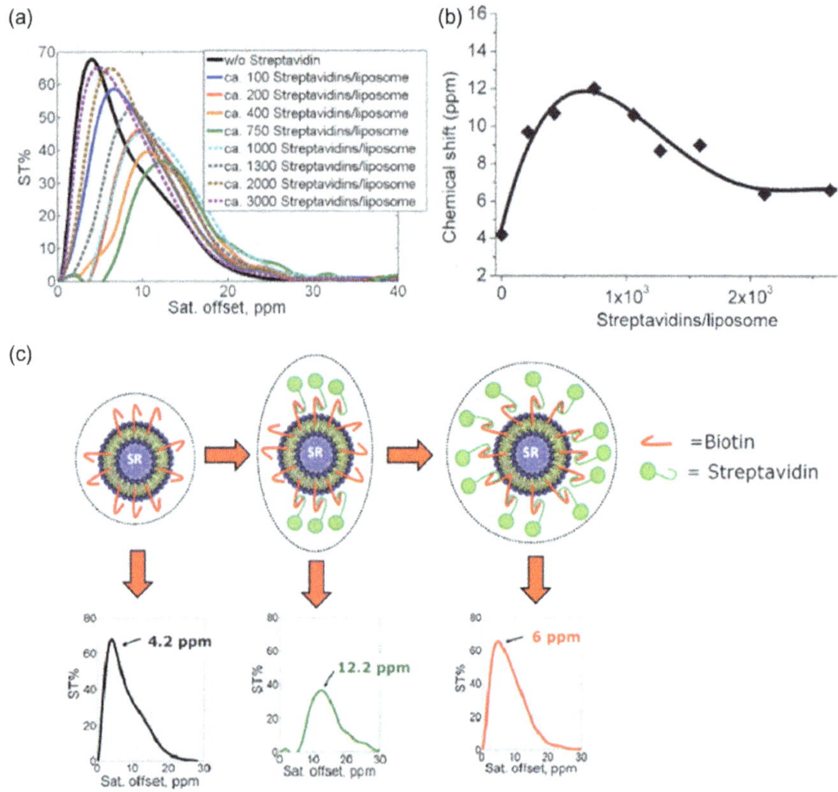


**FIGURE 7** (a) ST-spectra of LipoCESTs at different streptavidin/liposome ratios. (b) Dependence of the intraliposomal chemical shift offset on streptavidin/liposome ratio. (c) Schematic representation of the changes of liposome shape upon binding of streptavidin to exposed biotin moieties and corresponding ST spectra. (Adapted from Ref. [22].)

This method could be applied to the investigation of other enzymes and provided the insertion of the proper peptide on the surface of the LipoCEST agent. The water chemical shift is a parameter that does not depend on the actual concentration of the LipoCEST agent (the knowledge of which is not easily accessible *in vivo*), thus it may be a powerful candidate for the development of *in vivo* MRI-CEST responsive agents.

## 5 GIANT LipoCEST

The already remarkable sensitivity of LipoCEST agents has been recently challenged by the design of giant liposomes (giant unilamellar vesicles, GUVs) loaded with Ln(III) complexes [23]. GUVs are liposomes of micron size often used as cell mimicking systems. Their dimensions (ranging from 0.8 μm to 2 μm or even higher) depend on the methodology of preparation or on the membrane

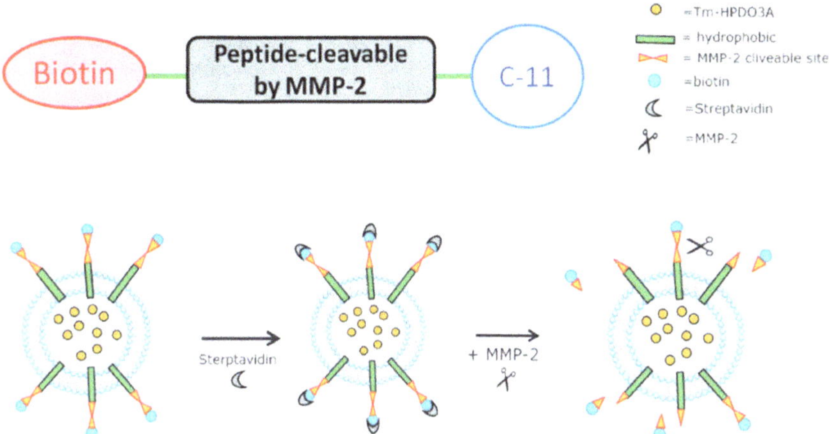


**FIGURE 8** Top: Sketched structure of the MMP-2 responsive biotinylated lipid structure incorporated in a LipoCEST agent. Bottom: Mechanism of action of the responsive probe. (Adapted from Ref. [22].)

composition. In the context of generating improved LipoCEST agents, GUVs and small unilamellar vesicles (SUVs) sharing the same formulation and bearing the same paramagnetic cargo were compared. The results could not be simply accounted in terms of the increase in size (from 100–150 nm to 1– 2 μm). Figure 9 shows a representative fluorescence microscopy image of a spherical GUV and the size distribution of a GUVs suspension.

Small and giant liposomes' suspensions containing 40 mM of the LSR [Tm-HPDO3A] in the aqueous cavity and 15% of an amphiphilic Tm-based complex in the membrane were prepared and resuspended in HEPES/NaCl 0.15 M isotonic buffer to induce an osmotic stress and form aspherical vesicles. The ST spectra presented in Figure 10 showed that the GUVs suspension outperformed SUVs, despite having a much lower concentration of nanovesicles (27 pM compared to 31 nM).

Interestingly, despite the same concentration of the entrapped Shift Reagent, the chemical shift of the intraliposomal water was quite different being 5.2 ppm for the GUV-LipoCEST and 14.0 ppm for the SUV-LipoCEST, respectively. This finding indicates that the intraliposomal shift of GUVs did not appear to be affected by the BMS contribution to the same extent as it was observed for the small ones. Confocal fluorescence microscopy images (Figure 11) showed that the osmotic shock on the GUV vesicles, while maintaining the overall spherical shape, yielded several invaginations.

Clearly, the osmotic stress induced various deformations, resulting in close contacts between opposite bilayers and extensive rearrangements. This, in turn, seems to lead to the formation of a multilamellar system. Despite the structural

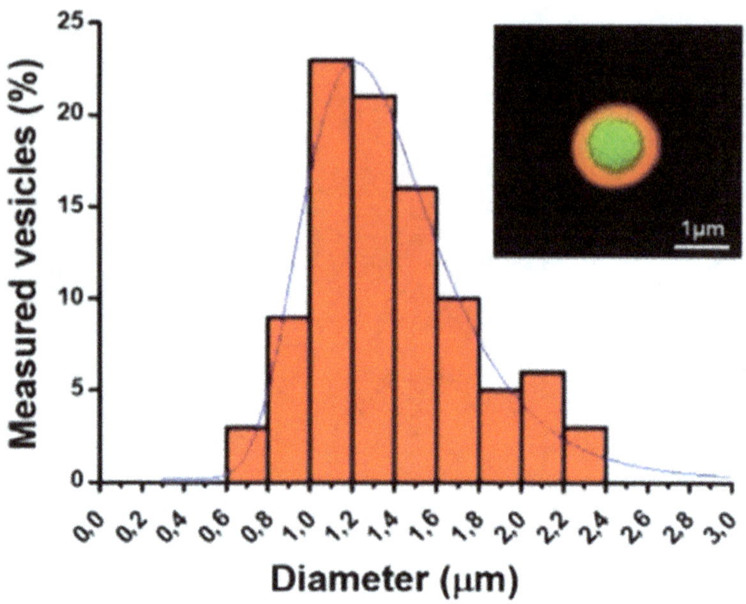


**FIGURE 9** Confocal fluorescence microscopy image of a dual-labelled GUV (inset, red: rhodamine phospholipid incorporated in the bilayer, green: carboxyfluorescein encapsulated in the core) and a typical size distribution of a GUVs sample. (Adapted from Ref. [23].)

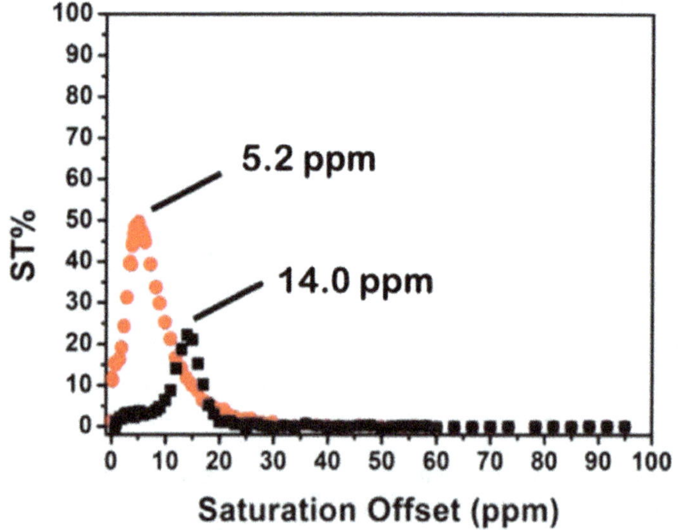


**FIGURE 10** ST spectra of a suspension of GUV-LipoCEST (red circles, 27 pM liposome concentration) and SUV-LipoCEST (black squares, 31 nM concentration) entrapping the same amount of a Tm-based SR and exposed to the same osmotic force. (Adapted from Ref. [23].)

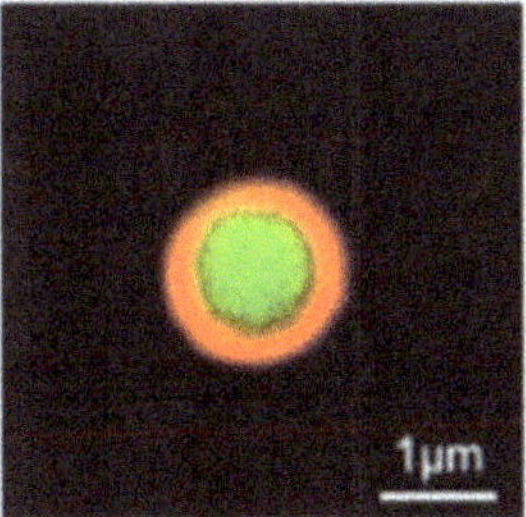

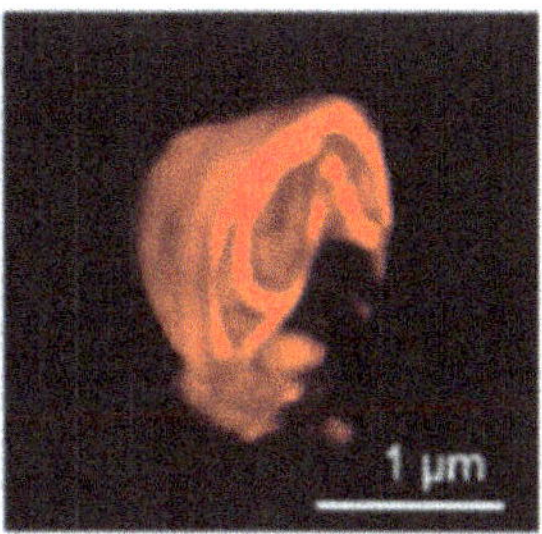


**FIGURE 11** Comparison between 2D confocal fluorescence microscopy images of a spherical GUV (left) and an osmotically stressed GUV (middle). Right: a 3D section of an osmotically stressed GUV. The bilayer of both nanovesicles was labelled with rhodamine-phospholipid (red) and their aqueous core was labelled with carboxyfluorescein (green). (Adapted from Ref. [23].)

differences, the "shrunken" GUV-LipoCEST systems exhibited the expected sensitivity enhancement, making them intriguing candidates for future applications.

## 6 Lipo-ParaCEST: LIPOSOMES ENCAPSULATING PARACEST AGENTS

Liposomes can also serve as carriers for paramagnetic CEST (paraCEST) agents. Different from LipoCEST, the source of the CEST contrast is not generated by the intraliposomal water protons, but it arises from the saturation of the typically highly shifted mobile protons of the encapsulated CEST probe (Scheme 2).

However, such probes may have interesting properties, because the CEST contrast may still be modulated by the water permeability of the nanocarrier, and it can be "activated" as a result of the release of the paraCEST agent [19]. When the water permeability of the nanocarrier is not a limiting factor on the contrast and the paraCEST agent is responsive to a given physico-chemical variable of the biological microenvironment, the liposome may have the advantage to deliver the probe to the desired site. As an example, Opina et al. reported a pH-sensitive Lipo-paraCEST probe encapsulating a 150 mM solution of the pH-responsive paramagnetic CEST agent $[\text{Tm-DOTA(gly)}_4]^-$ [24]. Lanthanide DOTA-tetraglycinate complexes contain four exchangeable amide protons, whose exchange with the water solvent is markedly pH-dependent [25]. Once encapsulated, the paraCEST complex maintained its pH dependence, although the dynamic range of the responsiveness was rather attenuated because of the additional exchange at the liposome bilayer. However, this drawback was compensated for by a consistent amplification in the detection sensitivity (liposome *vs.* metal complex concentration) of approximately four orders of magnitude.

This work was followed by many others essentially devoted to entrapping diamagnetic CEST agents, including sugars [26], drugs [27], iodinated agents [28], and hyperpolarized CEST probes [29], in liposomes.

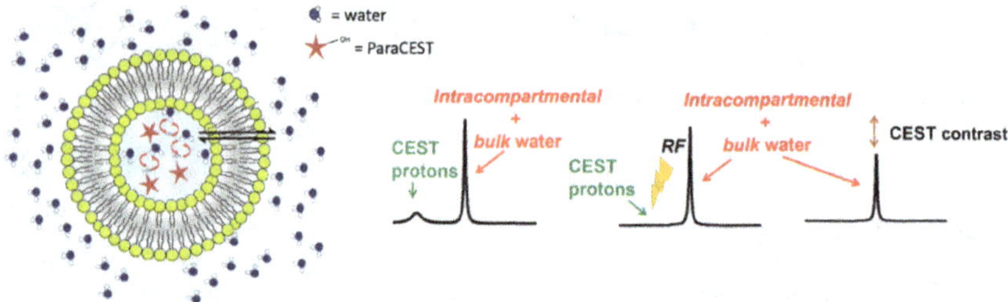


**SCHEME 2** General scheme describing the mechanism of action of a Lipo-ParaCEST agent. (Adapted from Ref. [4].)

Interestingly, very recently, it has also been reported that even label-free liposomes may produce a saturation transfer MRI contrast based on a relayed nuclear Overhauser effect (rNOE) at a chemical shift offset of –1.6 ppm that has been attributed to the choline methyl group on phosphatidylcholines [30].

## 7 CellCEST: LIVING COMPARTMENTS ENCAPSULATING LANTHANIDE SHIFT REAGENTS

The remarkable similarities between liposomes and cells inspired the development of CellCEST systems, where the CEST contrast is the result of the saturation of the intracellular water protons suitably shifted by a LSR (Scheme 3).

Red blood cells (RBCs) emerged as the candidates of choice for testing this possibility due to their: (1) good exchange level of water across their membranes, (2) efficient uptake of metal complexes through the application of the osmotic shock methodology, and (3) natural discoidal shape, which is expected to induce a significant shift in the intracellular water resonance via the BMS contribution [31,32].

First, RBCs were loaded with [Ln-HPDO3A] complexes (Ln=Eu, Gd, Dy, Tm, and Yb) by applying the hypotonic swelling procedure consisting in the incubation of the RBCs in a hypotonic medium containing the shift reagent [33]. The osmotic gradient caused the swelling of the erythrocytes with the formation of pores through which the metal complexes can enter the cytoplasmatic space. This step is followed by the return to iso-osmolar conditions thus allowing the RBCs to assume their usual discoidal shape (Figure 12).

Good CEST responses were obtained with all the tested complexes, with Dy(III)-complex showing the best performance. The CEST properties of RBCs loaded with [Dy-HPDO3A] ($3\times10^8$ Dy complexes/RBC corresponding to ca. 4–5 mM of metal complex in the cytoplasm) were quite different from those of the native, unlabelled RBCs, with a maximum ST effect (ca. 65%) observed at a chemical shift offset of about 6.5 ppm (Figure 13, left). To assess the detection threshold of such systems, the Dy-HPDO3A loaded RBCs suspension was progressively diluted (Figure 13, right) and a detectable ST contrast was observed at

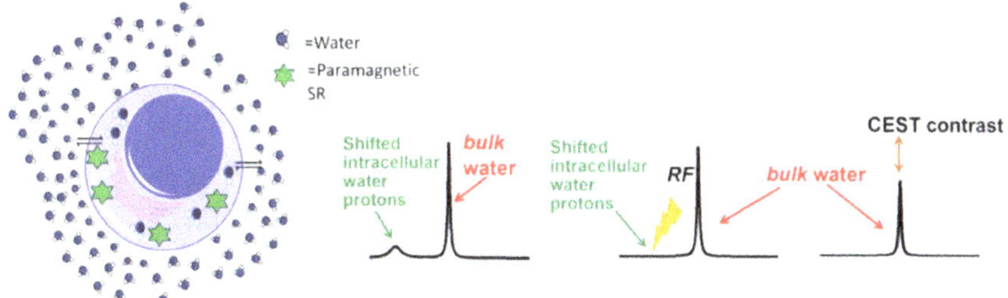


**SCHEME 3** General scheme describing the mechanism of action of a CellCEST system. (Adapted from Ref. [4].)

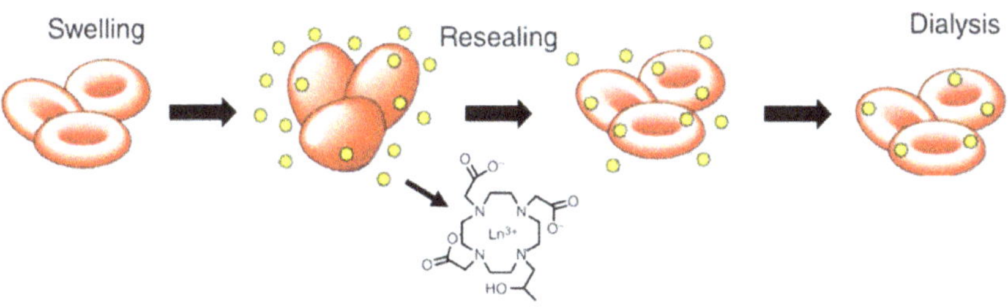


**FIGURE 12** Schematic representation of the procedure for labelling RBCs with [Ln-HPDO3A] complexes. (Adapted from Ref. [32].)

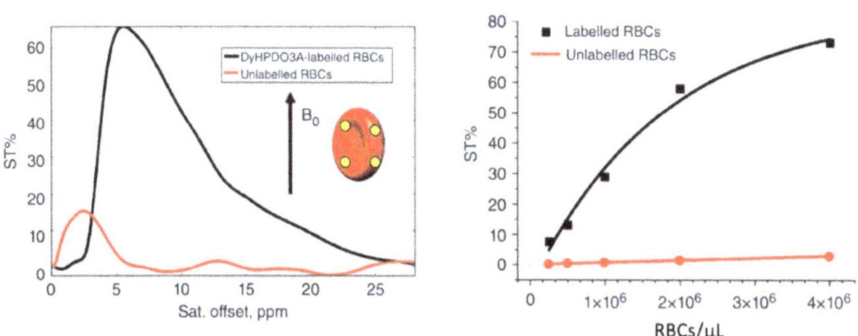


**FIGURE 13** Left: ST-spectra of Dy-labelled RBCs (black curve) and unlabelled RBCs (red curve). Right: CEST contrast dependence on the number of labelled RBCs/$\mu$L. (Adapted from Ref. [32].)

$2.5\times10^5$ labelled cells/$\mu$L, value corresponding to about 5% of the physiologically circulating RBCs in murine blood.

Next, a proof-of-concept for RBC-based CEST agents (dubbed erythroCEST) was reported in the assessment of tumor vasculature volume upon the administration of RBCs labelled with [Dy-HPDO3A] (Figure 14).

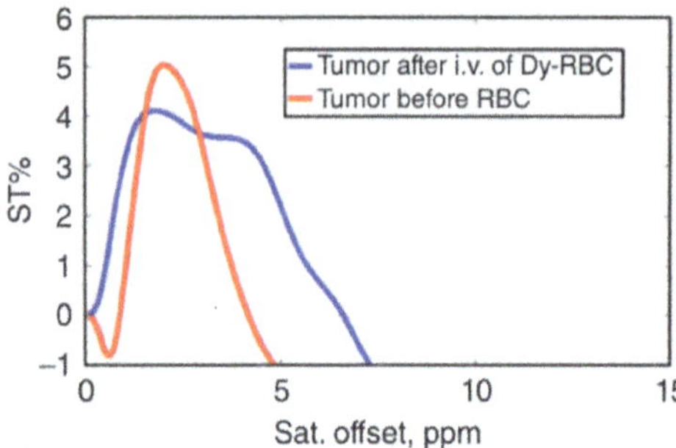


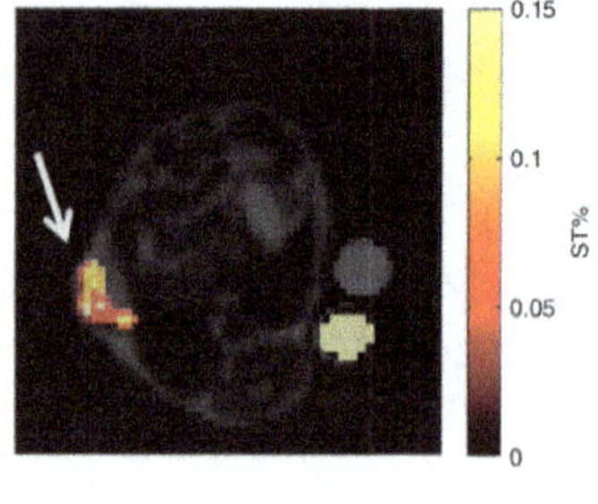


**FIGURE 14** Left: *In vivo* ST-spectra of tumor region before (red line) and after (blue line) the i.v. administration of Dy-labelled RBCs in mouse bearing subcutaneous TS/A breast cancer. Right: parametric CEST map upon irradiation at the frequency of [Dy-HPDO3A]-labelled RBCs. (Adapted from Ref. [32].)

As ErythroCEST is entirely retained in the vascular compartment, the ST% response can directly reflect the vascular volume. Comparing MRI CEST before and after the administration of labelled RBCs can emphasize the more vascularized areas within the tumor [34].

The concept that led to ErythroCEST was also successfully extended to label other cell types, including tumor and immune cell phenotypes [35].

The obtained results support the view that this approach has a general validity although the main problem could be related to the limited water exchange across the plasmalemma membrane. This drawback hampers the exploitation of large pools of compartmental exchanging water, thus limiting the sensitivity of the method.

## 8 COMBINED USE OF LipoCEST AND RBCs

The peculiar discoidal shape of RBCs has a key role to determine the chemical shift of the intracellular water resonance through the BMS term. An analogous effect was observed by anchoring Dy-based LipoCEST agents at the cell surface without encapsulating the LSR into the erythrocytes [31]. The "trick" consisted of using cationic Liposomes to promote an electrostatic binding to the negatively charged RBC membrane. As shown in Figure 15, the observed effect results in two CEST responses from: (i) the intracellular water protons shifted *via* the BMS contribution (positive offset), and (ii) the intraliposomal water protons shifted by the LSR (negative offset).

Quite surprisingly, despite the much larger volume of RBCs, the two contributions showed comparable ST effects. The observed behaviour was accounted in terms of a very high water permeability of the liposomal membrane due to the presence of the unsaturated cationic DOTAP component in the bilayer formulation; $k_{ex}$ was estimated to be ca. $700\,s^{-1}$ for liposomes and ca. $25\,s^{-1}$ for RBCs. Moreover, it is likely that the presence of liposomes on the surface of RBC may

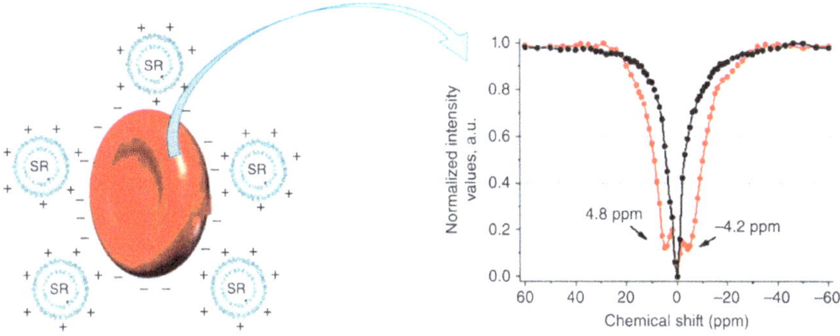


**FIGURE 15** Formation of LipoCEST/RBCs aggregates and Z-spectra of unlabelled RBCs free (black) or anchored to Dy-loaded cationic liposomes (red). (Adapted from Ref. [32].)

further hamper the water exchange in the latter systems. When [Dy-HPDO3A] was replaced by $[Dy\text{-}TTHA]^{3-}$ (complex with no metal coordinated water molecule), the BMS effect remained unaltered but, as expected, no CEST response assignable to the LipoCEST component was detected.

These observations pave the way to potential theranostic applications as the detachment of the LipoCEST from the surface of the RBCs causes the disappearance of the BMS contribution, whereas the integrity of the LipoCEST particles would permit their detection.

## 9 CONCLUSIONS

The field of liposome-imageable agents, drugs, and drug carriers is a quickly developing research area, and it is expected to bring important contributions to the applications in precision medicine. CEST MRI provides a unique opportunity to develop theranostic nano-drug delivery systems, paving a new pathway for constructing highly clinically translatable, image-guided therapeutic treatments.

LipoCEST and related agents arise from the remarkable effects of paramagnetic Ln(III) ions on the chemical shift of compartmentalized water molecules. The extent of the effect is associated with both the direct coordination of water molecules to the paramagnetic metal centre and the changes in the overall magnetic susceptibility term of the resonances of compartmentalized water molecules. In fact, the latter effect may be markedly larger. The MR CEST image readout is highly modulated by the water exchange rate across the phospholipid bilayer, which, for artificial liposomes, can be efficiently controlled through an active manipulation of the components of their membranes. In the case of CellCEST, the permeability of their membranes is an intrinsic property of the investigated cells, and its assessment can provide useful information on their patho-physiological state. Given the interesting properties demonstrated by CellCEST, one can envision new applications in various fields, such as the visualization of exosomes or in the area of CAR-T cells. More generally, the amplification of the magnetic response observed for LipoCEST and related systems may open new avenues for nanotechnology insights to explore fundamental aspects in life sciences as well as innovative diagnostic and therapeutic solutions.

## ABBREVIATIONS AND DEFINITIONS

| | |
|---|---|
| **BMS** | bulk magnetic susceptibility |
| **CellCEST** | cells loaded with SR as CEST agents |
| **CEST** | chemical exchange saturation transfer |
| **ErythroCEST** | RBCs loaded with SR as CEST agents |
| **GUV** | giant unilamellar vesicle |
| **LipoCEST** | liposomes loaded with SR as CEST agents |
| **LSR** | lanthanide shift reagent |
| **MLV** | multilamellar vesicle |
| **MRI** | magnetic resonance imaging |
| **ParaCEST** | paramagnetic CEST agents |
| **RBC** | red blood cells |
| **rNOE** | relayed nuclear overhauser effect |
| **SAR** | specific absorption rate |
| **SR** | shift reagent |
| **SUV** | small unilamellar vesicle |

## REFERENCES

1. K. M. Ward, A. H. Aletras, R. S. Balaban, *J. Magn. Reson.* **2000**, *143*, 79–87.
2. E. Terreno, D. Delli Castelli, S. Aime, *Contrast Media Mol. Imaging* **2010**, *5*, 78–98.
3. H. Hauser, *Biochim. Biophys. Acta* **1984**, *772*, 37–50.
4. G. Ferrauto, E. Terreno, *NMR Biomed.* **2023**, *36*, e4791.
5. E. Terreno, A. Sanino, C. Carrera, D. Delli Castelli, G. B. Giovenzana, A. Lombardi, R. Mazzon, L. Milone, M. Visigalli, S. Aime, *J. Inorg. Biochem.* **2008**, *102*, 1112–1119.
6. S. Aime, D. Delli Castelli, E. Terreno, *Angew. Chem. Int. Ed.* **2005**, *44*, 5513–5515.
7. J. M. Zhao, Y. Harel Y, M. T. McMahon, J. Zhou, A. D. Sherry, G. Sgouros, J. W. M. Bulte, P. C. M. van Zijl, *J. Am. Chem. Soc.* **2008**, *130*, 5178–5184.
8. B. Chahid, L. Vander Elst, J. Flament, F. Boumezbeur, C. Medina, M. Port, R. N. Muller, S. Lesieur, *Contrast Media Mol. Imaging.* **2014**, *9*, 391–399.
9. E. Terreno, C. Cabella, C. Carrera, D. Delli Castelli, R. Mazzon, S. Rollet, J. Stancanello, M. Visigalli, S. Aime, *Angew. Chem. Int. Ed.* **2007**, *46*, 966–968.
10. D. Delli Castelli, E. Terreno, C. Carrera, G. B. Giovenzana, R. Mazzon, S. Rollet, M. Visigalli, S. Aime, *Inorg. Chem.* **2008**, *47*, 2928–2930.
11. E. Terreno, D. Delli Castelli, E. Violante, H. M. H. F. Sanders, N. A. J. M. Sommerdijk, S. Aime, *Chem. Eur. J.* **2009**, *15*, 1440–1448.
12. E. Terreno, A. Barge, L. Beltrami, G. Cravotto, D. Delli Castelli, F. Fedeli, B. Jebasingh, S. Aime, *Chem. Commun.* **2008**, *5*, 600–602.
13. D. Delli Castelli, E. Terreno, D. Longo, S. Aime, *NMR Biomed.* **2013**, *26*, 839–849.
14. S. Aime, D. Delli Castelli, D. Lawson, E. Terreno, *J. Am. Chem. Soc.* **2007**, *129*, 2430–2431.

15. S. M. Abozeid, D. Asik, G. E. Sokolow, J. F. Lovell, A. Y. Nazarenko, J. R. Morrow, *Angew. Chem. Int. Ed.* **2020**, *59*, 12093–12097.
16. D. Burdinski, J. A. Pikkemaat, M. Emrullahoglu, F. Costantini, W. Verboom, S. Langereis, H. Grull, J. Huskens, *Angew. Chem. Int. Ed.* **2010**, *49*, 2227–2229.
17. J. Flament, F. Geffroy, C. Medina, C. Robic, J. F. Mayer, S. Mériaux, J. Valette, P. Robert, M. Port, D. Le Bihan, F. Lethimonnier, F. Boumezbeur, *Magn. Reson. Med.* **2013**, *69*, 179–187.
18. S. Langereis, J. Keupp, J. L. van Velthoven, I. H. de Roos, D. Burdinski, J. A. Pikkemaat, H. Grüll, *J. Am. Chem. Soc.* **2009**, *131*, 1380–1381.
19. D. Delli Castelli, C. Boffa, P. Giustetto, E. Terreno, S. Aime, *J. Biol. Inorg. Chem.* **2014**, *19*, 207–214.
20. D. Delli Castelli, W. Dastrù, E. Terreno, E. Cittadino, F. Mainini, E. Torres, M. Spadaro, S. Aime, *J. Control. Release* **2010**, *144*, 271–279.
21. H. Grüll, S. Langereis, L. Messager, D. Delli Castelli, A. Sanino, E. Torres, E. Terreno, S. Aime, *Soft Matter.* **2010**, *6*, 4847–4850.
22. G. Ferrauto, E. Di Gregorio, M. Ruzza, V. Catanzaro, S. Padovan, S. Aime, *Angew. Chem. Int. Ed.* **2017**, *56*, 12170–12173.
23. M. Tripepi, G. Ferrauto, P.O. Bennardi, S. Aime, D. Delli Castelli, *Angew. Chem. Int. Ed.* **2020**, *59*, 2279–2283.
24. A. C. Opina, K. B. Ghaghada, P. Zhao, G. Kiefer, A. Annapragada, A.D. Sherry, *PLoS One* **2011**, *6*, e27370.
25. S. Aime, A. Barge, D. Delli Castelli, F. Fedeli, A. Mortillaro, F. U. Nielsen, E. Terreno, *Magn. Reson. Med.* **2002**, *47*, 639–648.
26. E. Demetriou, H. E. Story, R. Bofinger, H. C. Hailes, A. B. Tabor, X. Golay, *J. Phys. Chem. B.* **2019**, *123*, 7545–7557.
27. K. W. Chan, T. Yu, Y. Qiao, Q. Liu, M. Yang, H. Patel, G. Liu, K. W. Kinzler, B. Vogelstein, J. W. Bulte, P. C. van Zijl, J. Hanes, S. Zhou, M. T. McMahon, *J. Control. Release* **2014**, *180*, 51–59.
28. L. H. Law, J. Huang, P. Xiao, Y. Liu, Z. Chen, J. H. C. Lai, X. Han, G. W. Y. Cheng, K. H. Tse, K. W. Y. Chan, *J. Control. Release* **2023**, *354*, 208–220.
29. J. O. Jost, L. Schröder, *NMR Biomed.* **2023**, *36*, e4714.
30. Q. X. Wu, H. Q. Liu, Y. J. Wang, T. C. Chen, Z. Y. Wei, J. H. Chang, T. H. Chen, J. Seema, E. C. Lin, *Biomedicines* **2022**, *10*, 1220.
31. G. Ferrauto, E. Di Gregorio, S. Baroni, S. Aime, *Nano Lett.* **2014**, *14*, 6857–6862.
32. G. Ferrauto, D. Delli Castelli, E. Di Gregorio, E. Terreno, S. Aime, *Wiley Interdiscip. Rev. Nanomed. Nanobiotechnol.* **2016**, *8*, 602–618.
33. E. Di Gregorio, G. Ferrauto, E. Gianolio, S. Aime, *Contrast Media Mol. Imaging* **2013**, *8*, 475–486.
34. G. Ferrauto, E. Di Gregorio, W. Dastrù, S. Lanzardo, S. Aime, *Biomaterials* **2015**, *58*, 82–92.
35. G. Ferrauto, E. Di Gregorio, D. Delli Castelli, S. Aime, *Magn. Reson. Med.* **2018**, *80*, 1626–1637.

# 9 Superparamagnetic Iron Oxide and Ferrite Nanoparticles for MRI

*Thomas Vangijzegem, Levy Van Leuven, and Dimitri Stanicki*
General, Organic and Biomedical Chemistry Unit, NMR and Molecular Imaging Laboratory, University of Mons,
B-7000 Mons, Belgium
thomas.vangijzegem@umons.ac.be,
levy.vanleuven@umons.ac.be,
dimitri.stanicki@umons.ac.be

*Robert N. Muller and Sophie Laurent*
General, Organic and Biomedical Chemistry Unit, NMR and Molecular Imaging Laboratory, University of Mons, B-7000 Mons, Belgium
Center for Microscopy and Molecular Imaging, Rue Adrienne Bolland, 8,
B-6041 Gosselies, Belgium
robert.muller@umons.ac.be,
sophie.laurent@umons.ac.be

## CONTENTS



DOI: 10.1201/9781003374688-9

**ABSTRACT**

Starting from the mid-1990s, colloidal suspensions of superparamagnetic nanoparticles, consisting of iron oxides, magnetite ($Fe_3O_4$), maghemite ($Fe_2O_3$) or other ferrites, were introduced as contrast agents for magnetic resonance imaging (MRI) due to their very large magnetic moments and their suitable biocompatibility for *in vitro* and *in vivo* applications. Within the MRI field, superparamagnetic iron oxide nanoparticles (SPION) are well known as $T_2$ contrast agents, producing significant signal loss on $T_2$- and $T_2$*-weighted images. Several SPION formulations have been designed and clinically evaluated as MRI contrast agents, initially as liver-specific contrast agents due to their fast clearance by the mononuclear phagocyte system, but also for other MRI applications such as blood-pool imaging, lymph node imaging or gastrointestinal imaging. In this chapter, we discuss the fundamental aspects of iron oxide nanoparticles with regard to their use as contrast agents for MRI. Specifically, the first section of this chapter describes the theoretical basis of proton relaxation induced by the presence of SPION and provides a detailed description of the superparamagnetic relaxation theory to understand the performances of SPION as contrast agents for MRI. The second part of this chapter focuses on clinically developed SPION-based contrast agents, with a particular emphasis on $T_2$ contrast agents and reasons for their withdrawal from clinical settings. We then discuss recent applications of SPION as efficient $T_1$ contrast agents, including the stringent physicochemical requirements which have to be fulfilled for their satisfactory use as clinically preferred $T_1$ contrast agents when gadolinium-based contrast agents are contraindicated for patients at risk with renal impairment.

## KEYWORDS

Magnetic Resonance Imaging (MRI); Superparamagnetic Iron Oxide Nanoparticles (SPION); Ferrite Nanoparticles; Contrast Agents; Longitudinal Relaxation; Transverse Relaxation

## 1 INTRODUCTION

Iron is the most abundant transition metal in the Earth's crust and, due to its high reactivity, reacts with water and oxygen to yield iron oxides (iron oxides, iron hydroxides and iron oxyhydroxides) [1]. Up to now, 16 phases of oxides, hydroxides and oxyhydroxides have been identified and classified according to their compositions and molecular structures [2]. Among the iron oxides subgroup, hematite ($\alpha$-$Fe_2O_3$), magnetite ($Fe_3O_4$) and maghemite ($\gamma$-$Fe_2O_3$) are the most naturally occurring iron oxide minerals, accompanied by wüstite (FeO). The properties of these three abundant iron oxides are listed in Table 1 [3].

**TABLE 1**
**Chemical and Structural Properties of Hematite ($\alpha$-$Fe_2O_3$), Magnetite ($Fe_3O_4$) and Maghemite ($\gamma$-$Fe_2O_3$) [3]**

| Property | Iron Oxide Phase | | |
|---|---|---|---|
| | Hematite | Magnetite | Maghemite |
| Molecular formula | $\alpha$-$Fe_2O_3$ | $Fe_3O_4$ | $\gamma$-$Fe_2O_3$ |
| Crystal structure | Rhombohedral, hexagonal | Cubic | Cubic |
| Density (g $cm^{-3}$) | 5.26 | 5.18 | 4.87 |
| Melting point (°C) | 1350 | 1597 | / |
| Boiling point (°C) | / | 2623 | / |
| Color | Red | Black | Reddish-brown |
| Hardness (Mohs scale) | 6.5 | 5.5 | 5 |
| Magnetic behavior | Weakly ferromagnetic or antiferromagnetic | Ferrimagnetic | Ferrimagnetic |
| Curie temperature (K) | 956 | 850 | 820–986[a] |
| $M_{SAT}$ at 300 K (A $m^2$ $kg^{-1}$) | 0.3 | 92–100 | 76 |
| Standard free energy of formation ($\Delta G_f^0$) (kJ $mol^{-1}$) | −742.7 | −1012.6 | −711.1 |
| Heat of decomposition (kJ $mol^{-1}$) | 461.4 | 605 | 457.6 |

[a] Range due to thermal conversion of maghemite to hematite at temperatures below the Curie temperature.

Magnetite and maghemite are the only two iron oxides exhibiting strong magnetic properties, i.e., high values of saturation magnetization ($M_S$). At the macroscale, both phases exhibit ferrimagnetic ordering characterized by the formation of organized sub-networks with uncompensated magnetic moments. Such arrangement leads to the formation of a multidomain structure showing magnetic hysteresis (Figure 1a; reprinted from Ref. [4]). This important characteristic leads to the presence of remanent magnetization ($M_R$) and coercivity ($H_C$) in the (de) magnetization curves of ferrimagnetic compounds (Figure 1a; Orange curve). However, when their size decreases to the nanoscale level, transition occurs from nanomaterials bearing a multidomain structure toward single-domain nanoparticles exhibiting a peculiar magnetic behavior called superparamagnetism. The absence of domains leads to the formation of uniformly magnetized nanoparticles without remanence or coercivity (Figure 1a; green curve). This transition toward single-domain nanoparticles exhibiting superparamagnetism usually occurs when decreasing their size below 25 nm (superparamagnetic size ($r_{SP}$); Figure 1b).

This unique form of magnetism has made iron oxide nanoparticles remarkable nanoscale candidates for numerous applications in the biomedical field as

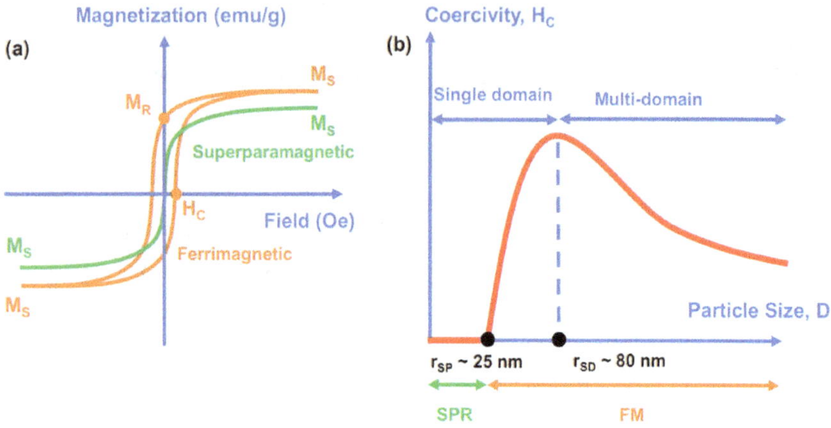


**FIGURE 1** (a) Magnetization *versus* applied field (*M–H*) curves for superparamagnetic (green curve) and ferrimagnetic (FM) (orange curve) materials and (b) relations between size, coercivity and magnetic behavior. (Reprinted with permission from Ref. [4]. Copyright 2021, MDPI Applied Sciences.)

they exhibit a rapid "on–off" switching capability, i.e., they reveal their magnetic properties only when subjected to an external magnetic field and retain no residual magnetism once the applied magnetic field is removed. Superparamagnetic iron oxide nanoparticles (SPION) behave similarly to paramagnets, except that the magnetic susceptibility of one particle is several orders of magnitude higher than the one of paramagnetic substances, hence resulting in higher magnetic properties [5]. Moreover, the return to an equilibrium non-magnetized state after removal of the magnetic field makes SPION unlikely to agglomerate when they are introduced into living systems, thus increasing their half-life circulation times and avoiding the threat of cardiovascular disorders such as thrombosis or blockage of blood capillaries due to particle agglomeration [6,7].

As a result of their superparamagnetic state, combined with their excellent toxicological and safety profiles [8], SPION have been used as probes for diagnosis and/or as therapeutic substances in various applications of the biomedical field such as magnetic resonance imaging, targeted drug delivery [9–13], cell labeling [14–17], magnetic fluid hyperthermia [18,19], gene therapy [20,21], tissue engineering [22,23], and more recently, magnetic particle imaging [24] and radiation therapy [25–27]. SPION have been an integral part in the development of MRI due to their ability to shorten the relaxation rates of surrounding water molecules and improve the contrast. This ability originates from local magnetic field inhomogeneities induced by SPION when submitted to an external magnetic field. Theoretical relaxation models have been developed to describe the influence of SPION on proton relaxation rates [28,29].

## 2 SUPERPARAMAGNETIC RELAXATION THEORY

Single-domain nanoparticles are characterized by magnetic anisotropy, i.e., the magnetic moments of SPION are aligned in energetically favorable positions called "easy axis" or "anisotropy axis". Considering spherical nanoparticles, anisotropy is often considered to be uniaxial, that is, with magnetic moments aligned either parallel or anti-parallel to a single anisotropy axis (Figure 2).

The anisotropy energy ($E_A$), defined as the energy required to flip from one easy axis to another, is proportional to the volume of the considered crystal according to the following equation:

$$E_A = K_A \cdot V \cdot \sin^2 \theta \qquad (1)$$

where $V$ is the volume of the crystal, $\theta$ is the angle between the magnetic moment vector and the easy axis and $K_A$ is an anisotropy constant depending on the physicochemical properties of the considered material.

The volume dependence of anisotropy energy involves that, when decreasing the nanoparticle size, the anisotropy energy decreases until being inferior to thermal agitation energy ($kT$). In this situation, thermal fluctuations are sufficient to induce flips of the magnetic moments between different anisotropy axes. This flipping phenomenon was initially described by Néel who described the time between two flips between the easy axes, known as the Néel relaxation time ($\tau_N$), given as follows:

$$\tau_N = \tau_0 e^{\frac{E_A}{kT}} \qquad (2)$$

where $\tau_0$ is the pre-exponential factor, $k$ is the Boltzmann constant ($1.380649 \times 10^{-23}$ m$^2$kg s$^{-2}$ K$^{-1}$), $T$ is the sample temperature ($K$) and $E_A$ is the anisotropy energy.

Besides Néel relaxation, SPION dispersed in a liquid medium are also subject to Brownian relaxation, corresponding to the complete rotation of SPION due to

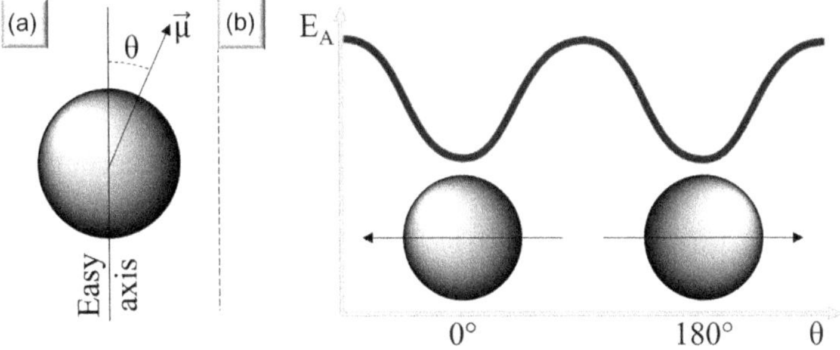


**FIGURE 2** (a) Uniaxial anisotropy axis of spherical nanoparticle, $\theta$ is the angle between the magnetic moment ($\vec{\mu}$) and the easy axis; (b) Evolution of anisotropy energy with the angle between easy axis and magnetic moment.

collisions with solvent molecules. The resulting Brownian relaxation time ($\tau_B$) is given in Equation 3:

$$\tau_B = \frac{3V\eta}{kT} \tag{3}$$

where $V$ is the particle volume and $\eta$ is the fluid viscosity.

The global relaxation rate of SPION dispersed in a liquid medium is dependent upon the two mechanisms of relaxation and it can be expressed as the sum of the two contributions according to Equation 4:

$$\frac{1}{\tau} = \frac{1}{\tau_N} + \frac{1}{\tau_B} \tag{4}$$

Depending on the size of the considered crystal, one of the two relaxation mechanisms will dominate the relaxation of the system. Néel relaxation time will be dominant for small nanoparticles (characterized by lower anisotropy energy), while larger particles will mainly relax through the Brownian contribution of relaxation.

This global relaxation mechanism is only related to the relaxation of the nanoparticle's magnetic moment and not directly the relaxation of water protons. However, dipolar coupling between the magnetic moment of SPION and the magnetic moments of water protons is greatly responsible for the modification of relaxation rates. Water protons in the presence of superparamagnetic nanoparticles experience magnetic fluctuations due to their free diffusion through magnetic inhomogeneities produced by the dipolar magnetic field (Figure 3). This interaction is modulated by water diffusion, but also by Néel and Brownian relaxation rates of the superparamagnetic nanoparticles [30].

## 2.1 Longitudinal Relaxation

The description of the superparamagnetic relaxation phenomena is based on theoretical model established by Roch, Muller, and Gillis (theoretical model known as the RMG model or superparamagnetic (SPM) model [28]). This model is built upon the original relaxation theory described for paramagnetic compounds. However, in the case of SPION, the model has been adapted to consider their much higher magnetic moment and their anisotropy. Inner-sphere relaxation does not contribute significantly to the proton relaxation which is rather defined by an outer-sphere mechanism where the dipolar interaction is modulated by both Néel relaxation and water diffusion, with the latter being defined by the translational diffusion time ($\tau_D = r^2/D$, where r is the crystal radius and $D$ the water diffusion coefficient).

The RMG model enables understanding of the proton longitudinal relaxation rate in aqueous suspensions of SPION. This theoretical approach assumes a uniform distribution of crystals with uniaxial anisotropy within pure water and is valid for the estimation of proton longitudinal relaxation rate in the presence of

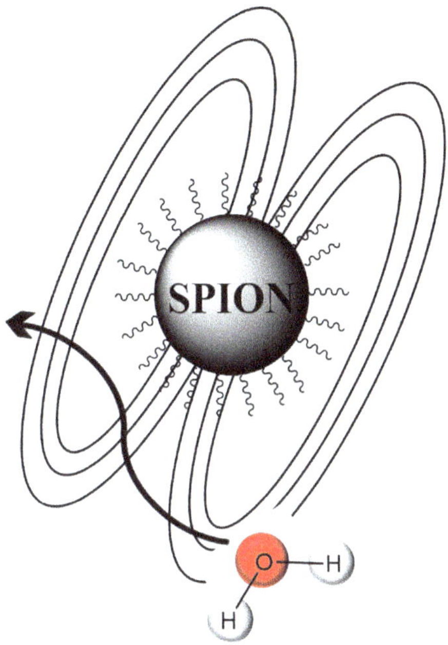


**FIGURE 3** Illustration of the diffusion of water molecules through magnetic field inhomogeneities responsible for modification of proton relaxation rates.

both large and small crystals (i.e., particle radius above and below 7.5 nm, respectively) respecting the Redfield condition [28,29]. Therefore, distinction should be made according to the size of the considered SPION and the relative strength of the external magnetic field.

**Large crystals** (particle radius >7.5 nm) are characterized by high anisotropy energy inducing the locking of the magnetic moment onto the anisotropy axis.

At low magnetic field, flips of the magnetic moment between different easy axes can occur; the dipolar interaction is therefore modulated by the Néel relaxation time and the translational diffusion time $\tau_D$. In this case, longitudinal relaxation rate is given by Equation 5; Néel relaxation and water diffusion are introduced thanks to the Freed spectral density [31] ($J_F$; Equation 6):

$$R_1 = \frac{1}{T_1} = \frac{32\pi}{405}\gamma^2\mu^2\left(\frac{N_A C}{r^3}\right)\left[10J_F(\omega_I, \tau_D, \tau_N)\right] \tag{5}$$

with

$$J_F(\omega_I, \tau_D, \tau_N) = R_e\left[\frac{1+\frac{1}{4}\Omega^{\frac{1}{2}}}{1+\Omega^{\frac{1}{2}}+\frac{4}{9}\Omega+\frac{1}{9}\Omega^{\frac{3}{2}}}\right] \text{ and } \Omega = i\omega_I\tau_D + \frac{\tau_D}{\tau_N} \tag{6}$$

where $\gamma$ is the proton gyromagnetic ratio, $\mu$ is the electron magnetic moment, $N_A$ is the Avogadro number, $C$ is the molar concentration of superparamagnetic compound, $r$ is the crystal radius, $\omega_I$ is the proton angular frequency and $\tau_D$ is the translational correlation time of water molecules (equal to the square of crystal radius ($r$) divided by the diffusion constant ($D$)).

Néel relaxation cannot occur at high magnetic field, magnetic moments remain locked onto the easy axis and Néel relaxation time is much longer. Relaxation is therefore only influenced by diffusion of water protons. In this case, Ayant spectral density [32] ($J_A$) can be used to describe the relaxation (Equation 7 and $J_A$; Equation 8):

$$R_1 = \frac{1}{T_1} = \frac{32\pi}{405}\gamma^2\mu^2\left(\frac{N_A C}{r^3}\right)\left[9L^2(\alpha)J_A\left(\sqrt{2\omega_I\tau_D}\right)\right] \tag{7}$$

with

$$J_A = \frac{1+\frac{5\mu}{8}+\frac{\mu^2}{8}}{1+\mu+\frac{\mu^2}{2}+\frac{\mu^3}{6}+\frac{4\mu^4}{81}+\frac{\mu^5}{81}+\frac{\mu^6}{648}} \tag{8}$$

At intermediate magnetic field, both contributions influence the global relaxation mechanism. The weighting given to each contributions is expressed using a linear combination of Equations 5 and 7. The Langevin function ($L(\alpha)$), which gives the average magnetization of the sample, is used to mathematically describe the combination of the low magnetic field and high magnetic field contributions (Equation 9).

$$R_1 = \frac{1}{T_1} = \frac{32\pi}{405}\gamma^2\mu^2\left(\frac{N_A C}{r^3}\right)\left\{\begin{array}{c}\left(\frac{L(\alpha)}{\alpha}\right)21J_F(\omega_I,\tau_D,\tau_N)\\ +9\left[1-L^2(\alpha)-2\left(\frac{L(\alpha)}{\alpha}\right)\right]J_F(\omega_I,\tau_D,\tau_N)\\ +9L^2(\alpha)J_A\left(\sqrt{2\omega_I\tau_D}\right)\end{array}\right\} \tag{9}$$

Using these field-dependent contributions, the RMG model enables the fitting of the relaxation curves for large nanoparticles. The field-dependent (NMRD profile) global relaxation rate resulting from each contribution is depicted in Figure 4 (Reproduced from Ref. [33]).

For **small crystals** (particle radius $<$ 7.5 nm), slight adaptation to the RMG model has been made to consider the smaller anisotropy energy of these materials. Smaller anisotropy energy results in attenuated locking of the magnetic

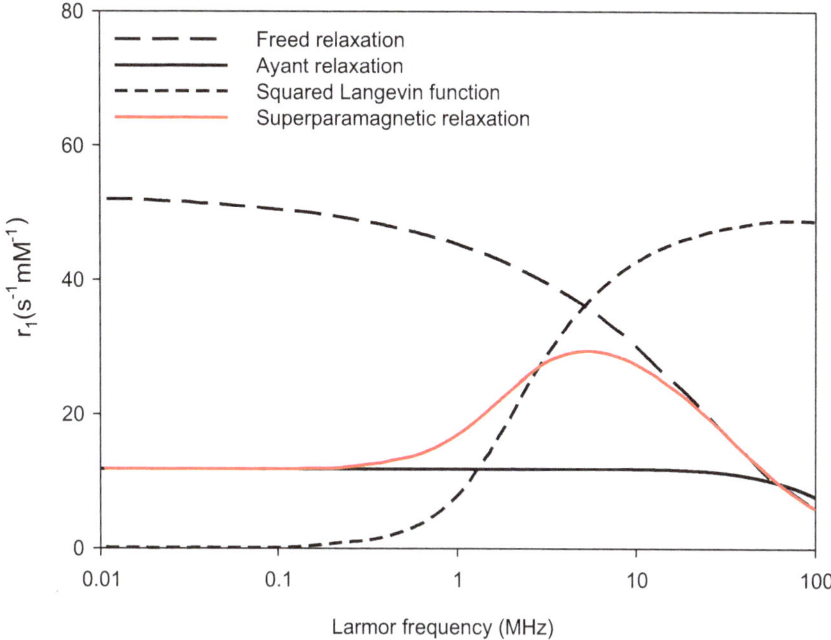


**FIGURE 4** Illustration of field-dependent relaxation processes contributing to the global superparamagnetic relaxation of large crystals. (Reproduced from Ref. [33].)

moments onto the anisotropy directions [34]. This low anisotropy results in a low-field dispersion in the field-dependent curve of small iron oxide nanoparticles. Adaptation of the theoretical model by considering the anisotropy as an empirical parameter (*P*; Equation 10) allowed a good agreement with experimentally observed NMRD curves.

$$R_1 = \frac{1}{T_1} = \frac{32\pi}{405}\gamma^2\mu^2\left(\frac{N_A C}{r^3}\right)\left\{\begin{matrix} \left(\frac{L(\alpha)}{\alpha}\right)21PJ_F(\omega_I,\tau_D,\tau_N)+21(1-P) \\ J_F(\omega_I,\tau_D,\tau_N) \\ +9\left[1-L^2(\alpha)-2\left(\frac{L(\alpha)}{\alpha}\right)\right] \\ J_F(\omega_I,\tau_D,\tau_N)+9L^2(\alpha)J_A\left(\sqrt{2\omega_I\tau_D}\right) \end{matrix}\right\} \quad (10)$$

A typical NMRD curve of crystals with low anisotropy energy is depicted in Figure 5. Fitting of the experimental curves with the theoretical model provides information concerning the physicochemical properties of SPION. The curve

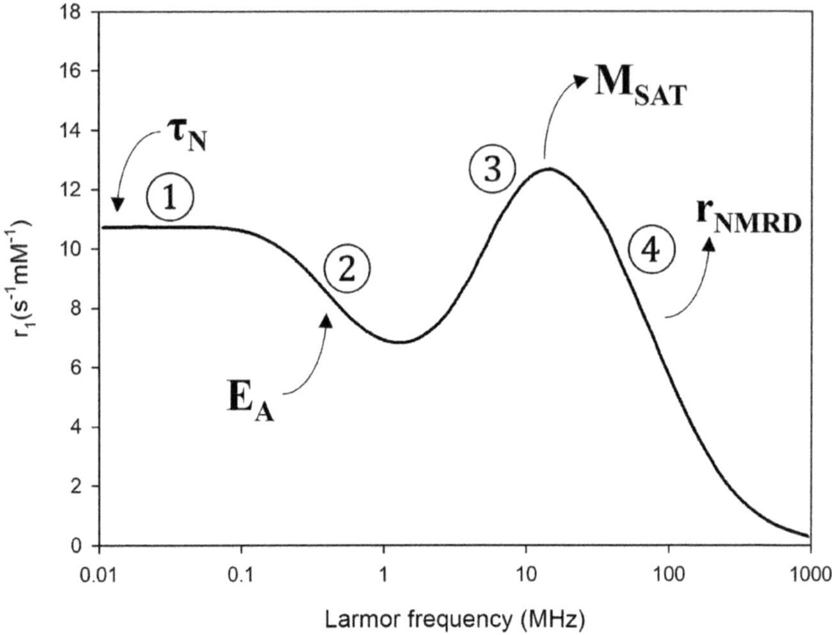


**FIGURE 5** Theoretical NMRD curve of SPION with low anisotropy energy obtained by the RMG model and region of influence of different physicochemical parameters of SPION.

depicted in Figure 5 can be split into four zones, each of which being associated with a physicochemical parameter of SPION: namely, their Néel relaxation time $\tau_N$, their anisotropy energy $E_A$, their average radius $r_{NMRD}$ and their saturation magnetization $(M_{SAT})$.

Fitting of experimental data using the RMG model is suitable for estimating the physicochemical parameters involved in the relaxation induced by superparamagnetic nanoparticles [35]. These parameters are listed below:

(1) An estimation of the Néel relaxation time $(\tau_N)$ can be obtained from the relaxivity plateau at low field. However, this value remains approximative and is usually used as qualitative information in addition to the other parameters (crystal size and saturation magnetization).

(2) The low-field dispersion observed for small crystals is an indication of their low anisotropy energy ($E_A$).

(3) The maximum value of longitudinal relaxivity reached at high field reflects the saturation magnetization ($M_{SAT}$) through the approximation $R_{MAX} \approx C \cdot M_{SAT}^2 \cdot \tau_D$, where $C$ is a constant and $R_{MAX}$ is the maximum relaxivity.

(4) At high magnetic field, an estimation of the nanoparticle radius $(r_{NMRD})$ can be obtained from the inflexion point of the NMRD curve. This point,

corresponding to the condition $\omega_I \cdot \tau_D \approx 1$, enables the determination of $\tau_D$ which subsequently allows the estimation of the nanoparticle radius (through $\tau_D = r_{NMRD}^2 / D$). Since the estimated value of particle size corresponds to the distance of closest approach of solvent molecules from the center of the particle, perfectly permeable coating is required to correctly estimate the particle size.

NMRD profiles, along with the RMG model enabling the theoretical fitting and the extraction of physical parameters of SPION, are powerful characterization tools to evaluate the efficacy of SPION as contrast agents for MRI. However, one must stress that the parameters extracted from the fitting are averaged values of the effective characteristics of ferrofluids. Moreover, NMRD profiles also offer significant information concerning the colloidal stability of nanoparticle suspensions, as the clustering of superparamagnetic nanoparticles highly influences their relaxometric properties.

## 2.2 Transverse Relaxation

The NMR dispersion of proton transverse relaxivity follows the same trend as longitudinal relaxivity at low field. However, its evolution with increasing magnetic field is different, with transverse relaxivity increasing toward reaching a residual value at high magnetic field, whereas longitudinal relaxivity always decreases toward zero (Figure 6). The residual value of transverse relaxivity is often referred as the "secular contribution" or "secular term" [28]. This contribution arises from local magnetic fields (magnetic inhomogeneities), caused by the magnetic nanoparticles, inducing a loss of coherence of the z-component of the magnetic field. The secular contribution therefore reflects the contribution of local fields causing spins to precess at different frequencies and therefore affecting the transverse relaxation rate [36].

Particle clustering highly affects the relaxation properties of iron oxide nanoparticles and their efficacy as contrast agents. *In vivo* formation of clusters or aggregates of nanoparticles results in unequal variations of both longitudinal and transverse relaxation rates. Longitudinal relaxivity tends to decrease throughout the clustering process [37] (Figure 7a). However, the transverse relaxation rate follows a bell curve in which it reaches a maximum (at approximately 20 nm) and then decreases (Figure 7b; Reprinted from Ref. [38]). The secular term is primarily responsible for the evolution of transverse relaxation. The initial increase of $R_2$ with agglomeration is provoked by the increase of cluster radius, while further increase of the clusters size leads to protons being almost motionless in the sample, resulting in a decrease of $R_2$ with aggregation [38].

Aggregated SPION can be regarded as single particles characterized by their own sizes and magnetic moments [38,39]. The marked effect of clustering on both relaxation rates gives rise to magnetic assemblies characterized by very high $r_2$ values and $r_2/r_1$ ratios that can reach up to several hundred depending on

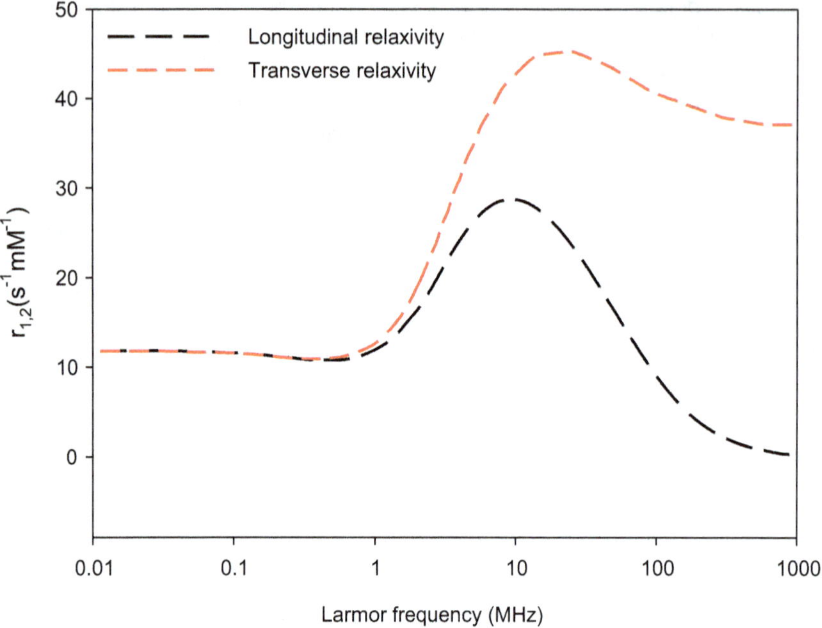


**FIGURE 6** NMRD profiles for longitudinal and transverse relaxations of 10 nm superparamagnetic iron oxide nanoparticles.

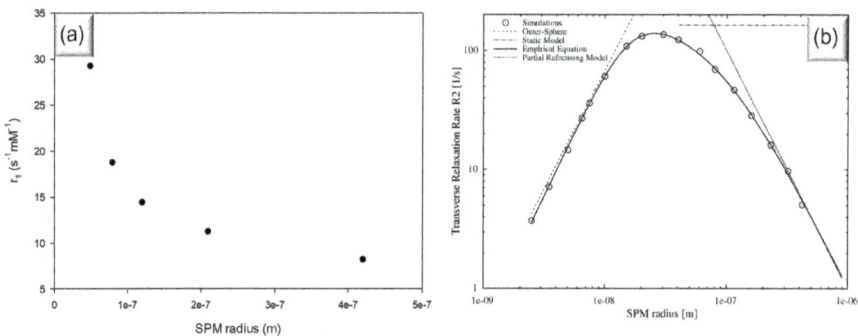


**FIGURE 7** (a) Evolution of proton longitudinal relaxivity (at magnetic field of 20 MHz) as a function of nanoparticle agglomeration. (Data extracted from Ref. [37]). (b) Simulation results and theoretical predictions of the transverse relaxation rate at high magnetic field using various models. (Reprinted with permission from Ref. [38]. Copyright 2011, Elsevier.)

the strength of the external magnetic field. Some of such SPION-based contrast agents have been clinically approved for $T_2$-*w* MRI; these agents and their applications are discussed in the next sections.

# 3 SPION-BASED CONTRAST AGENTS

Contrast agents based on superparamagnetic nanoparticles are made up of specific iron oxide cores and coating polymeric materials, either natural (such as dextran, carboxydextran, starch, and chitosan, etc.) or synthetic (polymers based on polyvinylpyrrolidone (PVP), polyethylene glycol (PEG), polyvinyl alcohol (PVA), etc.) [40]. Inorganic materials such as gold, silver, or silica have also been reported as stabilizers for SPION [41]. These hydrophilic coatings are designed to provide nanoparticles with enhanced stability in physiological media and immunological stealthiness.

## 3.1 Clinically-Developed SPION-Based Contrast Agents

Due to their very high $r_2$ values, iron oxide nanoparticles were initially designed for applications as $T_2$ contrast agents. Several formulations of SPION-based $T_2$ contrast agents have been developed by pharmaceutical companies and/or are in clinical trials; some of them have also been approved by regulatory bodies for clinical use [42]. Table 2 summarizes clinically developed SPION-based contrast agents and their utility as MRI contrast agents.

### 3.1.1 Hepatic MRI

Ferumoxides and Ferucarbotran are two liver-specific contrast agents which have been developed for the detection of focal liver lesions. Accumulation of the contrast agents in the normal liver parenchyma surrounding focal liver lesions increases the contrast/noise ratio (CNR) of these lesions, which appear more hyperintense on $T_2$-weighted images compared to the pre-contrast image [43].

**TABLE 2**
**Generic, Trade Names and Applications of SPION-Based $T_2$ Contrast Agents**

| Agent | Generic Name | Trade Name | Application |
|---|---|---|---|
| AMI-25 | Ferumoxide | Endorem®<br>Feridex® | Liver MRI |
| SHU 555A | Ferucarbotran | Resovist® | Liver MRI |
| AMI-227 | Ferumoxtran-10 | Sinerem®<br>Combidex® | Lymph node MRI |
| NC100150 | Feruglose | Clariscan® | Blood-pool MRI |
| VSOP C184 | Ferropharm | / | Blood-pool MRI |
| AMI-121 | Ferumoxsil | Lumirem® | Gastrointestinal MRI |
| OMP | Ferristene | Abdoscan® | Gastrointestinal MRI |
| / | Ferumoxytol | Feraheme® | Treatment of iron deficiency anemia |

**Ferumoxide** (Endorem® in Europe (Guerbet); Feridex® in the U.S. (Berlex Laboratories); AMI-25) was the first clinically approved SPION-based contrast agent. Particles of ferumoxides are composed of several iron oxide cores (core diameter: 5 nm) embedded in low-molecular-weight dextran coating. The size of the particles is generally between 120 and 180 nm. Ferumoxides are administered as drip infusions over 30 min to avoid potential side effects. Nanoparticles are rapidly taken up by the mononuclear phagocyte system (MPS) and phagocytosed by Kupfer cells (macrophages). Uptake by the liver (80%) and the spleen (6%–10%) occurs quickly and persists for several hours, allowing a broad window for effective MR imaging [44]. The accumulation of ferumoxides into liver tissues allows visualization of liver lesions or tumors due to the absence of Kupfer cells in these abnormal tissues. These products were initially designed for the detection of liver metastases or hepatocellular carcinomas (HCCs) [45,46]. Ferumoxides were withdrawn from the market due to lack of use by radiologists.

**Ferucarbotran** (Resovist® (Schering AG); SHU 555A) is another liver-specific contrast agent enabling the detection of small focal liver lesions. Resovist consists of 4.2 nm SPION embedded in a carboxydextran coating, with an overall hydrodynamic diameter of 62 nm [47]. Resovist® has a better safety profile compared to ferumoxides, enabling its intravenous injection by bolus. Resovist® was also assessed for dynamic MRI of the brain, but didn't outreach the diagnosis efficacy of gadopentetate dimeglumine [48].

### 3.1.2 Lymph Node Imaging

**Ferumoxtran-10** (Sinerem® in Europe (Guerbet); Combidex® in the U.S. (AMAG Pharmaceuticals); AMI-227) is a formulation containing smaller particles (5 nm SPION with low-molecular-weight dextran coating; hydrodynamic diameter of approximately 30 nm) [49]. Due to their small size, nanoparticles slowly leak into the intravascular space (Figure 8; [50]) from which they are subsequently taken up by macrophages and transported to lymph nodes by way of lymphatic vessels [51].

These were primarily designed for metastatic lymph node imaging due to their prolonged blood half-life. Visualization of lymph nodes was studied for staging pelvic cancers in a total of 271 patients. However, unconcordant results between three different blinded specialist readers led to abandoning the ferumoxtran-10 clinical development [52]. In 2013, the Radboud University Medical Center (Radboudumc) restarted the production of ferumoxtran-10 for clinical trials on patients with prostate cancer and succeeded in the detection of metastatic lymph nodes with size down to 2 mm [53]. Ferumoxtran-10 is now available under the brand name Ferrotran® (SPL Medical B.V., Nijmegen, The Netherlands).

### 3.1.3 Blood-Pool Imaging

**Feruglose** (Clariscan®; GE Healthcare; NC100150) and **Ferropharm** (VSOP C184) are SPION-based contrast agents which were designed for blood-pool MRI. Both are composed of single crystals: Feruglose consists of crystals (4–7 nm

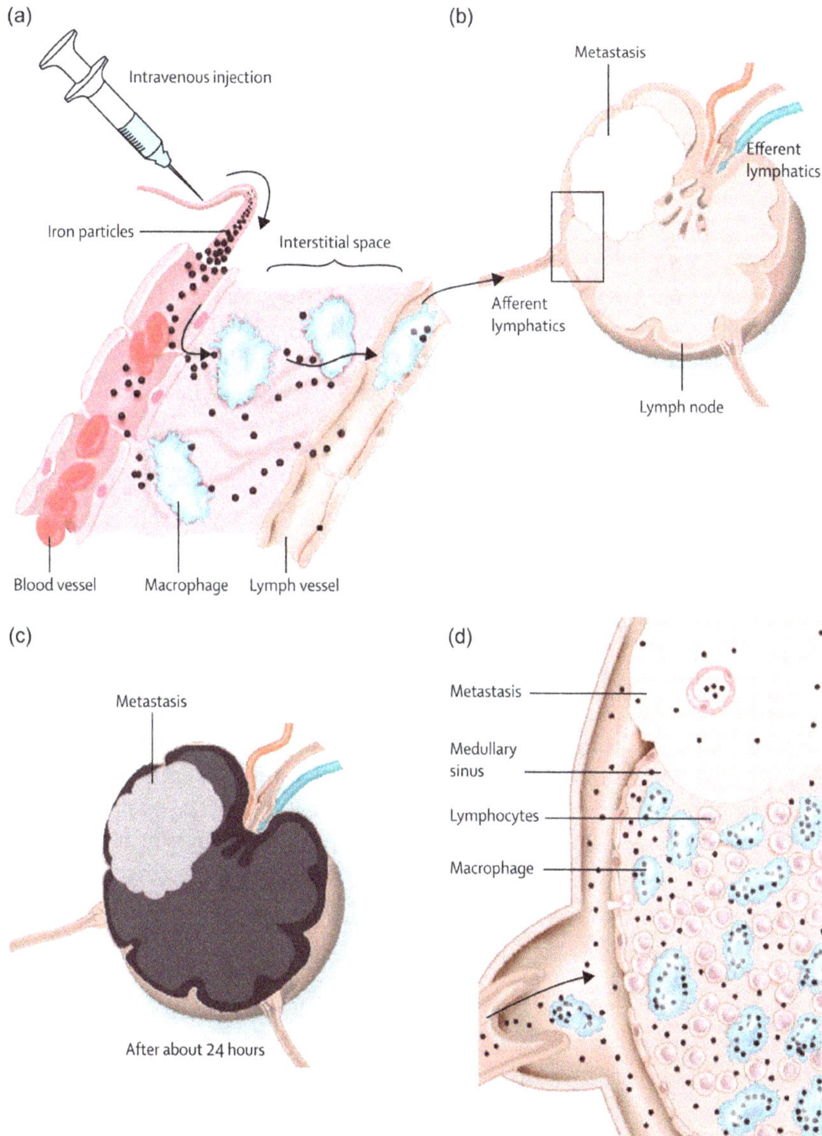


**FIGURE 8** (a) Infused iron-particles slowly extravasate from the vascular to the interstitial space and are internalized by macrophages. (b) and (c) Iron-loaded macrophages are transported to lymph nodes via lymphatic vessels and accumulate in normal-sized lymph node tissue. These iron-loaded macrophages cause low signal intensity on $T_2$*-weighted MR image. Box in b shows area depicted in d. (d) Disturbances of lymph flow or nodal architecture by metastases leads to less macrophages, depicted at MR imaging by higher signal intensity. (Reprinted with permission from Ref. [50]. Copyright 2008, Elsevier.)

SPION) with a carbohydrate PEG [54] coating, while Ferropharm is made of 4 nm SPIONs with a citrate layer [55]. Feruglose and Ferropharm have small hydrodynamic particle size of 12 and 7 nm, respectively [54,56], enabling them to be used for both $T_1$-*w* and $T_2$-*w* imaging. Both agents were shown to be well suited for first-pass magnetic resonance angiography (MRA) experiments [56,57] as well as for perfusion-weighted diagnosis of various organs [58,59]. However, although VSOP C184 and NC100150 have been tested in human clinical trials up to phase II [60–63], none of the formulations received regulatory approval [64].

**Ferumoxytol** (Feraheme®; AMAG Pharmaceuticals), initially designed and developed as an MRI contrast agent, is now used to treat iron deficiency anemia in adults with chronic kidney disease (CKD) [65,66]. Ferumoxytol is made up of 6.4 nm SPION surrounded by a carbohydrate shell (polyglucose sorbitol carboxymethyl ether). The colloidal particle size is approximately 30 nm [67]. Since its approval as a therapeutic iron supplement by the United State Food and Drug Administration (FDA), Ferumoxytol has seen renewed interest as a contrast agent with great potential for numerous applications in MRI where the use of Gadolinium-based contrast agents (GBCA) is contraindicated [68]. Ferumoxytol has favorable physicochemical characteristics enabling satisfactory contrast-enhanced MRI in different organs at various phases (Figure 9; [69]).

Following i.v. administration, imaging can be performed during the dynamic phase (directly after bolus injection) to provide valuable insight of the blood supply to the brain. During the blood-pool phase, Ferumoxytol allows high-resolution contrast-enhanced imaging of the intravascular space (arterial and venous systems, hence enabling the diagnosis of vascular diseases [70,71]. Finally, the disrupted blood–brain barrier of brain lesions allows the progressive leakage of nanoparticles resulting in signal changes occurring in the delayed phase around 24 h after injection [72]. For example, visualization of malignant brain tumors was demonstrated through delayed enhancement peaking at 24–28 h after administration of Ferumoxytol nanoparticles [73].

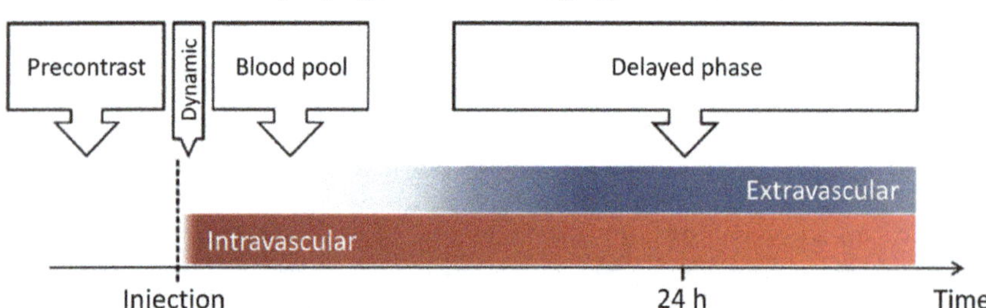


**FIGURE 9** Phases of Ferumoxytol enhancement. (Reprinted with permission from Ref. [69]. Copyright 2017, Elsevier.)

### 3.1.4 Gastrointestinal Imaging

**Ferumoxsil** (Lumirem®; Guerbet; AMI-121) and **Ferristene** (Abdoscan®; Amersham Health; OMP) are SPION-based oral formulations designed for gastrointestinal MRI. Ferumoxsil contains silica-coated particles containing multiple 10 nm SPION and the hydrodynamic diameter is approximately 300 nm, while Ferristene consists of micrometric particles (total particle size is 3.5 μm) coated with polystyrene and filled with SPION having diameters below 50 nm. Ferumoxsil and Ferristene were tested in clinical trials for the darkening of bowel loops; however, both agents were taken off the market due to lack of users [74].

### 3.1.5 $T_2$-Weighted MRI with SPION

Other than the clinically developed contrast agents, several other SPION-based compounds with various architectures have been studied for $T_2$-weighted MRI applications. SPION with larger sizes have higher $r_2$ values; non-spherical particles such as cubes, octapods or plates have been widely studied and showed enhanced $r_1$ and $r_2$ relaxivities [75]. Assembly of nanoparticles into magnetic nanostructures such as amphiphilic polymers, matrices of mesoporous silica, or liposomes has also been studied and has proven to be effective for attaining high relaxivity values [76,77].

However, despite the increasing amount of SPION-based nanosystems exhibiting remarkable efficacy as $T_2$ contrast agents, their transition to clinical trials has been very limited due to their inherent contrast mechanism, producing a signal decreasing effect. Many diseases (bleeding, calcification and metal deposits) naturally induce hypointense areas, which can be confused with the presence of $T_2$ contrast agents. Moreover, SPION generate magnetic inhomogeneities causing distortion of the magnetic field in their close vicinity. This effect, often referred as "blooming effect" or "blooming artifact", causes signal destruction and makes the diagnosis difficult when it comes to determining the exact state of a lesion [78,79], often resulting in an overestimation of the lesion size as depicted in Figure 10.

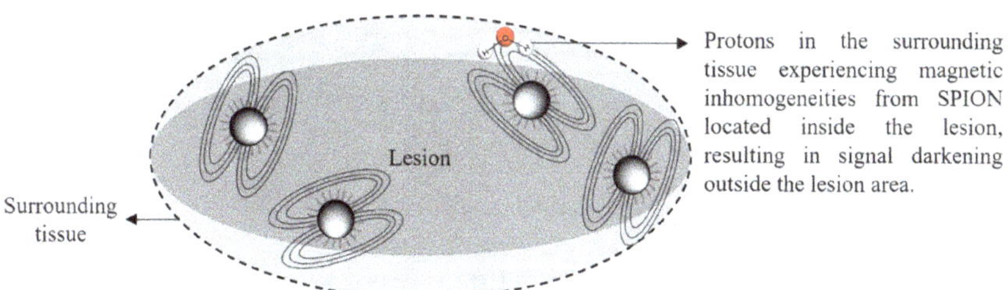


**FIGURE 10** Schematical illustration of the blooming effect, causing the overestimation of lesion size due to SPION located inside the lesion inducing relaxation of protons in the surrounding tissue.

## 3.2 SPION AS $T_1$ CONTRAST AGENTS

The major advantage of iron oxide nanoparticles is their biocompatibility which makes them very attractive for biomedical applications. This favorable feature, combined with the ongoing concerns about the toxicity of GBCA, whether in the case of patients at risk with nephrogenic systemic fibrosis [80–83] or more recently with findings of long-term retention of Gd in the brain of patients administered with GBCA [84–88], led researchers into developing SPION as $T_1$ contrast agents for MRI. To that end, several physicochemical prerequisites must be satisfied in order to reach optimal imaging efficacy and those features are described in the next section.

### 3.2.1 Parameters Affecting $T_1$ Contrast with SPION

Using iron oxide nanoparticles with low dimensions (size below 5 nm) that are capable of generating bright contrast on $T_1$-weighted images is a promising alternative [89]. The $r_2/r_1$ ratio can be used as an indicator of the suitability of a compound for $T_1$-$w$ or $T_2$-$w$ MRI (Figure 11; reprinted from Ref. [76]). As a rule of thumb, SPION with smaller core, i.e., smaller $r_2/r_1$ ratios (< 5), can be used as $T_1$ contrast agents, while bigger nanoparticles are used as $T_2$ contrast agents ($r_2/r_1 > 10$ at standard clinical field, i.e., 1.5 T) [90,91].

The link between reduction of $r_2/r_1$ ratio with the size of nanoparticles can be understood by the spin-canting effect. Magnetic moments of atoms constituting a nanoparticle have different configurations depending on whether they are located on the outer surface of the nanoparticles or inside the nanoparticles. Spins on the outside layer of the nanoparticle have random orientations due to structural disorder at the nanoparticle surface. As these spins are not aligned with the bulk spins of the nanoparticle, they form a magnetically dead layer on the nanoparticle's surface. The spin-canting effect is a finite-size effect; experimental data highlighted

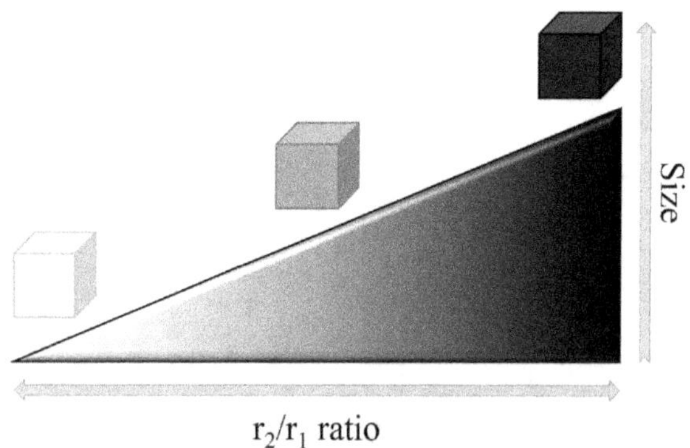


**FIGURE 11** Influence of the $r_2/r_1$ ratio on the efficacy of a contrast agent. (Reproduced from Ref. [76]. Originally published by and used with permission from Dove Medical Press Ltd.)

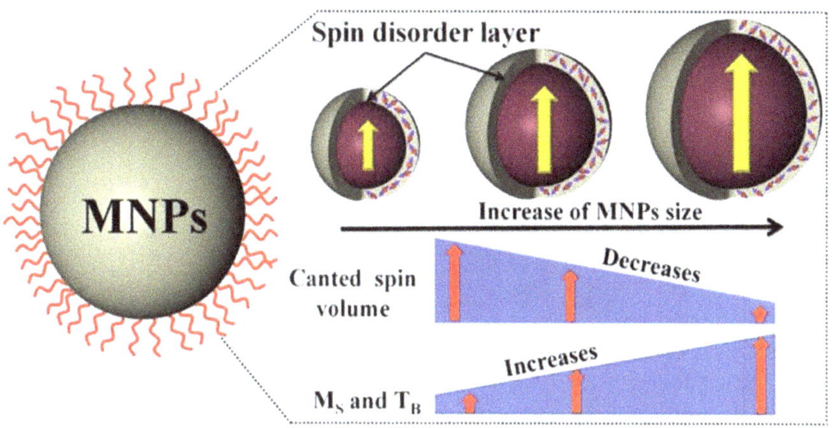


**FIGURE 12** Schematic representation of the variation of spin-canting effect, saturation magnetization and blocking temperature with the MNPs size. (Reprinted with permission from Ref. [94]. Copyright 2019, MDPI Materials.)

that the thickness of the magnetically dead layer is comprised between 0.5 and 0.9 nm depending on the type of nanoparticle [92,93]. Considering a magnetic core of 5 nm, roughly half of the nanoparticle contains randomly oriented spins constituting the magnetically dead region (Figure 12; [94]), assuming a 0.5 nm spin-canted layer [95]. Atoms inside the nanoparticles can also be subjected to spin-canting effects due to cation vacancies in the internal structure of the magnetic region. Such breakage of the crystal's internal structure disrupts their crystallinity and is also partially responsible for the reduced magnetization of small SPION [96].

Another essential prerequisite is to preserve the stability of SPION *in vivo*, as aggregation would lead to an increase of the $r_2/r_1$ ratio (i.e., Figure 7), resulting in a diminished efficacy of the particles as $T_1$ contrast agents.

### 3.2.2 $T_1$-Weighted MRI with SPION

For many years, and with the advent of synthetic methodologies enabling the preparation of small-sized and highly calibrated SPION [97,98], a great number of SPION-based contrast agents (Table 3) have been evaluated as potential $T_1$ contrast agents. The developed formulations differ in their synthetic procedures and in the various surface modifications used for the stabilization of the nanoparticles. Taupitz and co-workers initially explored preclinical studies about the potential of VSOP (Ferropharm) as $T_1$ contrast agents for MR angiography [56,57,61,99]. These pioneering studies demonstrated the potential of using SPION-based CAs for the investigation of coronary arteries. Moreover, VSOP was later shown to be suitable for tracking the evolution of atherosclerotic lesions at risk of destabilization [100]. Similar citrate coated 5 nm SPION were investigated by Taboada et al. [101]. These nanoparticles showed relaxivity values of $r_1 = 14.5\,mM^{-1}s^{-1}$ and $r_2 = 66.9\,mM^{-1}s^{-1}$ at 1.5 T.

**TABLE 3**
**Inorganic Diameter, Coating Material, Relaxivity Values and Ratios at Corresponding Magnetic Fields of Reported SPION-Based T1 Contrast Agents**

| Coating | $D_{TEM}$ (nm) | $r_1$ ($s^{-1}mM^{-1}$) | $r_2$ ($s^{-1}mM^{-1}$) | $r_2/r_1$ | Magnetic Field (T) | Reference |
|---|---|---|---|---|---|---|
| Citrate | 4.8±0.6 | 14.5 | 66.9 | 4.61 | 1.5 | [101] |
| $PEG_{1100}$ | 4 | 7.3 | 17.6 | 2.4 | 1.41 | [102] |
| $PEG_{600}$ | 5.4±0.7 | 19.7 | 39.5 | 2 | 1.5 | [103] |
| Phosphine oxide-PEG | 2.2±0.4 | 4.8 | 17.5 | 3.67 | 3 | [104,105] |
| | 3±0.5 | 4.8 | 29.2 | 6.12 | | |
| | 12 | 2.4 | 58.8 | 24.8 | | |
| PAA | 1.7 | 8.2 | 16.7 | 2.03 | 1.41 | [106] |
| | 2.2 | 6.1 | 28.6 | 4.65 | | |
| | 4.6 | 1.1 | 64.4 | 61.81 | | |
| Dextran | 7 | 13.3 | 40.9 | 3.07 | 1.43 | [107] |
| | | 6.8 | 39.7 | 5.84 | 3 | |
| Glutathione | 3.7±0.1 | 3.6 | 8.3 | 2.28 | 4.7 | [108] |
| $SiO_2$ | 4–5 | 1.2 | 7.8 | 6.5 | 3 | [109] |
| $SiO_2$-PEG | 10±2.5 | 6.9 | ~ 270 | 39.13 | 1.4 | [110] |
| Dextran | 2.5±0.2 | 6 | 27.9 | 4.7 | 1.5 | [111] |
| PAA-PEG | 3.6 | 12.7 | 23.5 | 1.9 | 0.5 | [112] |
| | | 8.8 | 22.7 | 2.6 | 1.5 | |
| | | 2.3 | 24.4 | 10.5 | 7 | |
| ZDS | 3.7±0.8 | 2.4 | 5.2 | 2.16 | 1 | [113] |
| ZDS | 7 | 0.2 | 0.9 | 4.5 | 0.5 | [114] |
| | 10 | 2.5 | 46.4 | 18.7 | | |
| BSA | 5.3±1.2 | 39.3 | 179.8 | 4.57 | 0.5 | [115] |
| PMAA-PTTM | 4.3±1.5 | 24.2 | 67.2 | 2.77 | | |
| PEG | 2.3 | 1.4 | 7.5 | 5.5 | 7 | [116] |
| $PEG_{750}$ | 3.5±0.6 | 12.7 | 19.7 | 1.55 | 0.5 | [118] |
| | | 11 | 28.1 | 2.55 | 1.5 | |
| $PEG_{2000}$ | | 7.9 | 13 | 1.64 | 0.5 | |
| | | 6.8 | 17.9 | 2.63 | 1.5 | |
| $PEG_{800}$ | 4.8±1 | 19.8 | 32.2 | 1.62 | 0.5 | |
| $PEG_{2000}$ | 4.8±0.9 | 20.1 | 32.8 | 1.63 | | [119] |
| $PEG_{5000}$ | 5±1 | 19.9 | 32.8 | 1.65 | | |

In 2009, Tromsdorf et al. developed PEGylated iron oxide nanoparticles with various PEG chain lengths [102]. The authors demonstrated that a minimum PEG chain length with a molecular weight of 500 g $mol^{-1}$ is required to provide stability to the nanoparticles in various buffer systems. SPION coated with $PEG_{1100}$

exhibited a low $r_2/r_1$ ratio making them suitable as $T_1$ contrast agents for MRI. The low cytotoxicity and low unspecific cellular uptake into cells from the MPS (J774 mouse cells macrophages) were further evidenced. Similar PEGylated iron oxide nanoparticles were also developed by Hu et al. which obtained SPION with higher relaxivities and smaller $r_2/r_1$ ratio, resulting in better contrast efficacy in both $T_1$-*w* and $T_2$-*w* images [103]. In 2011, Kim et al. investigated the magnetic and relaxometric properties of SPION with size varying between 1.5 and 3.7 nm as potential $T_1$ contrast agents. Size-dependent properties due to spin-canting effect were evaluated; small-sized SPION was characterized by low magnetic moments and almost paramagnetic field-dependent curves (Figure 13a and b). Their efficacy for high-resolution blood-pool imaging was demonstrated to be superior to Dotarem®, enabling the visualization of various blood vessels with size as low as 0.2 mm [104] (Figure 13c). In a later study, the same SPION were compared with manganese oxide nanoparticles and a clinically used GBCA (Gd-DTPA) [105]. The biosafety of each compound was evaluated on a mouse model. Results revealed a better safety profile for SPION resulting from their accumulation mostly occurring in the spleen, therefore inducing only stress-related effect and no other severe damage.

In a similar study, Wang et al. developed PAA-coated SPION of 1.7, 2.2, and 4.6 nm with respective $r_2/r_1$ ratios of 2.03, 4.65, and 61.81 (at 7 T). The 2.2 nm SPION were evaluated as dual $T_1$–$T_2$ contrast agents and displayed good contrast enhancement, long-term circulation and low toxicity, which make them attractive for various clinical applications such as diagnosis of myocardial infarction, renal failure, atherosclerotic plaque, thrombosis and angiogenesis of tumor cells [106]. Furthermore, in 2014, Jung et al. demonstrated the feasibility of dual contrast-enhanced imaging using iron oxide nanoparticles as both $T_1$ and $T_2$ contrast agents for detecting vascular signals in MR angiography [107].

The same year, glutathione-capped iron oxide nanoparticles developed by Liu et al. were shown to have excellent dual contrast ability and biocompatibility [108]. *In vivo* results showed a strong enhancement in mouse's superior sagittal sinus, enabling the detection of cerebral arterial occlusions. Metabolization of the nanoparticles by kidneys was assessed by the observed enhancement of the renal cortex, confirming the versatile potential of such systems for the study of renal disease or urinary tract tumor diagnosis.

Silica-coated iron oxide nanoparticles (SPION@$SiO_2$) as $T_1$ contrast agents were developed by Iqbal et al. in 2015 [109]. The biocompatibility of SPION was enhanced using silica, yielding SPION@$SiO_2$ displaying $r_1$ and $r_2$ values of 1.2 $s^{-1}mM^{-1}$ and 7.8 $s^{-1}mM^{-1}$, respectively. *In vivo*, strong positive signal enhancement was observed in heart, liver, kidney and bladder. Another study from Hurley et al. showed the potential of silica-coated SPION for both positive contrast enhancement (using a SWIFT sequence, i.e., SWept Imaging with Fourier Transform) and hyperthermia treatment using intra-tumoral injection of these nanoparticles in prostate cancer tumors in nude mice [110].

Bhavesh et al. developed fluorescent dextran-coated iron oxide nanoparticles (fdIONP) with the use of microwave technology [111]. These nanoparticles

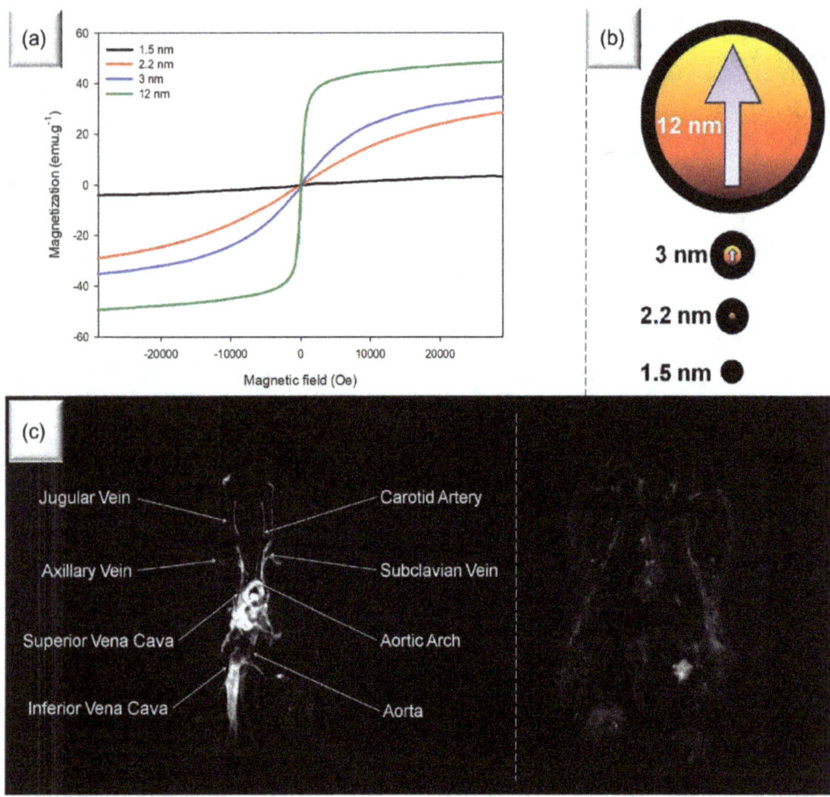


**FIGURE 13** (a) Field-dependent magnetization curves for SPION with various diameter. (b) Illustration of the size-dependent spin-canting effect in iron oxide nanoparticles with various diameters. (c) Comparison of *in vivo* blood-pool images obtained with a 3d-FLASH sequence using 3 nm SPION (left) and Dotarem (right). (Adapted with permission from Ref. [104]. Copyright 2011, American Chemical Society.)

showed relaxivity values of $r_1 = 6\,s^{-1}mM^{-1}$ and $r_2 = 27.9\,s^{-1}mM^{-1}$ (at 1.5 T) along with small hydrodynamic diameter of 21.5 nm. MRI and fluorescent phantom images were obtained with mouse adult fibroblasts incubated with increasing concentrations of iron (Figure 14a). The MRI positive contrast was further evidenced *in vivo* (Figure 14b) during CE-MRA experiments. A clear depiction of the main vascular architecture was maintained even 90 min post-injection.

In 2017, Shen et al. studied the relationship between the relaxivity values ($r_1$, $r_2$ and $r_2/r_1$ at 7 T) and the particle size [112]. SPION with various sizes were synthesized and a particle size of 3.6 nm was identified as the optimal one, combining ideal $r_2/r_1$ ratio and high $r_1$ value, for efficient positive contrast. A stimuli-responsive drug delivery system based on these SPION was then devised using dimeric RGD peptide ($RGD_2$) for tumor targeting, doxorubicin (DOX) as anticancer drug and PEG methyl ether (mPEG) grafted *via* an acid-labile-β-thiopropionate linker,

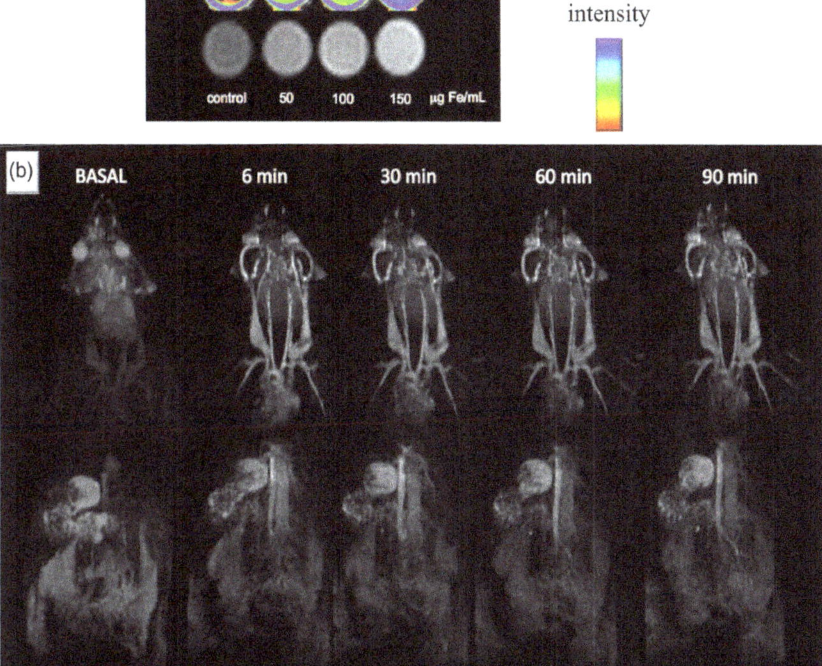


**FIGURE 14** (a) *In vitro* phantom images (top-row: fluorescence phantoms; bottom-row: MRI phantoms) of mouse adult fibroblasts incubated with increasing concentrations of fdIONP; (b) *In vivo* CE-MRA images before and after administration of fdIONP at increasing times. (Reprinted from Ref. [111]. Copyright 2015, MPDI Nanomaterials.)

enabling drug release in acidic tumor condition. $T_1$-weighted MR images recorded at different times after the administration of the nanoparticles *in vivo* demonstrated highest tumor signal with the specific systems compared to unspecific ones (Magnevist® and 3.6 nm SPION without targeting vector).

Liang et al. developed an innovative SPION-based $T_1$ contrast agent based on zwitterionic SPION [113]. Zwitterionic dopamine sulfonate (ZDS) was used as stabilizer enabling the direct preparation of SPION (named ZUIONs) with a core size of 3.7 nm and a hydrodynamic size of 7 nm. The authors demonstrated the significant contrast enhancement induced *in vivo* by these systems as well as persistent intravascular circulation (over 1 h), largely superior to the circulation time of Gd-DTPA, which was directly observed in the bladder 5 min post-injection (Figure 15). ZDS was more recently used by Liu et al. for the stabilization of hollow 7 nm and 10 nm SPION [114]. $T_1$ signal could be observed *in vivo* in a 4T1 xenograft mouse model on a 3 T scanner. Renal excretion of these ZDS-SPION

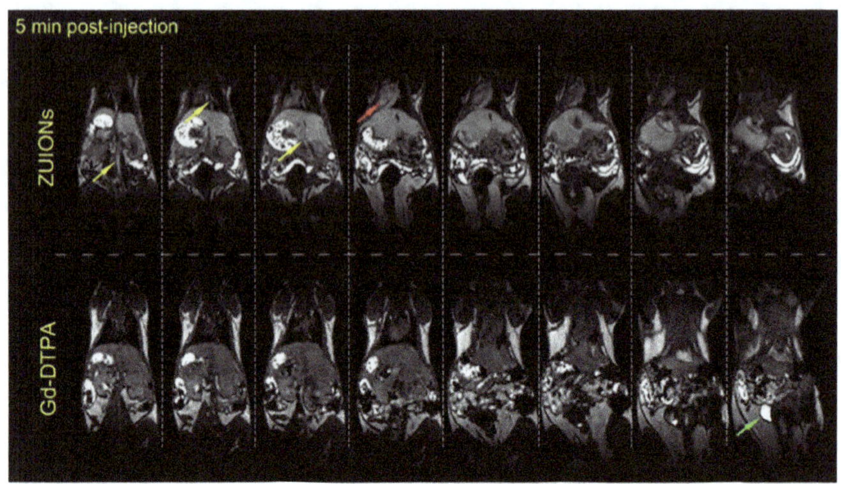


**FIGURE 15** MRI slices (slice thickness=0.8 mm) of mice at 5 min post-injection of ZUIONs (0.041 mmol Fe $kg^{-1}$) or Gd-DTPA (0.041 mmol Gd $kg^{-1}$) showing signal enhancement in blood vessels (yellow arrow), heart (red arrow) and bladder (green arrow), the latter one observed for the injection of Gd-DTPA only. (Reprinted with permission from Ref. [113]. Copyright 2018, American Chemical Society.)

was also assessed by both $T_1$ signal observed in the bladder and TEM analysis of the urine of mice showing the presence of the nanoparticles.

In 2019, Tao et al. performed a comparative study concerning the $T_1$ performance of SPION modified with either natural protein macromolecule (bovine serum albumin (BSA)) or an artificial molecule (poly(acrylic acid)-poly(methacrylic acid), PMAA-PTTM) [115]. Both systems displayed positive contrast enhancement *in vitro* on a 0.5 T MRI scanner. However, *in vivo* images showed opposite behaviors for the two types of SPION, with BSA-coated nanoparticles showing negative contrast and PMMA-PTTM-coated nanoparticles showing positive contrast. These observations, linked with differences in the stability of SPION, showed the importance of surface ligands on the performance of SPION as $T_1$ contrast agents.

Wang et al. demonstrated the feasibility of SPION-based contrast agent for ultrahigh field (UHF) (≥ 7 T) $T_1$-weighted dynamic CE-MRI [116]. UHF MRI is particularly interesting in the case of CE-MRA for vessel visualization with higher spatial resolution and signal-to-noise ratio (SNR) [117]. Using extremely small superparamagnetic iron oxide nanoparticles (2.3 nm SPION), clear visualization of microvascular anatomy was attained, providing a novel way for the medical diagnosis of vascular-related diseases. Our team also investigated the feasibility of using SPION for UHF $T_1$-weighted CE-MRI [118,119]. Sub-5 nm superparamagnetic iron oxide nanoparticles (SPIO-5) coated with PEG of different chain lengths (i.e., PEG-800, -2000 and -5000) were investigated as $T_1$ contrast agents at 9.4 T. Specifically, a follow-up of biodistribution and elimination

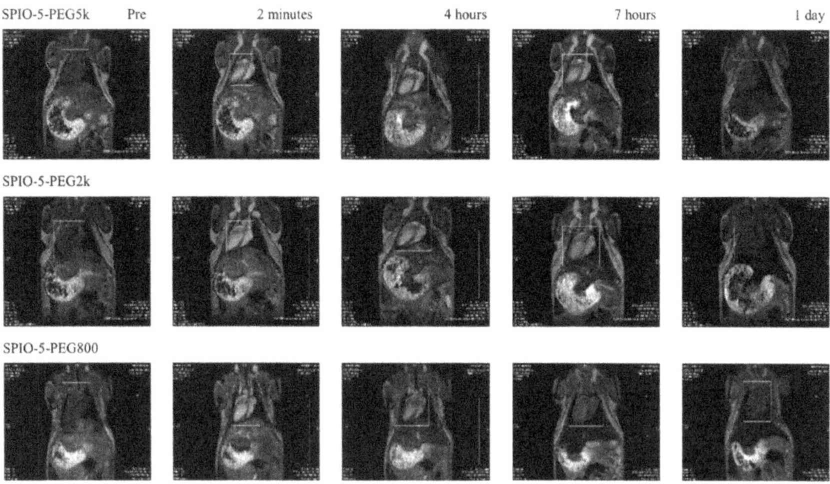


**FIGURE 16** Illustration of *in vivo* MRI data collected on mice treated with different types of SPIO-5 at a dose of 70 mmol Fe $kg^{-1}$. Positive signal enhancement in $T_1$-weighted FISP imaging before, at maximum contrast (2 min), after 2 h, after 4 h and one day after injection [119]. (Copyright 2021, *Journal of Materials Chemistry B.*)

pathways by $T_1$- and $T_2$-weighted MRI as well as by optoacoustic imaging through the presence of a near-infrared-emitting dye (NIR-dye) onto the particle surface. Interestingly, differences in the pharmacokinetic properties could be observed depending on the thickness of the PEG coating, SPIO-5 bearing longer chains of PEG showing a persistent $T_1$ signal in cardiac left ventricle for several hours, while SPIO-5-$PEG_{800}$ were washed out within an hour (Figure 16).

Other than pure iron oxide nanoparticles, hybrid nanoparticles containing various metals such as europium (Eu), copper (Cu), or manganese (Mn), having enhanced $T_1$ signal, have been developed in the past years. For instance, Fernandez-Barahona et al. developed Cu-doped SPION displaying very large longitudinal relaxivity values and demonstrated their suitability for *in vivo* use in angiography as well as targeted molecular imaging [120]. However, even if such strategy has proven to yield systems exhibiting enhanced positive contrast, this must be weighed against the additional toxicity potentially induced by the dopants as well as the increased complexity required to achieve their synthesis [95].

## 4 GENERAL CONCLUSIONS

Since the advent of MRI as a powerful imaging modality in clinical radiology practices, SPION have always held great promises as contrast agents due to a combination of favorable attributes, namely their superparamagnetic properties, biocompatibility and their ease of surface modification with various coatings. Overall, this chapter provides an overview of the theoretical basis (SPM model) of the use of SPION as MRI contrast agents and (pre)clinical applications. Although a substantial amount of work has been performed toward the translation of several SPION formulations into clinical applications, all the developed contrast agents didn't met the expected outcome and were withdrawn as a result of safety concern and/or lack of profit. Still, despite these pitfalls, SPION still show enormous potential as the next-generation MRI contrast agents.

For many years, a particular emphasis is placed on the use of SPION-based $T_1$ contrast agents as potential alternatives to conventional GBCA, as a consequence of issues faced by the latter regarding their safety. The literature has clearly evidenced the excellent imaging performance of SPION as $T_1$ contrast agents for applications such as CE-MRA or targeted molecular imaging. However, much research has yet to be done to advance beyond proof-of-concept studies and to reach suitable SPION-based $T_1$ contrast agent yet to be developed to a clinical setting.

## ABBREVIATIONS AND DEFINITIONS

| | |
|---|---|
| **CE-MRA** | contrast-enhanced magnetic resonance angiography |
| **CE-MRI** | contrast-enhanced magnetic resonance imaging |
| **GBCA** | gadolinium-based contrast agents |
| **MPI** | magnetic particle imaging |
| **MRI** | magnetic resonance imaging |
| **NMR** | nuclear magnetic resonance |
| **NMRD (profiles)** | nuclear magnetic resonance dispersion (profiles) |
| **PEG** | polyethylene glycol |
| **SNR** | signal-to-noise ratio |
| **SPION** | superparamagnetic iron oxide nanoparticle |
| $T_1$ | longitudinal relaxation time |
| $T_2$ | transverse relaxation time |
| **UHF** | ultra-high field |
| **ZDS** | zwitterionic dopamine sulfonate |

## REFERENCES


1. D. Huber, *Small* **2005**, *1*, 482–501.
2. D. Faivre, *Iron Oxides: From Nature to Applications*, 2nd completely rev. and extended ed., Wiley-VCH, Weinheim, New York, **2016**. ISBN 978-3-527-29669-9.
3. U. T. Lam, R. Mammucari, K. Suzuki, N. R. Foster, *Ind. Eng. Chem. Res.* **2008**, *47*, 599–614.
4. M. D. Nguyen, H.-V. Tran, S. Xu, T. R. Lee, *Appl. Sci.* **2021**, *11*, 11301. https://doi.org/10.3390/app112311301.
5. B. Issa, I. M. Obaidat, in *Magnetic Resonance Imaging*, Ed.: L. Manchev, IntechOpen, London, United Kingdom, **2019**.
6. J. Estelrich, E. Escribano, J. Queralt, M. Busquets, *Int. J. Mol. Sci.* **2015**, *16*, 8070–8101.
7. N. Wahajuddin, S. Arora, *Int. J. Nanomed.* **2012**, 3445–3471. https://doi.org/10.2147/IJN.S30320.

8. N. Malhotra, J.-S. Lee, R. A. D. Liman, J. M. S. Ruallo, O. B. Villaflores, T.-R. Ger, C.-D. Hsiao, *Molecules* **2020**, *25*, 3159. https://doi.org/10.3390/molecules25143159.
9. R. Tietze, J. Zaloga, H. Unterweger, S. Lyer, R. P. Friedrich, C. Janko, M. Pöttler, S. Dürr, C. Alexiou, *Biochem. Biophys. Res. Commun.* **2015**, *468*, 463–470.
10. K. Ulbrich, K. Holá, V. Šubr, A. Bakandritsos, J. Tuček, R. Zbořil, *Chem. Rev.* **2016**, *116*, 5338–5431.
11. T. Vangijzegem, D. Stanicki, S. Laurent, *Expert Opin. Drug Deliv.* **2019**, *16*, 69–78.
12. S. Palanisamy, Y.-M. Wang, *Dalton Trans.* **2019**, *48*, 9490–9515.
13. D. Stanicki, T. Vangijzegem, I. Ternad, S. Laurent, *Expert. Opin. Drug Deliv.* **2022**, *19*, 321–335. https://doi.org/10.1080/17425247.2022.2047020.
14. Jasmin, A. L. M. Torres, L. Jelicks, A. C. C. de Carvalho, D. C. Spray, R. Mendez-Otero, in *Nanoparticles in Biology and Medicine*, *Vol. 906*, Ed. M. Soloviev, Humana Press, Totowa, NJ, 2012, pp. 239–252.
15. L. Li, W. Jiang, K. Luo, H. Song, F. Lan, Y. Wu, Z. Gu, *Theranostics* **2013**, *3*, 595–615.
16. Y. Wang, C. Xu, H. Ow, *Theranostics* **2013**, *3*, 544–560.
17. D. Stanicki, S. Boutry, S. Laurent, L. Wacheul, E. Nicolas, D. Crombez, L. Vander Elst, D. L. J. Lafontaine, R. N. Muller, *J. Mater. Chem. B* **2014**, *2*, 387–397.
18. Z. Hedayatnasab, F. Abnisa, W. M. A. W. Daud, *Mater. Des.* **2017**, *123*, 174–196, https://doi.org/10.1016/j.matdes.2017.03.036.
19. I. Obaidat, B. Issa, Y. Haik, *Nanomaterials* **2015**, *5*, 63–89.
20. D. Kami, S. Takeda, Y. Itakura, S. Gojo, M. Watanabe, M. Toyoda, *Int. J. Mol. Sci.* **2011**, *12*, 3705–3722.
21. V. Mulens, M. del Puerto Morales, D. F. Barber, *ISRN Nanomater.* **2013**, *2013*, 1–14.
22. O. Ziv-Polat, S. Margel, A. Shahar, *Neural Regen. Res.* **2015**, *10*, 189–191.
23. T. Nanda, M. Alobaid, K. Rege, *Nano LIFE* **2021**, *11*, 2030001.
24. J. W. M. Bulte, *Adv. Drug Deliv. Rev.* **2019**, *138*, 293–301. https://doi.org/10.1016/j.addr.2018.12.007.
25. A. K. Hauser, M. I. Mitov, E. F. Daley, R. C. McGarry, K. W. Anderson, J. Z. Hilt, *Biomaterials* **2016**, *105*, 127–135. https://doi.org/10.1016/j.biomaterials.2016.07.032.
26. D. Kwatra, A. Venugopal, S. Anant, *Transl. Cancer Res.* **2013**, *2*, 13.
27. I. Ternad, S. Penninckx, V. Lecomte, T. Vangijzegem, L. Conrard, S. Lucas, A.-C. Heuskin, C. Michiels, R. N. Muller, D. Stanicki, S. Laurent, *Nanomaterials* **2023**, *13*, 201. https://doi.org/10.3390/nano13010201.
28. A. Roch, R. N. Muller, P. Gillis, *J. Chem. Phys.* **1999**, *110*, 5403–5411.
29. A. Roch, P. Gillis, A. Ouakssim, R. N. Muller, *J. Magn. Magn. Mater.* **1999**, *201*, 77–79.
30. Q. L. Vuong, P. Gillis, A. Roch, Y. Gossuin, *Wiley Interdiscip. Rev. Nanomed. Nanobiotechnol.* **2017**, 9, e1468.
31. J. H. Freed, *J. Chem. Phys.* **1978**, *68*, 4034–4037.
32. Y. Ayant, E. Belorizky, J. Aluzon, J. Gallice, *J. Phys.* **1975**, *36*, 991–1004.
33. G. De Marco, A. Roch, I. Peretti, Y. Gossuin, P. Lehmann, C. Menuel, J. Vallée, R. Muller, *Médecine Nucl.* **2006**, *30*, 645–658.
34. Y. Gossuin, P. Gillis, A. Hocq, Q. L. Vuong, A. Roch, *Wiley Interdiscip. Rev. Nanomed. Nanobiotechnol.* **2009**, *1*, 299–310.
35. S. Laurent, J.-L. Bridot, L. V. Elst, R. N. Muller, *Future Med. Chem.* **2010**, *2*, 427–449.
36. P. Gillis, S. H. Koenig, *Magn. Reson. Med.* **1987**, *5*, 323–345.
37. A. Roch, Y. Gossuin, R. N. Muller, P. Gillis, *J. Magn. Magn. Mater.* **2005**, *293*, 532–539.
38. Q. L. Vuong, P. Gillis, Y. Gossuin, *J. Magn. Reson.* **2011**, *212*, 139–148.
39. Y. Matsumoto, A. Jasanoff, *Magn. Reson. Imaging* **2008**, *26*, 994–998.

40. A. K. Gupta, M. Gupta, *Biomaterials* **2005**, *26*, 3995–4021.
41. A. Ali, H. Zafar, M. Zia, I. ul Haq, A. R. Phull, J. S. Ali, A. Hussain, *Nanotechnol. Sci. Appl.* **2016**, *9*, 49–67.
42. Y.-X. J. Wang, S. M. Hussain, G. P. Krestin, *Eur. Radiol.* **2001**, *11*, 2319–2331.
43. G. Morana, *Cancer Imaging* **2007**, *7*, S24–S27. https://doi.org/10.1102/1470-7330.2007.9001.
44. R. Weissleder, D. Stark, B. Engelstad, B. Bacon, C. Compton, D. White, P. Jacobs, J. Lewis, *Am. J. Roentgenol.* **1989**, *152*, 167–173.
45. O. Clément, N. Siauve, C.-A. Cuénid, G. Frija, *Top. Magn. Reson. Imaging* **1998**, *9*, 167–182.
46. P. R. Ros, P. C. Freeny, S. E. Harms, S. E. Seltzer, P. L. Davis, T. W. Chan, A. E. Stillman, L. R. Muroff, V. M. Runge, M. A. Nissembaum, *Radiology* **1995**, *196*, 481–488.
47. P. Reimer, T. Balzer, *Eur. Radiol.* **2003**, *13*, 1266–1276.
48. P. Reimer, G. Schuierer, T. Balzer, P. E. Peters, *Magn. Reson. Med.* **1995**, *34*, 694–697.
49. C. W. Jung, P. Jacobs, *Magn. Reson. Imaging* **1995**, *13*, 661–674.
50. R. A. M. Heesakkers, A. M. Hövels, G. J. Jager, *Lancet Oncol.* **2008**, *9*, 7.
51. M. G. Harisinghani, W. M. Deserno, *N. Engl. J. Med.* **2003**, *348*, 2491–2499.
52. C. Corot, D. Warlin, *Nanomed. Nanobiotechnol.* **2013**, *5*, 12.
53. A. S. Fortuin, R. Brüggemann, J. Linden, I. Panfilov, B. Israël, T. W. J. Scheenen, J. O. Barentsz, *WIREs Nanomed. Nanobiotechnol.* **2018**, *10*. https://doi.org/10.1002/wnan.1471.
54. K. E. Kellar, D. K. Fujii, W. H. H. Gunther, K. Briley-Sæbø, S. H. Koenig, *J. Magn. Reson. Imaging 11*, 488–494.
55. J. Schnorr, S. Wagner, H. Pilgrimm, B. Hamm, M. Taupitz, *Acad. Radiol.* **2002**, *9*, S307–S309.
56. J. Schnorr, S. Wagner, C. Abramjuk, I. Wojner, T. Schink, T. J. Kroencke, E. Schellenberger, B. Hamm, H. Pilgrimm, M. Taupitz, *Invest. Radiol.* **2004**, *39*, 546–553.
57. M. Taupitz, S. Wagner, H. Pilgrimm, B. Hamm, *J. Magn. Reson. Imaging* **2000**, *12*, 905–911.
58. R. Bachmann, R. Conrad, B. Kreft, O. Luzar, W. Block, S. Flacke, D. Pauleit, F. Träber, J. Gieseke, K. Saebo, H. Schild, *J. Magn. Reson. Imaging* **2002**, *16*, 190–195. https://doi.org/10.1002/jmri.10149.
59. H. Ittrich, K. Peldschus, N. Raabe, M. Kaul, G. Adam, *RöFo - Fortschritte Auf Dem Geb. Röntgenstrahlen Bildgeb. Verfahr.* **2013**, *185*, 1149–1166. https://doi.org/10.1055/s-0033-1335438.
60. C. Klein, E. Nagel, B. Schnackenburg, A. Bornstedt, S. Schalla, V. Hoffmann, A. Lehning, E. Fleck, *Magn. Reson. Mater. Phys.* **2000**, *11*, 65–67.
61. M. Taupitz, S. Wagner, J. Schnorr, I. Kravec, H. Pilgrimm, H. Bergmann-Fritsch, B. Hamm, *Invest. Radiol.* **2004**, *39*, 394–405.
62. T. Leiner, K. Y. J. A. M. Ho, V. B. Ho, G. Bongartz, W. P. T. M. Mali, W. Rasch, J. M. A. Van Engelshoven, *Eur. Radiol.* **2003**, *13*, 1620–1627. https://doi.org/10.1007/s00330-002-1791-6.
63. M. Wagner, S. Wagner, J. Schnorr, E. Schellenberger, D. Kivelitz, L. Krug, M. Dewey, M. Laule, B. Hamm, M. Taupitz, *J. Magn. Reson. Imaging* **2011**, *34*, 816–823.
64. Y.-X. J. Wang, *World J. Gastroenterol.* **2015**, *21*, 13400.
65. M. Lu, M. H. Cohen, D. Rieves, R. Pazdur, *Am. J. Hematol.* **2010**, *85*, 315–319.
66. A. Barton Pai, *J. Blood Med.* **2012**, *3*, 77–85.

67. V. S. Balakrishnan, M. Rao, A. T. Kausz, L. Brenner, B. J. G. Pereira, T. B. Frigo, J. M. Lewis, *Eur. J. Clin. Invest.* **2009**, *39*, 489–496.
68. M. R. Bashir, L. Bhatti, D. Marin, R. C. Nelson, *J. Magn. Reson. Imaging* **2015**, *41*, 884–898.
69. G. B. Toth, C. G. Varallyay, A. Horvath, M. R. Bashir, P. L. Choyke, H. E. Daldrup-Link, E. Dosa, J. P. Finn, S. Gahramanov, M. Harisinghani, *I. Macdougall, Kidney Int.* **2017**, *92*, 47–66.
70. E. D. Lehrman, A. N. Plotnik, T. Hope, D. Saloner, *Clin. Radiol.* **2019**, *74*, 37–50. https://doi.org/10.1016/j.crad.2018.02.021.
71. M. D. Hope, T. A. Hope, C. Zhu, F. Faraji, H. Haraldsson, K. G. Ordovas, D. Saloner, *Am. J. Roentgenol.* **2015**, *205*, W366–W373. https://doi.org/10.2214/AJR.15.14534.
72. Y. Huang, J. C. Hsu, H. Koo, D. P. Cormode, *Theranostics* **2022**, *12*, 796–816. https://doi.org/10.7150/thno.67375.
73. E. A. Neuwelt, C. G. Várallyay, S. Manninger, D. Solymosi, M. Haluska, M. A. Hunt, G. Nesbit, A. Stevens, M. Jerosch-Herold, P. M. Jacobs, J. M. Hoffman, *Neurosurgery* **2007**, *60*, 601–612. https://doi.org/10.1227/01.NEU.0000255350.71700.37.
74. Y. X. J. Wáng, J.-M. Idée, *Quant. Imaging Med. Surg.* **2017**, *7*, 88–122. https://doi.org/10.21037/qims.2017.02.09.
75. Z. Zhou, L. Yang, J. Gao, X. Chen, *Adv. Mater.* **2019**, *31*, 1804567.
76. J. Estelrich, M. J. Sánchez-Martín, M. A. Busquets, *Int. J. Nanomed.* **2015**, *10*, 1727–1741.
77. Y. Javed, K. Akhtar, H. Anwar, Y. Jamil, *J. Nanoparticle Res.* **2017**, *19*. https://doi.org/10.1007/s11051-017-4045-x.
78. H. B. Na, T. Hyeon, *J. Mater. Chem.* **2009**, *19*, 6267.
79. J. W. M. Bulte, D. L. Kraitchman, *NMR Biomed.* **2004**, *17*, 484–499.
80. T. Grobner, *Nephrol. Dial. Transplant.* **2006**, *21*, 1104–1108.
81. T. Grobner, F. Prischl, *Kidney Int.* **2007**, *72*, 260–264.
82. J.-M. Idée, M. Port, C. Medina, E. Lancelot, E. Fayoux, S. Ballet, C. Corot, *Toxicology* **2008**, *248*, 77–88.
83. C. Do, J. DeAguero, A. Brearley, X. Trejo, T. Howard, G. P. Escobar, B. Wagner, *Kidney360* **2020**, *1*, 561–568. https://doi.org/10.34067/KID.0000272019.
84. T. Kanda, K. Ishii, H. Kawaguchi, K. Kitajima, D. Takenaka, *Radiology* **2014**, *270*, 834–841.
85. E. Tedeschi, F. Caranci, F. Giordano, V. Angelini, S. Cocozza, A. Brunetti, *Radiol. Med. (Torino)* **2017**, *122*, 589–600.
86. T. Kanda, T. Fukusato, M. Matsuda, K. Toyoda, H. Oba, J. Kotoku, T. Haruyama, K. Kitajima, *S. Furui, Radiology* **2015**, *276*, 228–232.
87. V. Gulani, F. Calamante, F. G. Shellock, E. Kanal, S. B. Reeder, *Lancet Neurol.* **2017**, *16*, 564–570.
88. H. Malikova, M. Holesta, *J. Vasc. Access* **2017**, *18*, 1–7.
89. Y. Bao, J. Sherwood, Z. Sun, *J. Mater. Chem. C* **2018**, *6*, 1280–1290.
90. S. Caspani, R. Magalhães, J. P. Araújo, C. T. Sousa, *Materials* **2020**, *13*, 2586.
91. I. Fernández-Barahona, M. Muñoz-Hernando, J. Ruiz-Cabello, F. Herranz, J. Pellico, *Inorganics* **2020**, *8*, 28. https://doi.org/10.3390/inorganics8040028.
92. J. M. D. Coey, *Phys. Rev. Lett.* **1971**, *27*, 1140–1142.
93. S. Linderoth, P. V. Hendriksen, F. Bo/dker, S. Wells, K. Davies, S. W. Charles, S. Mo/rup, *J. Appl. Phys.* **1994**, *75*, 6583–6585.
94. J. Mohapatra, M. Xing, J. P. Liu, *Materials* **2019**, *12*, 3208. https://doi.org/10.3390/ma12193208.
95. M. Jeon, M. V. Halbert, Z. R. Stephen, M. Zhang, *Adv. Mater.* **2020**, 1906539. https://doi.org/10.1002/adma.201906539.

96. M. P. Morales, S. Veintemillas-Verdaguer, M. I. Montero, C. J. Serna, A. Roig, L. Casas, B. Martínez, F. Sandiumenge, *Chem. Mater.* **1999**, *11*, 3058–3064.
97. D. Stanicki, L. V. Elst, R. N. Muller, S. Laurent, *Curr. Opin. Chem. Eng.* **2015**, *8*, 7–14. https://doi.org/10.1016/j.coche.2015.01.003.
98. A. Avasthi, C. Caro, M. L. Garcia-Martin, M. Pernia Leal, *React. Chem. Eng.* **2023**. https://doi.org/10.1039.D2RE00516F.
99. M. Taupitz, J. Schnorr, S. Wagner, C. Abramjuk, H. Pilgrimm, D. Kivelitz, T. Schink, J. Hansel, G. Laub, H. Hünigen, B. Hamm, *Radiology* **2002**, *222*, 120–126.
100. M. Taupitz, S. Wagner, A. Ludwig, V. Stangl, M. Ebert, B. Hamm, *Int. J. Nanomed.* **2013**, *8*, 767–779. https://doi.org/10.2147/IJN.S38702.
101. E. Taboada, E. Rodríguez, A. Roig, J. Oró, A. Roch, R. N. Muller, *Langmuir* **2007**, *23*, 4583–4588.
102. U. I. Tromsdorf, O. T. Bruns, S. C. Salmen, U. Beisiegel, H. Weller, *Nano Lett.* **2009**, *9*, 4434–4440.
103. F. Hu, Q. Jia, Y. Li, M. Gao, *Nanotechnology* **2011**, *22*. https://doi.org/10.1088/0957-4484/22/24/245604.
104. B. H. Kim, N. Lee, H. Kim, K. An, Y. I. Park, Y. Choi, K. Shin, Y. Lee, S. G. Kwon, H. B. Na, Park, J. G. *J. Am. Chem. Soc.* **2011**, *133*, 12624–12631.
105. R. Chen, D. Ling, L. Zhao, S. Wang, Y. Liu, R. Bai, S. Baik, Y. Zhao, C. Chen, T. Hyeon, *ACS Nano* **2015**, *9*, 12425–12435.
106. G. Wang, X. Zhang, A. Skallberg, Y. Liu, Z. Hu, X. Mei, K. Uvdal, *Nanoscale* **2014**, *6*, 2953–2963.
107. H. Jung, B. Park, C. Lee, J. Cho, J. Suh, J. Park, Y. Kim, J. Kim, G. Cho, H. Cho, *Nanomed. Nanotechnol. Biol. Med.* **2014**, *10*, 1679–1689.
108. C.-L. Liu, Y.-K. Peng, S.-W. Chou, W.-H. Tseng, Y.-J. Tseng, H.-C. Chen, J.-K. Hsiao, P.-T. Chou, *Small* **2014**, *10*, 3962–3969.
109. M. Z. Iqbal, X. Ma, T. Chen, L. Zhang, W. Ren, A. Wu, *J. Mater. Chem. B* **2015**, *3*, 5172–5181.
110. K. R. Hurley, H. Ring, M. L. Etheridge, J. Zhang, Q. Shao, N. D. Klein, V. Szlag, C. Chung, M. Reineke, M. G. Garwood, *Mol. Pharm.* **2016**, *13*, 2172–2183.
111. R. Bhavesh, A. V. Lechuga-Vieco, J. Ruiz-Cabello, F. Herranz, *Nanomaterials* **2015**, *5*, 1880–1890.
112. Z. Shen, T. Chen, X. Ma, W. Ren, Z. Zhou, G. Zhu, A. Zhang, Y. Liu, J. Song, Z. Li, H. Ruan, *ACS Nano* **2017**, *11*, 10992–11004.
113. G. Liang, J. Han, Q. Hao, *ACS Appl. Bio Mater.* **2018**, *1*, 1389–1397.
114. W. Liu, G. Deng, D. Wang, M. Chen, Z. Zhou, H. Yang, S. Yang, *J. Mater. Chem. B* **2020**, *8*, 3087–3091.
115. C. Tao, Q. Zheng, L. An, M. He, J. Lin, Q. Tian, S. Yang, *Nanomaterials* **2019**, *9*, 11. https://doi.org/10.3390/nano9020170.
116. J. Wang, Y. Jia, Q. Wang, Z. Liang, G. Han, Z. Wang, J. Lee, M. Zhao, F. Li, R. Bai, D. Ling, *Adv. Mater.* **2020**, *33*. https://doi.org/10.1002/adma.202004917.
117. P. Fries, J. N. Morelli, F. Lux, O. Tillement, G. Schneider, A. Buecker, *Wiley Interdiscip. Rev. Nanomed. Nanobiotechnol.* **2014**, *6*, 559–573.
118. T. Vangijzegem, D. Stanicki, S. Boutry, Q. Paternoster, L. Vander Elst, R. N. Muller, S. Laurent, *Nanotechnology* **2018**, *29*, 265103. https://doi.org/10.1088/1361-6528/aabbd0.
119. D. Stanicki, L. Larbanoix, S. Boutry, T. Vangijzegem, I. Ternad, S. Garifo, R. N. Muller, S. Laurent, *J. Mater. Chem. B* **2021**, *9*, 5055–5068. https://doi.org/10.1039/d1tb00573a.
120. I. Fernández-Barahona, L. Gutiérrez, S. Veintemillas-Verdaguer, J. Pellico, M. D. P. Morales, M. Catala, M. A. Del Pozo, J. Ruiz-Cabello, F. Herranz, *ACS Omega* **2019**, *4*, 2719–2727. https://doi.org/10.1021/acsomega.8b03004.

# 10 Ln-Based Particles as MRI Contrast Agents

*Joop A. Peters and Kristina Djanashvili*
Department of Biotechnology, Delft University of Technology, Van der Maasweg 9, 2629 HZ Delft, The Netherlands
j.a.peters@tudelft.nl, k.djanashvili@tudelft.nl

## CONTENTS



### Abstract

Lanthanides have proven effective in enhancing the contrast of magnetic resonance images by modulating the intensity of the proton signal of water in close proximity. However, the low sensitivity of this imaging modality has stimulated the search for new MRI probes along with the already established small paramagnetic complexes. This is where nanoparticulate systems came into play, owing to their immense morphological diversity and their ability to accommodate and deliver high payloads of lanthanides to target sites through facile surface grafting with various functional molecules. Decades of dedicated research in this direction yielded fascinating examples of lanthanide-based nanoparticles with enhanced capabilities. These include responsiveness to diverse endogenous stimuli, multimodal imaging, and most importantly, the integration of imaging and therapeutic modalities. In this

DOI: 10.1201/9781003374688-10

chapter, we review some examples of solid and porous lanthanide-based nanoparticles and the corresponding relaxation enhancement mechanisms, and discuss their applications, potential, limitations, and prospects.

## KEYWORDS

Nanoparticles; Relaxation; AGuIX; Ho-microspheres; Image-Guided Therapy; Stimuli Responsive Nanoparticles; Mesoporous Silica-Based Nanoparticles; Zeolites

## 1 INTRODUCTION

The current clinical Ln-based MRI contrast agents (CAs) are low-molecular-weight organic ligand complexes of $Gd^{3+}$ that distribute non-selectively in the extracellular space and are quickly excreted renally. They enhance $T_1$-weighted MRI scans, aiding in the visualization of abnormalities and structures in the body, particularly for diseases like cancer, cardiovascular disorders, multiple sclerosis, and inflammatory conditions.

Advances in molecular sciences have identified disease-specific biomarkers, enabling the development of targeted and thus more precise CAs [1]. However, the low sensitivity of typical receptors (on the order of $10^5$ per cell) results in local concentrations of CA in the nM–$\mu$M range based on cell volume [1–3], while clinical CAs require at least about 0.1 mM for detectable contrast enhancement. Both soft and hard nanoparticles (NPs) containing a high payload of $Ln^{3+}$ ions have been explored to overcome this limitation and amplify the relative sensitivity, with further enhancements through selective targeting [4,5].

Soft NPs involve grafting $Ln^{3+}$ chelates onto organic polymers or encapsulating them in various structures, while hard NPs are Ln-oxides, Ln-salts, Ln-loaded (alumino)silicates, zeolites, mesoporous materials, fullerenes, carbon nanodots, nanotubes, and fibers. To prevent clumping and settling of NPs in aqueous suspensions, coatings are applied, often incorporating targeting groups, functions for other imaging modalities, or therapeutics.

The properties of NPs, such as size, shape, charge, and coating, significantly influence their behavior in the body, impacting factors like blood residence time, biodistribution, and cellular uptake. Spherical NPs under 5.5 nm are rapidly excreted, while larger particles may be taken up by organs or trapped in lung alveoli [6–8]. Delivery of the NPs to the site of interest can be achieved either passively through the enhanced permeability and retention (EPR) effect [9,10] typical for the leaky neovasculature of tumors or actively by exploiting targeting vectors attached to the NPs. Surface PEGylation (PEG = polyethylene glycol) can be used to reduce nonspecific binding to blood proteins and macrophages leading to increased blood circulation times [11]. Properties of the NPs such as composition, morphology, and size can also determine their cytotoxicity [12].

Some $Ln^{3+}$ ions are also luminescent probes with applications in bio-imaging due to their narrow transition bands and long luminescence lifetimes [13–15].

$Ln^{3+}$-doped NPs can upconvert long-wavelength IR or near-IR excitation radiation to emit shorter wavelengths. Furthermore, lanthanides also have radioisotopes that are suitable for diagnostic imaging (PET and SPECT) and radiotherapy [16,17].

This chapter explores selected $Ln^{3+}$-containing hard particles, specifically their relationships between structure and relaxivity. Our primary focus is on the aspects of enhancing MRI contrast. Two particulate systems that have been translated to clinical application will be discussed in more detail.

## 2 LANTHANIDE OXIDES AND SALTS AS MRI CONTRAST AGENTS

Small $Gd^{3+}$-based NPs (size ≤10 nm) have usually $r_2/r_1$ <10, and therefore, may be used as positive CAs, whereas NPs of larger sizes and those based on paramagnetic $Ln^{3+}$ from the second half of the Ln-series ($Dy^{3+}$, $Ho^{3+}$, $Er^{3+}$, and $Tm^{3+}$) are usually considered as negative CAs. The characteristics of NPs, such as pharmacodynamics, pharmacokinetics, and relaxivity, are highly dependent on their size and morphology, necessitating well-defined NPs for potential applications. Various inorganic materials containing $Ln^{3+}$ ions are being considered for nanoparticulate MRI CAs, including $Ln_2O_3$ [18–20], $Ln_2O_2S$ [21], $LnF_3$ [22], $NaLnF_4$ [23,24], and $LnPO_4$ [25,26]. This involved the development of production methodologies for $Ln^{3+}$-based oxides and salts with well-defined shapes and particle sizes [27–30], such as sol–gel, co-precipitation, thermal decomposition, hydrothermal, microemulsion, and microwave-assisted syntheses.

Among these NPs, the Ln-oxides have the highest possible $Ln^{3+}$ density. For example, a spherical NP of $Gd_2O_3$ with a diameter of 5 nm contains approximately 1660 $Gd^{3+}$ ions. Coating NPs with a diamagnetic layer generally does not affect the magnetization ($M$) of the core, though ultrasmall NPs (<10 nm) may display slightly reduced $M$ values due to a spin canting effect by surface $Ln^{3+}$ ions [31].

The concentration of paramagnetic $Ln^{3+}$ (Ln≠La, Lu) in suspensions can be determined from the bulk magnetic susceptibility (BMS) shift as measured with an NMR experiment and using Equation 1 [32,33], where $c$ is the $Ln^{3+}$-concentration in mM, $\Delta_\chi$ is the bulk magnetic susceptibility shift in ppm, and $\mu_{\text{eff}}$ is the effective magnetic moment of the lanthanide under study.

$$c = 1.926\frac{\Delta_\chi T}{\mu_{\text{eff}}^{\,2}} \tag{1}$$

The relaxivity of paramagnetic $Ln^{3+}$-based NPs ($r_i$, $i = 1, 2$) has contributions from the inner sphere through exchange between bulk and inner-sphere water molecules ($r_{i,\text{IS}}$), from the second sphere through water molecules held close to surface $Ln^{3+}$ ions longer than the diffusion correlation time ($r_{i,\text{SS}}$), and from the outer sphere through water molecules that diffuse along the NP without being bound ($r_{i,\text{OS}}$) (Equation 2).

$$r_i = r_{i,\text{IS}} + r_{i,\text{SS}} + r_{i,\text{OS}} \quad i = 1,2 \tag{2}$$

The relative importance of $r_{i,OS}$ with respect to $r_{i,IS}$ and $r_{i,SS}$ can be evaluated by relaxation rate measurements of a compound that does not coordinate to $Ln^{3+}$ ions [34,35] or by performing relaxation measurements in methanol–water mixtures [36]. If the relaxation rate enhancement of the OH and CH protons are about equal, $r_{i,OS}$ dominates. If not, $r_{i,IS}$ and/or $r_{i,SS}$ dominate(s).

## 2.1 Gd Oxides and Inorganic Gd Salts as $T_1$ MRI Contrast Agents

$Gd^{3+}$ has an isotropic electronic ground state ($^8S_{7/2}$), leading to relatively long electronic relaxation times ($10^4$ ps) [37]. For such a slow electronic relaxation, the inner- and second-sphere relaxation contributions can be modeled with a combination of the Luz–Meiboom equations for longitudinal relaxation rates in exchanging systems and the Solomon–Bloembergen–Morgan (SBM) equations for the dipolar relaxation (see Chapter 1) [38–43]. Parameters that govern the relaxation are the number of water molecules in the first and second coordination sphere of $Gd^{3+}$ ($q_{IS}$ and $q_{SS}$), the electronic relaxation times ($\tau_{ie}$), the rotational correlation time ($\tau_R$), and the average residence time of a water molecule in the first or second coordination sphere of $Gd^{3+}$ ($\tau_M$ or $\tau_{M,SS}$). The conditions under which the SBM model can be applied are no longer valid for slowly rotating systems. More complex models have been developed to treat them, although the simpler classical SBM model is usually still adequate to evaluate data for $B_0 > 3$ T [39,44,45].

The relaxivities of many Gd-based NPs with different sizes and coatings measured under different experimental conditions (temperature and magnetic field strength) have been reported. Comparing these data is challenging because of the numerous parameters that influence $r_{1,IS}$ and/or $r_{1,SS}$ [46], which can be highly correlated.

The paramagnetic effect, caused by the exchange between first- and second-sphere water molecules and the bulk water molecules, originates almost entirely from the $Gd^{3+}$ ions located at the surface of the NP. A water proton bound at a $Gd^{3+}$ at the surface of an NP is always at a short distance (< 4.7 Å) of neighboring $Gd^{3+}$ ions, which therefore may have an additional effect on the relaxation enhancement of that water proton. DFT calculations suggested that water molecules may also bridge between two adjacent Gd-ions at a surface [47]. As $r_{i,IS}$ and $r_{i,SS}$ are proportional to $r_{GdH}^{-6}$ and $r_{GdH,SS}^{-6}$, respectively (where $r_{GdH}$ and $r_{GdH,SS}$ are the distances between the $Gd^{3+}$ ion and a water H-atoms of interest), the relaxation enhancing effect diminishes rapidly with increasing distance. $Gd^{3+}$ ions below the surface of an NP hardly contribute to the relaxivity. Consequently, $r_{1,IS}$ and $r_{1,SS}$ depend on the surface/volume ratio of the NPs, which decreases rapidly for spherical NPs as they grow larger. Additionally, when considering NPs of equal volume but different shapes, the surface/volume ratio is the smallest for a sphere. Hence, it is reasonable to expect that any shape other than a sphere will result in a larger $r_{1,IS}$ and/or $r_{1,SS}$. The reported data often show the expected trend in the dependence of $r_1$ as a function of the NP size, but generally, the magnitudes of the size-related effects are smaller than expected (see for example Figure 1a) [48,49]. It should be noted that the tumbling time ($\tau_R$) of the NPs is usually larger than $\tau_{ie}$

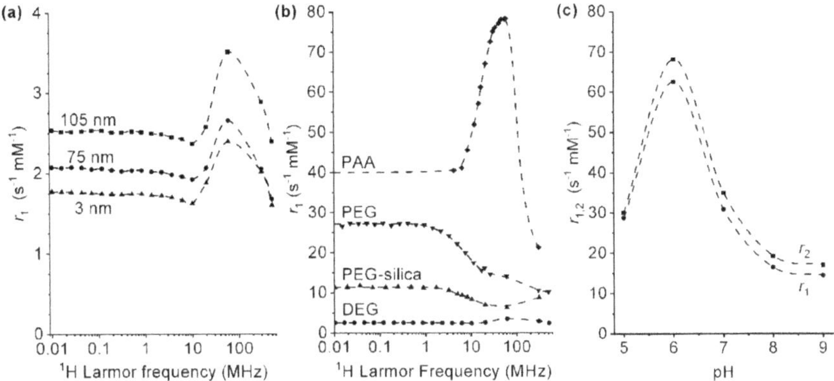


**FIGURE 1** Relaxivities of aqueous suspensions of $Gd_2O_3$ NPs: (a) DEG coated with core sizes between 3 and 105 nm at 310 K. (Constructed with data from Ref. [48].) (b) NPs with 1–2 nm diameter with different coatings at 310 K. (Constructed with data from Refs. [48,52,53].) (c) NPs with 1.9 nm diameter coated with $PAA_{100}$, ratio PAA/Gd = 0.89, at 298 K, 0.5 T. (Constructed with data from Ref. [54].) The dashed curves are guides to the eye.

or $\tau_M$, and therefore, it has almost no effect on $r_1$ [50]. However, the shape of the surface may increase the residence time of second-sphere waters with respect to the diffusion lifetime of bulk water and consequently increase $r_1$ [51].

It has been suggested that lower crystallinity results in higher relaxivities [54], while for nanoplates, the relaxivity even appears to be dependent on the orientation of the crystal planes that are exposed to the water molecules [47]. A detailed study of citrate-coated NPs (6–8 nm size) showed that cubic crystals induced $r_1$ and $r_2$ twice as large as their hexagonal analogs [55]. This was attributed to a combination of differences in hydrodynamic radius and effective magnetic moments. Therefore, control of size and morphology is an important challenge for the production of Ln-based NPs.

It is noteworthy that $Gd_2O_3$ NPs with a core diameter smaller than 10 nm have relatively high $r_1$ and $r_2$ values (typically 10–20 s$^{-1}$mM$^{-1}$ at 1.5 T and 310 K) compared with classical CAs such as Gd-DOTA, although it can be expected that for the NPs, only the $Gd^{3+}$ ions at the surface can cause a significant enhancement of the relaxation. This can be ascribed to both the relatively large rotational correlation times ($\tau_R$) of the NPs and the second-sphere contributions by exchangeable protons at the coatings of the NPs or water molecules bound to the coatings through hydrogen bonds. The latter is demonstrated in Figure 1b, which compares the reported nuclear magnetic relaxation dispersion (NMRD) profiles for $Gd_2O_3$ NPs with small cores (1.3–3 nm) that are coated with different materials. The magnitude of $r_1$ at low Larmor frequencies (LF < 1 MHz) is usually determined mainly by $\tau_{ie}$, $q_{IS}$, and $q_{SS}$ [38]. Assuming that $\tau_{ie}$ and $q_{IS}$ in these systems are approximately the same, these NMRD profiles suggest that $q_{SS}$ varies significantly for different coatings. Apparently, PEG attached directly to $Gd_2O_3$ keeps more water molecules close to the surface than the less polar diethylene glycol (DEG). The NMRD profile for $Gd_2O_3$ with a PEG-coating separated from the core by a polysiloxane shell suggests that this shell is porous, which allows water to approach the core [56]. Most NMRD profiles in Figure 1a and b show a local maximum at high LF (> 10 MHz), which is a characteristic of slowly rotating systems [38]. However, such a maximum is absent in the system with PEG directly attached to $Gd_2O_3$. Simulations with the SBM model suggest that this might be attributed to either very slow ($\tau_M > 10^{-6}$ s) or very fast exchange of first- and/or second-sphere water molecules ($\tau_M < 10^{-10}$ s).

Carniato et al. have analyzed the $^1H$ and $^{17}O$ relaxivity of citrate-coated $GdF_3$ NPs (core diameter < 5 nm) by considering only the $Gd^{3+}$ ions at the surface [57]. The $r_1$ value of the local maximum in the $^1H$ NMRD profile calculated in this way was at about 40 $s^{-1}mM^{-1}$ (surface $Gd^{3+}$ ions) at about 100 MHz and 310 K. Analysis of the NMRD profile with the SBM model suggested that each surface $Gd^{3+}$ ion is directly bound to 2 water molecules and has 4 water molecules in the second coordination sphere. Similar results were obtained for $NaGdF_4$ coated with EDTA derivatives [58].

Many types of coatings have been investigated including PEG, dextran, dextran cross-linked with epichlorohydrin, polyvinylpyrrolidone (PVP), polycarboxylates, polysaccharides, and silica. The grafting degree of the coating, its conformation, and hydrophilicity are examples of parameters that may influence $r_{1,SS}$ [46]. In general, the relaxivity increases with increasing hydrophilicity of the coating. This has been demonstrated by $r_1$ and $r_2$ values measured for NPs (diameter 13–21 nm) coated with polyols of increasing length (DEG, tri(tetra)ethylene glycol, $PEG_{200}$) that increased from 1.1 to 5.8 $s^{-1}mM^{-1}$ and 13.5 to 28.7 at 3 T, respectively [59]. Very high relaxivities have been obtained with ultrasmall $Gd_2O_3$, $NaGdF_4$, and GdOF NPs coated with PAA [53,60]. For example, relaxivities up to about 80 $s^{-1}mM^{-1}$ 1.4 T have been observed for ultrasmall $Gd_2O_3$ NPs (core diameter 1.9 nm) coated with polyacrylic acid ($PAA_{1800}$, molar ratio PAA/Gd 0.89) (Figure 1b) [53]. After conjugation with RGD dimer as a targeting group for integrin $\alpha_v\beta_3$, these NPs were successfully tested for *in vitro* and *in vivo* $T_1$-weighted imaging of tumors. The relaxivities of a series of $Gd_2O_3$-based NPs with different coatings have been correlated with their isoelectric points (p*I*s) by Liu et al. [54]. Of the series investigated, $Gd_2O_3$ NPs with a 1.9 nm sized core and a coating with $PAA_{5000}$ showed the highest relativity had the lowest isoelectric point (p*I*=2.40), whereas corresponding NPs covered with aminated silica had the lowest $r_1$ value and the highest isoelectric point (p*I*=6.26). The pH dependence (Figure 1c) suggests a relatively fast prototrophic exchange at pH 7.4 on these carboxylate-decorated NPs.

Ultrahigh relaxivities were also achieved by encapsulation of oleic acid-coated $NaGdF_4$ NPs (core diameter 3 nm) in micelles by intercalation of the oleic acid chains with the PEGylated phospholipid DSPE-$PEG_{2000}$ (1,2-distearoyl-*sn*-glycero-3-phosphoethanolamine-*N*-[methoxy(polyethyleneglycol)$_{2000}$]) [49]. The $r_1$ value appeared to increase with the hydrodynamic diameter. The maximum

($r_1$=78.2 $s^{-1}$ $mM^{-1}$ at 1.41 T and 310 K) was achieved for a hydrodynamic diameter of 4.2 nm, which is still below the threshold for effective renal clearance.

## 2.2 Polysiloxanes Grafted with Ln-Chelates (AGuIX) for Image-Guided Therapy

Several NPs of the types mentioned above have demonstrated promising performance during *in vitro* cell studies and preliminary animal experiments while exhibiting no signs of toxicity, which makes them potential candidates for future clinical applications. However, none yet made it to the clinical stage. A significant obstacle to translation into clinical use is their hydrodynamic size ($D_h$), which often exceeds the renal clearance limit of approximately 5.5 nm resulting in prolonged residence times within the body. This raises concerns about toxicities due to the potential release of free $Gd^{3+}$ ions from these NPs. In addition, the efficiency of $Gd_2O_3$ NPs as $T_1$ contrast agents is limited by the fact that only the $Gd^{3+}$ ions at the surface of the cores contribute to the $r_1$ relaxivity. Furthermore, the extremely high cost of research and development poses another major challenge in bringing these novel medicines to the market. For instance, the average cost of developing a new cancer medicine has been estimated to be around € 3.7 billion [61].

A Gd-based nanoparticulate system that has successfully advanced to clinical trials (phase I and II) is a Gd-chelated polysiloxane matrix-based NP named AGuIX (activation and guiding of irradiation by X-ray) [62–66]. The AGuIX NPs were prepared by an elegant top-down approach [62,67] starting from ultrasmall $Gd_2O_3$ NPs ($D_h$=3.5 nm) synthesized with the polyol method using DEG as the solvent. Coating with a thin polysiloxane layer (0.5 nm) by TEOS and APTES followed by conjugation of the amino groups present on the silica coating with DOTAGA and transfer into water resulted in hollow spheres due to dissolution of the $Gd_2O_3$. Rearrangement and fragmentation of the polysiloxane and finally collapse produced AGuIX (Figure 2). These NPs have a $D_h$ of only 3±0.1 nm and contain 10 DOTA units, with 7 coordinated with $Gd^{3+}$ ions and 3 remaining unbound.

By adding extra $Gd^{3+}$ ions, the remaining free DOTA groups can also be coordinated to obtain a potent MRI CA with a relaxivity higher than the classical Gd-DOTA (Figure 3). Alternatively, coordination with radioisotopes (e.g., $^{111}In$) enables potential multimodal imaging (SPECT, PET) or theranostic applications. The free DOTA groups can also be used for attachment of targeting peptide groups [68–76]. The silica can be utilized for doping with a chromophore for luminescence, and free amino functions at the silica introduced by the APTES during the synthesis can be applied as anchors for other functions such as porphyrins for photodynamic therapy. Recently, a more flexible one-pot synthesis has been developed, in which DOTAGA was first connected to APTES. This method could also be used to prepare NPs with larger sizes (3–15 nm) [77].

AGuIX has a zeta potential of 17 mV at physiological pH and consequently shows high colloidal stability [67]. It accumulates in tumors by the EPR effect quickly and the tumor retention time is relatively long, while it is rapidly washed

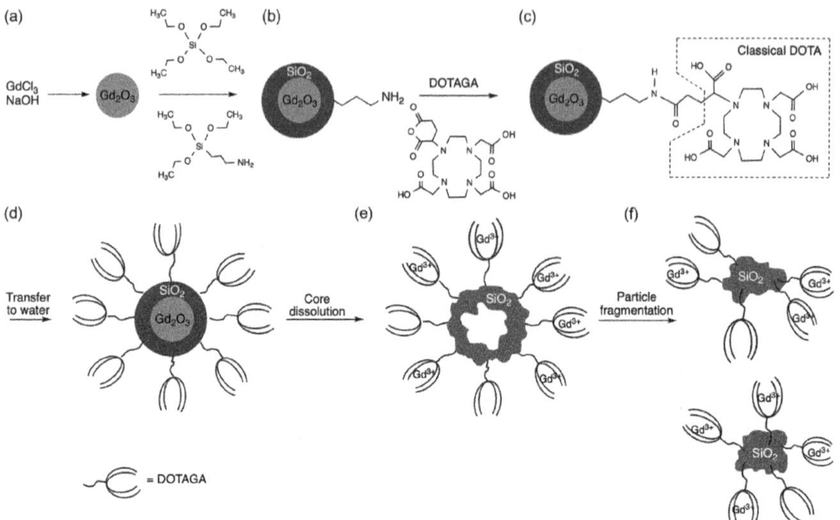


**FIGURE 2** Schematic representation of the AGuIX synthesis: (a) core synthesis, (b) polysiloxane shell synthesis, (c) DOTAGA grafting, (d) transfer into water, (e) core dissolution, and (f) polysiloxane fragmentation. (Reproduced with permission from Ref. [67]. Copyright 2013, Wiley-VCH Verlag GmbH&Co.)

out from healthy parts of the body by renal elimination [62,79]. This dual behavior limits systemic toxicity and simultaneously provides extremely selective MRI contrast enhancement of tumors and metastases. A low dose of AGuIX NPs accumulated in tumors is sufficient for the imaging by MRI, primarily due to the presence of 10 Gd ions (each with a high $r_1$, see Figure 3) within the rigid framework of an NP. AGuIX is biodegradable by hydrolysis of the polysiloxane into low-molecular-weight fragments with Gd-DOTA that eliminate renally very rapidly without the formation of free $Gd^{3+}$ ions [80,81].

Once the optimal accumulation of AGuIX in tumors is achieved, the Gd ions in the NPs can serve as highly effective sensitizers in subsequent MRI-guided localized therapy using different types of irradiations (radiotherapy, neutron therapy, or hadron therapy) [64,82–85].

A comprehensive meta-analysis of literature data focused on the pharmacokinetics and biodistribution of AGuIX, and other tumor-targeting NPs revealed that the core composition of NPs predominantly influences their pharmacokinetics and biodistribution, rather than the active targeting function [65]. However, the analysis also demonstrated that employing active targeting strategies enhances the tumor uptake of AGuIX NPs and prolongs their residence time in tumors compared to passively targeted AGuIX NPs. By optimizing an active targeting approach, it becomes feasible to enhance the accumulation of AGuIX NPs in tumors, thereby potentially minimizing the frequency of administrations while maximizing their effectiveness.

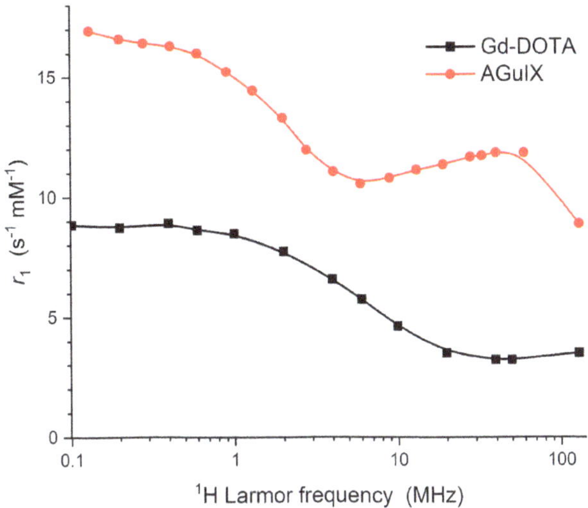


**FIGURE 3** NMRD profiles of AGuIX and Gd-DOTA. (Constructed with data from Refs. [67,78].)

After extensive *in vitro* and animal preclinical evaluation, utilizing a variety of tumor models and administration methods (intravenous, intertumoral, intraperitoneal, intratracheal, intradermal, and pulmonary) [63,64,86–91], AGuIX has progressed to several phase I and II clinical trials. These trials aim to assess the MRI-guided treatment efficacy of AGuIX for brain metastases, locally advanced cervical cancer, centrally located lung tumors, pancreatic cancer, and glioblastoma [92–94]. Encouragingly, initial results demonstrate that AGuIX NPs accumulate and enhance MRI contrast in all types of brain metastases (see Figure 4), and they effectively reduce tumor volume when combined with radiation therapy sensitized by the NPs [95,96]. Moreover, no acute adverse effects have been observed.

## 2.3 $Dy_2O_3$, $Ho_2O_3$, $NaDyF_4$, AND $NaHoF_4$ AS $T_2$ CONTRAST AGENTS

Paramagnetic $Ln^{3+}$ ions (Ln≠Gd) have much shorter electronic relaxation times ($\tau_e$=0.3–0.5 ps at 298 K) than $Gd^{3+}$ ($\tau_e$=$10^4$ ps at 298 K) [97], while, e.g., $Dy^{3+}$-$Tm^{3+}$ ions also have high magnetic moments ($\mu_{eff}$=7.9–10.6 Bohr magneton). Typically, the $r_1$ values of NPs based on these ions are quite small (< 0.5 s$^{-1}$mM$^{-1}$),

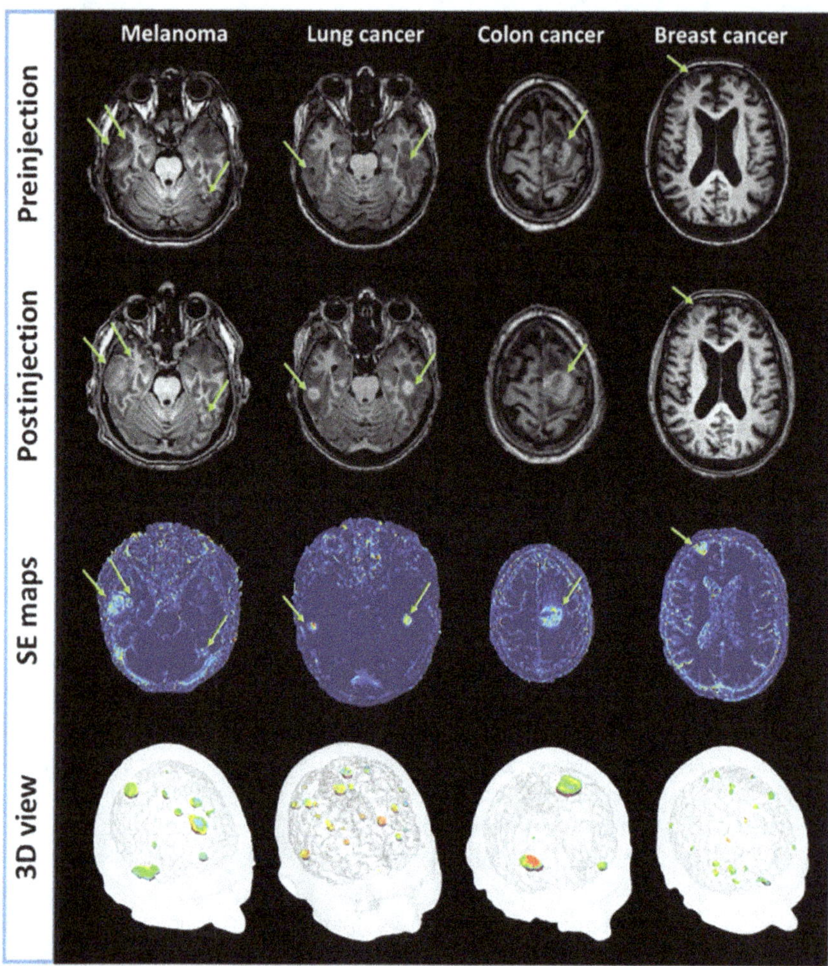


**FIGURE 4** MRI signal enhancement in four types of brain metastases. The first- and second-row images are obtained pre/post-administration of Gd-based NPs using a three-dimensional (3D) $T_1$-weighted imaging sequence. The green arrows point to highlighted metastases. The third-row images are corresponding signal enhancement (SE) maps with a conspicuous local increase of intensity (light blue to orange color) in all different types of brain metastases. The fourth row shows a 3D visualization of all metastases with SE. (Reprinted with permission from Ref. [95]. Copyright 2020, American Association for the Advancement of Science.)

in contrast to their $Gd^{3+}$-based counterparts, which generally are an order of magnitude greater [19,33]. The Curie spin relaxation dominates over the dipolar relaxation for these NPs, resulting in much higher $r_{2,IS}$ values compared to $r_{1,IS}$ ones. However, inner-sphere contributions for $Dy^{3+}$-$Tm^{3+}$ are usually still small as shown by relaxation measurements of $Dy_2O_3$ in water–methanol mixtures ([98], see above).

The outer-sphere transverse relaxation enhancement arises from magnetic field inhomogeneities induced by these paramagnetic Ln-based NPs, resulting in the loss of phase coherence among the precessing nuclear spins. The relaxation rate enhancements ($R_2$ and $R_2$*) are modulated by the diffusion correlation time ($\tau_D = d^2/4D$) and the reciprocal of the Larmor frequency shift induced at the NP's equator ($\Delta\omega^{-1}$). For spherical NPs, $\Delta\omega$ is related to $M$, the magnetization of an NP, through Equation 3, where $\gamma_H$ is the proton magnetogyric ratio, $\mu_0$ is the permeability of a vacuum, $n$ represents the number of $Ln^{3+}$ ions within the NP, $v$ denotes the volume of the NP, $k$ is the Boltzmann constant, and $T$ stands for temperature. For small particles with $\tau_D\Delta\omega < 1$, the motion averaging regime (MAR) holds, where diffusive motions occur faster than the spatial variations in the local magnetic field generated by individual NPs. Under this condition, the relaxation enhancements at LF > 1 T can be modeled by Equation 4, where $f$ is the volume fraction occupied by the NPs.

$$\Delta\omega = \frac{\gamma_H \mu_0 M}{3} = \frac{n\gamma_H \mu_0 \mu_{\text{eff}}^2 B}{3vkT} \tag{3}$$

$$R_2^* = \frac{16}{45} f\tau_D (\Delta\omega)^2 \tag{4}$$

It is crucial to emphasize that both $R_2$ and $R_2$* are intricately connected to $\Delta\omega$ and, consequently to $M$, the collective contribution of all paramagnetic $Ln^{3+}$ ions within the NP. This stands in contrast to $R_1$, where the dominant influence stems from the $Ln^{3+}$ ions located on the surfaces of the NPs. It can be concluded that upon an increase in size of the NPs in the MAR, $R_2$ and $R_2$* increase linearly with $\tau_D$ (or $d^2$).

Furthermore, for particles of the same size, $R_2$ and $R_2$* are linearly dependent on $B^2$ and $\mu_{\text{eff}}^4$ in the MAR. Since $Dy^{3+}$ and $Ho^{3+}$ have the largest $\mu_{\text{eff}}$, $T_2$-CAs based on these $Ln^{3+}$ ions are the most effective. It is worth noting that as the magnetic field strength increases, the value of $r_2$ also rises consistently, while $r_1$ decreases steadily once the magnetic field strength surpasses 2 T. If a diamagnetic coating is applied to a paramagnetic NP, its magnetization remains unchanged, but $R_2$ and $R_2$* are affected by the resulting increased size of the NP. Taking into account the distance-dependent properties of $\Delta\omega$, $f$, and $\tau_D$ leads to the following scaling factors for spherical NPs coated with a water-impermeable layer:

$$\Delta\omega_{\text{imp}} = \Delta\omega_{\text{core}} \left(\frac{d_{\text{core}}}{d_{\text{imp}}}\right)^3 ; f_{\text{imp}} = f_{\text{core}} \left(\frac{d_{\text{imp}}}{d_{\text{core}}}\right)^3 ; \tau_{\text{D,imp}} = \tau_{\text{D,core}} \left(\frac{d_{\text{imp}}}{d_{\text{core}}}\right)^2 \tag{5}$$

Here, "imp" and "core" subscripts refer to parameters for the coated and bare NPs, respectively. Application of these factors indicates that coating with an impermeable layer reduces $R_2$ and $R_2$* within the MAR regime [99].

$$R^*_{2,\text{imp}} = R_{2,\text{core}}\left(\frac{d_{\text{core}}}{d_{\text{imp}}}\right) \quad (6)$$

Coatings that allow water to penetrate can present some challenges; simply examining the coating's size pre- and post-application cannot provide an accurate calculation. In addition, the pores of the coating may experience a slower rate of water diffusion compared to bulk water. This can lead to a localized increase in $\tau_D$ [100,101], resulting in higher $R_2$ and $R_2$*. However, if the exchange of water between the interior of the coating and the bulk is too slow, the opposite effect may occur.

In general, the observations reported in the literature are consistent with the predictions described above. An example is the behavior of suspensions containing $Dy_2O_3$ NPs coated with dextran that was crosslinked by epichlorohydrin. The results showed a linear relationship between $r_2$ as a function of $\tau_D$ up to 1.5 μs, which corresponds to particle sizes of 120–140 nm ($B_0$=7–17.6 T) [102], which is a behavior characteristic of the MAR. Surprisingly, $r_2$ (as measured with the CPMG pulse sequence) was somewhat smaller than $r_2$*. This can be attributed to the water molecules in the polydextran coating being hindered from diffusing freely during the time between the refocusing $\pi$-pulses of the CPMG sequence ($\tau_{CP}$). Consequently, the dephasing of these water protons is fully recovered by the spin-echo pulses and these protons do not contribute to $r_2$. This effect is similar to that of an impermeable coating.

Notably, ultrasmall NPs ($d$<5 nm), though occasionally displaying slightly lower magnetization, often manifest much higher $r_2$ values than those predicted by the MAR model (see Table 1). Conversely, larger NPs tend to have their experimental $r_2$ values more accurately aligned with this model. A possible explanation for this apparent disparity lies in the dominance of the inner sphere's contribution to $r_2$ in ultrasmall NPs. Since only the $Ln^{3+}$ atoms located on the NP's surface make substantial contributions, the impact of the inner sphere diminishes swiftly as the size increases. In contrast, the outer-sphere effect is influenced by all $Ln^{3+}$ ions within an NP, thereby gaining prominence in larger NPs.

## 2.4 Gd-Based NPs as Dual $T_{1,2}$ MRI Contrast Agents

The ability to simultaneously influence both $T_1$ and $T_2$ with a single CA is of great interest, particularly for the mitigation of artifacts in MR images [110,111]. To achieve optimal performance as a dual $T_1$–$T_2$ probe, CAs with high values for both $r_1$ and $r_2$ are necessary, while keeping the $r_2/r_1$ ratio close to 1. As the $r_1$ values decrease with the size of Gd-based NPs while the opposite is true for $r_2$, manipulating the NP size can be a useful strategy to adjust the $r_2/r_1$ ratio at the applied magnetic field strength. For $Gd^{3+}$-based NPs, the magnitude of the $r_1$ and $r_2$ values strongly depends on the nature of the coating. Another way to create dual modality $T_1$–$T_2$ CAs is to construct a multi-layer system, such as doping the surface of ultrasmall $Gd_2O_3$ NPs with MnO and then coating the resulting $Gd_2O_3$@

**TABLE 1**
**Comparison of Selected Reported Experimental $r_2$ Data of Coated Spherical Dy- and Ho-Based NPs with Values as Computed from the Reported Magnetizations Using Equations 3 and 4**

| Core | Coating | $d$[a] (nm) | $D_h$[b] (nm) | $M$[c] ($10^4$ A $m^2kg^{-1}$) | $B_0$ (T) | $r_{2,calc}$ ($s^{-1}mM^{-1}$) | $r_{2,exp}$ ($s^{-1}mM^{-1}$) | Reference |
|---|---|---|---|---|---|---|---|---|
| $Dy_2O_3$ | Poly-dextran[d] | 2 | 40 | 3.12 | 7 | 0.1 | 15.7 | [102] |
| | | 74 | 150 | 3.12 | 7 | 74.9 | 77.6 | |
| | | 86 | 188 | 3.12 | 7 | 101.1 | 104.3 | |
| | | 118 | 134 | 3.12 | 7 | 190.3 | 159.1 | |
| | | 190 | 278 | 3.12 | 7 | 493.5 | 129.7 | |
| $Dy_2O_3$ | GlcA[e] | 2.2 | 6.7 | 5.20 | 3 | 0.2 | 37.5 | [103] |
| $NaDyF_4$ | PMAO-PEG[f] | 20.3 | 33.7 | 3.18 | 3 | 10.8 | 15.8 | [104] |
| | | 9.8 | 26.3 | 2.86 | 3 | 2.1 | 5.8 | |
| | | 5.4 | 18.6 | 2.31 | 3 | 0.4 | 4.3 | |
| $Ho_2O_3$ | GlcA[e] | 1.9 | 6.7 | 6.00 | 3 | 0.2 | 35.2 | [105] |
| $Ho_2O_3$ | $PAA_{1800}$[g] | 1.7 | 12.7 | 6.27 | 3 | 0.1 | 1.4 | [106] |
| $Ho_2O_3$ | $PEI_{1200}$[h] | 2.1 | 30.1 | 6.53 | 3 | 0.2 | 13.1 | [107] |
| $Ho_2O_3$ | $PEI_{6000}$ | 1.9 | 52.5 | 6.74 | 3 | 0.2 | 9.9 | [107] |
| $Ho_2O_3$ | $PEGD_{250}$[i] | 2.1 | 8.7 | 6.55 | 3 | 0.2 | 30.4 | [108] |
| $Ho_2O_3$ | $PEGD_{600}$ | 2.1 | 13.5 | 6.22 | 3 | 0.2 | 11.3 | [108] |
| $NaHoF_4$ | DSPE-$PEG_{5000}$[j] | 3.2 | 12.9 | 2.25 | 3 | 0.1 | 24.0 | [109] |
| | | 7.4 | 19.0 | 2.59 | 3 | 0.9 | 10.6 | |
| | | 13.2 | 22.7 | 2.78 | 3 | 3.5 | 36.0 | |

[a] Diameter of the core.
[b] Hydrodynamic diameter as measured by DLS.
[c] As estimated from reported *M-H* curves.
[d] Dextran crosslinked by epichlorohydrin.
[e] GlcA = glucuronic acid.
[f] PMAO-PEG = poly(maleic anhydride-*alt*-1-octadecene)-polyethylene glycol.
[g] PAA = polyacrylic acid.
[h] PEI = polyethylenimine.
[i] PEGD = poly(ethylene glycol) diacid.
[j] DSPE-PEG = 1,2-distearoyl-*sn*-glycero-3-phosphoethanolamine-*N*-[methoxy (polyethylene glycol).

MnO (diameter of 1–2 mm) with a 1 nm thick layer of lactobionic acid [112]. This process has produced NPs with $r_1 = 12.8\,s^{-1}mM^{-1}$ and $r_2 = 26.6\,s^{-1}mM^{-1}$ at 1.5 T. Table 2 shows a selection of $Gd_2O_3$ systems with $r_2/r_1$ ratios close to 1 at 1.5 T, which may have potential as $T_1$–$T_2$ CA.

**TABLE 2**
**Selected Potential $Gd_2O_3$-Based Dual Modality $T_1$–$T_2$ CAs for Application at 1.5 T**

| Coating | $d$ (nm) | $a$ (nm) | $r_1$ ($s^{-1}mM^{-1}$) | $r_2$ ($s^{-1}mM^{-1}$) | Reference |
|---|---|---|---|---|---|
| $PAA_{5100}$ | 2.0 | 6.3 | 31 | 37.4 | [113] |
| $PAA_{1800}$/PAA-rhodamine | 1.5 | –[a] | 22.6 | 29.5 | [114] |
| HOOC-$(CH_2)_5$-COOH | 2 | –[a] | 21.1 | 21.7 | [115] |
| DEG | | 5.9 | 13.3 | 11.8 | [116] |

[a] Not reported

## 2.5 Ho-Based Microspheres for Image-Guided Embolization Therapy

Lanthanides offer a wide array of radioisotopes, holding immense significance in both diagnostic and therapeutic applications. Among these, $^{158}Sm$, $^{165}Dy$, $^{166}Ho$, and $^{177}Lu$ are noteworthy examples [16,17]. In particular, $^{166}Ho$ stands out due to its ease of production through thermal neutron capture by $^{165}Ho$. The radioisotope $^{166}Ho$ decays with a favorable half-life to the stable isotope $^{166}Er$, under the emission of $\beta$-particles and $\gamma$-radiation [17,117]. The maximum range of the emitted $\beta$-particles in tissue is 8.7 mm with a mean of 2.5 mm. Its beta energy (1.77 and 1.85 MeV, $I_\beta$=48.7 and 50.0%, respectively) is very favorable for selective radiotherapy. The half-life is 26.8 h, which means that more than 90% of the radiation is delivered within the first 4 days following the administration. Additionally, its $\gamma$-emission is valuable for $\gamma$-scintigraphy and SPECT, and $^{166}Ho$ shows excellent contrast properties in computed tomography (CT) and $T_2$- or $T_2$*-weighted MRI. Consequently, compounds and materials based on $Ho^{3+}$ appear to be highly suitable for developing multimodality probes for both diagnostic and therapeutic purposes. In this context, they surpass $^{90}Y$-based materials, often used in therapy but unsuitable for diagnostic applications due to their diamagnetic nature and lack of $\gamma$-emission [17].

Microspheres (MSs) with diameters of 25–35 µm and composed of $^{166}Ho$-acetylacetonate complex embedded in poly-(L-lactate) have reached the market and are applied clinically in intra-arterial radioembolization treatment of liver tumors [118–130]. These radioactive MSs can be prepared by neutron activation of their $^{165}Ho$-analogs with preservation of the integrity of the MSs [122,128,131,132]. The intra-arterial radioembolization technique exploits different blood supply patterns in liver malignancies compared to normal liver parenchyma, mainly arterial *vs* portal, respectively. As a result, these MSs spread into the liver through the bloodstream and eventually become entrapped in the arterioles in the periphery or surface of the tumor. No leaching of $Ho^{3+}$ ions was observed in *in vivo* rat MRI experiments followed by neutron activation analysis of bone tissue after 14 months [118]. Achieving a high local dose of radioactivity at the tumor site and minimal radiation-induced toxicity is critical for efficacy and requires a personalized-image guided approach. The treatment efficacy and biodistribution rely on patient selection, pre-imaging with a low dose, dose determination, monitoring, and follow-up evaluations. This requires visualization and quantification using MRI and/or SPECT-CT. During and after the administration, dynamic insight into the distribution and uptake of the $^{166}$Ho-MSs can be obtained by 3-dimensional dose distributions on a voxel-level (in-plane resolution of $2\times2\,mm^2$), which are constructed based on 3D SPECT-CT and/or multislice MR images [133,134]. The administration could ideally be performed within a clinical MRI scanner [135]. To enable accurate delivery, a catheter has been designed with a $Dy_2O_3$ marker attached to its tip, allowing for clear visibility during MRI-guided navigation [127].

Given the size of these MSs, the MAR model for transverse relaxation enhancement is not valid, because the magnetic field experienced by each proton becomes time independent, rendering diffusion negligible [123,136]. In this case, $R_2$* primarily relies on $\Delta\omega$ and can be approximated by Equation 7 [137,138]. Moreover, complete refocusing after applying a $\pi$-pulse in a $T_2$-weighted imaging procedure may become impossible for large $\tau_D$-values, leading to a substantially lower $R_2$ than $R_2$*. Therefore, MR images are preferably acquired with $T_2^*$-weighting, which can be achieved using multigradient spin echo pulse sequences [139]. For MRI-based dosimetry by quantification of the biodistribution of the Ho-MSs, voxel-wise $R_2$* maps are being used [140–142].

$$R_2 = R_2^* = \frac{2\pi}{3\sqrt{3}} f\Delta\omega \tag{7}$$

A multigradient sampling of the free induction decay was found to be superior to multigradient sampling of a spin echo envelope because the latter technique appears to be sensitive to diffusion [136,143]. Measuring and post-processing techniques have been developed to further enhance the accuracy of the quantification [133]. Phantom studies have revealed that the Ho-MSs can be detected at a minimal concentration of 5.5 $\mu g\ mL^{-1}$ at 1.5 T and 25°C [130], which corresponds to a concentration of 5.7 μM Ho, given their 17% Ho content (w/w). The $r_2$* relaxivity in phantoms was measured to be $92.6\,s^{-1}mg^{-1}L$ Ho-MSs ($\approx 90\,s^{-1}mM^{-1}$ $Ho^{3+}$) [130,140].

The safety and the efficacy of the Ho-MS for the treatment of unresectable liver malignancies have been demonstrated in extensive preclinical tests, animal tests [123,124,127,144–146], and human clinical trials (phase I and II) [134,140,147–154]. The Ho-MSs have also been shown to be promising for application in the embolization treatment of various other types of tumors including cerebral tumors [155], recurrent head and neck squamous cell carcinoma [156–158], and kidney tumors [159]. Furthermore, several other $^{166}$Ho particulates have been investigated for brachytherapy such as $^{166}Ho_2O_3$, $^{166}HoPO_4$ [160,161], $^{166}$Ho-alginate [162–166], $^{166}Ho(acac)_3$ [167–169], and $^{166}$Ho-labeled chitosan [170].

# 3 $Ln^{3+}$-LOADED POROUS SILICATES

Loading $Ln^{3+}$ ions into silica-based scaffolds for MRI purposes has been extensively explored in recent decades, primarily due to their inherent physicochemical properties, such as biocompatibility, stability, and adjustable surface characteristics [4,171]. An intriguing feature of silicates lies in their porosity, giving rise to unique frameworks with interconnected cavities that allow for drug delivery in addition to an MRI signal [172]. The design of such probes is strongly reliant on pore sizes, classified as macro- (> 50 nm), meso- (2–50 nm), and micropores (< 2 nm). Further noteworthy advantages include: (1) tunability of pore sizes/shapes, expanding the range of drugs that can be incorporated in the formed cavities, (2) control over drug release kinetics by leveraging various stimuli [173] or molecular gatekeepers [174], and (3) facile surface functionalization using reactive silanol-groups to conjugate various bioactive molecules, such as targeting vectors [175,176] and/or 'stealth' polymers [177]. At the same time, development of the synthetic methods for reducing particle size to the nanometer scale [178,179] not only endows silicates with new properties but, most importantly, determines their advancement toward the clinic.

## 3.1 $Ln^{3+}$-Loaded Mesoporous Systems

Mesoporous silica nanoparticles (MSNs) are composed of an amorphous matrix formed by organosilanes undergoing acid/base-mediated hydrolysis/condensation reactions. The presence of templating surfactants during the synthesis determines the final morphology of the MSNs, resulting in 2D structures with hexagonal or cubic pore arrays (e.g., MCM-41, SBA-15), 3D- (MCM-48, SBA-1) or lamellar (MCM-50) frameworks with a large surface area (> 1000 $m^2/g$) and pores ranging from 2 to 10 nm, with a narrow size distribution. The rigid framework of MSNs serves as an excellent platform to confine high payloads of paramagnetic ions, impeding their rotation, while benefiting from the unrestricted movement of water molecules through the pores, which accelerates the relaxation rate of the surrounding water and enhances the $T_1$ contrast. One of the straightforward methods to achieve this is to add Gd-precursors to the Si-source during the synthesis of MSNs. Mou and co-workers demonstrated this by simple mixing of $GdCl_3$ with long-chain cationic surfactants during the synthesis of Gd-MCM-41 [180] or DTPA-3-aminopropylsilane subsequently complexed with $Gd^{3+}$ ions to yield Gd-MS-nanorods [181]. Both 2D frameworks exhibited increased $r_1$- and $r_2$-relaxivities without evident Gd-leaching. However, XRD patterns indicated some distorted structural order and a minor occurrence of collapsing pores. Co-precipitation of Gd-oleate along with long-chain templating molecules caused even more prominent disordered porosity leading to smaller surface area and abundant Gd-O bonds in the framework, which consequently reduced the degree of water coordination and yielded the $r_1$ relaxivity that was not higher than that of Magnevist [182].

3D frameworks are generally attractive scaffolds to generate high relaxivities due to their open water-permeable network. Incorporation of $Gd^{3+}$ ions by ion-exchange in 3D mesoporous aluminosilicate TUD-1 was attempted, but the stability of the probe was not sufficient for its further development [183]. Alternatively, insertion of $Gd^{3+}$ ions via the incipient wetness technique and calcination at 700°C resulted in $GdSi_xO_y$-MSNs (size <200 nm and pores diameter ±4 nm) with both 3D (MCM-48) and 2D (MCM-41) frameworks [184] with $r_1$-values (at 1.5 T and 37°C) of 18.54 and 11.99 $s^{-1}mM^{-1}$, respectively. Interestingly, increasing the Gd-loading from 2 to 5 mmol Gd/g $SiO_2$ in both systems led to a significant decrease in $r_1$ to about 2.50 $s^{-1}mM^{-1}$, attributed to the restricted water diffusion caused by the reduced porosity. A similar effect was observed after the soft metalorganic grafting of 2D (MCM-41, SBA-15) and 3D (SBA-1) framework surfaces with $Gd[N\text{-}(SiHMe_2)_2]_3(THF)_2$: the $r_1$-values of the materials with 1 wt% Gd were dramatically higher than those for the analogs with higher Gd-loadings (up to 12 wt%) [185]. Growing a mesoporous $SiO_2$-layer around $Gd_2O_3$ nanoparticles [186] or their inclusion into the mesopores [187,188] represents another strategy that takes advantage of a high Gd-payload in combination with favorable water exchange through the pores.

An alternative approach involves post-synthetic grafting of MSNs with the Gd complexes via the abundant silanol-groups. The eventual anchoring location depends on whether the templating molecules are still present in the pores or have been removed by calcination [189]. For the latter scenario, the final position of the complex is dictated by the pore diameter, as shown for MCM-41 (3.3 nm) and SBA-15 (8.5 nm), with the Gd-DOTA-like complex anchored to the outer surface and both inner and outer surfaces, respectively. As a result, four times lower $r_1$-relaxivity was found for SBA-15 due to the limited access of water to the mesopores [190]. The anchoring process requires modification of the chelates either with triethoxysilane derivatives that can be easily attached to the surface [181,191] or with activated pendant carboxylates ready to react with the surface silanols pre-functionalized with amino groups [190]. Complexation with the $Gd^{3+}$ ions should preferably occur before the attachment of Gd complexes to the MSNs surface to avoid problems with slow or incomplete complexation and removal of the excess metal ions trapped in the pores [192].

In a more recent study, the inner surface of MSNs was impregnated with the chemotherapeutic drug mitoxantrone, while the outer surface was functionalized with PEG-molecules conjugated to azadibenzocyclooctyne to exploit biorthogonal click chemistry to target cancer cells after their metabolic labeling with azido-mannose [193]. Even though the *in vitro* click-reaction did not lead to a significant increase in affinity to the cancer cells, the MRI data demonstrated contrast enhancement as high as 200% due to the Gd-DOTAGA complexes grafted to the outer surface of MSNs. Finally, the efficacy of MSNs for controlled drug delivery was showcased by loading Gd-DTPA into the pores, sequentially capped with PEG. Exposure of this system to a high-intensity focused ultrasound induced the release of the MR-imageable cargo by increasing the temperature by only 4°C, which could be seen in a $T_1$ contrast effect [194]. The interesting aspect of such systems lies in the drug delivery perspective, where timing and dose can be controlled upon modulation of exposure time and power levels.

## 3.2 $Ln^{3+}$-LOADED ZEOLITES

Zeolites are microporous aluminosilicates composed of $SiO_4$ and $AlO_4^-$ tetrahedra arranged in a unique crystalline 3D framework. Interconnected through oxygen atoms, these tetrahedra form 4–8 membered rings (e.g., sodalite cages) that are further assembled in various cavities (supercages) and channels with pore sizes ranging from 0.4 to 1.2 nm filled with water molecules. The negatively charged framework of zeolites (Figure 5a) can serve as an MRI probe scaffold after ion-exchange with trivalent lanthanides replacing the balancing mono/divalent alkali/alkaline earth metals.

A $Gd^{3+}$-exchanged micro-sized Na-Y was the first zeolite-based MRI CA proposed for oral [196,197] and later for intravenous use [198] upon its availability in the size range of 80–100 nm. The supercages with a diameter of 11.8 Å and a pore opening of 7.4 Å accommodate $Gd^{3+}$ ions coordinating seven water molecules ($q$) with a residence time ($\tau_m$) and longitudinal relaxation time $T_{1m,}$ as defined by the SBM theory [38,41,42]. This arrangement inside the zeolite was conceptualized as a concentrated aqueous $Gd^{3+}$ solution, with the longitudinal relaxation rate ($1T_{1zeo}$) governed by Equation 8, where $w$ is the number of free water molecules per $Gd^{3+}$ ion.

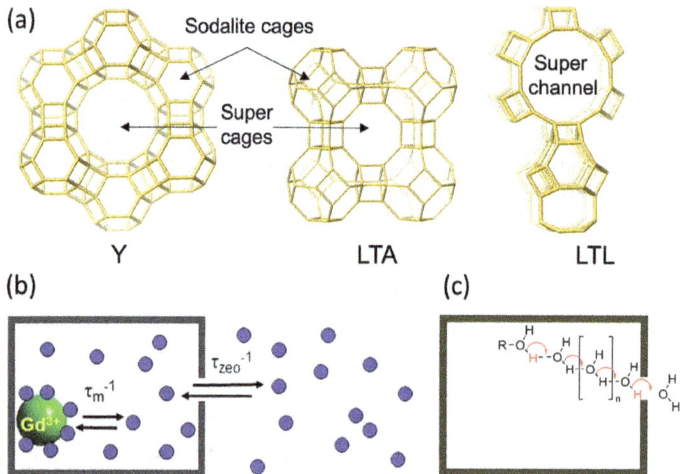


**FIGURE 5** Illustration of the concepts in design of MRI CAs based on zeolites: framework structures of the representative examples (a), the two-step relaxation mechanism (b), and the proton exchange mechanism (c). (Figures (b) and (c) are reprinted with permission from Ref. [195]. Copyright 2015, American Chemical Society.)

$$\frac{1}{T_{1zeo}} = \frac{q/w}{\tau_m + T_{1m}\left(1+\frac{q}{w}\right)} \tag{8}$$

At the same time, the cavity water is in exchange with the bulk water outside of the nanoparticles, thus considering the residence time of the water protons inside the zeolite ($\tau_{zeo}$), the overall relaxivity could be derived by Equation 9. This entire process (Figure 5b) was formulated as a two-step relaxation mechanism [198], which later proved useful for other $Gd^{3+}$-loaded zeolites.

$$r_1 = \frac{w+q}{55500}\left(\frac{1}{T_{1zeo} + \tau_{zeo}}\right) \tag{9}$$

The importance of the pore size in the second step was illustrated in a study with $Gd^{3+}$-loaded zeolite-LTA [199] with a comparable supercage diameter (11.4 Å) but a smaller pore opening of 4.1 Å, which is less efficient for unhindered water diffusion. As a result, the $r_1$-relaxivity of Gd(1.5 wt%)-LTA of 8 $s^{-1}mM^{-1}$ was significantly lower compared to 38 $s^{-1}mM^{-1}$ of Gd(1.3 wt%)-Y obtained at 60 MHz and 25°C due to the larger pore opening (7.4 Å) of the latter. In both cases, a higher Gd-loading (up to 5.4 wt%) led to an overall relaxivity reduction by about 20% based on a decreased number of water molecules inside the cavities. This water fraction, and subsequently, the relaxivity could be increased by dealumination of the framework using a mild ($(NH_4)_2SiF_6$) or a severe (0.5 M HCl) method [199]. Changing the Si/Al ratio of Gd-Y from 1.6 (no dealumination) to 2.6/3.4 (mild/severe dealumination) did not affect the immobilization effectiveness of the $Gd^{3+}$ ions in the framework. The relaxivities remained independent of $\tau_R$ and relied on $\tau_m$, which was reduced upon dealumination (5.6 *vs* 1.6 ns).

A series of studies was conducted with nanozeolite LTL, whose framework is built from cancrinite cages, connected via the upper/lower 6-membered rings to form columns. These oxygen-bridged neighboring columns create expansive undulating 12-membered ring channels with the largest diameter of 12.6 Å and pore opening of 7.1 Å. Initially, the $K^+$ ions in the big channels were ion-exchanged with a cationic dye serving as a drug prototype. In the second step, the surface was functionalized via an APES (3-aminopropyl triethoxysilane) linker with DOTA-chelates that were subsequently complexed with $Eu^{3+}$ and $Gd^{3+}$ ions for optical- and MR-imaging, respectively [200]. In another study, the framework itself was exploited for the complexation of $Gd^{3+}$ ions and at the same time absorbed 5.27 $cm^3$ of oxygen per gram of LTL for MRI-guided tumor reoxygenation purposes [201]. However, the full advantage of the peculiar LTL framework for multimodal imaging was exploited in another study, where $Eu^{3+}$ ions were loaded into the big LTL-channels, after which the nanoparticles were calcined, washed, and reloaded with $Gd^{3+}$ ions [202]. This phenomenon of $Ln^{3+}$ ions migrating from the supercages to the small sodalite cavities upon calcination at ±400°C was shown earlier for Eu-Y [203]. In the LTL-probe, the $Eu^{3+}$ ions were irreversibly confined in the small cavities ($q=1$), while $Gd^{3+}$ ions were placed in the big channels ($q=6$)

enabling simultaneous optical and MR imaging. An exceptionally high $r_1$ relaxivity of 37.8 $s^{-1}mM^{-1}$ as obtained with Gd(5.2 wt%)-LTL compared to 9 $s^{-1}mM^{-1}$ of Gd(5.4 wt%)-Y despite similar pore opening (7.1 *vs* 7.4 Å) was found to be independent of water exchange [195]. Furthermore, the exchange mode exhibited pH sensitivity at a physiologically relevant region, with a steep decrease in $r_1$ from 38 to 8 $s^{-1}mM^{-1}$ and $r_2$ from 98 to 50 $s^{-1}mM^{-1}$ (300 MHz and 25°C) between pH 4 and 9. This could be explained by a prototrophic exchange regime (Figure 5c), where water protons were hopping through the linear channels at pH < 7, while at higher pH, the whole water molecules were exchanging with a lower rate between the interior at the bulk. Remarkably, both high relaxivity and pH-sensitivity were retained after grafting the surface of the nanoparticles with long-chain PEG-molecules < 6.2 wt% [204]. The biodistribution of the probe was recently studied by PET imaging after injection of the Gd-LTL radiolabeled with $^{89}Zr$ into control mice, demonstrating the potential of nanozeolites for multimodal imaging, possibly combined with radionuclide therapy [205].

## 4 GENERAL CONCLUSIONS

The general advancement of nanotechnology inevitably leads to the introduction of NPs into medicine, opening enormous possibilities for their use in imaging and therapy, both individually and in combination. This immensity, mainly due to the structural diversity of the NPs, becomes even more pronounced when using lanthanides with their concurrent chemical similarity and physical versatility. This chapter provides an overview of the most striking examples of Ln-based nano- and microsized particles for MRI. The presented examples show how the proton relaxation theory, originally formulated for small molecular paramagnetic complexes, can still be applied to design particulate systems.

The first argument for the use of particles in medical MRI is the possibility to deliver high payloads of paramagnetic ions to the site of interest, ultimately resulting in increased local contrast. The longitudinal and transverse proton relaxivities of such systems are typically tens of times higher than those of conventional MRI CAs. Since $T_1$-weighted contrast of particles relies on the efficient exchange between water molecules coordinated with the $Gd^{3+}$ ions and water in the surrounding tissues, the efficacy will exclusively depend on the paramagnetic ions that are exposed on the surface. To overcome this inefficient exploitation of $Gd^{3+}$ ions, many interesting architectures have been designed, including nanocarriers grafted with $Gd^{3+}$ complexes, various types of core/shell NPs with different metal ions in the compartments, and ultrasmall Ln-based NPs with high surface-to-volume ratios. Application of hydrophilic coatings may significantly increase the relaxivities. Of particular interest are NPs that provide dual $T_1$- and $T_2$-weighted contrast, an option that is valued by radiologists for its effectiveness in distinguishing between the real signals and artifacts in MR images. For particles containing $Ln^{3+}$ ions with electronic relaxation times shorter than those of $Gd^{3+}$ ions, such as $Dy^{3+}$ or $Ho^{3+}$ ions, the prevalent process shifts from dipolar relaxation (determined almost exclusively by $Ln^{3+}$ ions at the surface) to of Curie spin relaxation (to which all paramagnetic $Ln^{3+}$ ions not just those at the surface contribute). This transition results in increased transverse relaxivities, facilitating $T_2$-weighted contrast. Moreover, the transverse relaxation acceleration increases with magnetic field strength, in contrast to the longitudinal relaxation acceleration, which then actually decreases.

For theranostic applications, particles may be complemented by doping with radioisotopes (e.g., $^{166}Ho$). Porous NPs offer perhaps the most elegant solutions to achieve high $r_1$- and $r_2$-relaxivities by incorporating paramagnetic $Ln^{3+}$ ions into their versatile frameworks. In addition, their various cavities can serve either as sites for $Ln^{3+}$ ions or as drug reservoirs, which can be designed to release the contents by applying appropriate stimuli.

As research on Ln-based NPs continues, new functional materials with attractive imaging/therapeutic properties are emerging. However, translating NP-based technologies into clinical applications remains challenging due to the complexities associated with the intricate interactions of NPs *in vivo*. These include issues like aggregation, immune responses, premature bloodstream clearance, and undesired organ uptake/storage. Together, these factors pose potential risks of long-term toxicity. Consequently, it is necessary to develop advanced solutions that enable the translation of chemical and technical innovations in the field of $Ln^{3+}$-based nano- and microparticles to clinical applications. This necessitates the establishment of standardized methodologies to ensure the safety, efficacy, and reproducibility of emerging technologies. The first two particle systems that have already reached clinical application (AGuIX and Ho-MSs) demonstrate that image-guided therapy is emerging as an important field. Delivery of the agents to the relevant sites without collateral damage is crucial. Therefore, ultra-small targeting NPs with rapid clearance and MSs designed for embolization or brachytherapy may hold the most potential in this area.

## ABBREVIATIONS AND DEFINITIONS

| | |
|---|---|
| **AGuIX** | activation and guiding of irradiation by X-ray |
| **APTES** | (3-aminopropyl)triethoxysilane |
| $\boldsymbol{B_0}$ | strength of the external magnetic field |
| ***c*** | concentration |
| **CA** | contrast agent |
| **CT** | computed tomography |
| ***d*** | diameter (as determined by TEM) |
| **CPMG** | Carr–Purcell–Meiboom–Gill (pulse sequence) |
| ***D*** | diffusion coefficient |
| $\boldsymbol{D_h}$ | hydrodynamic size (as determined by DLS) |
| **DEG** | diethylene glycol |
| **DFT** | density-functional theory |
| **DOTA** | 1,4,7,10-tetraazacyclododecane-1,4,7,10-tetraacetate |

| | |
|---|---|
| **DOTAGA** | 1,4,7,10-tetraazacyclodocecane,1-(glutaric acid)-4,7,10-triacetic acid |
| **DTPA** | diethylenetriamine pentaacetic acid |
| **DSPE-PEG** | 1,2-distearoyl-*sn*-glycero-3-phosphoethanolamine-*N*-[methoxy(polyethylene glycol)] |
| **EDTA** | ethylenediaminetetraacetic acid |
| **EPR** | enhanced permeability and retention |
| $f$ | volume fraction |
| **GlcA** | glucuronic acid |
| $H$ | magnetic field strength |
| $k$ | Boltzmann constant |
| **LF** | Larmor frequency |
| $M$ | magnetization |
| **MAR** | motional averaging regime |
| **MRI** | magnetic resonance imaging |
| **MS** | microsphere |
| **MSN** | mesoporous silica nanoparticles |
| $n$ | number of $Ln^{3+}$ ions within the NP |
| **NMRD** | nuclear magnetic relaxation dispersion |
| **NP** | nanoparticle |
| **PAA** | polyacrylic acid |
| **PEG** | polyethylene glycol |
| **PET** | positron emission tomography |
| $q_{IS}$ **and** $q_{OS}$ | number of inner- and second-sphere water molecules |
| p$I$ | isoelectric point |
| **PMAO-PEG** | poly(maleic anhydride-*alt*-1-octadecene)-polyethylene glycol |
| **PVP** | polyvinylpyrrolidone |
| $R_1$ | longitudinal relaxation rate |
| $R_2$ | transverse relaxation rate |
| $R_2$* | observed rate constant of the free induction decay |
| $r_1$ | longitudinal relaxivity ($s^{-1}mM^{-1}$ paramagnetic ion) |
| $r_2$ | transverse relaxivity ($s^{-1}mM^{-1}$ paramagnetic ion) |
| $r_{i,OS}$, $r_{i,IS}$, **and** $r_{i,SS}$ ($i = 1,2$) | inner-, outer-, and second-sphere relativity |
| $r_{GdH}$ **and** $r_{GdH,SS}$ | the distances between the $Ln^{3+}$ ion and a water H-atoms of first- and second-sphere water molecules |
| **RGD** | arginylglycylaspartic acid |
| **SBM** | Solomon–Bloembergen–Morgan |
| **SPECT** | single-photon emission computed tomography |
| $T_1$ | longitudinal relaxation time |
| $T_2$ | transverse relaxation time |
| $T_2$* | observed time constant of the free induction decay |
| **TEM** | transmission electron microscopy |
| **TEOS** | tetraethoxysilane |

| | |
|---|---|
| $\gamma_H$ | gyromagnetic ratio of $^1H$ |
| $\Delta\chi$ | bulk magnetic susceptibility shift in ppm |
| $\Delta\omega$ | the Larmor frequency of water protons at the particle's surface as compared to those at infinity |
| $\mu_{eff}$ | effective magnetic moment |
| $\mu_0$ | permeability of a vacuum |
| $\tau_D$ | diffusion correlation time |
| $\tau_{ie}$ ($i$ =1,2) | electronic relaxation times |
| $\tau_M$ **and** $\tau_{M,SS}$ | average residence time of water molecules in the first and second coordination sphere of a $Ln^{3+}$ ion |
| $\tau_R$ | the rotational correlation time |
| $v$ | volume of the NP |

## REFERENCES


1. P. Caravan, Z. Zhang, in *The Chemistry of Contrast Agents in Medical Magnetic Resonance Imaging*, 2nd edn., Eds.: A. Merbach, L. Helm, É. Tóth, John Wiley & Sons Ltd., Chichester, UK, **2013**, pp. 311–342.
2. L. M. De Leon-Rodriguez, A. J. M. Lubag, C. R. Malloy, G. V. Martinez, R. J. Gillies, A. D. Sherry, *Acc. Chem. Res.* **2009**, *42*, 948–957.
3. S. Iwaki, K. Hokamura, M. Ogawa, Y. Takehara, Y. Muramatsu, T. Yamane, K. Hirabayashi, Y. Morimoto, K. Hagisawa, K. Nakahara, T. Mineno, T. Terai, T. Komatsu, T. Ueno, K. Tamura, Y. Adachi, Y. Hirata, M. Arita, H. Arai, K. Umemura, T. Nagano, K. Hanaoka, *Org. Biomol. Chem.* **2014**, 8611–8618.
4. J. A. Peters, K. Djanashvili, *Eur. J. Inorg. Chem.* **2012**, *2012*, 1961–1974.
5. M. Botta, L. Tei, *Eur. J. Inorg. Chem.* **2012**, *2012*, 1945–1960.
6. H. S. Choi, W. Liu, P. Misra, E. Tanaka, J. P. Zimmer, B. I. Ipe, M. G. Bawendi, J. V. Frangioni, *Nat. Biotechnol.* **2007**, *25*, 1165–1170.
7. N. Hoshyar, S. Gray, H. Han, G. Bao, *Nanomedicine* **2016**, *11*, 673–692.
8. M. Longmire, P. L. Choyke, H. Kobayashi, *Nanomedicine* **2008**, *3*, 703–717.
9. R. Sun, J. Xiang, Q. Zhou, Y. Piao, J. Tang, S. Shao, Z. Zhou, Y. H. Bae, Y. Shen, *Adv. Drug Deliv. Rev.* **2022**, *191*, 114614.
10. Y. Matsumura, H. Maeda, *Cancer Res.* **1986**, *46*, 6387–6392.
11. J. S. Suk, Q. Xu, N. Kim, J. Hanes, L. M. Ensign, *Adv. Drug Deliv. Rev.* **2016**, *99*, 28–51.
12. T. Kihara, Y. Zhang, Y. Hu, Q. Mao, Y. Tang, J. Miyake, *J. Biosci. Bioeng.* **2011**, *111*, 725–730.
13. J.-C. G. Bünzli, *J. Lumin.* **2016**, *170*, 866–878.
14. K. Malhotra, D. Hrovat, B. Kumar, G. Qu, J. V. Houten, R. Ahmed, P. A. E. Piunno, P. T. Gunning, U. J. Krull, *ACS Appl. Mater. Interfaces* **2023**, *15*, 2499–2528.
15. Y. Li, C. Chen, F. Liu, J. Liu, *Mikrochim. Acta* **2022**, *189*, 109.
16. F. Rösch, E. Forssell-Aronsson, *Met. Ions Biol. Syst.* **2004**, *42*, 77–108.
17. J. F. W. Nijsen, G. C. Krijger, A. D. van het Schip, *Anti-Cancer Agents Med. Chem.* **2007**, *7*, 271–290.
18. M. Norek, J. A. Peters, *Prog. Nucl. Magn. Reson. Spectrosc.* **2011**, *59*, 64–82.
19. Q. L. Vuong, S. Doorslaer, J.-L. Bridot, C. Argante, G. Alejandro, R. Hermann, S. Disch, C. Mattea, S. Stapf, Y. Gossuin, *Magn. Reson. Mater. Phys., Biol. Med.* **2012**, *25*, 467–478.

20. J. Pellico, C. M. Ellis, J. J. Davis, *Contrast Media Mol. Imaging* **2019**, 1845637.
21. S. A. Osseni, S. Lechevallier, M. Verelst, P. Perriat, J. Dexpert-Ghys, D. Neumeyer, R. Garcia, F. Mayer, K. Djanashvili, J. A. Peters, E. Magdeleine, H. Gros-Dagnac, P. Celsis, R. Mauricot, *Nanoscale* **2014**, *6*, 555–564.
22. F. Evanics, P. R. Diamente, F. C. J. M. Van Veggel, G. J. Stanisz, R. S. Prosser, *Chem. Mater.* **2006**, *18*, 2499–2505.
23. X. Jin, F. Fang, J. Liu, C. Jiang, X. Han, Z. Song, J. Chen, G. Sun, H. Lei, L. Lu, *Nanoscale* **2015**, *7*, 15680–15688.
24. N. J. J. Johnson, W. Oakden, G. J. Stanisz, R. S. Prosser, F. C. J. M. van Veggel, *Chem. Mater.* **2011**, *23*, 4877.
25. H. Hifumi, S. Yamaoka, A. Tanimoto, D. Citterio, K. Suzuki, *J. Am. Chem. Soc.* **2006**, *128*, 15090–15091.
26. C. Frangville, M. Gallois, Y. Li, H. H. Nguyen, N. Lauth-de Viguerie, D. R. Talham, C. Mingotaud, J.-D. Marty, *Nanoscale* **2016**, *8*, 4252–4259.
27. B. Ortega-Berlanga, L. Betancourt-Mendiola, C. del Angel-Olarte, L. Hernández-Adame, S. Rosales-Mendoza, G. Palestino, *Crystals* **2021**, *11*, 1094.
28. F. Mayer, J. A. Peters, K. Djanashvili, *Chem. Eur. J.* **2012**, *18*, 8004–8007.
29. W. Zhang, J. Martinelli, F. Mayer, C. S. Bonnet, F. Szeremeta, K. Djanashvili, *RSC Adv.* **2015**, *5*, 69861–69869.
30. L. Wu, X. Lu, Y. Lu, M. Shi, S. Guo, J. Feng, S. Yang, W. Xiong, Y. Xu, C. Yan, Z. Shen, *Small* **2023**, e2308547.
31. B. Xavier, L. Amílcar, *J. Phys. D: Appl. Phys.* **2002**, *35*, 201.
32. D. M. Corsi, C. Platas-Iglesias, H. van Bekkum, J. A. Peters, *Magn. Reson. Chem.* **2001**, *39*, 723–726.
33. M. Norek, G. A. Pereira, C. F. G. C. Geraldes, A. Denkova, W. Zhou, J. A. Peters, *J. Phys. Chem. C* **2007**, *111*, 10240–10246.
34. J. A. Peters, H. van Bekkum, W. M. M. J. Bovée, *Tetrahedron* **1982**, *38*, 331–335.
35. J. A. Peters, J. Huskens, D. J. Raber, *Prog. Nucl. Magn. Reson. Spectrosc.* **1996**, *28*, 283–350.
36. Y. Gossuin, A. Roch, R. N. Muller, P. Gillis, *J. Magn. Reson.* **2002**, *158*, 36–42.
37. G. Parigi, E. Ravera, C. Luchinat, *Prog. Nucl. Magn. Reson. Spectrosc.* **2019**, *114–115*, 211–236.
38. É. Tóth, L. Helm, A. Merbach, in *Chemistry of Contrast Agents in Medical Magnetic Resonance Imaging*, 2nd ed., Eds.: É. Tóth, L. Helm, A. Merbach, John Wiley & Sons Ltd., Chichester, UK, **2013**, pp. 25–81.
39. L. Helm, *Prog. Nucl. Magn. Reson. Spectrosc.* **2006**, *49*, 45–64.
40. Z. Luz, S. Meiboom, *J. Chem. Phys.* **1964**, *40*, 2686–2692.
41. I. Solomon, *Phys. Rev.* **1955**, *99*, 559–565.
42. N. Bloembergen, L. O. Morgan, *J. Chem. Phys.* **1961**, *34*, 842–850.
43. N. Bloembergen, *J. Chem. Phys.* **1957**, *27*, 572–573.
44. C. S. Bonnet, P. H. Fries, A. Gadelle, S. Gambarelli, P. Delangle, *J. Am. Chem. Soc.* **2008**, *130*, 10401–10413.
45. P. H. Fries, E. Belorizky, In *Chemistry of Contrast Agents in Medical Magnetic Resonance Imaging*, 2nd ed, Eds A. E. Merbach, L. Helm, É. Tóth, John Wiley & Sons Ltd., Chichester, UK, **2013**, , pp. 277–309.
46. Z. Zhou, L. Yang, J. Gao, X. Chen, *Adv. Mater.* **2019**, *31*, e1804567.
47. Z. Zhou, R. Hu, L. Wang, C. Sun, G. Fu, J. Gao, *Nanoscale* **2016**, *8*, 17887–17894.
48. L. Faucher, Y. Gossuin, A. Hocq, M.-A. Fortin, *Nanotechnol.* **2011**, *22*, 295103/1–295103/10, S295103/1–S295103/5.

49. N. J. J. Johnson, S. He, V. A. Nguyen Huu, A. Almutairi, *ACS Nano* **2016**, *10*, 8299–8307.
50. A. E. Merbach, L. Helm, É. Tóth, *The Chemistry of Contrast Agents in Medical Magnetic Resonance Imaging*, 2nd edn., Eds A. E. Merbach, L. Helm, É. Tóth, John Wiley & Sons, Ltd, Chichester, UK, **2013**.
51. M. W. Rotz, K. S. B. Culver, G. Parigi, K. W. MacRenaris, C. Luchinat, T. W. Odom, T. J. Meade, *ACS Nano* **2015**, *9*, 3385–3396.
52. L. Faucher, M. Tremblay, J. Lagueux, Y. Gossuin, M.-A. Fortin, *ACS Appl. Mater. Interfaces* **2012**, *4*, 4506–4515.
53. Z. Shen, W. Fan, Z. Yang, Y. Liu, V. I. Bregadze, S. K. Mandal, B. C. Yung, L. Lin, T. Liu, W. Tang, L. Shan, Y. Liu, S. Zhu, S. Wang, W. Yang, L. H. Bryant, D. T. Nguyen, A. Wu, X. Chen, *Small* **2019**, 1903422.
54. Y. Liu, Y. Dai, H. Li, D. Duosiken, N. Tang, K. Sun, K. Tao, *Nanoscale Adv.* **2022**, *4*, 95–101.
55. N. Liu, R. Marin, Y. Mazouzi, G. O. Cron, A. Shuhendler, E. Hemmer, *Nanoscale* **2019**, *11*, 6794–6801.
56. J.-L. Bridot, A.-C. Faure, S. Laurent, C. Rivière, C. Billotey, B. Hiba, M. Janier, V. Josserand, J.-L. Coll, L. Vander Elst, R. Muller, S. Roux, P. Perriat, O. Tillement, *J. Am. Chem. Soc.* **2007**, *129*, 5076–5084.
57. F. Carniato, K. Thangavel, L. Tei, M. Botta, *J. Mater. Chem. B* **2013**, *1*, 2442–2446.
58. F. Carniato, L. Tei, S. Phadngam, C. Isidoro, M. Botta, *ChemPlusChem* **2015**, *80*, 503–510.
59. A. Guleria, P. Pranjali, M. K. Meher, A. Chaturvedi, S. Chakraborti, R. Raj, K. M. Poluri, D. Kumar, *J. Phys. Chem. C* **2019**, *123*, 18061–18070.
60. X.-Y. Zheng, K. Zhao, J. Tang, X.-Y. Wang, L.-D. Li, N.-X. Chen, Y.-J. Wang, S. Shi, X. Zhang, S. Malaisamy, L.-D. Sun, X. Wang, C. Chen, C.-H. Yan, *ACS Nano* **2017**, *11*, 3642–3650.
61. O. J. Wouters, M. McKee, J. Luyten, *J. Am. Med. Assoc.* **2020**, *323*, 844–853.
62. F. Lux, A. Mignot, P. Mowat, C. Louis, S. Dufort, C. Bernhard, F. Denat, F. Boschetti, C. Brunet, R. Antoine, P. Dugourd, S. Laurent, E. L. Vander, R. Muller, L. Sancey, V. Josserand, J.-L. Coll, V. Stupar, E. Barbier, C. Remy, A. Broisat, C. Ghezzi, D. G. Le, S. Roux, P. Perriat, O. Tillement, *Angew. Chem. Int. Ed.* **2011**, *50*, 12299–12303, S12299/1–S12299/16.
63. L. Sancey, F. Lux, S. Kotb, S. Roux, S. Dufort, A. Bianchi, Y. Cremillieux, P. Fries, J.-L. Coll, C. Rodriguez-Lafrasse, M. Janier, M. Dutreix, M. Barberi-Heyob, F. Boschetti, F. Denat, C. Louis, E. Porcel, S. Lacombe, D. G. Le, E. Deutsch, J.-L. Perfettini, A. Detappe, C. Verry, R. Berbeco, K. Butterworth, S. McMahon, K. Prise, P. Perriat, O. Tillement, *Br. J. Radiol.* **2014**, 20140134.
64. F. Lux, V. L. Tran, E. Thomas, S. Dufort, F. Rossetti, M. Martini, C. Truillet, T. Doussineau, G. Bort, F. Denat, F. Boschetti, G. Angelovski, A. Detappe, Y. Crémillieux, N. Mignet, B. T. Doan, B. Larrat, S. Meriaux, E. Barbier, S. Roux, P. Fries, A. Müller, M. C. Abadjian, C. Anderson, E. Canet-Soulas, P. Bouziotis, M. Barberi-Heyob, C. Frochot, C. Verry, J. Balosso, M. Evans, J. Sidi-Boumedine, M. Janier, K. Butterworth, S. McMahon, K. Prise, M. T. Aloy, D. Ardail, C. Rodriguez-Lafrasse, E. Porcel, S. Lacombe, R. Berbeco, A. Allouch, J. L. Perfettini, C. Chargari, E. Deutsch, G. Le Duc, O. Tillement, *Br. J. Radiol.* **2019**, *92*, 20180365.
65. L. Carmes, M. Banerjee, P. Coliat, S. Harlepp, X. Pivot, O. Tillement, F. Lux, A. Detappe, *Adv. Ther. (Weinheim, Ger.)* **2023**, *6*, 2300019
66. F. Lux, L. Sancey, A. Bianchi, Y. Cremillieux, S. Roux, O. Tillement, *Nanomedicine* **2015**, *10*, 1801–1815.

67. A. Mignot, C. Truillet, F. Lux, L. Sancey, C. Louis, F. Denat, F. Boschetti, L. Bocher, A. Gloter, O. Stephan, R. Antoine, P. Dugourd, D. Luneau, G. Novitchi, L. C. Figueiredo, M. P. C. de, L. Bonneviot, B. Albela, F. Ribot, L. L. Van, I. Dechamps-Olivier, F. Chuburu, G. Lemercier, C. Villiers, P. N. Marche, D. G. Le, S. Roux, O. Tillement, P. Perriat, *Chem. Eur. J.* **2013,** 19, 6122–6136.
68. H. Benachour, A. Seve, T. Bastogne, C. Frochot, R. Vanderesse, J. Jasniewski, I. Miladi, C. Billotey, O. Tillement, F. Lux, M. Barberi-Heyob, *Theranostics* **2012**, *2*, 889–904.
69. M. Dentamaro, F. Lux, L. Vander Elst, N. Dauguet, S. Montante, A. Moussaron, C. Burtea, R. N. Muller, O. Tillement, S. Laurent, *Contrast Media Mol. Imaging* **2016**, *11*, 381–395.
70. D. Bechet, F. Auger, P. Couleaud, E. Marty, L. Ravasi, N. Durieux, C. Bonnet, F. Plenat, C. Frochot, S. Mordon, O. Tillement, R. Vanderesse, F. Lux, P. Perriat, F. Guillemin, M. Barberi-Heyob, *Nanomedicine* **2015**, *11*, 657–670.
71. A. Moussaron, S. Vibhute, A. Bianchi, S. Guenduez, S. Kotb, L. Sancey, V. Motto-Ros, S. Rizzitelli, Y. Cremillieux, F. Lux, N. K. Logothetis, O. Tillement, G. Angelovski, *Small* **2015**, *11*, 4900–4909.
72. C. Truillet, P. Bouziotis, C. Tsoukalas, J. Brugiere, M. Martini, L. Sancey, T. Brichart, F. Denat, F. Boschetti, U. Darbost, I. Bonnamour, D. Stellas, C. D. Anagnostopoulos, V. Koutoulidis, L. A. Moulopoulos, P. Perriat, F. Lux, O. Tillement, *Contrast Media Mol. Imaging* **2015**, *10*, 309–319.
73. M. Plissonneau, J. Pansieri, L. Heinrich-Balard, J.-F. Morfin, N. Stransky-Heilkron, P. Rivory, P. Mowat, M. Dumoulin, R. Cohen, E. Allemann, E. Toth, M. J. Saraiva, C. Louis, O. Tillement, V. Forge, F. Lux, C. Marquette, *J. Nanobiotechnol.* **2016**, *14*, 60/1–60/15.
74. P. Bouziotis, D. Stellas, E. Thomas, C. Truillet, C. Tsoukalas, F. Lux, T. Tsotakos, S. Xanthopoulos, M. Paravatou-Petsotas, A. Gaitanis, L. A. Moulopoulos, V. Koutoulidis, C. D. Anagnostopoulos, O. Tillement, *Nanomedicine* **2017**, *12*, 1561–1574.
75. A. Detappe, M. Reidy, C. Mathieu, I. M. Ghobrial, P. P. Ghoroghchian, Y. Yu, F. Lam, H. V. T. Nguyen, J. A. Johnson, T. P. Coroller, P. Jarolim, P. Harvey, A. Protti, Q.-D. Nguyen, Y. Cremillieux, O. Tillement, *Nanoscale* **2019**, *11*, 20485–20496.
76. V. Thakare, V.-L. Tran, M. Natuzzi, E. Thomas, M. Moreau, A. Romieu, B. Collin, A. Courteau, J.-M. Vrigneaud, C. Louis, S. Roux, F. Boschetti, O. Tillement, F. Lux, F. Denat, *RSC Adv.* **2019**, *9*, 24811–24815.
77. V.-L. Tran, V. Thakare, F. Rossetti, A. Baudouin, G. Ramniceanu, B.-T. Doan, N. Mignet, C. Comby-Zerbino, R. Antoine, P. Dugourd, F. Boschetti, F. Denat, C. Louis, S. Roux, T. Doussineau, O. Tillement, F. Lux, *J. Mater. Chem. B* **2018**, *6*, 4821–4834.
78. C. Verry, S. Dufort, B. Lemasson, S. Grand, J. Pietras, I. Tropres, Y. Cremillieux, F. Lux, S. Meriaux, B. Larrat, J. Balosso, G. Le Duc, E. L. Barbier, O. Tillement, *Sci. Adv.* **2020**, *6*, eaay5279.
79. G. Bort, F. Lux, S. Dufort, Y. Cremillieux, C. Verry, O. Tillement, *Theranostics* **2020**, *10*, 1319–1331.
80. L. Labied, P. Rocchi, T. Doussineau, J. Randon, O. Tillement, H. Cottet, F. Lux, A. Hagège, *Anal. Chim. Acta* **2021**, *1185*, 339081.
81. L. Sancey, S. Kotb, C. Truillet, F. Appaix, A. Marais, E. Thomas, B. van der Sanden, J.-P. Klein, B. Laurent, M. Cottier, R. Antoine, P. Dugourd, G. Panczer, F. Lux, P. Perriat, V. Motto-Ros, O. Tillement, *ACS Nano* **2015**, *9*, 2477–2488.
82. G. Le Duc, I. Miladi, C. Alric, P. Mowat, E. Bräuer-Krisch, A. Bouchet, E. Khalil, C. Billotey, M. Janier, F. Lux, T. Epicier, P. Perriat, S. Roux, O. Tillement, *ACS Nano* **2011**, *5*, 9566–9574.

83. S. Dufort, G. Le Duc, M. Salome, V. Bentivegna, L. Sancey, E. Brauer-Krisch, H. Requardt, F. Lux, J.-L. Coll, P. Perriat, S. Roux, O. Tillement, *Sci. Rep.* **2016**, *6*, 29678.
84. C. Verry, S. Dufort, E. L. Barbier, O. Montigon, M. Peoc'h, P. Chartier, F. Lux, J. Balosso, O. Tillement, L. Sancey, G. Le Duc, *Nanomedicine* **2016**, *11*, 2405–2417.
85. A.-S. Wozny, M.-T. Aloy, G. Alphonse, N. Magne, M. Janier, O. Tillement, F. Lux, M. Beuve, C. Rodriguez-Lafrasse, *Nanomedicine* **2017**, *13*, 2655–2660.
86. A. Bianchi, S. Dufort, F. Lux, A. Courtois, O. Tillement, J.-L. Coll, Y. Crémillieux, *Magn. Reson. Mater. Phys., Biol. Med.* **2014**, *27*, 303–316.
87. G. Le Duc, S. Roux, A. Paruta-Tuarez, S. Dufort, E. Brauer, A. Marais, C. Truillet, L. Sancey, P. Perriat, F. Lux, O. Tillement, *Cancer Nanotechnol.* **2014**, *5*, 4/1–4/14, 14 pp.
88. E. Porcel, S. C. Le, S. Li, O. Tillement, F. Lux, P. Mowat, N. Usami, K. Kobayashi, Y. Furusawa, S. Lacombe, *Nanomedicine* **2014**, *10,* 1601–1608
89. S. Dufort, A. Bianchi, M. Henry, F. Lux, G. Le Duc, V. Josserand, C. Louis, P. Perriat, Y. Cremillieux, O. Tillement, J.-L. Coll, *Small* **2015**, *11*, 215–221.
90. N. Tassali, A. Bianchi, F. Lux, G. Raffard, S. Sanchez, O. Tillement, Y. Cremillieux, *Contrast Media Mol. Imaging* **2016**, *11*, 396–404.
91. Y. Cremillieux, N. Pinaud, V. Pham, Y. Montigaud, J. Pourchez, C. Bal, S. Perinel, S. Perinel, M. Natuzzi, F. Lux, O. Tillement, N. Ichinose, B. Zhang, *Magn. Reson. Med.* **2019**, *83*, 1774–1782.
92. C. Verry, L. Sancey, S. Dufort, G. Le Duc, C. Mendoza, F. Lux, S. Grand, J. Arnaud, J. L. Quesada, J. Villa, O. Tillement, J. Balosso, *BMJ Open* **2019**, *9*, e023591.
93. C. Verry, S. Dufort, J. Villa, M. Gavard, C. Iriart, S. Grand, J. Charles, B. Chovelon, J. L. Cracowski, J. L. Quesada, C. Mendoza, L. Sancey, A. Lehmann, F. Jover, J. Y. Giraud, F. Lux, Y. Crémillieux, S. McMahon, P. J. Pauwels, D. Cagney, R. Berbeco, A. Aizer, E. Deutsch, M. Loeffler, G. Le Duc, O. Tillement, J. Balosso, *Radiother. Oncol.* **2021**, *160*, 159–165.
94. TherAguix SA, *Proof of concept to be generated by 5 clinical trials.* https://nhtheraguix.com/pipeline/ (last visited December 18, 2023).
95. C. Verry, S. Dufort, B. Lemasson, S. Grand, J. Pietras, I. Troprès, Y. Crémillieux, F. Lux, S. Mériaux, B. Larrat, J. Balosso, G. Le Duc, E. L. Barbier, O. Tillement, *Sci. Adv.* **2020**, *6*, eaay5279.
96. A. Lavielle, F. Boux, J. Deborne, N. Pinaud, S. Dufort, C. Verry, S. Grand, I. Troprès, C. Vecco-Garda, G. Le Duc, S. Mornet, Y. Crémillieux, *J. Magn. Reson. Imaging* **2023**, *58*, 313–323.
97. G. Parigi, C. Luchinat, in *Paramagnetism in Experimental Biomolecular NMR*, The Royal Society of Chemistry, 2018 ebook collection, **2018**, pp. 1–41.
98. Y. Gossuin, A. Hocq, Q. L. Vuong, S. Disch, R. P. Hermann, P. Gillis, *Nanotechnology* **2008**, *19*, 475102 (8 pp).
99. S. L. C. Pinho, G. A. Pereira, P. Voisin, J. Kassem, V. Bouchaud, L. Etienne, J. A. Peters, L. Carlos, S. Mornet, C. F. G. C. Geraldes, J. Rocha, M.-H. Delville, *ACS Nano* **2010**, *4*, 5339–5349.
100. D.-X. Chen, N. Sun, H.-C. Gu, *J. Appl. Phys.* **2009**, *106*, 063906.
101. H. W. de Haan, C. Paquet, *Magn. Reson. Med.* **2011**, *66*, 1759–1766.
102. M. Norek, E. Kampert, U. Zeitler, J. A. Peters, *J. Am. Chem. Soc.* **2008**, *130*, 5335–5340.
103. S. Marasini, H. Yue, S. L. Ho, K.-H. Jung, J. A. Park, H. Cha, A. Ghazanfari, M. Y. Ahmad, S. Liu, Y. J. Jang, X. Miao, K.-S. Chae, Y. Chang, G. H. Lee, *Eur. J. Inorg. Chem.* **2019**.

104. G. K. Das, N. J. J. Johnson, J. Cramen, B. Blasiak, P. Latta, B. Tomanek, F. C. J. M. van Veggel, *J. Phys. Chem. Lett.* **2012**, *3*, 524–529.
105. K. Kattel, C. R. Kim, W. Xu, T. J. Kim, J. W. Park, Y. Chang, G. H. Lee, *J. Nanosci. Nanotechnol.* **2015**, *15*, 7311–7316.
106. S. Marasini, H. Yue, S. L. Ho, J. A. Park, S. Kim, K.-H. Jung, H. Cha, S. Liu, T. Tegafaw, M. Y. Ahmad, A. Ghazanfari, K.-S. Chae, Y. Chang, G. H. Lee, *Nanomaterials* **2021**, *11*, 1355.
107. S. Liu, H. Yue, S. L. Ho, S. Kim, J. A. Park, T. Tegafaw, M. Y. Ahmad, S. Kim, A. K. A. A. Saidi, D. Zhao, Y. Liu, S.-W. Nam, K. S. Chae, Y. Chang, G. H. Lee, *Nanomaterials* **2022**, *12*, 1588.
108. S. Liu, T. Tegafaw, H. Yue, S. L. Ho, S. Kim, J. A. Park, A. Baek, M. Y. Ahmad, S. H. Yang, D. W. Hwang, S. Kim, A. K. Ali Al Saidi, D. Zhao, Y. Liu, S.-W. Nam, K. S. Chae, Y. Chang, G. H. Lee, *Mater. Adv.* **2022**, *3*, 5857–5870.
109. D. Ni, J. Zhang, W. Bu, C. Zhang, Z. Yao, H. Xing, J. Wang, F. Duan, Y. Liu, W. Fan, X. Feng, J. Shi, *Biomaterials* **2016**, *76*, 218–225.
110. T.-H. Shin, J.-s. Choi, S. Yun, I.-S. Kim, H.-T. Song, Y. Kim, K. I. Park, J. Cheon, *ACS Nano* **2014**, *8*, 3393–3401.
111. F. Hu, Y. S. Zhao, *Nanoscale* **2012**, *4*, 6235–6243.
112. E. Sook Choi, J. Young Park, M. Ju Baek, W. Xu, K. Kattel, J. Hyun Kim, J. Jun Lee, Y. Chang, T. Jeong Kim, J. Eun Bae, K. Seok Chae, K. Jin Suh, G. Ho Lee, *Eur. J. Inorg. Chem.* **2010**, *2010*, 4555–4560.
113. X. Miao, S. L. Ho, T. Tegafaw, H. Cha, Y. Chang, I. T. Oh, A. M. Yaseen, S. Marasini, A. Ghazanfari, H. Yue, K. S. Chae, G. H. Lee, *RSC Adv.* **2018**, *8*, 3189–3197.
114. S. L. Ho, H. Cha, I. T. Oh, K.-H. Jung, M. H. Kim, Y. J. Lee, X. Miao, T. Tegafaw, M. Y. Ahmad, K. S. Chae, Y. Chang, G. H. Lee, *RSC Adv.* **2018**, *8*, 12653–12665.
115. J. Y. Park, S. J. Kim, G. H. Lee, S. Jin, Y. Chang, J. E. Bae, K. S. Chae, J. *Korean Phys. Soc.* **2015**, *66*, 1295–1302.
116. G. Azizian, N. Riyahi-Alam, S. Haghgoo, H. R. Moghimi, R. Zohdiaghdam, B. Rafiei, E. Gorji, *Nanoscale Res. Lett.* **2012**, *7*, 549.
117. N. J. M. Klaassen, M. J. Arntz, A. Gil Arranja, J. Roosen, J. F. W. Nijsen, *EJNMMI Radiopharm. Chem.* **2019**, *4*, 19.
118. S. W. Zielhuis, J. F. W. Nijsen, J.-H. Seppenwoolde, C. J. G. Bakker, G. C. Krijger, H. F. J. Dullens, B. A. Zonnenberg, P. P. van Rijk, W. E. Hennink, A. D. van het Schip, *Biomaterials* **2007**, *28*, 4591–4599.
119. J. F. W. Nijsen, B. A. Zonnenberg, J. R. W. Woittiez, D. W. Rook, I. A. Swildens-van Woudenberg, P. P. van Rijk, A. D. van Het Schip, *Eur. J. Nucl. Med.* **1999**, *26*, 699–704.
120. F. Nijsen, D. Rook, C. Brandt, R. Meijer, H. Dullens, B. Zonnenberg, J. de Klerk, P. van Rijk, W. Hennink, F. van het Schip, *Eur. J. Nucl. Med.* **2001**, *28*, 743–749.
121. J. F. W. Nijsen, M. J. van Steenbergen, H. Kooijman, H. Talsma, L. M. J. Kroon-Batenburg, M. van de Weert, P. P. van Rijk, A. de Witte, A. D. van het Schip, W. E. Hennink, *Biomaterials* **2001**, *22*, 3073–3081.
122. J. F. W. Nijsen, A. D. van het Schip, M. J. van Steenbergen, S. W. Zielhuis, L. M. J. Kroon-Batenburg, M. van de Weert, P. P. van Rijk, W. E. Hennink, *Biomaterials* **2002**, *23*, 1831–1839.
123. J. F. W. Nijsen, J.-H. Seppenwoolde, T. Havenith, C. Bos, C. J. G. Bakker, A. D. van het Schip, *Radiology* **2004**, *231*, 491–499.
124. J.-H. Seppenwoolde, J. F. W. Nijsen, L. W. Bartels, S. W. Zielhuis, A. D. van het Schip, C. J. G. Bakker, *Magn. Reson. Med.* **2005**, *53*, 76–84.
125. S. W. Zielhuis, J. F. W. Nijsen, R. Figueiredo, B. Feddes, A. M. Vredenberg, A. D. van het Schip, W. E. Hennink, *Biomaterials* **2005**, *26*, 925–932.

126. S. W. Zielhuis, J. F. W. Nijsen, J. H. Seppenwoolde, B. A. Zonnenberg, C. J. G. Bakker, W. E. Hennink, P. P. van Rijk, A. D. van het Schip, *Curr. Med. Chem.: Anti-Cancer Agents* **2005**, *5*, 303–313.
127. J.-H. Seppenwoolde, L. W. Bartels, R. van der Weide, J. F. W. Nijsen, A. D. van het Schip, C. J. G. Bakker, *J. Magn. Reson. Imaging* **2006**, *23*, 123–129.
128. S. W. Zielhuis, J. F. W. Nijsen, R. de Roos, G. C. Krijger, P. P. van Rijk, W. E. Hennink, A. D. van het Schip, *Int. J.Pharm.* **2006**, *311*, 69–74.
129. S. W. Zielhuis, J.-H. Seppenwoolde, V. A. P. Mateus, C. J. G. Bakker, G. C. Krijger, G. Storm, B. A. Zonnenberg, A. D. van het Schip, G. A. Koning, J. F. W. Nijsen, *Cancer Biother. Radiopharm.* **2006**, *21*, 520–527.
130. P. R. Seevinck, J.-H. Seppenwoolde, T. C. de Wit, J. F. W. Nijsen, F. J. Beekman, A. D. van het Schip, C. J. G. Bakker, *Anti-Cancer Agents Med. Chem.* **2007**, *7*, 317–334.
131. M. A. D. Vente, J. F. W. Nijsen, R. de Roos, M. J. van Steenbergen, C. N. J. Kaaijk, M. J. J. Koster-Ammerlaan, P. F. A. de Leege, W. E. Hennink, A. D. van het Schip, G. C. Krijger, *Biomed. Microdevices* **2009**, *11*, 763–772.
132. S. W. Zielhuis, J. F. W. Nijsen, G. C. Krijger, A. D. Van het Schip, W. E. Hennink, *Biomacromolecules* **2006**, *7*, 2217–2223.
133. J. Roosen, M. W. M. van Wijk, L. E. L. Westlund Gotby, M. J. Arntz, M. J. R. Janssen, D. Lobeek, G. H. van de Maat, C. G. Overduin, J. F. W. Nijsen, *Med. Phys.* **2023**, *50*, 935–946.
134. J. Roosen, L. E. L. Westlund Gotby, M. J. Arntz, J. J. Fütterer, M. J. R. Janssen, M. W. Konijnenberg, M. W. M. van Wijk, C. G. Overduin, J. F. W. Nijsen, *Eur. J. Nucl. Med. Mol. Imaging* **2022**, *49*, 4705–4715.
135. J. Roosen, M. J. Arntz, M. J. R. Janssen, S. F. de Jong, J. J. Fütterer, C. G. Overduin, J. F. W. Nijsen, *Cancers* **2021**, *13*, 5462.
136. P. R. Seevinck, J.-H. Seppenwoolde, J. J. M. Zwanenburg, J. F. W. Nijsen, C. J. G. Bakker, *Magn. Reson. Med.* **2008**, *60*, 1466–1476.
137. D. A. Yablonskiy, E. M. Haacke, *Magn. Reson. Med.* **1994**, *32*, 749–763.
138. H. W. de Haan, *Magn. Reson. Med.* **2011**, *66*, 1748–1758.
139. S. Shin, S. D. Yun, N. J. Shah, *Sci. Rep.* **2023**, *13*, 1138.
140. G. H. van de Maat, P. R. Seevinck, M. Elschot, M. L. Smits, H. de Leeuw, A. D. van Het Schip, M. A. Vente, B. A. Zonnenberg, H. W. de Jong, M. G. Lam, M. A. Viergever, M. A. van den Bosch, J. F. Nijsen, C. J. Bakker, *Eur. Radiol.* **2013**, *23*, 827–835.
141. M. L. Smits, M. Elschot, M. A. van den Bosch, G. H. van de Maat, A. D. van het Schip, B. A. Zonnenberg, P. R. Seevinck, H. M. Verkooijen, C. J. Bakker, H. W. de Jong, M. G. Lam, J. F. Nijsen, *J. Nucl. Med.* **2013**, *54*, 2093–2100.
142. P. R. Seevinck, G. H. van de Maat, T. C. de Wit, M. A. Vente, J. F. Nijsen, C. J. Bakker, *Int. J. Radiat. Oncol. Biol. Phys.* **2012**, *83*, e437–e444.
143. G. H. van de Maat, H. de Leeuw, P. R. Seevinck, M. A. van den Bosch, J. F. Nijsen, C. J. Bakker, *Magn. Reson. Med.* **2015**, *73*, 273–283.
144. W. Bult, M. A. Vente, E. Vandermeulen, I. Gielen, P. R. Seevinck, J. Saunders, A. D. van Het Schip, C. J. Bakker, G. C. Krijger, K. Peremans, J. F. Nijsen, *Brachytherapy* **2013**, *12*, 171–177.
145. M. A. D. Vente, T. C. de Wit, M. A. A. J. van den Bosch, W. Bult, P. R. Seevinck, B. A. Zonnenberg, H. W. A. M. de Jong, G. C. Krijger, C. J. G. Bakker, A. D. van het Schip, J. F. W. Nijsen, *Eur. Radiol.* **2010**, *20*, 862–869.
146. M. A. D. Vente, J. F. W. Nijsen, T. C. de Wit, J. H. Seppenwoolde, G. C. Krijger, P. R. Seevinck, A. Huisman, B. A. Zonnenberg, T. S. G. A. M. van den Ingh, A. D. van het Schip, *Eur. J. Nucl. Med. Mol. Imaging* **2008**, *35*, 1259–1271.

147. M. T. M. Reinders, K. J. van Erpecum, M. L. J. Smits, A. J. A. T. Braat, J. de Bruijne, R. Bruijnen, D. Sprengers, R. A. de Man, E. Vegt, J. N. M. IJzermans, A. Moelker, M. G. E. H. Lam, *J. Nucl. Med.* **2022**, *63*, 1891–1898.
148. P. Hendriks, D. D. D. Rietbergen, A. R. van Erkel, M. J. Coenraad, M. J. Arntz, R. J. Bennink, A. E. Braat, A. Crobach, O. M. van Delden, T. van der Hulle, H. J. Klümpen, R. W. van der Meer, J. F. W. Nijsen, C. S. P. van Rijswijk, J. Roosen, B. N. Ruijter, F. Smit, M. K. Stam, R. B. Takkenberg, M. E. Tushuizen, F. H. P. van Velden, L. F. de Geus-Oei, M. C. Burgmans, *Cardiovasc. Intervent. Radiol.* **2022**, *45*, 1057–1063.
149. J. F. Prince, M. van den Bosch, J. F. W. Nijsen, M. L. J. Smits, A. F. van den Hoven, S. Nikolakopoulos, F. J. Wessels, R. C. G. Bruijnen, M. Braat, B. A. Zonnenberg, M. Lam, *J. Nucl. Med.* **2018**, *59*, 582–588.
150. A. J. A. T. Braat, J. F. Prince, R. van Rooij, R. C. G. Bruijnen, M. A. A. J. van den Bosch, M. G. E. H. Lam, *Eur. Radiol.* **2018**, *28*, 920–928.
151. J. F. Prince, M. L. Smits, G. C. Krijger, B. A. Zonnenberg, M. A. van den Bosch, J. F. Nijsen, M. G. Lam, *J. Vasc. Interv. Radiol.* **2014**, *25*, 1956–1963.e1.
152. M. L. J. Smits, J. F. W. Nijsen, M. A. A. J. van den Bosch, M. G. E. H. Lam, M. A. D. Vente, W. P. T. M. Mali, A. D. van het Schip, B. A. Zonnenberg, *Lancet Oncol.* **2012**, *13*, 1025–1034.
153. M. L. Smits, J. F. Nijsen, M. A. van den Bosch, M. G. Lam, M. A. Vente, W. P. Mali, A. D. van Het Schip, B. A. Zonnenberg, *Lancet Oncol.* **2012**, *13*, 1025–1034.
154. M. L. J. Smits, J. F. W. Nijsen, M. A. A. J. van den Bosch, M. G. E. H. Lam, M. A. D. Vente, J. E. Huijbregts, A. D. van het Schip, M. Elschot, W. Bult, H. W. A. M. de Jong, P. C. W. Meulenhoff, B. A. Zonnenberg, *J. Exp. & Clinical Cancer Res.* **2010**, *29*, 70–81.
155. F. de Lange, M. Dieleman Jan, L. A. Blezer Erwin, J. F. Houston Ralph, C. J. Kalkman, J. F. W. Nijsen, *J. Neurosci. Methods* **2009**, *176*, 152–156.
156. R. C. Bakker, R. J. J. van Es, A. Rosenberg, S. A. van Nimwegen, R. Bastiaannet, H. de Jong, J. F. W. Nijsen, M. Lam, *Nucl. Med. Commun.* **2018**, *39*, 213–221.
157. R. J. van Es, J. F. Nijsen, A. D. van het Schip, H. F. Dullens, P. J. Slootweg, R. Koole, *Int. J. Oral Maxillofac. Surg.* **2001**, *30*, 407–413.
158. R. J. van Es, J. F. Nijsen, H. F. Dullens, M. Kicken, A. van der Bilt, W. Hennink, R. Koole, P. J. Slootweg, *J. Craniomaxillofac. Surg.* **2001**, *29*, 289–297.
159. W. Bult, S. G. Kroeze, M. Elschot, P. R. Seevinck, F. J. Beekman, H. W. de Jong, D. R. Uges, J. G. Kosterink, P. R. Luijten, W. E. Hennink, A. D. van het Schip, J. L. Bosch, J. F. Nijsen, J. J. Jans, *PLoS One* **2013**, *8*, e52178.
160. A. G. Arranja, W. E. Hennink, C. Chassagne, A. G. Denkova, J. F. W. Nijsen, *Mater. Sci. Eng. C. Mater. Biol. Appl.* **2020**, *106*, 110244.
161. A. G. Arranja, W. E. Hennink, A. G. Denkova, R. W. A. Hendrikx, J. F. W. Nijsen, *Int. J. Pharm.* **2018**, *548*, 73–81.
162. M. van Elk, B. Ozbakir, A. D. Barten-Rijbroek, G. Storm, F. Nijsen, W. E. Hennink, T. Vermonden, R. Deckers, *PLoS One* **2015**, *10*, e0141626/1–e0141626/19.
163. M. van Elk, C. Lorenzato, B. Ozbakir, C. Oerlemans, G. Storm, F. Nijsen, R. Deckers, T. Vermonden, W. E. Hennink, *Eur. Polym. J.* **2015**, *72*, 620–631.
164. C. Oerlemans, P. R. Seevinck, M. L. Smits, W. E. Hennink, C. J. G. Bakker, M. A. A. J. van den Bosch, J. F. W. Nijsen, *Int. J. Pharm.* (*Amsterdam, Neth.*) **2015**, *482*, 47–53.
165. C. Oerlemans, P. R. Seevinck, d. M. G. H. van, H. Boulkhrif, C. J. G. Bakker, W. E. Hennink, J. F. W. Nijsen, *Acta Biomater.* **2013**, *9*, 4681–4687.
166. S. W. Zielhuis, J. H. Seppenwoolde, C. J. G. Bakker, U. Jahnz, B. A. Zonnenberg, A. D. van het Schip, W. E. Hennink, J. F. W. Nijsen, *J. Biomed. Mater. Res., Part A* **2007**, *82A*, 892–898.

167. W. Bult, H. de Leeuw, O. M. Steinebach, M. J. van der Bom, H. T. Wolterbeek, R. M. Heeren, C. J. Bakker, A. D. van Het Schip, W. E. Hennink, J. F. Nijsen, *Pharm. Res.* **2012**, *29*, 827–836.
168. W. Bult, R. Varkevisser, F. Soulimani, P. R. Seevinck, H. de Leeuw, C. J. G. Bakker, P. R. Luijten, A. D. van het Schip, W. E. Hennink, J. F. W. Nijsen, *Pharm. Res.* **2010**, *27*, 2205–2212.
169. W. Bult, P. R. Seevinck, G. C. Krijger, T. Visser, L. M. J. Kroon-Batenburg, C. J. G. Bakker, W. E. Hennink, A. D. van Het Schip, J. F. W. Nijsen, *Pharm. Res.* **2009**, *26*, 1371–1378.
170. J. K. Kim, K.-H. Han, J. T. Lee, Y. H. Paik, S. H. Ahn, J. D. Lee, K. S. Lee, C. Y. Chon, Y. M. Moon, *Clin. Cancer Res.* **2006**, *12*, 543–548.
171. F. Carniato, L. Tei, M. Botta, *Eur. J. Inorg. Chem.* **2018**, **2018**, 4936–4954.
172. M. Khodadadi Yazdi, P. Zarrintaj, H. Hosseiniamoli, A. H. Mashhadzadeh, M. R. Saeb, J. D. Ramsey, M. R. Ganjali, M. Mozafari, *J. Mater. Chem. B* **2020**, *8*, 5992–6012.
173. P. Yang, S. Gai, J. Lin, *Chem. Soc. Rev.* **2012**, *41*, 3679–3698.
174. J. Wen, K. Yang, F. Liu, H. Li, Y. Xu, S. Sun, *Chem. Soc. Rev.* **2017**, *46*, 6024–6045.
175. F. Danhier, O. Feron, V. Préat, *J. Controlled Release* **2010**, *148*, 135–146.
176. S. Z. Hosseinabadi, S. Safari, M. Mirzaei, E. Mohammadi, S. M. Amini, B. Mehravi, *Adv. Nat. Sci.: Nanosci. Nanotechnol.* **2020**, *11*, 045010.
177. K. E. Wheeler, A. J. Chetwynd, K. M. Fahy, B. S. Hong, J. A. Tochihuitl, L. A. Foster, I. Lynch, *Nat. Nanotechnol.* **2021**, *16*, 617–629.
178. J. E. Lee, N. Lee, T. Kim, J. Kim, T. Hyeon, *Acc. Chem. Res.* **2011**, *44*, 893–902.
179. L. Tosheva, V. P. Valtchev, *Chem. Mater.* **2005**, *17*, 2494–2513.
180. Y.-S. Lin, Y. Hung, J.-K. Su, R. Lee, C. Chang, M.-L. Lin, C.-Y. Mou, *J. Phys. Chem. B* **2004**, *108*, 15608–15611.
181. C.-P. Tsai, Y. Hung, Y.-H. Chou, D.-M. Huang, J.-K. Hsiao, C. Chang, Y.-C. Chen, C.-Y. Mou, *Small* **2008**, *4*, 186–191.
182. G. Liu, N. M. K. Tse, M. R. Hill, D. F. Kennedy, C. J. Drummond, *Aust. J. Chem.* **2011**, *64*, 617–624.
183. M. Norek, I. C. Neves, J. A. Peters, *Inorg. Chem.* **2007**, *46*, 6190–6196.
184. R. Guillet-Nicolas, J.-L. Bridot, Y. Seo, M.-A. Fortin, F. Kleitz, *Adv. Funct. Mater.* **2011**, *21*, 4653–4662.
185. H. Skar, Y. Liang, E. S. Erichsen, R. Anwander, J. G. Seland, *Microporous Mesoporous Mater.* **2013**, *175*, 125–133.
186. Y. Z. Shao, L. Z. Liu, S. Q. Song, R. H. Cao, H. Liu, C. Y. Cui, X. Li, M. J. Bie, L. Li, *Contrast Media Mol. Imaging* **2011**, *6*, 110–118.
187. Y. Shen, Y. Shao, H. He, Y. Tan, X. Tian, F. Xie, L. Li, *Int. J. Nanomed.* **2013**, *8*, 119–127.
188. Z. Li, J. Guo, M. Zhang, G. Li, L. Hao, *Front. Chem.* **2022**, *10*, 837032.
189. D. Brühwiler, *Nanoscale* **2010**, *2*, 887–892.
190. F. Carniato, L. Tei, W. Dastrù, L. Marchese, M. Botta, *Chem. Commun.* **2009**, 1246–1248.
191. K. M. L. Taylor, J. S. Kim, W. J. Rieter, H. An, W. Lin, W. Lin, *J. Am. Chem. Soc.* **2008**, *130*, 2154–2155.
192. M. Laprise-Pelletier, M. Bouchoucha, J. Lagueux, P. Chevallier, R. Lecomte, Y. Gossuin, F. Kleitz, M.-A. Fortin, *J. Mater. Chem. B* **2015**, *3*, 748–758.
193. F. Carniato, D. Alberti, A. Lapadula, J. Martinelli, C. Isidoro, S. Geninatti Crich, L. Tei, *J. Mat. Chem. B* **2019**, *7*, 3143–3152.
194. C.-A. Cheng, W. Chen, L. Zhang, H. H. Wu, J. I. Zink, *J. Am. Chem. Soc.* **2019**, *141*, 17670–17684.

195. W. Zhang, J. A. Peters, F. Mayer, L. Helm, K. Djanashvili, *J. Phys. Chem. C* **2015**, *119*, 5080–5089.
196. I. Bresinska, K. J. Balkus, Jr., *J. Phys. Chem.* **1994**, *98*, 12989–12994.
197. S. W. Young, F. Qing, D. Rubin, K. J. Balkus, Jr., J. S. Engel, J. Lang, W. C. Dow, J. D. Mutch, R. A. Miller, *J. Magn. Reson. Imaging* **1995**, *5*, 499–508.
198. C. Platas-Iglesias, I. VanderElst, W. Z. Zhou, R. N. Muller, C. F. G. C. Geraldes, T. Maschmeyer, J. A. Peters, *Chem. Eur. J.* **2002**, *8*, 5121–5131.
199. É. Csajbók, I. Bányai, I. VanderElst, R. N. Muller, W. Z. Zhou, J. A. Peters, *Chem. Eur. J.* **2005**, *11*.
200. M. Tsotsalas, M. Busby, E. Gianolio, S. Aime, L. De Cola, *Chem. Mater.* **2008**, *20*, 5888–5893.
201. A. Amedlous, C. Hélaine, R. Guillet-Nicolas, O. Lebedev, S. Valable, S. Mintova, *Inorg. Chem. Front.* **2023**, *10*, 2665–2676.
202. F. Mayer, W. Zhang, T. Brichart, O. Tillement, C. S. Bonnet, É. Tóth, J. A. Peters, K. Djanashvili, *Chem. Eur. J.* **2014**, *20*, 3358–3364.
203. S. Lee, H. Hwang, P. Kim, D.-J. Jang, *Catal. Lett.* **1999**, *57*, 221–226.
204. W. Zhang, J. Martinelli, J. A. Peters, J. M. A. van Hengst, H. Bouwmeester, E. Kramer, C. S. Bonnet, F. Szeremeta, É. Tóth, K. Djanashvili, *ACS Appl. Mat. Interfaces* **2017**, *9*, 23458–23465.
205. S. Lacerda, W. Zhang, R. T. M. de Rosales, I. Da Silva, J. Sobilo, S. Lerondel, É. Tóth, K. Djanashvili, *ACS Appl. Mater. Interfaces* **2022**, *14*, 32788–32798.

# 11 Manganese- and Iron-Based Nanoparticles as MRI Contrast Agents

*Fabio Carniato and Lorenzo Tei*
Dipartimento di Scienze e Innovazione Tecnologica, Università del Piemonte Orientale "A. Avogadro", I-15121 Alessandria, Italy
fabio.carniato@uniupo.it, lorenzo.tei@uniupo.it

*Carlos F. G. C. Geraldes*
Department of Life Sciences and Coimbra Chemistry Center - Institute of Molecular Sciences (CQC-IMS), University of Coimbra, 3004-535 Coimbra, Portugal
Department of Chemistry, University of Coimbra, 3004-535 Coimbra, Portugal
geraldes@ci.uc.pt

**CONTENTS**



DOI: 10.1201/9781003374688-11

**Abstract**

Manganese- and Iron(III)-based contrast agents for magnetic resonance imaging (MRI) have attracted significant attention over the past two decades as an alternative to Gd-based compounds. Various nanostructures have been proposed for potential applications in *in vivo* diagnostics and theranostics. While different types of Mn-based nanoparticles (NPs) obtained at the +2, +3, and +4 oxidation states, including $Mn_xO_y$ NPs, are discussed in this chapter, in case of iron, only the Fe(III)-based NPs, excluding Fe-oxides, are presented. For MRI, multimodal imaging, or theranostic applications, a diverse range of NPs has been utilized, encompassing amphiphilic polymer-based NPs, mixed metal NPs, and silica-, carbohydrate-, and polyphenolic-based NPs containing Mn(II, III, IV) or Fe(III) ions or chelates. Both Mn and Fe showcase favorable magnetic properties, good biocompatibility, and an improved toxicity profile with respect to Gd(III); these properties position them as the paramagnetic ions of preference for the next generation of nanoscale MRI and theranostic contrast agents. An examination of selected examples published in the past decade demonstrates the potential for enhancing relaxivity and MR-contrast effects of these Mn- or Fe-based NPs pushing them to effectively rival Gd(III)-based nanosystems.

## KEYWORDS

Functionalized Nanoparticles; Contrast Agents; Manganese; Iron(III); Metal Oxides; Magnetic Resonance Imaging; Multimodal Imaging; Theranostics

## 1 INTRODUCTION

Over the past two decades, functionalized nanoparticles (NPs) have become a valuable tool for diagnostic and therapeutic biomedical applications. They offer several advantages: (1) NPs possess sizable compartments (large interior loading volume and/or surface area) that enable them to incorporate or conjugate with numerous imaging or therapeutic agents; (2) they can integrate multiple types of imaging or therapeutic agents, making them versatile nanoplatforms suitable for diagnostic or theranostic purposes; (3) NPs can be modified on their surface with polyethylene glycol (PEG) or hydrophilic polymers, enhancing their stealth properties and prolonging their circulation time. Additionally, they can be equipped with targeting molecules to guide the nanoprobes to specific disease sites for drug delivery and imaging. These characteristics make multifunctional NPs highly promising as innovative diagnostic and therapeutic probes in the field of medicine [1–5].

Magnetic resonance imaging (MRI) stands out as a non-invasive diagnostic technique that offers exceptional contrast capabilities, as well as high spatial resolution and no limitation in tissue penetration. However, the technique's limited sensitivity and low temporal resolution has hindered its molecular imaging clinical application and has driven the advancement of nanostructured MRI contrast agents (CAs). These agents have the potential to significantly enhance the sensitivity of MRI and reduce the required injected dose for signal enhancement. Also, the combination in a single MRI nanoplatform with other modalities with complementary advantages, such as the high sensitivity provided by positron emission tomography (PET) and optical imaging (despite the low penetration depth of the latter), can lead to improved functional and anatomical imaging information.

The most extensively studied nanosystems for $T_1$ MRI involve the utilization of gadolinium(III)-based agents. These agents consist of various Gd(III) complexes incorporated into or conjugated to nanocarriers, or $Gd^{3+}$ ions being an integral part of the inorganic nanostructure [6,7]. The incorporation of numerous active paramagnetic centers within the nanoprobes enhances their efficiency, particularly within the magnetic field range of 0.5–2 T. This improvement is attributed to the increased molecular volume of the nanoprobe, which in turn leads to a significant reduction in the global tumbling motion rate ($1/\tau_R$) [8]. Here, $\tau_R$ refers to the correlation time for the rotational tumbling of the system. A large number of Gd-based nanosystems have been investigated, ranging from organic-based to inorganic and hybrid NPs, and there are several excellent and comprehensive reviews that examined in detail the preparation, properties, and applications of these nanosized Gd-based CAs [6–9]. However, while Gd(III) chelates used in clinical practice are generally considered safe and well-tolerated by patients, there have been two clinical concerns in recent years that have prompted research efforts to explore safer alternatives to Gd-based CAs [10]. Thus, although investigations into nanosystems for MRI are still in the preclinical phase, there is an emerging shift in research toward exploring the use of other paramagnetic ions, such as Mn(II/III) and Fe(III), even within the realm of nanoprobes.

Both Fe(III) and Mn(II) are essential elements found in all known living organisms and are involved in many metabolic processes. They are essential cofactors for many proteins involved in the normal function of the brain. However, it must be highlighted that exposure to elevated levels of these metal ions can lead to adverse health effects: in fact, while high levels of Fe have been associated with diseases such as Alzheimer, Parkinson, and Huntington, excess Mn is correlated to manganism, attention deficit hyperactivity disorder, depression, and hepatic encephalopathy [11].

## 2 MANGANESE-BASED NANOPARTICLES AS MRI CONTRAST AGENTS

Manganese is a transition metal with a $[Ar]3d^5 4s^2$ electronic configuration that displays several possible oxidation states ranging from 0 to +7. The number of unpaired electrons distributed in its five *d* orbitals determines the paramagnetic character of the Mn ions. In its divalent, high-spin state, Mn(II), it is a very efficient paramagnetic relaxation agent with electronic spin quantum number $S=5/2$, high effective magnetic moment ($\mu_{eff}=5.92$ $\mu_B$/atom), long electronic relaxation time (in the 0.1–1 ns range), and quite fast exchange rates of the coordinated water molecules. Such interesting paramagnetic properties grant the application of Mn(II) chelates as MRI contrast agents in alternative to Gd(III) chelates. Typical donor atoms for hexacoordinate Mn(II) (ionic radius=0.83 Å) are carboxylate oxygen and amino nitrogen atoms, but when it comes to the harder and smaller $Mn^{3+}$ ion (0.64 Å), it prefers strongly electron-releasing ligands such as phenolate or catecholate groups. The Mn(III) oxidation state, with $3d^4$ configuration, usually provide high-spin $S=2$ paramagnetic compounds with lower effective magnetic moment ($\mu_{eff}=4.9$ $\mu_B$/atom). However, Mn(III) compounds are less efficient relaxation agents due to the shorter electronic relaxation times (ca. 10 ps), although the water exchange is also fast. Moreover, Mn easily forms multiple magnetic ions and complexes reacting for example with oxygen to form Mn oxides such as MnO, $Mn_3O_4$, $Mn_2O_3$, and $MnO_2$, in which Mn shows +2, mixed +2/+3, +3, and +4 oxidation states, respectively. Since Mn(IV) is $3d^3$ metal ion with three unpaired electrons and lower paramagnetism ($S=3/2$, $\mu_{eff}=3.87$ $\mu_B$/atom), the contrast enhancement in $MnO_2$ like nanoparticles is provided by the release of free $Mn^{2+}$ after dissolution of the particle triggered by endogenous stimuli [12].

Considering Mn(II)-based nanosystems, there are two main classes of nanoparticles that have been investigated in the literature: different types of Mn(II)-chelates anchored on or embedded into soft and/or hard nanostructures and $T_1$-based Mn-oxide (MnO) or other inorganic nanoparticles (i.e., $KMnF_3$ and $MnWO_4$).

## 2.1 Non-Mn(II) Oxide Systems

### 2.1.1 Non-MnO Mn(II) Systems-Based Nanoparticles

The focus of this section on non-MnO Mn(II) systems is mainly on Mn(II)-chelate-based and hybrid organic/inorganic Mn(II)-based NPs. It should be emphasized that certain manganese-based NPs can exhibit long-term release of $Mn^{2+}$ ions, which may result in the accumulation of $Mn^{2+}$ in the body and potential toxic effects, as demonstrated by several toxicological studies [13–16]. Therefore, the use of stable and inert Mn(II) complexes with polydentate polyaminocarboxylate ligands is crucial in order to prevent any release of free $Mn^{2+}$ within the body [17,18]. These chelates are commonly employed in the construction of various types of paramagnetic nanoparticles, ranging from silica to lipid-based systems. In case of new Mn(II)-chelates used in NP preparation, the investigation of the thermodynamic stability and kinetic inertness is essential to ensure the safe *in vivo* application of the final nanoprobe [18]. Additionally, the selection of the Mn-chelate for conjugation to the macromolecular or nanosized system depends on the functional groups present on the nano-object and the bifunctional chelating agent [19]. The bifunctional chelating agent should possess strong coordination capability toward the $Mn^{2+}$ ion and include a functional group capable of forming a stable covalent link with the carrier through a wide range of chemical bonds. The ligands for Mn(II) complexation, as well as the stability and relaxometric properties of the Mn(II) complexes, have been recently reviewed and will not be extensively discussed here [17].

In this section, we start focusing our attention to different types of Mn(II)-chelate-based nanoparticles which have been used for MRI, multimodal, and theranostic applications. Three main examples of Mn(II)-based NPs will be covered: (1) lipid-based NPs consisting of supramolecular interactions of self-assembled amphiphilic units; (2) Mn(II)-chelates anchored on polysaccharide NPs or nanogels (NGs); and (3) silica or carbon NPs with Mn-chelates anchored on the surface or confined in porous structures. The main characteristics of most of the described nanosystems are summarized in Table 1.

#### *2.1.1.1 Mn(II)-Based Micelles or Liposomes*

Amphiphilic Gd(III) complexes composed of a hydrophobic and a hydrophilic component that spontaneously self-assemble into micelles or liposomes in aqueous solutions have been extensively used for the development of high-relaxivity Gd-based nanosystems [20,21]. Additional self-assembled nanosystems, including block-copolymers like polymersomes [22] or dendritic molecules such as dendrimersomes [23], have also been proposed for nanomedicine applications. Generally, micellar aggregates possess a distinctive core–shell architecture with sizes typically ranging from 5 to 50 nm, while liposomes are nanosized vesicles with diameters spanning from 50 to 1000 nm. The controlled structural characteristics of these nanosystems facilitate the modulation of their pharmacokinetic properties by modifying size, surface charge, or membrane composition, as well as the attachment of specific targeting ligands for molecular imaging applications [24].

The first example of Mn-based micelle was a dextran amphiphilic polymer conjugated to a semirigid pyridine-based linear Mn-complex via azide-alkyne click chemistry reaction (Figure 1a) [25]. The paramagnetic micelles display a size of ca. 85 nm and a relaxivity of 13.3 $mM^{-1}s^{-1}$ at 1.5 T and 298 K, which is 2.8 times higher than that of the mononuclear complex. A contrast enhanced Magnetic Resonance Angiography (MRA) study on rats at 3 T showed an evident signal enhancement in the blood vessels at a relatively low manganese dosage (0.1 Mn mmol $kg^{-1}$ BW). Then, amphiphilic MnEDTA-like or Mn(1,4-DO2A)-like complexes incorporating one or two aliphatic chains on the ethylene backbone were reported to aggregate in micellar structures in water or by mixing them with pegylated phospholipids [26]. In particular, Mn(1,4-DO2A) shows a relaxivity lower than $Mn(EDTA)^{2-}$ (2.1 *vs* 3.3 $mM^{-1}s^{-1}$), but its stability and kinetic inertness in physiological conditions are much higher ($\log K_{Mn(1,4\text{-}DO2A)} = 15.68$; $t_{1/2}$ at pH 7.4 = 57 h) [27]. The nanosized micellar systems obtained with the amphiphilic complexes based on EDTA and 1,4-DO2A bearing one (or two) $C_{12}$ or $C_{16}$ aliphatic chain(s) exhibit sizes in the range 13–84 nm and interesting relaxivity values, especially for the systems containing two aliphatic chains that

**TABLE 1**

**Summary of Basic Properties of Some of the Examples of Non-oxide Mn(II)-Based NPs as MRI CAs Discussed in this Section**

| NPs Components | Mn(II)-chelate | Relaxivity ($mM^{-1}s^{-1}$), 298 K[a], Bo (T) | Imaging/ Therapeutic Modalities | Cell/Animal Model | Reference |
|---|---|---|---|---|---|
| Amphiphilic dextran conjugated to Mn-chelate | Mn(Pyridine-bis-pipecolinate) | $r_1$=13.3 (1.5 T) | MRA | Sprague-Dawley (SD) rats | [25] |
| Amphiphilic Mn-EDTA or Mn(1,4-DO2A) | Mn(HCDTA) Mn(DD-DO2A) Mn(DH-DO2A) | Mn(HCDTA) NPs $r_1$=18.4; Mn(DD-DO2A) NPs $r_1$=15.3 (0.5 T) | MRI | --- | [26] |
| Amphiphilic chelates embedded in liposomes | Mn(HCDTA) Mn(DD-DO2A) | Mn(HCDTA) NPs $r_1$=17.3; Mn(DD-DO2A) NPs $r_1$=16.3 (0.5 T) | MRI | --- | [28] |
| mPEG-p(MnL-a-HDMI)-mPEG amphiphilic polymer | Mn-pyridine-bis-hydroxyproline | $r_1$=23.2 (0.5 T) $r_1$=9.3 (3 T) | MRA | Sprague-Dawley (SD) rats | [29] |
| Mn-polydopamine on cellulose nanocrystals | Mn-polydopamine | $r_1$=38.0 (3 T) for $(Mn\text{-}PDA)_4$@CNCs | MRI/ phototherapy | --- | [30] |
| Mn-polydopamine on cellulose nanocrystals | MnDTPA-amide | $r_1$=57 (3 T) for MnDTPA@DCNC-8 | MRI/ phototherapy | RAW 264.7 and HUVEC cell lines | [31] |

*(Continued)*

**TABLE 1 (*Continued*)**
**Summary of Basic Properties of Some of the Examples of Non-oxide Mn(II)-Based NPs as MRI CAs Discussed in This Section**

| NPs Components | Mn(II)-chelate | Relaxivity ($mM^{-1}s^{-1}$), 298 K[a], Bo (T) | Imaging/ Therapeutic Modalities | Cell/Animal Model | Reference |
|---|---|---|---|---|---|
| Carboxymethylchitosan-(Mn-DTPA)$_n$ | MnDTPA-amide | $r_1$=15.42 (3 T) | MRI | Sprague-Dawley (SD) rats | [32] |
| Alginate-PDA Ca/Mn NGs | Mn-alginate-PDA | $r_1$=12.54 (7 T)[b] | MRI | BALB/c mice | [33] |
| MnCDTA-bisamide-Chitosan nanogels | Mn-*t*-CDTA-amide | $r_1$=30.5 (0.5 T) | MRI | Breast tumors on BALB/c male mice | [34] |
| MSNs-MnDTPA | MnDTPA-monoamide | $r_1$ =7.18 (1 T)[b] | MRI | C57BL/6j mice | [36] |
| SNPs-MnCDTA | MnCDTA | $r_1$=18.6 (0.5 T) | MRI | C57BL/6j mice | [37] |
| Porous silica nanospheres | Mn(pyridine-bis-picolinate) | $r_1$=8.46 (1.41 T) | MRI | *In vitro* HeLa cells | [38] |
| Mn@CC carbon NPs | Mn-gluconate and aspartic acid treated at 180°C | $r_1$=42.9 (1.5 T)[b] | MRI and bioluminescence | Orthotopic brain tumors in mice | [39] |
| Mn@CD carbon NPs | Mn-gluconate and aspartic acid treated at 180°C | $r_1$=10.8 (3.0 T)[b] | CE-MRI | Acute kidney injury | [40] |

[a] Unless otherwise stated.
[b] Temperature not indicated.

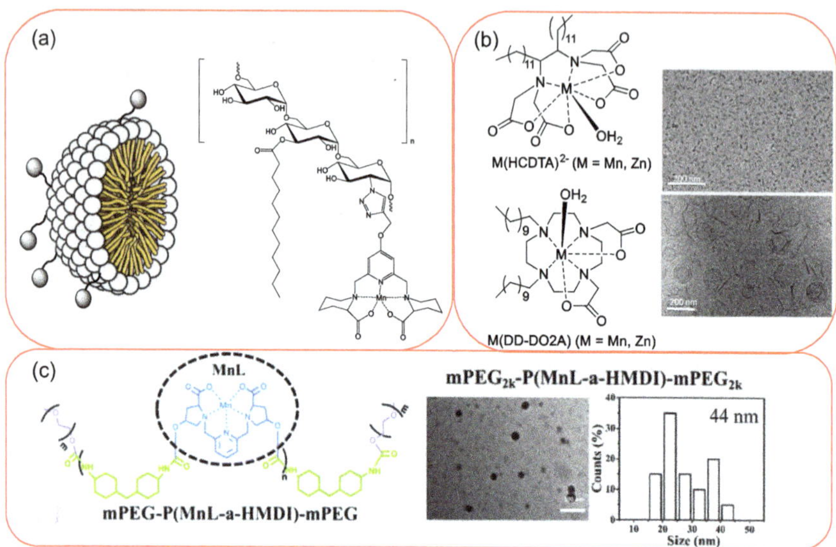


**FIGURE 1** (a) Mn (II) complexes conjugated to dextran-based micelles [25]. (b) Chemical structures of the amphiphilic EDTA or 1,4-DO2A-based ligands with Cryo-TEM images of phospholipid-based nanoparticles entrapping 20% $Zn(HCDTA)^{2-}$ and 10% Zn(DD-DO2A). (c) Mn-pyridine-bis-hydroxyproline complex conjugated to HMDI and different mPEGs (mPEG-P(L-a-HMDI)-mPEG) used for the preparation of nanomicelles. (Adapted from Ref. [28] (b) and [29] (c).)

partially block the local rotation of the chelate once the system is embedded in the lipidic nanoparticle ($r_1 = 18.4\,mM^{-1}s^{-1}$ for $Mn(HCDTA)^{2-}$ and $12.6\,mM^{-1}s^{-1}$ for MnDD-DO2A, at 298 K, 0.5 T) [26]. The same Mn complexes were also incorporated in the lipid bilayer of liposomes, and it was found that the length and position of the aliphatic chains control the formation of either vesicular liposomes (size ≈110 nm), non-vesicular bicelles (size ≈14 nm), or a mixture of both (Figure 1b) [28]. The relaxivity (at 298 K and 0.5 T) of the nanoaggregates was in line with that found for the previously reported micelles, i.e., $17.3\,mM^{-1}s^{-1}$ for $Mn(HCDTA)^{2-}$ (bicelles) and $16.3\,mM^{-1}s^{-1}$ for MnDD-DO2A (liposomes).

Finally, a series of amphiphilic polyurethane-based polymers made by a two-step polycondensation reaction between hexamethylene diisocyanate (HMDI), the ligand pyridine-bis-hydroxyproline, and different molecular weights of mPEG (mPEG-P(L-a-HMDI)-mPEG) were recently proposed (Figure 1c) [29]. The corresponding Mn(II) complexes (mPEG-P(MnL-a-HMDI)-mPEG) formed an aggregated micellar system with 48 nm size and high relaxivity (in the case of mPEG2k system, $r_1 = 23.2$, 14.4, and $9.7\,mM^{-1}s^{-1}$ at 0.5 T, 1.5 T, and 3.0 T, respectively). This micellar system was used to perform a MRA study *in vivo* and showed excellent $T_1$ contrast-enhanced performance including relatively long time-window vascular imaging and stenosis detection with no accumulation of Mn(II) in the brain.

#### 2.1.1.2 Mn(II) Chelates on Polysaccharides NPs

The use of functionalized biopolymers such as polysaccharides or cellulose nanocrystals is common for the preparation of biocompatible and biodegradable Mn-based nanoprobes [30–34]. In one example, Mn-dopamine complexes were reacted with cellulose nanocrystals (CNCs) at different ratios to form nanoparticles with $Mn^{2+}$ embedded into polydopamine (PDA)-coated CNCs [$(Mn\text{-}PDA)_{1-6}$@CNCs] for MRI/photothermal therapy (PTT) theranostic applications (Figure 2a) [30]. These NPs had a rod-shaped morphology (size ca. 150–200 × 10 nm) and $r_1$ of 38 $mM^{-1}s^{-1}$ at 298 K and 3 T for $(Mn\text{-}PDA)_4$@CNCs. On the other hand, the increase of the (Mn-PDA)/CNCs ratio increases the photothermal conversion efficiency ($\eta$) of PDA from 17.7% to 44.4% making these NPs suitable for tumor therapy and thus integrating tumor diagnosis and treatment. In another paper, different amounts of DTPA-bis-anhydride were conjugated to CNCs in order to form MnDTPA@CNSs nanosystems with relaxivity values that increase with the loading of DTPA-units on the surface of the nanocrystals (Figure 2b) [31].

DTPA was also conjugated to carboxymethyl chitosan (CMCS) to obtain a Mn-based macromolecular MRI contrast agent with good biocompatibility, controlled biodegradability, and good water solubility [32]. CMCS-$(Mn\text{-}DTPA)_n$ showed a DTPA grafting ratio on chitosan of 14.5%, a relaxivity at 3 T of 15.42 $mM^{-1}s^{-1}$, and the injection in SD rats generated a considerable $T_1$ signal intensity increase in liver and kidneys. Furthermore, nanogels (NGs) made by alginate–polydopamine or alginate–dopamine complexed by Ca(II)/Mn(II) [AlgPDA/DA(Ca/Mn)] were prepared and functionalized with PEG molecules to increase the blood circulation half-life [33]. These nanogels had a negative surface charge with size between 66 and 135 nm and showed a $r_1$ at 7 T of 12.54 $mM^{-1}s^{-1}$ for AlgPDA(Ca/Mn) and 10.13 $mM^{-1}s^{-1}$ for AlgDA(Ca/Mn) NGs.

Recently, a new type of nanogel based on chitosan crosslinked by the stable and inert Mn-*trans*-CDTA-bisamide complexes was reported (Figure 2c) [34]. These complexes serve as both paramagnetic and crosslinking agents, resulting in nanosystems with high relaxivities and exceptional stability. These nanogels showed a $r_1$ of 30.5 $mM^{-1}s^{-1}$ at 0.5 T, 298 K, which is seven times higher than that of typical monohydrated Mn(II) chelates. This improvement is attributed to the restricted mobility of the complex combined with fast exchange of the coordinated water molecule. Furthermore, the Mn-NGs demonstrated excellent long-term stability in both aqueous solutions and biological fluids, eliminating concerns of metal leakage or particle aggregation. A comprehensive $^{1}H$ and $^{17}O$ NMR relaxometric study on the Mn-NGs and on model Mn-CDTA-mono and bis-glucosamide complexes allowed to confirm that the Mn(II) ion is in a monohydrated state ($q = 1$) and exhibits fast water exchange dynamics (Figure 2d). *In vivo* experiments were performed using mice with subcutaneous breast cancer showing successful extravasation and accumulation of Mn-NGs in the tumor, leading to approximately a 20% signal enhancement at 1 T, 24 h after injection (Figure 2e).

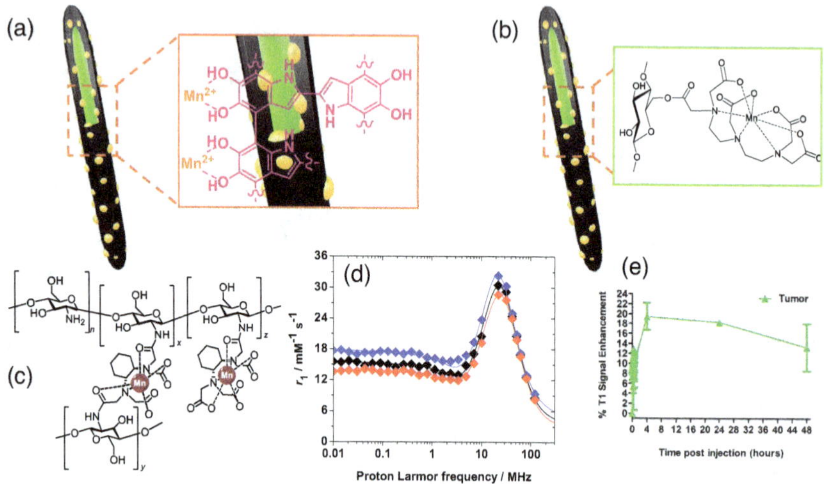


**FIGURE 2** (a) Schematic illustration of (Mn-PDA)@CNC nanoparticles; (b) MnDTPA@CNC nanoparticles; (c) MnCDTA-amide nanogels; (d) $^1$H NMRD profiles Mn-NGs at 283 (blue), 298 (black), and 310 K (red) at neutral pH; and (e) % $T_1$ SE measured in the tumor at different time points post-intravenous injection of the MnCDTA-amide NGs. (Adapted from Ref. [30] (a), [31] (b), and [34] (c, d and e).)

### *2.1.1.3 Mn(II) Chelates Anchored on Silica or Carbon NPs*

The conjugation of paramagnetic chelates to silica nanoparticles, either mesoporous or amorphous silicas, has been largely exploited for Gd-chelates in the last 15 years [35], but in the case of Mn chelates, only few examples are reported. The first one consisted of an organo-modified silica functionalized on the surface with amino groups for the conjugation of DTPA ligands for $Mn^{2+}$ complexation (Figure 3a) [36]. Although MnDTPA is a $q=0$ complex with only outer-sphere contribution, the nanospheres showed a $r_1$ value at 1 T of ca. 7.2 mM$^{-1}$ s$^{-1}$, and after administration in mice, they exhibited marked liver-specific $T_{1w}$ MRI contrast. Another more recent example considered similar amino-functionalized silica NPs conjugated to CDTA-like ligands. In particular, a bisamide derivative of MnCDTA was used for conjugation, and the final nanoprobes showed a spherical shape with size of ca. 90 nm and a $r_1$ of 18.6 mM$^{-1}$ s$^{-1}$ (at 0.5 T and 298 K) (Figure 3b). Preliminary *in vivo* studies on healthy mice at 1 T demonstrated the efficiency of the novel Mn(II)-based silica nanoparticle as $T_{1w}$ MRI probes, resulting in about 60% contrast enhancement in the liver [37].

A hexadentate pyridine-picolinate ligand was used to complex Mn(II) and confine it into porous silica nanospheres in a non-covalent fashion (Figure 3c). A reverse-microemulsion method was used to form first nanosized aqueous droplets containing the Mn-chelates and then creating the silica nanospheres around the water droplets, thus entrapping the complexes inside. The paramagnetic nanosystem showed a size of 14 nm and a $r_1$ of 8.46 mM$^{-1}$ s$^{-1}$ (1.41 T and 298 K), a good

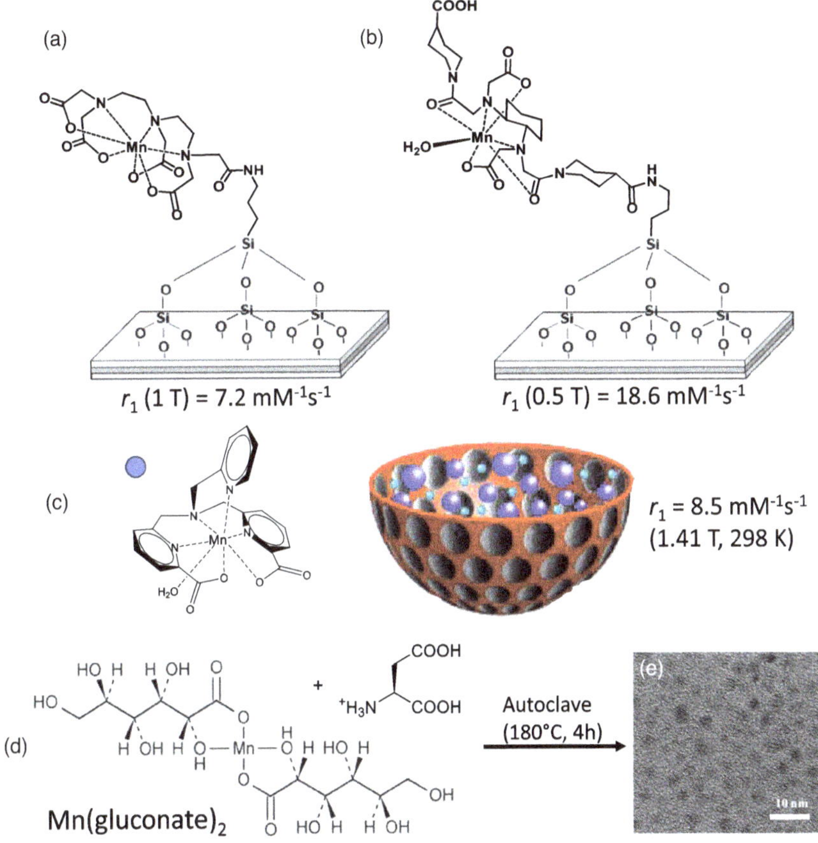


**FIGURE 3** (a) Chemical structure of the Mn-DTPA-like complex conjugated to silica nanoparticles [36]; (b) Mn-CDTA-like complex conjugated to SiNPs [37]; (c) Mn(II)-pyridine-picolinate ligand confined within a porous silica nanosphere [38]; (d) scheme of the preparation of Mn@CC nanoparticles; and (e) TEM image of the 5 nm NPs [39,40].

biocompatibility on HeLa cells, and stability in the presence of excess $Zn^{2+}$ ions and physiologically relevant anions [38].

Another approach was reported by Qin at al. who used complexes of Mn(II) with gluconate and aspartic acid and carbonized them in autoclave at 180°C for 4 h to obtain Mn-NPs (Mn@CCs) with hydrodynamic diameter of 11.6 nm and near neutral electric charge, which provide good colloidal stability in physiological media and in serum (Figure 3d and e) [39]. Mn@CCs exhibited quite high relaxivity at different magnetic field strengths (31.1 mM$^{-1}$ s$^{-1}$, 7 T) and high cellular viability in normal cell lines. Moreover, Mn@CCs were able to cross the intact blood–brain barrier (BBB) in normal health mice and are efficiently excreted from the host at 4 h post-injection through renal clearance. *In vivo* MRI and light-sheet fluorescence microscopy (LSFM) imaging highlighted that Mn@CCs preferentially target the glioma tissues and distribute homogeneously with high penetration in an intracranial mouse model, thereby delineating clear brain tumor margins and enabling precise MR imaging of microscopic orthotopic single-nodule brain tumors and multinodular liver tumors in mice [39]. Similar Mn-doped carbon dots (Mn-CDs) allowed non-invasive diagnosis of acute kidney injury (AKI) [40]. Mn-CDs have a size of 5 nm, good biocompatibility, and a high $r_1$ at 3.0 T (10.8 $mM^{-1}s^{-1}$). Thus, they were able to produce good signal enhancement in the kidneys and show the fine structures of the kidneys including the cortex, outer medulla, and inner medulla.

### 2.1.2 Non-$Mn_3O_4$ Mn(III) Systems-Based Nanoparticles

Three classes of Mn(III)-containing NPs have been studied: Mn-12 ($[Mn_{12}O_{12}(O_2CCH_3)_{16}(H_2O)_4]$) clusters and derivatives, $Mn_3O_4$-based NPs, and Mn(III)-porphyrin NPs. The use of ($[Mn_{12}O_{12}(O_2CCH_3)_{16}(H_2O)_4]$) and derivatives and $Mn_2O_3$-based NPs will be discussed in Section 2.2.2. In this section, we will briefly focus on Mn(III)porphyrin NPs, as several recent reviews cover this subject in detail [41–43].

Many studies have reported NPs incorporating Mn(III) porphyrins as MRI CAs [41]. Mn(III)-porphyrins bound to or incorporated into soft NPs (micelles, polymers, liposomes, virus capsids, and microbubbles) have been used as MRI CAs and as versatile platforms for MRI-guided theranostics, such as chemotherapy (ChT), PTT, photodynamic therapy (PDT), radiotherapy (RT), and sonodynamic therapy (SDT). Here, we will describe only one example.

Mn(III)-labeled nanobialys were prepared as toroidal-shaped NPs (190 nm diameter) by self-assembly of amphiphilic branched polyethyleneimine (PEI) and linoleic acid, forming inverted micelles, and including, at their surface, Mn(III)-protoporphyrin (PP) and biotin-caproyl-DSPE. The longitudinal relaxivity values reflected the water exposure of the Mn(III)-PP. Since each NP contained about 165,000 Mn(III) chelates, they had very high particulate relaxivities ($r_{1part}$=612,307 mmol [nanobialy])$^{-1}$s$^{-1}$ and $r_{1part}$=866,989 mmol [nanobialy])$^{-1}$s$^{-1}$) and could be targeted *in vitro* to fibrin clots after attaching biotin as a targeting vector, using avidin–biotin interactions and biotinylated fibrin-specific monoclonal antibodies. MRI images (3 T) of the clots showed the fibrin-targeted Mn(III) nanobialys with a positive contrast enhancement. The potential utility of nanobialys loaded with both hydrophilic (doxorubicin, DOX) and hydrophobic (camptothecin) chemotherapeutic agents for theranostic applications for microthrombi in ruptured atherosclerotic plaques was demonstrated by their *in vitro* release of the drugs [44].

### 2.1.3 Non-$MnO_2$ Mn(IV) Systems-Based Nanoparticles

Not all Mn(IV)-based NPs for MRI-guided tumor theranostics are based on $MnO_2$. One example consists of structures (Mn(IV)-HDCL) based on a hexahydrazide clathrochelate ligand (HDCL), with a 2D sheet structure, obtained by scanning tunneling microscopy (STM). The Mn(IV) ion is trapped within the cavity of the cage-like ligand, as shown by single crystal X-ray diffraction (XRD) analysis. Crystalline Mn-HDCL is deep black and is essentially non-transmissive to visible light. The UV–vis–NIR spectrum, with a strong broadband absorption (200–1000 nm), overlapped well with that of the solar spectrum. It showed a high photothermal conversion (PC) efficiency (ca. 71%) in aqueous solution under NIR 730 nm laser photo-irradiation, comparable to single-wall carbon nanotubes in the solid state. Mn-HDCL showed very low cytotoxicity toward 4T1 mouse breast cancer and HUVEC cell lines, which became very high at 1.0 mM under 730 nm laser irradiation, where the cell viability decreased to 5%, as seen by fluorescence microscopic images, consistent with high PC and photothermal cell ablation. Mn-HDCL is paramagnetic with a very low $r_1$, providing a modest increase in the $T_{1w}$ MRI contrast both *in vitro* and *in vivo* for a BALB/c mouse upon i.v. injection of Mn-HDCL. Mn-HDCL targeted tumors passively, defining tumor margins *in vivo* and had an efficient laser-triggered PTT effect, showing potential as a tumor-targeting photothermal sensitizer [45].

## 2.2 Mn Oxides

### 2.2.1 MnO Nanoparticles

As there are several reviews regarding the preparation, characterization, MR-imaging, and toxicological properties of MnO [1,12,46,47], in this chapter, ultrasmall, biocompatible MnO NPs employed as $T_1$ MRI contrast agents will be only briefly covered. The main characteristics of most of the described systems are summarized in Table 2.

In the case of MnO NPs, only the $Mn^{2+}$ ions exposed on the surface in contact with bulk water molecules can be responsible for contrast enhancement and their number is highly related to the surface-to-volume ratio of the NPs. Thus, in recent years, much work has been focused on engineering the surface of MnO nanoparticles to achieve higher $T_1$ relaxivities and accurate diagnosis [1]. Unfortunately, suffering from relatively low $T_1$ relaxivity, high dosage MnO NPs are often necessary to assess lesion detection, increasing the potential risk of adverse effect and limiting the application of MnO nanoparticles in clinic diagnosis [46–49].

MnO NPs of sub-10 nm size may display crystalline surface, highly related to the morphology, and surface coating ligands that can be tuned to increase the amount of water around the nanoparticles. Thus, since nanostructures with anisotropic morphology may show significantly higher surface-to-volume ratio than spherical morphology, MnO NPs of different shapes and sizes have been synthesized ranging from spheres [50] to cubes [51] and many other non-spherical morphologies [52], with sizes typically lower than 10 nm to obtain a low $r_2/r_1$ ratio and thus $T_1$ contrast ability. However, it must be highlighted that non-spherical nanoparticles have often shown rather undesirable $r_2/r_1$ values, and a correct analysis of surface–water interactions is needed to elucidate the relationship between crystal surface and $T_1$ relaxivity and optimize the contrast enhancement ability of the NPs [53]. Furthermore, surface coating is typically needed to improve the *in vivo* pharmacokinetics of NPs; thus, MnO NPs have been functionalized with PEG [54,55], coated with silica [56], or modified by phospholipids [57].

**TABLE 2**
**Summary of Basic Properties of Some MnO-Based NPs as MRI CAs Discussed in This Section**

| NPs Components | Mn(II)-chelate | Relaxivity ($mM^{-1}s^{-1}$), 298 K[a], Bo (T) | Imaging/ Therapeutic Modalities | Cell/Animal Model | Reference |
|---|---|---|---|---|---|
| MnO-PEG-Aptamer | MnO-PEG | $r_1=12.9$ (3 T)[b] | Targeted MRI | 786-0 cells on BALB/c mice | [54] |
| Hollow porous MnO octapod NPs+DOXO | MnO ($Mn^{2+}$ release with pH) | $r_1$ of free $Mn^{2+}$ ions | MRI/OI/ Chemotherapy | SMMC-7721 cells H22 tumor bearing mice | [57] |
| MnM48SNs | MnO-silicates | $r_1=8.4$ (1.5 T, 310 K) | MRI | *In vitro* leukemia cells | [58] |
| MnO-PAA-FA | MnO-PAA | $r_1=9.3$ (3 T)[b] | Targeted MRI | U87MG brain tumor | [59] |
| MnO-PEG-RGD | MnO-PEG | $r_1=12.1$ (4.7 T)[b] | Targeted MRI | M21 tumor cells on Balb/c mice | [60] |
| MnO-PEG-catechol-RGD | MnO-PEG-catechol | $r_1=10.2$ (3 T, 310 K) | Targeted MRI | A549 tumor-bearing mice | [61] |

Hollow porous MnO NPs, obtained by acid or basic etching, exhibit a higher surface-to-volume ratio compared to their solid equivalents and therefore increased relaxivities. As an example, Gao et al. reported small-sized octapod-shaped hollow porous MnO NPs functionalized by zwitterionic dopamine sulfonate that can act as a versatile platform to load organic dyes or chemotherapeutic drugs with high loading efficiency [58]. The contrast enhancement was achieved by the release of free $Mn^{2+}$ ions in mild acidic conditions for real-time monitoring of the cargo delivery *in vivo*, especially in tumor microenvironment and lysosome.

Another example was reported by Fortin et al., who showed that PEGylated bis-phosphonate dendrons grafted on the surface of MnO NPs could be used as $T_1$ vascular contrast agents due to a fast excretion capacity of the NPs [55]. The same group described Mn-silica nanohybrid materials by confining Mn into the porous network of MCM-48 silica nanoparticles (MnM48SNs) [59]. No significant evidence of $Mn^{2+}$ release was found in Mn–M48SNs in acidic water (pH 6), although leaching of $Mn^{2+}$ was observed in acetate buffer at pH 5. Longitudinal relaxivity values of $r_1=8.4\,mM^{-1}s^{-1}$ were reported (1.5 T and 37°C), with a low $r_2/r_1$ ratio ($r_2/r_1=2$). Finally, leukemia cells were labeled with Mn–M48SNs and the MRI contrast enhancement provided was quantified at 1 T.

Targeting agents were also conjugated to the surface of MnO NPs for cancer imaging applications: for example, Lee et al. reported folic acid-conjugated, polyacrylic acid (PAA)-coated MnO NPs with an average particle diameter of 2.7 nm [60]. While PAA provided good colloidal stability and low cellular cytotoxicity, folic acid conferred cancer-targeting ability, confirmed by $T_1$ contrast enhancement at U87MG brain cancer site upon intravenous administration to mice tails. Further, the nanoparticles exhibited a quite high $r_1$ value of 9.3 $mM^{-1}s^{-1}$ ($r_2/r_1 = 2.2$) at 3.0 T. Then, Long and co-workers synthesized PEG-coated MnO NPs functionalized with RGD peptide to target tumors via $\alpha_v\beta_3$ integrin [61]. In this example, a $r_1$ value of 12.1 $mM^{-1}s^{-1}$ with a $r_2/r_1$ ratio close to 1 was measured at 4.7 T. They also studied the effect of two different PEG chain lengths (600 and 5000 Da) in the performance of MnO NPs *in vivo* showing that PEG 5000-modified nanoparticles underwent higher tumor accumulation. Furthermore, Wang et al. also described MnO NPs PEGylated via catechol-Mn chelation and conjugated with cRGD as active tumor targeting moiety [62]. They reported a larger size of ca. 100 nm after coating and RGD conjugation, a slightly high $r_2/r_1$ ratio of ca. 6 ($r_1 = 10.2$ $mM^{-1}s^{-1}$ at 3.0 T and 37°C), and an efficient accumulation in tumor verified by an *in vivo* biodistribution study. Finally, PEG-coated MnO NPs with a hydrodynamic diameter of ca. 15 nm were conjugated to the AS1411 aptamer which allowed targeting 786-0 renal cancer tumor cells and prolonging the accumulation time of the probe in tumor cells. At 3.0 T, this nanoprobe had a $r_1$ value of 12.9 $mM^{-1}s^{-1}$ and a $r_2/r_1$ ratio of 4.66 [54].

### 2.2.2 $Mn_3O_4$ Nanoparticles

As several recent reviews describing the synthesis, characterization, chemical properties, and performance of $Mn_3O_4$-based NPs as targeted, responsive, uni- and multimodal MRI CAs, as well as their theranostic applications [1,12,47,63,64], only some representative examples of each type of applications will be presented in this chapter.

- *$Mn_3O_4$-Based NPs as MRI CAs*

  $Mn_3O_4$ NPs contain $Mn^{2+}$ and $Mn^{3+}$ ions in a spinel structure in which O atoms are closely packed with $Mn^{2+}$ in the tetrahedral sites and $Mn^{3+}$ in the octahedral sites ($Mn^{2+}[Mn^{3+}]_2O_4^{2-}$). However, $Mn^{3+}$ ions can be reduced to $Mn^{2+}$ in an intracellular reducing environment by glutathione (GSH), increasing $r_1$ and producing redox-activated $T_{1w}$ MRI CAs. Only the surface ions of $Mn_3O_4$ NPs in contact with bulk water molecules are responsible for $T_1$ contrast enhancement, with a relative number that decreases with the surface-to-volume ratio (*S/V*) of spherical NPs, but also depends on their shape. Hyeon et al. first reported an easy, low-temperature synthesis of $Mn_3O_4$ nanocrystals of various shapes and sizes in nonpolar solvents in the presence of surfactants, which were fully characterized using transmission electron microscopy (TEM) and XRD. The size and shape (spherical, nanoplates and asymmetric nanowires and nanocytes) of the nanocrystals was controlled by varying the experimental conditions such as precursors, surfactants, and injection temperature of water. Water-dispersible 9 nm sized $Mn_3O_4$ nanoplates showed very small $r_1$ and $r_2$ ($r_2/r_1$=4.2 at 1.5 T) that, despite this, demonstrated their potential application as $T_{1w}$ MRI CAs [65].

The synthetic conditions for the formation of MnO, $Mn_3O_4$, or both simultaneously by pyrolysis of Mn(II) acetate in 1-octadecene in the presence of oleylamine and oleic acid have been studied on the basis of the kinetics of coordination between oleic acid/oleylamine and $Mn^{2+}$ [66]. The oxidation of MnO NPs (MONs) into hollow $Mn_3O_4$ (HMONs), used as $T_1$ and $T_2$ MRI CAs and as vehicles for drug (DOX) cell labeling, has been reported. *In vivo* $T_{1w}$- and $T_{2w}$-MRI of a mouse brain upon local injection of the HMONs showed a positive and negative contrast on the injection site, demonstrating the potential of these NPs as a dual contrast agent for both $T_1/T_2$-weighted MRI [67].

$Mn_3O_4$-based NPs have been designed as MRI CAs, some of which for tumor diagnosis and MRI imaging-guided therapy, as well as for MRI-based bimodal and multimodal imaging. As this area has recently been extensively reviewed [47,63,64], here only a few examples of each type of applications will be described, whose properties are summarized in Table 3.

- *Hybrid $Mn_3O_4$-based NPs as MRI CAs*

Many $Mn_3O_4$-based NPs surface-coated or modified with organic and inorganic molecules have been developed for MRI as $T_{1w}$ CAs. The coating avoids colloidal instability in water and improves biocompatibility and their $T_1$-relaxivity, making them usable as *in vivo* MRI CAs. Encapsulation of hydrophobic NPs by polymers, such as PEG, PEI, PAA, PDA, and polystyrene sulfonate (PSS), is the most common modification strategy, but silica inorganic coatings are also used.

For instance, Deka et al. designed $Mn_3O_4$ NPs encapsulated in a mesoporous 3D carbon framework (CF) functionalized by PEG ($Mn_3O_4$@CF NPs) as a potential *in vivo* $T_{1w}$ MRI CA [68]. Another commonly used type of polymer is PEI. For example, Luo et al. reported monodispersed acylated PEI-coated $Mn_3O_4$ NPs sequentially conjugated with fluorescein isothiocyanate (FI), folic acid (FA)-linked PEG and PEG monomethyl ether, PEGylated FA and PEG monomethyl ether ($Mn_3O_4$-PEI-Ac-FI-mPEG-PEG-FA NPs) as tumor targeted CA for *in vivo* MRI [69]. Although both PEG and PEI modification strategies are common, they create a thick hydrophobic hydrocarbon coating shell that usually hinders water access to the $Mn_3O_4$ core, resulting in relatively low $r_1$ values. An alternative strategy is to use hydrophilic coatings, such as alginate (Alg)-coated nanogels loaded with $Mn_3O_4$-PEI NPs (Alg/PEI-$Mn_3O_4$) [70].

**TABLE 3**
**Summary of Basic Properties of the Examples of MRI CAs and Theranostic Nanosystems Containing Hybrid $Mn_2O_3$ NPs Discussed in This Section**

| NPs Components | Imaging Modalities | Relaxivity ($mM^{-1}s^{-1}$), 298 K, Bo (T)[a] | Therapeutic Modalities | Cell/Animal Model | Reference |
|---|---|---|---|---|---|
| $Mn_3O_4$@CF | MRI | $r_1$=3.5 (1.41 T) | – | *In vitro* RAW 264.7 cells | [66] |
| $Mn_3O_4$-PEI-Ac-FI-mPEG-PEG-FA | MRI | $r_1$=0.57 (a) | – | *In vitro* KB cancer cells<br>*In vivo* KB tumor mice | [67] |
| AG/PEI-$Mn_3O_4$ | MRI | $r_1$=26.12 (0.5 T) | – | *In vivo* U87MG tumor mice | [68] |
| $Mn_3O_4$ | MRI | $r_1$=8.26 (3.0 T) | – | *In vivo* NPC CNE-2 tumor Balb/c nude mice | [69] |
| GQD-PDA-$Mn_3O_4$ | MRI/OI | $r_1$=3.5 (3 T) | PDT | *In vivo* A549 tumor mice | [70] |
| FA(DOX)-$Mn_3O_4$ @PDA@PEG | MRI | $r_1$=14.47 (3 T) | CT (DOX), PTT | *In vitro* MCF-7 cells<br>*In vivo* MCF-7 and H22 tumor mice | [71] |
| $Mn_3O_4$-BSA-EDTA | $T_1/T_2$ MRI | $r_1$=8.75; $r_2$=40.09 (0.5 T) | PTT | *In vivo* HCT116 tumor mice | [72] |
| $^{64}$Cu(NOTA)/PEG–FA-FI-PEI– $Mn_3O_4$ | MRI/PET | $r_1$=0.996 (0.5 T) | – | *In vitro* HeLa cells<br>*In vivo* HeLa tumor mice | [73] |
| $Mn_3O_4$@$SiO_2$ (RBITC)–FA | MRI/Fl | $r_1$=0.50 mM (0.5 T)<br>$r_1$=0.47 (3.0 T), | – | *In vitro* Hela cells | [74] |
| $Mn_3O_4$@ PEG-Cy7.5 | MRI/Fl | $r_1$=0.53 (7 T) | SDT | *In vitro* PC-3, A549, and HEPG2 cells<br>*In vivo* HEPG2 tumor Balb/c mice | [75] |

[a] Not indicated.

- *$Mn_3O_4$-based NPs as MRI CAs for imaging-guided tumor therapy*
  The development of theranostics methods, combining diagnostic and therapeutic tools ideally in a single nanoplatform, is of great importance for the simultaneous diagnostic and specific therapy of cancer. Many research groups have used a variety of nanotechnology approaches to contribute to this field, including $Mn_3O_4$-based NPs.
  Xiao et al. prepared ligand-free $Mn_3O_4$ NPs with a 9 nm mean particle diameter using laser ablation in aqueous media, as shown by TEM. These uncoated $Mn_3O_4$ NPs, with a high $r_1$ value, did not significantly affect the *in vitro* viability of L929, 293, NP69 (normal nasopharyngeal epithelium), and CNE-2 (human nasopharyngeal carcinoma, NPC) cells owing to their low toxicity. $T_{1w}$ MRI (3 T) of Balb/c nude mice with the CNE-2 NPC showed high positive contrast enhancements 30 min after i.v. injection of the NPs, demonstrating their potential as a $T_{1w}$ MRI CA [71].
  Ternary hybrid nanoprobes, consisting of $Mn_3O_4$ NPs conjugated to graphene quantum dots (GQD) through PDA and thiol-amine, (GQD-PDA-$Mn_3O_4$ NPs) worked as dual (MRI/fluorescence) imaging-guided PDT agents [72]. Another ternary nanotheranostic system, consisting of $Mn_3O_4$ NPs surface covered with PDA conjugated to drug (DOX)-loaded PEG-FA (FA(DOX)-$Mn_3O_4$@PDA@PEG), was designed for MRI-guided combinatorial cancer ChT/PTT [73].
  Liu et al. developed a novel, inorganic nanomaterial for $T_1/T_2$-MRI-guided PTT consisting of EDTA- and bovine serum albumin (BSA)-capped $Mn_3O_4$ ($Mn_3O_4$-BSA-EDTA) NPs, which could be degraded in a controlled way in the presence of ascorbic acid [74].
- *$Mn_3O_4$-based NPs as CAs for multimodal imaging and theranostics*
  Zhu et al. developed a nanoplatform consisting of PEG-FA-FI-PEI-coated $Mn_3O_4$ NPs surface labeled with $^{64}$Cu(NOTA) chelates as a targeted PET/$T_{1w}$-MRI bimodal imaging probe ($^{64}$Cu(NOTA)/PEG–FA-FI-PEI–$Mn_3O_4$). The NPs were stable in aqueous solution, with a low $r_1$, no *in vitro* cytotoxicity and target specificity to folate receptors (FR) overexpressed in HeLa tumor cells. *In vivo* microPET images of the nude mice bearing HeLa xenografted tumors showed good tumor tracer uptake and positive tumor contrast in $T_{1w}$ MRI images (1.5 T). Thus, the nanoplatform is effective as a tumor-targeted bimodal PET/MRI probe for animal studies (Figure 4) [75].
  Yang et al. developed monodisperse core–shell silica-coated $Mn_3O_4$ NPs with the silica shell aminated through silanization with 3-aminopropyltriethoxysilane (APS) and covalently conjugated to a fluorescent dye, rhodamine B isothiocyanate (RBITC), and FA. The formed $Mn_3O_4$@$SiO_2$(RBITC)–FA core–shell nanocomposites showed to be useful for bimodal imaging ($T_1$-weighted MRI and fluorescence) in biological systems [76].

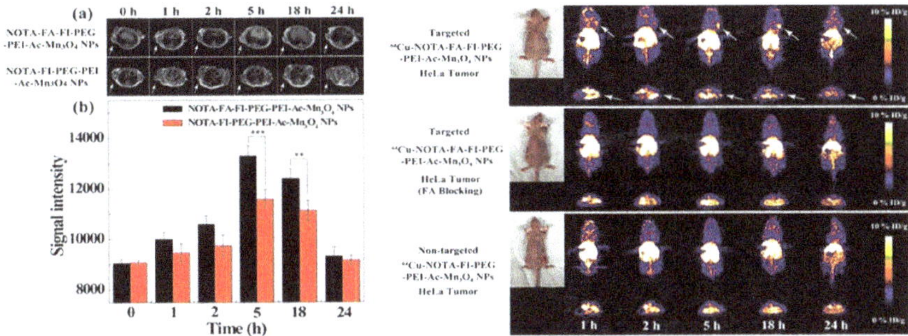


**FIGURE 4** Left: *In vivo* $T_1$-weighted MR images (a) and signal intensity (b) of HeLa tumors after intravenous injection of the NOTA–FA–FI– PEG–PEI–Ac–$Mn_3O_4$ NPs or the NOTA–FI–PEG–PEI–Ac–$Mn_3O_4$ NPs (500 mg Mn, in 0.2 mL saline) at different time points. Tumors are indicated by white arrows; Right: micro-PET images of the nude mice bearing HeLa xenografted tumors at different time points post-i.v. injection of the targeted NPs, the $^{64}$Cu–NOTA–FA–FI–PEG–PEI–Ac–$Mn_3O_4$ NPs with FA blocking, and the non-targeted NPs. The whole-body coronal (top) and transverse (bottom) micro-PET images of nude mice bearing HeLa xenografted tumors are shown. Tumors are indicated by arrows. (Reprinted with permission from Ref. [75]. Copyright 2010, American Chemical Society.)

Zhan et al. developed an optical/MRI dual-mode imaging probe for *in vivo* bimodal imaging to guide sentinel lymph node (SLN) mapping, which can be used to estimate the metastatic stage of a tumor. It was based on PEG-coated $Mn_3O_4$ NPs conjugated with the Cy7.5 dye ($Mn_3O_4$@PEG-Cy7.5 NPs). The NPs were characterized using several *in vitro* techniques, which are validated by *in vivo* fluorescence imaging (FI) and MRI (7 T) in healthy BALB/c mice [77].

### 2.2.3 $MnO_2$ Nanoparticles

Most Mn(IV)-based MRI active nanosystems have been reported as $MnO_2$-based NPs for MRI-guided tumor theranostics. As several extensive reviews of this field have been published recently [47,64,78–80], only a few examples will be described here. Moreover, as pointed out in the introduction, this type of NPs functions by releasing $Mn^{2+}$ ions that are known to be toxic, and therefore, the authors discourage the use of $MnO_2$-based NPs unless they are used for theranostic applications.

- *$MnO_2$-based NPs as MRI CAs of non-tumor diseases*

  BSA–$MnO_2$ nanoparticles (BM NPs) were developed for the permeability imaging of the BBB and hemorrhage transformation (HT) prediction in the acute ischemic stroke *in vivo*. The presence of BSA in the spherical (3 nm size) NPs led to a high $r_1$ value and a good potential as *in vivo* $T_1$ MRI CAs. Their BBB permeability ability was demonstrated in middle cerebral artery occlusion (MCAO) rats with reperfusion after acute ischemic stroke through the enhanced intensity in $T_1$ MRI images observed in the infarcted areas, allowing the visualization of the BBB damage in stroke, as the NPs cannot pass through the intact BBB. Thus, these BM NPs show high potential as an alternative biocompatible MRI CA for non-invasive BBB permeability imaging *in vivo*, with impact on the clinical treatment of stroke patients [81].

- *$MnO_2$-based NPs for MRI-based tumor theranostics*

  The MRI imaging and therapeutic approaches to tumors are challenging due to their abnormal blood vessel structure, proliferation, and metabolic patterns of tumor cells. Consequently, the tumor microenvironment (TME) has vascular abnormalities, severe hypoxia, high $H_2O_2$ and glutathione (GSH) concentrations and low pH (4.5–5.0), high lactate levels and glucose deprivation due to the upregulated glycolytic metabolism generating lactic acid. These properties could facilitate the growth of tumor cells, tumor progression, metastasis, and multidrug resistance (MDR). The heterogeneity of the tumor tissue prevents monotherapy from achieving their complete eradication, allowing recurrence and development of metastasis. The development of TME-responsive smart $MnO_2$-based NPs combining several therapeutic modalities has shown significant potential to increase the efficacy of cancer treatments due to their tunable structures/morphologies, pH responsive degradation, and excellent catalytic activities. Several strategies can modulate the TME, including relief of tumor hypoxia, depletion of excess GSH, glucose consumption, and moderation of tumor immunosuppressive microenvironment. The $MnO_2$-based TME modulation benefits cancer therapies, such as PDT, PTT, RT, SDT, ChT, starvation therapy, and immunotherapy [82].

  $MnO_2$ NPs integrated in nanoplatforms have recently allowed important progress in cancer MRI-based theranostics. They can integrate two or more types of therapies operating synergistically to minimize MDR and radioresistance. They work as stimulus (pH or redox) responsive CAs, as $MnO_2$ responds to the TME by generating $Mn^{2+}$ ions upon reacting with $H^+/H_2O_2$ or glutathione (GSH), which promote strong $T_1$ relaxation [83,84]. The $r_1$ of $MnO_2$ NPs is very low ($< 0.1\ mM^{-1} s^{-1}$) due to the low access of most of its $Mn^{4+}$ ions to bulk water. However, once exposed to acidic $H_2O_2$ or GSH in the TME, they are reduced to large amounts of $Mn^{2+}$ ions, resulting in drastic (>50-fold) $r_1$ increase and enhancement of $T_{1w}$ MRI contrast within the tumor. This real "OFF/ON" behavior is an important feature of $MnO_2$-containing NPs as responsive CAs.

  In this section, some representative recent examples of $MnO_2$-based theranostic nanoplatforms are described, classified according to the imaging techniques, besides MRI, in which they are active, and their main characteristics are summarized in Table 4.

**TABLE 4**
**Summary of Basic Properties of the Examples of MRI-Guided Theranostic Nanosystems Containing Hybrid $MnO_2$ NPs Discussed in the Section**

| Imaging Modalities | NPs Components | $MnO_2$ Reducing Agent in TME | TME Modulation Mode | Therapeutic Modalities | Animal Cancer Model | Reference |
|---|---|---|---|---|---|---|
| MRI | $PEG_{5000}$-$MnO_2$/Dox nanosheets | $H^+/H_2O_2$ and/or GSH | Tumor hypoxia relief/ GSH depletion | CT (DOX) | *In vivo* 4T1 tumor nude mice | [85] |
| MRI | AS1411/Ce6–LPMSNs–$MnO_2$ | $H^+/H_2O_2$ | Tumor hypoxia relief | PDT | *In vitro* HeLa cells | [86] |
| MRI/CT | PLGA/AuNR/DTX@$MnO_2$ | GSH/$H^+$ | GSH depletion | PTT/CT(DTX) | *In vitro* MCF-7 cells; *In vivo* S-180 tumor Kunming mice | [87] |
| MRI/NIR-Fl/PAI | DOX-Ce6-$MnO_2$@PLCA-PEG-PCLA NPs (CDM NPs) | $H^+/H_2O_2$ | Tumor hypoxia relief | PDT/CT(DOX) | *In vitro* MCF-7 cells; *In vivo* MCF-7 Balb/c tumor mice | [88] |

#### 2.2.3.1 Unimodal Systems

The first examples illustrate the use of $MnO_2$-based theranostic nanosystems using unimodal MRI imaging combined with one or more therapeutic modalities. A theranostic platform based on highly dispersed 2D $PEG_{5000}$-$MnO_2$ exfoliated nanosheets (NSs), surface loaded with the chemotherapeutic DOX drug via electrostatic interactions and Mn–N coordinate bonds ($PEG_{5000}$-$MnO_2$/DOX NS), was designed for pH-responsive MRI and drug delivery. Incorporation of the NSs in the acidic TME led to their break-up and disintegration, pH-responsive controlled release of DOX, and reduction of $MnO_2$ units from the NSs efficiently endocytosed into the endosomes and lysosomes (pH ca. 5.0–5.5) of the tumor cells to release $Mn^{2+}$ ions, which was responsible for a large increase of $r_1$ and strong *in vivo* $T_{1w}$ tumor contrast in nude mice 4T1 cancer xenograft after i.v. injection of the NSs. The capability of DOX-loaded $MnO_2$ NSs of intracellular drug delivery and the corresponding therapeutic efficiency were also evaluated in DOX-resistant MCF-7/ADR cancer cells [85].

Another reported unimodal system is a cancer cell-specific, endogenous TME $H^+/H_2O_2$-activated, $MnO_2$-containing the nanocomposite (AS1411/Ce6–LPMSNs–$MnO_2$), where the $MnO_2$ NPs were grown within the pores of large pore silica nanoparticles (LPMSNs), and the highly efficient photosensitizer chlorin e6 (Ce6) was covalently linked to LPMSNs. Then, the AS1411 aptamer, with high affinity to the overexpressed nucleolin on the plasma membrane of most cancer cells, was anchored onto the surface of LPMSNs. *In vitro* studies in cultured HeLa cells showed that, upon cancer cell internalization provided by the aptamer targeting vector, this biocompatible nanocomposite relieved tumor hypoxia, enhanced PDT resulting from production of singlet oxygen ($^1O_2$) produced from 660 nm laser irradiation of Ce6, and showed activated $T_1$ MRI contrast resulting from free $Mn^{2+}$ release [86].

#### 2.2.3.2 Multimodal Systems

The use of multimodal NPs capable of optimizing the advantages and disadvantages of MRI with those of other modalities is well known [64]. Fabrication procedures of multimodal $MnO_2$-based theranostic nanoplatforms have recently been developed, with high specific surface area, controllable size and morphologies, and easy surface modification, allowing targeted and controllable drug delivery. Two examples illustrating the use of $MnO_2$-based theranostic nanosystems using multimodal MRI imaging combined with one or more therapeutic modalities are now presented.

Poly(lactic-co-glycolic acid) (PLGA) NPs were loaded with gold nanorods (AuNRs) suitable for RF hyperthermia, and docetaxel (DTX) for chemotherapy and then surface coated with ultrathin $MnO_2$ nanofilms by the reduction of $KMnO_4$ to construct the PLGA/AuNR/DTX@$MnO_2$ drug delivery system. This theranostic nanosystem was successfully tested *in vitro* (MCF-7 human breast cancer cells) and *in vivo* on a subcutaneous S-180 tumor Kunming mouse model for combined chemotherapy and radiofrequency (RF) hyperthermia. AuNRs-induced

RF hyperthermia and the degradation of $MnO_2$ in the TME, resulting from the combined effect of GSH and acidic pH, promoted the controlled release of DTX in the tumor region and produced free $Mn^{2+}$ providing efficient $T_{1w}$ MRI contrast. This, combined with the presence of AuNRs, led to an efficient X-ray CT imaging CA, making the nanosystem a bimodal MRI/CT agent. Injection of the NPs in the tail vein of the tumor mouse showed clear positive MRI contrast and CT contrast imaging at the tumor site (Figure 5) [87].

A $MnO_2$-based theranostic nanosystem for trimodal imaging (MRI/NIR-Fl/PAI) and combined therapy (PDT/CT) was developed. Due to the low efficacy of PDT in the treatment of solid tumors, which is severely affected by hypoxia, $O_2$-generating theranostic NPs were prepared by adding aqueous DOX and colloidal $MnO_2$ to a Ce6 dispersed poly(ε-caprolactone-co-lactide)-β-poly (ethylene glycol)-β-poly (ε-caprolactone-co-lactide) (PLCA-PEG-PCLA co-polymer) ethyl

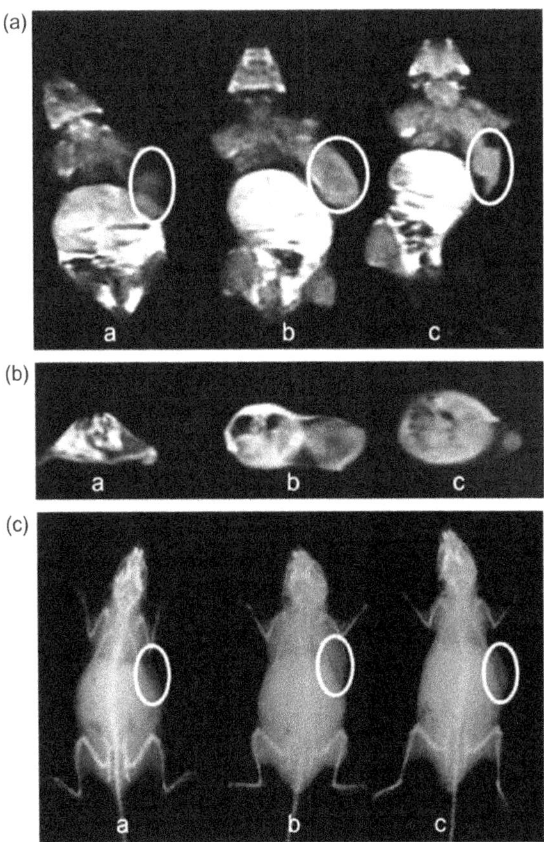


**FIGURE 5** Dual-mode imaging *in vivo*. Notes: (a and b) $T_1$-weighted MR images; (c) X-ray CT images: (a) control; (b) 4 h; and (c) 8 h. (Reprinted from Ref. [87].)

acetate solution followed by sonication, to obtain co-loaded DOX-Ce6-$MnO_2$@PLCA-PEG-PCLA NPs (CDM NPs). MCF-7 Balb/c tumor-bearing mice were injected with CDM NPs and subjected to *in vivo* whole body florescence imaging, photoacoustic imaging (PAI) using a multispectral optoacoustic tomography (MSOT) small animal scanner, and MRI (7 T). These studies showed that Ce6 fluorescence, Ce6 concentration-dependent PAI contrast ability at 680 nm, and positive MRI contrast occurred in the tumor area due to the accumulation of the NPs. The $MnO_2$ triggered the decomposition of excessive endogenous $H_2O_2$ in the TME to generate $O_2$ oxygen, which relieved tumor hypoxia and significantly improved the PDT effect under 606 nm laser irradiation. This effect, together with that of DOX, dramatically improved the combined CT-PDT efficacy of CDM NPs in the tumor-bearing mouse model [88].

- *$MnO_x$ NPs for MRI-based theranostics*

  The oxidation state of manganese oxide can be obtained using classical material science characterization techniques such as XRD, X-ray photoelectron spectroscopy (XPS), X-ray absorption near-edge spectroscopy (XANES), and Raman spectroscopy. However, when the size of the material decreases to the nanoscale, this characterization becomes extremely difficult due to the broadening and weakening of the corresponding peak. Therefore, to avoid inaccuracies, some researchers describe the nanostructures simply as $MnO_x$ [12].

  Several examples of $MnO_x$ NPs for MRI-based tumor diagnosis and therapy have been reported. Ren et al. prepared $MnO_x$-coated SPION-capped camptothecin (CPT)-loaded MSN ($MnO_x$-SPION@MSN@CPT) as TME-responsive $T_1/T_2$ dual-mode MRI-guided pancreatic cancer chemotherapy using controlled CPT drug release from the mesoporous silica. The NPs showed high magnetization and $T_2$ MRI contrast due to large $r_2$ resulting from to the density of SPION onto the surface of MSN. The $MnO_x$ shell degradation in acidic pH typical of the TME provided $T_1$ MRI contrast due to the release of $Mn^{2+}$ ions. This was confirmed in *in vivo* $T_{2w}$ and $T_{1w}$ MRI images of pancreatic tumor-bearing mice after the injection of a $MnO_x$-SPION@MSN solution through tail vein. Thus, this nanoplatform achieved its goal both *in vitro* and *in vivo* (Figure 6) [89]. The characteristics of this and other representative systems [90–93] are summarized in Table 5.

## 3 MRI $T_1$ CONTRAST AGENTS BASED ON IRON(III) NANOSYSTEMS

Iron shows a 3+ oxidation state, with $3d^5$ high-spin configuration and is an essential element with a concentration in biological serum in the 100–200 μg $L^{-1}$ range. Iron(III) tends to form stable high-spin chelates with typical octahedral coordination, although some heptacoordinate exceptions exist. For high-spin

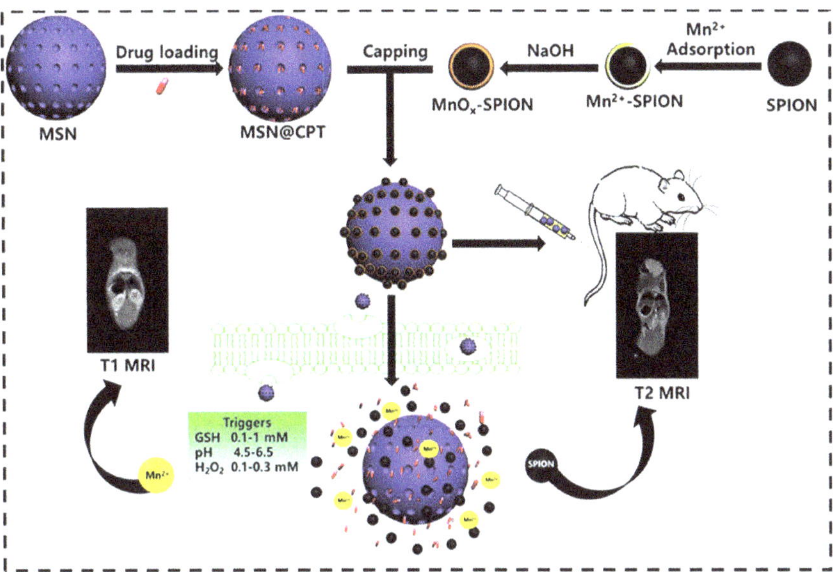


**FIGURE 6** Schematic illustration for the synthesis of $MnO_x$ -SPION capped MSN and controlled drug release in response to tumor microenvironment; TME release of SPION and $Mn^{2+}$ with *in vivo* $T_2$ and $T_1$ contrast, respectively. (Reprinted with permission from Ref. [89]. Copyright 2018, American Chemical Society.)

complexes, the magnetic moments are close to 5.9 BM and the electron relaxation times have been measured in the $10^{-11}$ – $10^{-9}$s range. These properties render iron(III) chelates valid candidate as $T_1$ MRI probes at clinical magnetic fields. Over the past decade, in addition to the search for new efficient iron chelates suitable for MRI applications, various nanosystems based on iron(III) complexed by low-molecular-weight chelating agents (siderophores) have been developed as potential alternatives to gadolinium-based compounds for MR-imaging applications. The chelators are typically based on hydroxamate, functionalized catechol, or derivative ligands and combine a high degree of specificity and huge thermodynamic stability of the final iron chelates. For these reasons, many cases reported in the literature describe the use of Fe(III)-containing nanostructures with these types of ligands as effective $T_1$ MRI contrast agents [94]. Some examples will be discussed in this chapter and summarized in Table 6.

## 3.1 Nanoparticles Based on Amphiphilic Polymeric Matrices

Amphiphilic polymeric nanoparticles are one of the first examples of nanostructures containing Fe(III)-chelates. In 2015, Grull et al. [95] investigated the use of a novel nanoplatform ($^{89}$Zr/Fe-DFO-micelles) for dual-modality PET/MR imaging. The nanoplatform showed promise in accurately detecting and localizing

**TABLE 5**

**Summary of Basic Properties of the Examples of $MnO_x$-Based NPs for MRI-Based Theranostics Discussed in the Section**

| Imaging Modalities | NPs Components | $MnO_x$ Reducing Agent in TME | TME Modulation Mode | Therapeutic Modalities | Animal Cancer Model | Reference |
|---|---|---|---|---|---|---|
| $T_1/T_2$ MRI | $MnO_x$-SPION@MSN @CPT | $H_2O_2/H^+$ and GSH | Tumor hypoxia relief/ GSH depletion | CT (CPT) | *In vitro*: pancreatic Panc-1 cancer cells; *In vivo*: pancreatic Panc-1 tumor mice | [89] |
| MRI | $MnO_x$-HMCNs | $H_2O_2/H^+$ | Tumor hypoxia relief | CT (DOX, CPT) | *In vitro*: MDA-MB-231 and 4T1 cells; *In vivo*: 4T1 tumor mice | [90] |
| MRI/SPECT | $^{99m}Tc$ -DTPA-$MnO_x$-MSNs-PEG | $H_2O_2/H^+$ | Tumor hypoxia relief | CT (DOX) | *In vitro*: MDA-MB-231 cells; *In vivo*: MBA-MD-231 female BALB/c nude mice and male Kunming mice | [91] |
| MRI/CT/PAI | $MnO_x/Ta_4C_3$-SP | $H_2O_2/H^+$ and GSH | Tumor hypoxia relief/ GSH depletion | PTT | *In vitro*: 4T1 cells; *In vivo*: 4T1 tumor male Balb/c nude mice | [92] |
| $^{19}F$/ $T_1/T_2$ $^1H$ MRI | PFOB@$MnO_x$-PEI-PEG (PM-CS) | $H_2O_2/H^+$ and GSH | Tumor hypoxia relief/ GSH depletion | Overcome hypoxia-induced ferroptosis resistance | *In vitro*: 4T1 cells; *In vivo*: 4T1 tumor mice | [93] |

**TABLE 6**

**Some Examples of Fe(III)-Based Nanosystems as MRI and Theranostics Probes Discussed in This Work**

| Nanoparticles | Imaging Modalities | Relaxivity ($mM^{-1}s^{-1}$) $T/B_0$ | Reference |
|---|---|---|---|
| $^{89}Zr$/ Fe-DFO micelles | $T_{1w}$ MRI/PET | $r_1$=2.7, $r_2$=3.1<br>310 K, 1.41 T | [95] |
| Fe-PLGA NPs | $T_{1w}$ MRI | $r_1$=10.6 (3.0 T)<br>$r_1$=3.0 (14.1 T)<br>293 K | [96] |
| Fe-rhodamine/ Chlorambucil@RD CH PG BN TP polymer | $T_{1w}$ MRI/ Fluorescence imaging | $r_1$=14.91; $r_2$=65.95<br>298 K, 11.7 T | [97] |
| Fe-CMN micellar systems | $T_{1w}$ MRI | $r_1$=7.9; $r_2$=11.1<br>310 K, 1.41 T | [98] |
| Fe-SMNP NPs | $T_{1w}$ MRI | $r_1$=8<br>310 K, 1.88 T | [99] |
| $Fe^{3+}$@PDOPA10-b-PSar50 | $T_{1w}$ MRI/MRI angiography | $r_1$=5.6<br>293 K, 3.0 T | [100] |
| Fe-CPNDs | $T_{1w}$ MRI | $r_1$=1.5; $r_2$=2.9<br>298 K, 1.5 T | [101] |
| GA-Fe@BSA | $T_{1w}$ MRI | $r_1$=0.89; $r_2$=0.95<br>298 K, 0.5 T | [103] |
| GA-Fe@BSA-PTX | $T_{1w}$ MRI | $r_1$=1.20; $r_2$=1.95<br>298 K, 0.5 T | [104] |
| Iron-tannic acid NPs | $T_{1w}$ MRI | $r_1$=3.14<br>298 K, T[a] | [105] |
| Fe-NGDA-dendritic-PEG oligomer NPs | $T_{1w}$ MRI | $r_1$=3.95<br>298 K, 7 T | [107] |
| PEG-$GdF_3$:Fe NPs | $T_{1w}$/$T_{2w}$ MRI/X-ray CT | $r_1$=3.3; $r_2$=36.0<br>298 K, T[a] | [109] |
| Gd-Fe(7:3)-TA NPs | $T_{1w}$ MRI | $r_1$=9.3; $r_2$=11.7<br>298 K, 0.94 T | [110] |
| Au@$SiO_2$-MDOTA-Cy7 @Au-mPEG M= Fe(III), Gd(III) | $T_{1w}$ MRI/ Fluorescence imaging | $r_1$=5.9, $r_2$=25.1 (Fe) at 1.5 T;<br>$r_1$=22, $r_2$=54.7 (Gd) at 4.7 T;<br>298 K | [111] |
| Fe(III)-TOB @ β-glucan particles | $T_{1w}$ MRI | $r_1$=0.21; $r_2$=3.45<br>310 K, 4.7 T | [112] |
| Fe-CDs | $T_{2w}$ MRI | $r_1$=0.13; $r_2$=9.9<br>298 K, 9.4 T | [113] |
| Fe-PFHA | $^{19}F$ $T_{1w}$ MRI | $r_1$=1.01; $r_2$=1.56<br>298 K, 9.4 T | [116] |

[a] Not reported.

targets using both PET and MR imaging techniques. The new $^{89}Zr$/Fe-DFO micelles can be used for $T_1$-weighted MRI experiments, unlike other PET/MRI and SPECT/MRI systems that rely on radiolabeled iron oxides for $T_2$-specific contrast agents. Micelles were formed by mixing polybutadiene-b-polyethylene oxide (PBD-b-PEO) and polybutadiene-b-polyacrylic acid (PBD-b-PAA) polymers, which were then functionalized with acrylic acid residues (AA) and able to promote the incorporation of $^{89}Zr$-DFO (for PET imaging) and $Fe^{3+}$-DFO (for MRI). The $r_1$ values collected on Fe-DFO-micelles indicated that these particles have a modest ionic $r_1$ at a temperature of 37°C and 1.41 T. The $r_1$ for Fe-DFO-micelles was found to be 2.7 $mM^{-1}s^{-1}$ ($r_1$ per micelle = 1620 $mM^{-1}s^{-1}$), whereas the $r_2$ was found to be 3.1 $mM^{-1}s^{-1}$. The efficacy of these micelles as $T_1$ probes was comparable to that of analogous systems containing Gd(III).

More recently, a novel formulation composed of NPs of poly(lactic-co-glycolic) acid (PLGA) and chelated iron(III) (Fe-PLGA) was also investigated [96]. The NPs showed a monodisperse distribution with size centered at 93 nm, spherical morphology, and high stability in PBS and serum. Experiments demonstrated strong chelation of Fe(III) ions with only 3.1% of iron release over 72 h in simulated body fluid. The $r_1$ values of the NPs at neutral pH were 10.6 and 3.0 $mmol^{-1}s^{-1}$ at 3.0 and 14.1 T, respectively. The effectiveness of Fe-PLGA NPs was evaluated in a live mouse model using dynamic MR imaging. The NPs were injected, and pre-contrast and post-injection images were obtained at various time intervals up to 3 h on a 3 T scanner. Thus, Fe-PLGA NPs demonstrated significant angiographic MRI contrast enhancement, providing a successful nanosystem for MR imaging.

High relaxivity values were also achieved with a dual-imaging polymer nanostructure functionalized with rhodamine labeled with Fe(III) ions [97]. The amphiphilic nature of the polymer allowed it to self-assemble into nanospheres of approximately 200 nm diameter, which were readily taken up by cancer cells via receptor-mediated endocytosis. Further, the nanospheres accumulated rapidly in mitochondria due to strong electrostatic interactions. The NPs displayed a high efficacy in enhancing the relaxation rate of the solvent protons. At 298 K and 500 MHz, the sample exhibited relaxivity values of $r_1 = 14.91\ mM^{-1}s^{-1}$ and $r_2 = 65.95\ mM^{-1}s^{-1}$.

### 3.2 Nanosystems Based on Polyphenolic Units

Researchers also developed amphiphilic triblock copolymers containing Fe(III)-catecholate complexes that generated spherical or cylindrical micellar nanoparticles (SMN and CMN, respectively), which act as $T_1$-weighted MRI [98]. The nanostructure was synthesized through post-polymerization functionalization. The segment size of each block was carefully controlled and adjusted to achieve the desired properties and obtain nanoparticles with different morphologies. Nuclear magnetic relaxation dispersion (NMRD) profiles were used to measure the $1/T_1$ values of nanosystems at varying temperatures and frequencies. The experiments showed that $r_1$ values of these systems were higher than those of

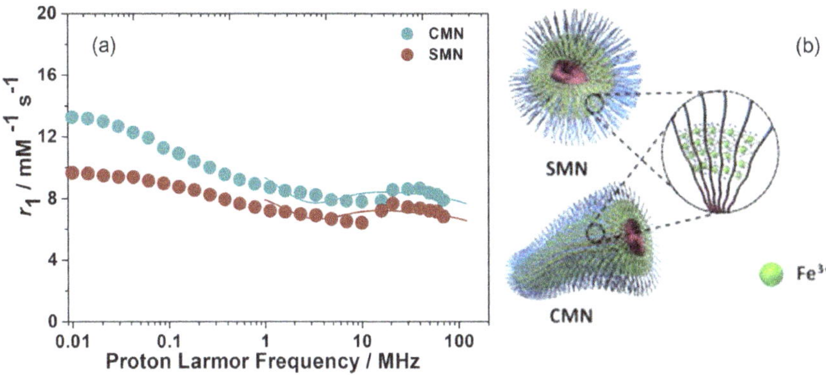


**FIGURE 7** (a) $^1$H NMRD profiles of both CMN and SMN nanoparticles. (b) Illustration of the copolymers containing Fe(III)-catecholate with spherical or cylindrical morphology. (Reprinted with permission from Ref. [98]. Copyright 2016, Wiley.)

Gd(III) contrast agents at multiple magnetic field strengths [8]. The high $r_1$ values of 7.1 and 7.9 mM$^{-1}$ s$^{-1}$ (at 1.5 T and 310 K) for SMN and CMN were attributed to an important second-sphere contribution of water molecules in interaction with the catechol groups (Figure 7). Furthermore, SMN and CMN were found to be stable and safe in biological environments, demonstrating high biocompatibility and low toxicity toward living cells.

Synthetic melanins loaded with Fe(III) are another example of nanoparticles containing polyphenolic groups. A team of researchers developed a method for improving and controlling the iron loading of synthetic melanin nanoparticles (SMNPs) with spherical morphology and subsequently conducted a detailed analysis on their structure–property relationship [99]. This study aimed to understand the behavior of SMNPs and how their physical properties were affected by the amount of iron incorporation. Interestingly, the iron load in SMNPs did not increase their relaxivity values proportionally at all frequencies. Instead, there is a limit of doping (iron load of 5.86%) beyond which the relaxivity decreases. SMNPs with this iron load exhibited high ionic $r_1$ values of approximately 7–8 mM$^{-1}$ s$^{-1}$ in the 20–80 MHz range.

Another example of NPs based on catechol ligands was proposed by Miao et al. [100]. These micellar NPs called $Fe^{3+}$@PDOPA-b-PSar were efficient as $T_{1w}$ MRI CAs due to the strong chelation of Fe(III) cations by catechol ligands (Figure 8). Optimal properties were achieved by balancing the hydrophilic and hydrophobic ratios of the PDOPA and PSar segments required for the NPs preparation. This involves searching for the best ratios of these properties to enhance the overall efficacy of the nanoparticles. The most stable $Fe^{3+}$@PDOPA10-b-PSar50 NPs, labeled as NPDS1, were formed using PDOPA10-b-PSar50. These NPs, with size below 50 nm, exhibited good results with $r_1$ of 5.6 mM$^{-1}$ s$^{-1}$ at 3.0 T and 293 K and good MRI contrast *in vivo* (Figure 8). In particular, *in vivo* MRA was conducted

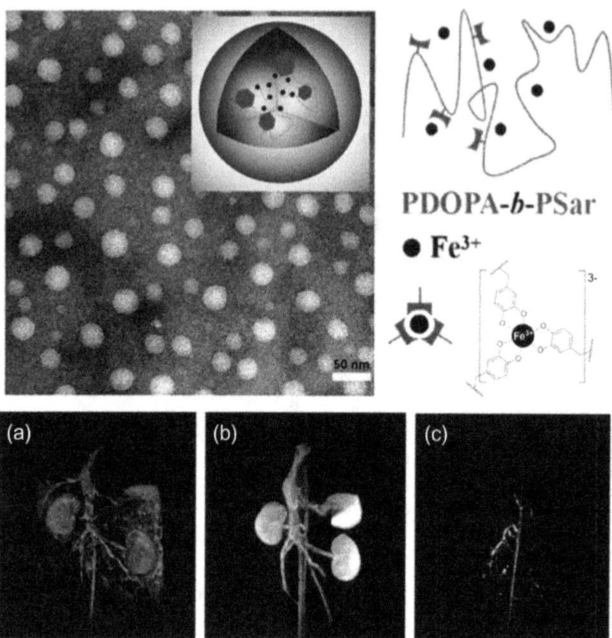


**FIGURE 8** Above: Schematic structure and TEM micrograph of the $Fe^{3+}$@PDOPA-b-PSar NPs. Below: (a) Volume rendering MR image of abdominal aorta and inferior vena cava. (b) Maximal intensity projection (MIP) MR image of abdominal aorta and inferior vena cava. (c) MIP image of the separated abdominal aorta and its branch. (Reprinted with permission from Ref. [100]. Copyright 2018, American Chemical Society.)

on rabbits by administering 10 mL of NPDS1 solution. The MRI images exhibited strong and consistent enhancement of the kidneys and vessels for over 120 min after the injection (Figure 8). This indicates that the NPDS1 solution has the potential to be used as a contrast agent for imaging in medical applications.

Liu and co-workers created ultrasmall nanodots (NDs) that can be activated at different pH levels. The Fe-CPNDs, composed of particles with size close to 5.3 nm, were developed using gallic acid (GA), which is optimal for Fe(III) ions complexation, and poly(vinylpyrrolidone) at room temperature [101]. Fe-CPNDs have an electrically neutral surface which reduces interactions with serum proteins and other blood components. These nanoparticles displayed pH-activatable MRI contrast and highly effective photothermal activity. The coordination between Fe(III) and GA is pH-dependent, with the more hydrated species dominant at lower pH, resulting in higher relaxivity. This activation of Fe-CPNDs by the low pH of tumors therefore can improve the MRI contrast [102]. In addition, PTT was used to test the effectiveness of Fe-CPNDs in a mouse model with xenograft tumor. The results showed that a concentration of 0.2 mg $kg^{-1}$ completely stopped the tumor growth. This suggested that Fe-CPNDs could be a promising new class for cancer diagnosis and treatment.

Furthermore, nanosystems made of Fe(III) and coated with protein matrices were investigated in the last years. For example, Mu et al. developed a method for synthesizing gallic acid-Fe(III) coordination polymer NPs using BSA as a stabilizing agent [103]. The GA-Fe@BSA NPs are biocompatible and ultrasmall (3.5 nm), making them ideal as MRI probes for $T_{1w}$ images. Their low $r_2/r_1$ ratio (1.06) further confirms their suitability for MRI imaging. Additionally, they exhibited strong absorption in the visible to near-infrared regions. The same researchers used paclitaxel to promote the assembling of BSA-coated ultrasmall GA-Fe@BSA nanoparticles through the hydrophobic effect. These NPs were then combined to form large, self-assembled multifunctional theranostic nanoparticles called GA-Fe@BSA-PTX [104]. The sample showed important advantages in terms of excellent MRI $T_1$ contrast and good therapeutic effect associated with a relevant tumor uptake. The $r_1$ value was calculated to be 1.2 $mM^{-1}s^{-1}$ at 0.5 T. The effectiveness of GA-Fe@BSA-PTX self-assembled nanoparticles in enhancing MRI performance and tumor accumulation was confirmed *in vivo*. The NPs showed a quite fast clearance from the body, with initial accumulation in the liver and subsequent clearance through the bloodstream to the kidney, bladder, and urine.

Nanoparticles based on iron–tannic acid complexes (FT NPs) were also investigated in the literature [105]. These NPs possessed beneficial properties, including small size, high MRI contrast ability, stability, solubility in water, and high efficiency in being taken up by tumor cells. A recent study utilized the same nanoparticles for a quantitative assessment of liver function using MRI and improving liver parenchyma clearance function [106]. The highest enhancement in MRI signal was observed after 30 min following injection in $T_{1w}$ experiments. The signal then steadily declined as a consequence of the FTs metabolism by the liver.

Very recently, Li et al. [107] synthesized a new iron-based nanosized contrast agent (IBCA) composed of telodendritic molecules. These molecules consist of a hydrophilic PEG tail conjugated to 3,4-dihydroxyhydrocinnamic acid and a hydrophobic dendritic oligomer based on nordihydroguaiaretic acid. These IBCA NPs showed high contrast in MRI with a $r_1$ value of ca. 4 $mM^{-1}s^{-1}$ at 7 T and exhibited good biocompatibility *in vitro*.

## 3.3 Nanoparticles Containing Mixed Metals

Mixed nanosystems containing a combination of Gd(III) and Fe(III) ions were also suggested as multimodal diagnostic probes. The two ions can provide excellent $T_1$ and $T_2$ relaxation effects. Moreover, Gd(III) has a high X-ray absorption coefficient, which makes it ideal for contrast enhancement in X-ray computed tomography (CT) [108]. As an example, in 2017, the incorporation of Fe(III) ions in the structure of $GdF_3$ nanoparticles stabilized by PEG molecules was investigated in detail [109]. These nanoparticles (PEG-$GdF_3$:Fe) were monodisperse measuring about 50 nm in length and 30 nm in width (Figure 9). They were found to accumulate in the tumor due to the enhanced permeability and retention (EPR)

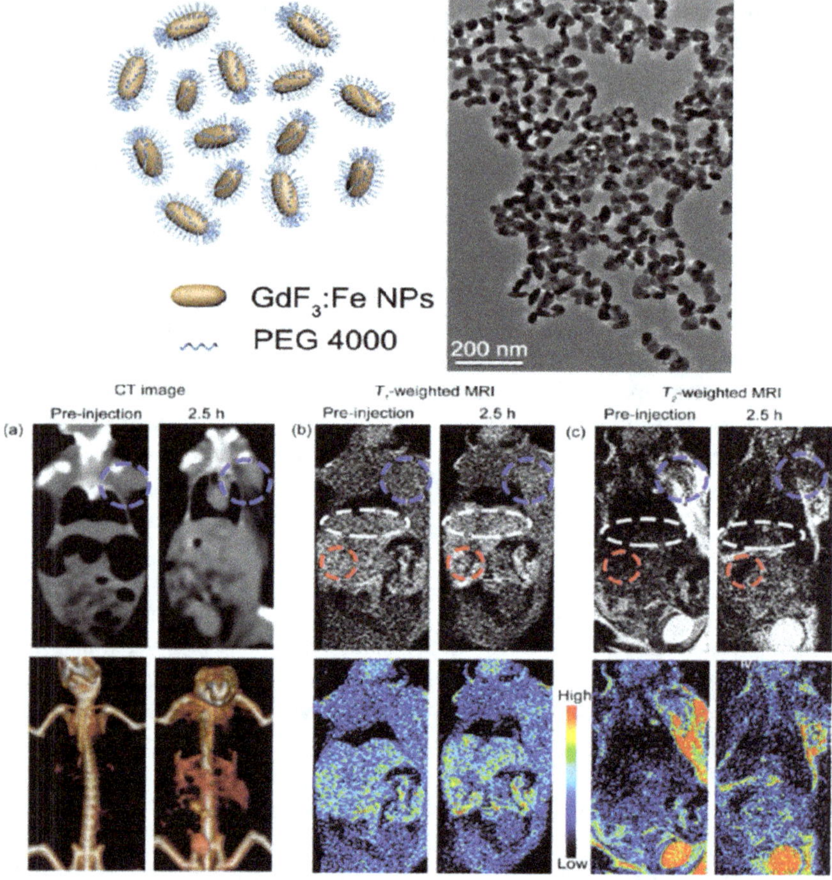


**FIGURE 9** Above: Schematic view and TEM image of PEG-$GdF_3$:Fe sample. Below: *In vivo* CT images (a), $T_1$-weighted MRI (b), and $T_2$-weighted MRI (c) of tumor before and after administration of the nanoparticles: (tumor (blue circle), liver (white ellipse), and spleen (red circles)). (Reprinted with permission from Ref. [109]. Copyright 2017, American Chemical Society.)

effect. The tumor tissues showed an increase in $T_1$ signal after 1.5 h of nanoparticle injection, which disappeared completely after 2.5 h. Furthermore, PEG-$GdF_3$:Fe showed high ability to enhance the X-ray contrast in *in vivo* analyses at a concentration lower than that used with the commercially available CT contrast agents (Figure 9). A preliminary *in vivo* study showed that these nanostructures exhibited a strong $T_1$-MRI signal and fluorescence, indicating its efficacy and potential as dual imaging probe.

NPs created by using a controllable process that uses $Gd(NO_3)_3$, $FeSO_4$, and polyphenols (tannic acid) as a ligand were also studied [110]. These bimetallic-phenolic nanoparticles showed sizes close to 23 nm, good stability, $r_1$ value of 9.3 $mM^{-1}s^{-1}$ at 40 MHz, and high photothermal conversion efficiency. The particles were tested *in vivo* in EMT-6 tumor-bearing mice, thus showing high signal enhancement and appreciable photothermal therapy ability.

The use of Fe(III) complexes as MRI probes has also been combined with near-infrared resonant gold nanoparticles to create hybrid nanostructures that can serve both imaging and therapy purposes. The authors, in this work, prepared a multilayer system composed of a gold core, a silica layer containing fluorescent dyes (Cy-7) and SCN-DOTA ligand able to chelate Fe(III) or Gd(III) ions, and a second gold shell [111]. At a magnetic field strength of 4.7 T, the two samples, containing Gd(III) (GdNM) or Fe(III) (FeNM) NPs, demonstrated significantly increased relaxivity compared to their respective counterparts, GdDOTA and FeDOTA. The $r_1$ relaxivity of GdNM NPs was found to increase by eight times, while that of FeNM NPs increased by 15 times (Figure 10). A preliminary study on animals indicated that a newly engineered structure was effective for dual-modality imaging. In a mouse model, both $T_1$-MRI and fluorescence signals were detected strongly, indicating the potential for multimodal imaging.

## 3.4 Miscellaneous Nanoparticles

Yeast-derived β-glucan particles (GPs) are commonly utilized to transport and distribute various pharmaceuticals. These particles are highly effective and versatile, allowing for efficient drug delivery. In this field, Morrow and colleagues successfully encapsulated macrocyclic Fe(III) chelates in graphene particles for MR imaging capabilities. The Fe(III) complexes were based on the 1,4,7-triazacyclononane (TACN) ring and efficiently incorporated into the graphene particles through electrostatic interactions [112]. Fe-labeled NPs have lower relaxivity than the free complex due to an alteration of second-sphere water molecules network. Nevertheless, a relaxation enhancement could be achieved through the release

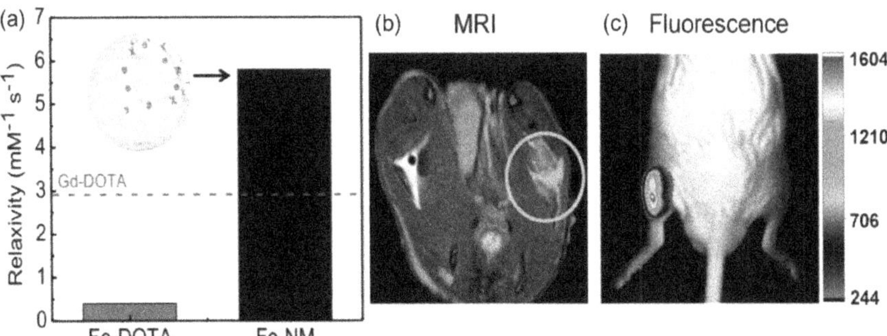


**FIGURE 10** Left: $r_1$ values calculated for Fe-NM (4.7 T) and Fe-DOTA (1.5 T); right: MRI (4.7 T) and fluorescence images of a mouse model after administration of bimodal FeCy7–NM nanoparticles (b and c, respectively). (Reprinted with permission from Ref. [111]. Copyright 2020, American Chemical Society.)

of Fe complexes from particles by acidifying the solution or by adding maltol bidentate ligand.

More recently, Fe-doped carbon nanozyme dots (Fe-CDs), characterized by peroxidase-like activity, were also synthesized and characterized in detail [113]. The sample was synthesized via hydrothermal reaction, resulting in uniform nanoparticles approximately 3.8 nm in size. The relaxivities of Fe-CDs were measured at 9.4 T with $r_1$ and $r_2$ values of 0.13 $mM^{-1} s^{-1}$ and 9.9 $mM^{-1} s^{-1}$, respectively. The high $r_2/r_1$ ratio of ca. 76 indicates that Fe-CDs may have potential as $T_{2w}$ MRI CAs. Researchers conducted an *in vivo* MRI test on U87MG tumor-bearing mice monitoring the $T_2$ signal at 9.4 T. The results showed a negative signal intensity in the tumor region, which was particularly pronounced approximately 1 h after injection. These findings provide valuable insights into the potential of Fe-CDs as a contrast agent in MRI imaging for tumor detection.

In the last decade, some efforts were also addressed to the use of $^{19}F$ MRI probes based on perfluorocarbon (PFC), although the long relaxation times of $^{19}F$ are a widely recognized barrier in the advancement of these systems [114]. A possible strategy to overcome this limit is based on the incorporation of paramagnetic centers in the same nanostructure. For instance, Ahrens and co-workers conjugated paramagnetic metal-binding β-diketones known as "FETRIS" to the end groups of a linear perfluoropolyether (PFPE) [115]. The sample, characterized by low cytotoxicity, significantly reduces the $^{19}F$ $T_1$ values by up to ten times. The effectiveness of this reduction appears to be dependent on the level of iron concentration present. The same group some years later developed other fluorinated nanoparticles functionalized with Fe(III) chelates (Fe-(perfluoroheptanoyl) acetone) (PFHE) [116]. The $^{19}F$ MRI SNR enhancement at 9.4 T of the nanoemulsion (NE) was increased 2.6 times after confinement of Fe-PFHA chelates (30 mM).

## 4 GENERAL CONCLUSIONS

The use of Mn(II/III) and Fe(III) ions as an alternative to Gd(III) in the design of nanoprobes for $T_1$ MR imaging and theranostics is constantly growing, and in this chapter, only a selection of nanosystems has been shown. Considering the lower toxicity of Fe(III) and its ability to form extremely stable complexes, Fe-based NPs might be preferable to Mn-based NPs that, however, have been much more explored in the last two decades. Despite the promising and interesting advances highlighted in this chapter, there are still more in-depth studies to be performed such as *in vivo* experiments aimed at investigating long-term toxicity, biodegradation, elimination mechanisms, and biodistribution since the reported data are still limited.

Moreover, it should be emphasized that for a correct analysis of the efficiency of these Fe/Mn-based nanoprobes, it is extremely important to measure the values of the relaxation rates $R_1$ (and $R_2$) at multiple frequencies, preferably between 2 and 5 T, and at multiple temperatures. This will enable to gain a deeper and comprehensive understanding of the correlation between the molecular parameters governing relaxivity and the structural features of nanoprobes. Ultimately, this knowledge will facilitate the maximum utilization of the immense potential of these Fe/Mn-based NPs for imaging studies.

## ACKNOWLEDGMENTS

C.F.G.C.G. acknowledges the Coimbra Chemistry Centre (UID/QUI/00313/2019 and POCI-01–0145-FEDER-027996) of the University of Coimbra. L.T. and F.C. acknowledge the financial support of the Ministero dell'Università e della Ricerca (PRIN 2017A2KEPL project).

## ABBREVIATIONS AND DEFINITIONS

**Alg** alginate
**BBB** blood–brain barrier
**BSA** bovine serum albumin
**CA** contrast agent
**CDTA** cyclohexanediamenetetraacetate
**ChT** chemotherapy
**CT** X-ray computed tomography
**DFO** deferoxamine
**1,4-DO2A** 1,4,7,10-tetraazacyclodocecane-1,4-diacetate
**DD-DO2A** 1,4,7,10-tetraazacyclodocecane-7,10-didocecyl-1,4-diacetate
**DOTA** 1,4,7,10-tetraazacyclododecane-1,4,7,10-tetraacetate
**DOX** doxorubicin
**DSPE** 1,2-distearoyl-sn-glycero-3-phosphoethanolamine
**DTPA** diethylene triamine pentaacetate
**DTX** docetaxel
**EDTA** ethylenediamietetraacetate
**FA** folic acid
**FI** fluorescein isothiocyanate
**FR** folate receptor
**GSH** glutathione
**HIFU** high intensity focused ultrasound
**MDR** multidrug resistance
**MRA** magnetic resonance angiography
**MRI** magnetic resonance imaging
**NG** nanogel
**NOTA** 1,4,7-triazacyclononane-N,N′,N″-triacetate
**NP** nanoparticle
**PAA** polyacrylic acid
**PAI** photoacoustic imaging
**PC** photothermal conversion
**PDA** polydopamine
**PDT** photodynamic therapy

| | |
|---|---|
| **PEG** | polyethylene glycol |
| **PEI** | polyethyleneimine |
| **PET** | positron emission tomography |
| **PFHA** | perfluoroheptanoylacetone |
| **PLGA** | poly(lactic-co-glycolic) acid |
| **PP** | protoporphyrin |
| **PSS** | polystyrene sulfonate |
| **PTT** | photothermal therapy |
| **PTX** | paclitaxel |
| $q$ | number of bound water molecules |
| **RT** | radiotherapy |
| $r_1$ | longitudinal relaxivity |
| $r_2$ | transverse relaxivity |
| **SDT** | sonodynamic therapy |
| **SPION** | superparamagnetic iron oxide nanoparticle |
| **TACN** | 1,4,7-triazacyclononane |
| **TEM** | transmission electron microscopy |
| **TME** | tumor microenvironment |
| $T_{1w}$ | $T_1$-weighted |
| $T_{2w}$ | $T_2$-weighted |
| **XANES** | X-ray absorption near-edge spectroscopy |
| **XPS** | X-ray photoelectron spectroscopy |
| **XRD** | X-ray diffraction |

## REFERENCES


1. R. Antwi-Baah, Y. Wang, X. Chen, Y. Kui, *Adv. Mater. Interfaces* **2022**, *9*, 2101710.
2. X. Mao, J. Xu, H. Cui, *Wiley Interdiscip Rev. Nanomed. Nanobiotechnol.* **2016**, *8*, 814–841.
3. Z. Gao, T. Ma, E. Zhao, D. Docter, W. Yang, R. H. Stauber, M. Gao, *Small* **2016**, *12*, 556–576.
4. T. L. Doane, C. Burda, *Chem. Soc. Rev.* **2016**, *41*, 2885–2911.
5. B. R. Smith, S. S. Gambhir, *Chem. Rev.* **2017**, *117*, 901–986.
6. A. Fatima, M. W. Ahmad, A. K. A. Al Saidi, A. Choudhury, Y. Chang, G. H. Lee, *Nanomaterials* **2021**, *11*, 2449.
7. C.-T. Yang, P. Padmanabhana, B. Z. Gulyás, *RSC Adv.* **2016**, *6*, 60945–60966.
8. M. Botta, L. Tei, *Eur. J. Inorg. Chem.* **2012**, 1945–1960.
9. A. J. L. Villaraza, A. Bumb, M. W. Brechbiel, *Chem. Rev.* **2010**, *110*, 2921–2959.
10. M. R. Rudnick, I. M. Wahba, A. K. Leonberg-Yoo, D. Miskulin, H. I. Litt, *Am. J. Kidney Dis.* **2021**, *77*, 517–528.
11. P. Chen, M. Totten, Z. Zhang, H. Bucinca, K. Erikson, A. Santamaría, A. B. Bowman, M. Aschner, *Exp. Rev. Neurotherap.* **2019**, *19*, 243–260.
12. L. García-Hevia, M. Bañobre-López, J. Gallo, *Chem. Eur. J.* **2019**, *25*, 431–441.
13. A. Sharma, L. Feng, D. F. Muresanu, S. Sahib, Z. R. Tiane, J. V. Lafuente, A. D. Buzoianu, R. J. Castellani, A. Nozari, L. Wiklund, H. S. Sharma, *Progress Brain Res.* **2021**, *265*, 385–406.
14. M. Parvaneh, S. Samaneh, F. Hashemi, M. Ale-Ebrahim, *J. Biol. Macromol.* **2018**, *125*, 674–682.

15. J. A. Roth, B. Ganapathy, A. J. Ghio, *Toxicol. In Vitro* **2012**, *26*, 1143–1149.
16. D. Ding, J. Roth, R. Salvi, *Neurotoxicology* **2011**, *32*, 233–41.
17. M. Botta, F. Carniato, D. Esteban-Gomez, C. Platas-Iglesias, L. Tei, *Future Med. Chem.* **2019**, *11*, 1461–1483.
18. R. Uzal-Varela, F. Pérez-Fernández, L. Valencia, A. Rodríguez-Rodríguez, C. Platas-Iglesias, P. Caravan, D. Esteban-Gómez, *Inorg. Chem.* **2022**, *61*, 14173–14186.
19. L. Lattuada, A. Barge, G. Cravotto, G. B. Giovenzana, L. Tei, *Chem. Soc. Rev.* **2011**, *40*, 3019–3049.
20. S. Langereis, T. Geelen, H. Grull, G. J. Strijkers, K. Nicolay, *NMR Biomed.* **2013**, *26*, 728–744.
21. M. Filippi, D. Remotti, M. Botta, E. Terreno, L. Tei, *Chem. Commun.* **2015**, *51*, 17455–17458.
22. M. Elsabahy, G. S. Heo, S. M. Lim, G. Sun, K. L. Wooley, *Chem. Rev.* **2015**, 115, 10967–11011.
23. M. Filippi, J. Martinelli, G. Mulas, M. Ferraretto, T. Eline, M. Botta, L. Tei, E. Terreno, *Chem. Commun.* **2014**, *50*, 3453–3456.
24. E. Terreno, D. Delli Castelli, C. Cabella, W. Dastru, A. Sanino, J. Stancanello, L. Tei, S. Aime, *Chem. Biodiversity* **2008**, *5*, 1901–1912.
25. C. Wu, D. Li, L. Yang, B. Lin, H. Zhang, Y. Xu, Z. Cheng, C. Xia, Q. Gong, B. Song, H. Ai, *J. Mater. Chem. B*, **2015**, *3*, 1470–1473.
26. G. A. Rolla, V. De Biasio, G. B. Giovenzana, M. Botta, L. Tei, *Dalton Trans.* **2018**, *47*, 10660–10670.
27. G.A. Rolla, C. Platas-Iglesias, M. Botta, L. Tei, L. Helm, *Inorg. Chem.* **2013**, 52, 3268–3279.
28. G. Mulas, G. A. Rolla, C. F. G. C. Geraldes, L. W. E. Starmans, M. Botta, E. Terreno, L. Tei, *ACS Appl. Bio Mater.* **2020**, *3*, 2401–2409.
29. X. Liu, S. Fu, C. Xia, M. Li, Z. Cai, C Wu, F. Lu, J. Zhu, B. Song, Q. Gong, H. Ai, *J. Mater. Chem. B* **2022**, *10*, 2204–2214.
30. Y. Shen, X. Li, H. Huang, Y. Lan, L. Gan, J. Huang, *Carbohydr. Polym.* **2022**, 297, 120061.
31. Y. Shen, X. Li, Y. Lan, M. Zu, X. Liu, H. Huang, N. Zhou, R. Duan, L. Gan, J. Huang, *Cellulose* **2021**, *28*, 2905–2916.
32. X. Wang, L. Xu, Z. Ren, M. Fan, J. Zhang, H. Qi, M. Xu, *Colloids Surf. B: Biointerfaces* **2019**, *183*, 110452.
33. K. D. Addisu, B. Z. Hailemeskel, S. L. Mekuria, A. T. Andrgie, Y.-C. Lin, H.-C. Tsai, *ACS Appl. Mater. Interfaces* **2018**, *10*, 5147–5160.
34. F. Carniato, M. Ricci, L. Tei, F. Garello, C. Furlan, E. Terreno, E. Ravera, G. Parigi, C. Luchinat, M. Botta, *Small* **2023**, 2302868.
35. F. Carniato, L. Tei, M. Botta, *Eur. J. Inorg. Chem.* **2018**, *46*, 4936–4954.
36. M. Palmai, A. Petho, L. N. Nagy, S. Klebert, Z. May, J. Mihaly, A. Wacha, K. Jemnitz, Z. Veres, I. Horvath, K. Szigeti, D. Mathe, Z. Varga, *J. Colloid Interface Sci.* **2017**, *498*, 298–305.
37. D. Lalli, G. Ferrauto, E. Terreno, F. Carniato, M. Botta, *J. Mater. Chem. B* **2021**, *9*, 8994–9004.
38. R. Mallik, M. Saha, C. Mukherjee, *ACS Appl. Bio Mater.* **2021**, *4*, 8356–8367.
39. R. Qin, S. Li, Y. Qiu, Y. Feng, Y. Liu, D. Ding, L. Xu, X. Ma, W. Sun, H. Chen, *Nat. Commun.* **2022**, *13*, 1938.
40. X. Huang, Z. Wang, S. Li, S. Lin, L. Zhang, Z. Meng, X. Zhang, S.-K. Sun, *Biomater. Sci.* **2023**, *11*, 4289–4297.
41. S. Shao, V. Rajendiran, J.F. Lovell, *Coord. Chem. Rev.* **2019**, *379*, 99–120.
42. K. Choi, D.-H. Lee, W.-D. Jang, *Korean Chem. Soc.* **2010**, *31*, 639–644.

43. C. F. G. C. Geraldes, M. M. C. A. Castro, J. A. Peters, *Coord. Chem. Rev.* **2021**, *445*, 214069.
44. D. Pan, S. D. Caruthers, G. Hu, A. Senpan, M. J. Scott, P. J. Gaffney, S. A. Wickline, G. M. Lanza, *J. Am. Chem. Soc.* **2008**, *130*, 9186–9187.
45. Y. Xu, C. Li, X. Wu, M.-X. Li, Y. Ma, H. Yang, Q. Zeng, J. L. Sessler, Z.-X. Wang, *J. Am. Chem. Soc.* **2022**, *144*, 18834–18843.
46. B. Y. W. Hsu, G. Kirby, A. Tan, A. M. Seifalian, X. Li, J. Wang, *RSC Adv.* **2019**, *6*, 45462–45474.
47. X. Cai, Q. Zhu, Y. Zeng, Q. Zeng, X. Chen, Y. Zhan, *Int. J. Nanomed.* **2019**, *14*, 8321–8344.
48. A. A. Gilad, P. Walczak, M. T. McMahon, H. B. Na, J. H. Lee, K. An, T. Hyeon, P. C. van Zijl, J. W. Bulte, *Magn. Reson. Med.* **2008**, *60*, 1–7.
49. R. Chen, D. Ling, L. Zhao, S. Wang, Y. Liu, R. Bai, S. Baik, Y. Zhao, C. Chen, T. Hyeon, *ACS Nano* **2015**, *9*, 12425–12435.
50. A. Banerjee, G. E. Bertolesi, C.-C. Ling, B. Blasiak, A. Purchase, O. Calderon, B. Tomanek, S. Trudel, *ACS Appl. Mater. Interfaces* **2019**, *11*, 13069–13078.
51. M. Lei, C. Fu, X. Cheng, B. Fu, N. Wu, Q. Zhang, A. Fu, J. Cheng, J. Gao, Z. Zhao, *Adv. Funct. Mater.* **2017**, *27*, 1700978.
52. A. Banerjee, W. Zeng, M. Taheri, B. Blasiak, B. Tomanek, S. Trudel, *AIP Adv.* **2019**, *9*, 125031.
53. L. Yang, L. Wang, G. Huang, X. Zhang, L. Chen, A. Li, J. Gao, Z. Zhou, L. Su, H. Yang, J. Song, *Theranostics* **2021**, *11*, 6966–6982.
54. J. Li, C. Wu, P. Hou, M. Zhang, K. Xu, *Biosens Bioelectron.* **2018**, *102*, 1–8.
55. P. Chevallier, A. Walter, A. Garofalo, I. Veksler, J. Lagueux, S. Begin- Colin, D. Felder-Flesch, M.-A. Fortin, *J. Mater. Chem. B* **2014**, *2*, 1779–1790.
56. B. Y. W. Hsu, M. Wang, Y. Zhang, V. Vijayaragavan, S. Y. Wong, A. Y.-C. Chang, K. K. Bhakoo, X. Li, J. Wang, *Nanoscale* **2014**, *6*, 293–299.
57. M. Costanzo, L. Scolaro, G. Berlier, A. Marengo, S. Grecchi, C. Zancanaro, M. Malatesta, S. Arpicco, *Int. J. Pharmaceut.* **2016**, *508*, 83–91.
58. R. Wei, X. Gong, H. Lin, K. Zhang, A. Li, K. Liu, H. Shan, X. Chen, J. Gao, *Nano Lett.* **2019**, *19*, 5394–5402.
59. R. Guillet-Nicolas, M. Laprise-Pelletier, M. M. Nair, P. Chevallier, J. Lagueux, Y. Gossuin, S. Laurent, F. Kleitz, M.-A. Fortin, *Nanoscale* **2013**, *23*,11499-511.
60. S. Marasini, H. Yue, S.-L. Ho, J.-A. Park, S. Kim, J.-U. Yang, H. Cha, S. Liu, T. Tegafaw, M. Y. Ahmad, A. K. A. Al Saidi, D. Zhao, Y. Liu, K.-S. Chae, Y. Chang, G.-H. Lee, *Appl. Sci.* **2021**, *11*, 2596.
61. J. Gallo, I. S. Alam, I. Lavdas, M. Wylezinska-Arridge, E. O. Aboagye, N. J. Long, *J. Mater. Chem. B* **2014**, *2*, 868–876.
62. H. Huang, T. Yue, K. Xu, J. Golzarian, J. Yua, J. Huang, *Colloids Surf. B* **2015**, *131*, 148–154.
63. X. Qian, X. Han, L. Yu, T. Xu, Y. Chen, *Adv. Funct. Mater.* **2020**, *30*, 1907066.
64. B. Ding, P. Zheng, P. A. Ma, J. Lin, *Adv. Mater.* **2020**, *32*, 1905823.
65. T. Yu, J. Moon, J. Park, Y. I. Park, H. B. Na, B. H. Kim, I. C. Song, W. K. Moon, T. Hyeon, *Chem. Mater.* **2009**, *21*, 2272–22797.
66. K. An, M. Park, J. H. Yu, H. B. Na, N. Lee, J. Park, S. H. Choi, I. C. Song, W. K. Moon, T. Hyeon, *Eur. J. Inorg. Chem.* **2012**, *12*, 2148–2155.
67. H. Zhang, L. Jing, J. Zeng, Y. Hou, Z. Li, M. Gao, *Nanoscale* **2014**, *6*, 5918–5925.
68. J. Shin, R. M. Anisur, M. K. Ko, G. H. Im, J. H. Lee, I. S. Lee, *Angew. Chem. Int. Ed.* **2009**, *48*, 321–324.
69. K. Poon, Z. Lu, Y. De Deene, Y. Ramaswamy, H. Zreiqat, G. Singh, *Nanoscale Adv.* **2021**, *3*, 4052–4061.

70. K. Deka, A. Guleria, D. Kumar, J. Biswas, S. Lodha, S. D. Kaushik, S. Dasgupta, P. Deb, *Colloids Surf. A* **2018**, *539*, 229–236.
71. J. Xiao, X. M. Tian, C. Yang, P. Liu, N. Q. Luo, Y. Liang, H. B. Li, D. H. Chen, C. X. Wang, L. Li, G. W. Yang, *Sci. Rep.* **2013**, *3*, 3424.
72. M. Nafiujjaman, M. Nurunnabi, S.-H Kang, G. R. Reeck, H. A. Khan, Y.-K. Lee, *J. Mater. Chem. B* **2015**, *3*, 5815–5823.
73. X. Ding, J. Liu, J. Li, F. Wang, Y. Wang, S. Song, H. Zhang, *Chem. Sci.* **2016**, *7*, 6695–6700.
74. Y. Liu, G. Zhang, Q. Guo, L. Ma, Q. Jia, L. Liu, J. Zhou, *Biomaterials* **2017**, *112*, 204–217.
75. J. Zhu, H. Li, Z. Xiong, M. Shen, P. S. Conti, X. Shi, K. Chen, *ACS Appl. Mater. Interfaces* **2018**, *10*, 34954–34964.
76. H. Yang, Y. Zhuang, H. Hu, X. Du, C. Zhang, X. Shi, H. Wu, S. Yang, *Adv. Funct. Mater.* **2010**, *20*, 1733–1741.
77. Y. Zhan, W. Zhan, H. Li, X. Xu, X. Cao, S. Zhu, J. Liang, X. Chen, *Molecules* **2017**, *22*, 2208.
78. G. Yang, J. Ji, Z. Liu, *WIREs Nanomed. Nanobiotech.* **2021**, *13*, e1720.
79. Y. Li, K. Song, Y. Cao, C. Peng, G. Yang, *ACS Appl. Mater. Interfaces* **2018**, *10*, 26039–26045.
80. J. Estelrich, M. J. Sánchez-Martín, M. A. Busquets, *Int. J. Nanomed.* **2015**, *10*, 1727–1741.
81. Y. Chen, D. Ye, M. Wu, H. Chen, L. Zhang, J. Shi, L. Wang, *Adv. Mater.* **2014**, *26*, 7019–7026.
82. X. Liu, Y. Zhou, W. Xie, S. Liu, Q. Zhao, W. Huang, *Small Methods* **2020**, *4*, 2000566.
83. Y. Gao, Z. Yin, Q. Ji, J. Jiang, Z. Tao, X. Zhao, S. Sun, A. Wu, L. Zeng, *J. Mater. Chem. B* **2021**, *9*, 314–321.
84. Z. He, Y. Xiao, J.-R. Zhang, P. Zhang, J.-J. Zhu, *Chem. Commun.* **2018**, *54*, 2962–2965.
85. Y. Hao, B. Zhang, C. Zheng, M. Niu, H. Guo, H. Zhang, J. Chang, Z. Zhang, L. Wang, Y. Zhang, *Colloids Surf. B: Biointerfaces* **2017**, *151*, 384–393.
86. C. Fu, X. Duan, M. Cao, S. Jiang, X. Ban, N. Guo, F. Zhang, J. Mao, T. Huyan, J. Shen, L. M. Zhang, *Adv. Healthcare Mater.* **2019**, *8*, 1900047.
87. L. Wang, D. Li, Y. Hao, M. Niu,Y. Hu, H. Zhao, J. Chang, H Zhang, Y. Zhang, *Int. J. Nanomed.* **2017**, *12*, 3059–307.
88. D. Hu, L. Chen, Y. Qu, J. Peng, B. Y. Chu, K. Shi, Y. Hao, L. Zhong, M. Y. Wang, Z. Y. Qian, *Theranostics* **2018**, *8*, 1558–1574.
89. S. S. Ren, J. Yang, L. Ma, X. Li, W. Wu, C. Liu, J. He, L. Miao, *ACS Appl. Mater. Interfaces* **2018**, *10*, 31947–31958.
90. S. J. Zhang, X. Q. Qian, L. L. Zhang, W. J. Peng, Y. Chen, *Nanoscale* **2015**, *7*, 7632–7643.
91. H. Gao, X. Liu, W. Tang W, D. Niu, B. Zhou, H. Zhang, W. Liu, B. Gu, X. Zhou, Y. Zheng, Y. Sun, X. Jia, L. Zhou, *Nanoscale* **2016**, *8*, 19573–19580.
92. C. Dai, Y. Chen, X. X. Jing, L. Xiang, D. Yang, H. Lin, Z. Liu, X. Han, R. Wu, *ACS Nano* **2017**, *11*, 12696–12712.
93. Z. Dong, P. Liang, G. Guan, B. Yin, Y. Wang, R. Yue, X. Zhang, G. Song, *Angew. Chem. Int. Ed.* **2022**, *61*, e202206074.
94. M. Botta, C. F. G. C. Geraldes, L. Tei, *WIREs Nanomed. Nanobiotechnol.* **2023**, *15*, e1858.
95. L. W. E. Starmans, M. A. P. M. Hummelink, R. Rossin, E. C. M. Kneepkens, R. Lamerichs, K. Donato, K. Nicolay, H. Grüll, *Adv. Healthc. Mater.* **2015**, *4*, 2137–2145.

96. R. Marasini, S. Rayamajhi, A. Moreno-Sanchez, S. Aryal, *RSC Adv.* **2021**, *11*, 32216–32226.
97. D. Patra, P. Kumar, T. Samanta, I. Chakraborty, R. Shunmugam, *ACS Appl. Bio Mater.* **2022**, *5*, 1284–1296.
98. Y. Li, Y. Huang, Z. Wang, F. Carniato, Y. Xie, J. P. Patterson, M. P. Thompson, C. M. Andolina, T. B. Ditri, J. E. Millstone, J. S. Figueroa, J. D. Rinehart, M. Scadeng, M. Botta, N. C. Gianneschi, *Small* **2016**, *12*, 668–677.
99. Y. Li, Y. Xie, Z. Wang, N. Zang, F. Carniato, Y. Huang, C. M. Andolina, L. R. Parent, T. B. Ditri, E. D. Walter, M. Botta, J. D. Rinehart, N. C. Gianneschi, *ACS Nano* **2016**, *10*, 10186–10194.
100. Y. Miao, F. Xie, J. Cen, F. Zhou, X. Tao, J. Luo, G. Han, X. Kong, X. Yang, J. Sun, J. Ling, *ACS Macro Lett.* **2018**, *7*, 693–698.
101. F. Liu, X. He, H. Chen, J. Zhang, H. Zhang, Z. Wang, *Nat. Commun.* **2015**, *6*, 8803–8811.
102. P. Mi, D. Kokuryo, H. Cabral, H. Wu, Y. Terada, T. Saga, I. Aoki, N. Nishiyama, K. Kataoka, *Nat. Nanotechnol.* **2016**, *11*, 724–730.
103. X. Mu, C. Yan, Q. Tian, J. Lin, S. Yang, *Int. J. Nanomed.* **2017**, *12*, 7207–7223.
104. L. An, C. Yan, X. Mu, C. Tao, Q. Tian, J. Lin, S. Yang, *ACS Appl. Mater. Interfaces* **2018**, *10*, 28483–28493.
105. T. Phatruengdet, J. Intakhad, M. Tapunya, A. Chariyakornkul, C.B. Hlaing, R. Wongpoomchai, C. Pilapong, *RSC Adv.* **2020**, *10*, 35419–35425.
106. T. Phatruengdet, P. Khuemjun, J. Intakhad, S. Krunchanuchat, A. Chariyakornkul, R. Wongpoomchai, C. Pilapong, *Nanotheranostics* **2022**, *6*, 195–204.
107. X. Xue, R. Bo, H. Qu, B. Jia, W. Xiao, Y. Yuan, N. Vapniarsky, A. Lindstrom, H. Wu, D. Zhang, L. Li, M. Ricci, Z. Ma, Z. Zhu, T. Lin, A.L. Louie, Y. Li, *Biomaterials* 2020, *257*, 120234–120245.
108. M. Ahmad, W. Xu, S.J. Kim, J.S. Baeck, Y. Chang, J.E. Bae, K.S. Chae, J.A. Park, T.J. Kim, G.H. Lee, *Sci. Rep.* **2015**, *5*, 8549–8559.
109. L. Dong, P. Zhang, P. Lei, S. Song, X. Xu, K. Du, J. Feng, H. Zhang, *ACS Appl. Mater. Interfaces* **2017**, *9*, 20426–20434.
110. J. Qin, G. Liang, Y. Feng, B. Feng, G. Wang, N. Wu, Y. Zhao, J. Wei, *Nanoscale* 2020, *12*, 6096–6103.
111. L. Henderson, O. Neumann, C. Kaffes, R. Zhang, V. Marangoni, M. K. Ravoori, V. Kundra, J. Bankson, P. Nordlander, N. J. Halas, *ACS Nano* 2018, *12*, 8214–8223.
112. A. Patel, D. Asik, E. M. Snyder, A. E. Dilillo, P. J. Cullen, J. R. Morrow, *ChemMedChem* 2020, *15*, 1050–1057.
113. R. Qin, Y. Feng, D. Ding, L. Chen, S. Li, H. Deng, S. Chen, Z. Han, W. Sun, H. Chen, *ACS Appl. Bio Mater.* **2021**, *4*, 5520–5528.
114. I. Tirotta, V. Dichiarante, C. Pigliacelli, G. Cavallo, G. Terraneo, F. B. Bombelli, P. Metrangolo, G. Resnati, *Chem. Rev.* **2015**, *115*, 1106–1129.
115. A. A. Kislukhin, H. Xu, S. R. Adams, K. H. Narsinh, R. Y. Tsien, E. T. Ahrens, *Nat. Mater.* **2016**, *15*, 662–668.
116. C. Wang, S. R. Adams, H. Xu, W. Zhu, E. T. Ahrens, *ACS Appl. Bio Mater.* **2019**, *2*, 3836–3842.

# 12 Advances in PET/MRI and Probe Development for Biomedical Precision Imaging Applications

*Jan Kretschmer*
Department of Preclinical Imaging and Radiopharmacy, Werner Siemens Imaging Center, Eberhard Karls University Tübingen, Röntgenweg 13/1, D-72076 Tübingen, Germany
Cluster of Excellence iFIT (EXC 2180) "Image-Guided and Functionally Instructed Tumor Therapies", Eberhard Karls University Tübingen, Röntgenweg 13/1, D-72076 Tübingen, Germany

*Juan Pellico*
School of Biomedical Engineering & Imaging Sciences, King's College London, 4th Floor, Lambeth Wing, St. Thomas Hospital, London, SE1 7EH, UK

*Angelina Prytula-Kurkunova*
Department of Preclinical Imaging and Radiopharmacy, Werner Siemens Imaging Center, Eberhard Karls University Tübingen, Röntgenweg 13/1, D-72076 Tübingen, Germany
Cluster of Excellence iFIT (EXC 2180) "Image-Guided and Functionally Instructed Tumor Therapies", Eberhard Karls University Tübingen, Röntgenweg 13/1, D-72076 Tübingen, Germany

*Rafael Torres Martin De Rosales*
School of Biomedical Engineering & Imaging Sciences, King's College London, 4th Floor, Lambeth Wing, St. Thomas Hospital, London, SE1 7EH, UK
rafael.torres@kcl.ac.uk

DOI: 10.1201/9781003374688-12

*Andre Ferreira Martins*

Department of Preclinical Imaging and Radiopharmacy, Werner Siemens Imaging Center, Eberhard Karls University Tübingen, Röntgenweg 13/1, D-72076 Tübingen, Germany

Cluster of Excellence iFIT (EXC 2180) "Image-Guided and Functionally Instructed Tumor Therapies", Eberhard Karls University Tübingen, Röntgenweg 13/1, D-72076 Tübingen, Germany

German Cancer Consortium (DKTK), partner site Tübingen, German Cancer Research Center (DKFZ), Im Neuenheimer Feld 280, D-69120 Heidelberg 69120, Germany

andre.martins@med.uni-tuebingen.de

## CONTENTS

**Abstract**

This chapter provides a detailed exploration of advanced imaging techniques, with a specific focus on simultaneous PET/MRI applications. It begins by highlighting the distinctive features of various imaging modalities, emphasizing the unique capabilities of PET/MRI in integrating functional and anatomical information. The discussion moves into the realm of multimodal imaging, particularly molecular imaging, and its significant role in precision medicine, coupled with the integration of artificial intelligence. The focus narrows down to the technical intricacies of PET/MRI systems, outlining their design, challenges, and clinical applications. A dedicated section explores the fundamental concepts of PET, MRI, and contrast agents, followed by an in-depth analysis of the rationale and challenges associated with bimodal PET/MRI contrast agents, comparing small molecules and nanoparticles. The chapter concludes with a detailed examination of current preclinical applications of PET/MRI contrast agents in cardiovascular scenarios, infection and inflammation imaging, oncology, and neuroimaging. Examples and breakthroughs in each area underscore the potential impact of bimodal PET/MRI agents in advancing diagnostics and therapeutic monitoring. In summary, this chapter serves as a comprehensive guide to advanced imaging, offering valuable insights for researchers and clinicians in the dynamic landscape of PET/MRI applications in medicine.

## KEYWORDS

PET/MRI probes; PET/MRI applications; PET/MRI theranostics; PET/MRI technology

## 1 ADVANCED IMAGING TECHNIQUES

Advanced imaging techniques enable the acquisition of anatomical and/or functional information of a subject by responding to specific internal or external triggers. Imaging modalities such as computed tomography (CT), ultrasound (US), and magnetic resonance imaging (MRI) harness intrinsic mechanisms to generate imaging contrast, eliminating the need for external substances (contrast agents), unlike nuclear imaging techniques such as positron emission tomography (PET), single-photon emission computed tomography (SPECT), or optical imaging (OI), which rely on exogenous substances.

Each of these imaging techniques possesses distinct strengths and limitations concerning signal acquisition and post-processing mechanisms, which in turn determine their spatial resolution, depth penetration, and sensitivity—key properties for every imaging technique. Our focus in this chapter will be on PET/MRI. Nevertheless, Table 1 shows a summary of the properties of the current advanced imaging modalities, along with their relative costs. Furthermore, a brief description on the status of each technique is provided in the next section.

**TABLE 1**
**Summary of the Properties of the Main Imaging Modalities**

| Imaging Technique | Spatial Resolution | Depth Penetration | Sensitivity | Relative Cost |
|---|---|---|---|---|
| MRI | ≤0.1 mm (PC)<br>1–2 mm (C) | No limit | μM–mM | €€€ |
| CT | ≤0.2 mm (PC)<br>0.5–1 (C) | No limit | mM | € |
| US | 1–2 mm (PC)<br>≤0.1 mm (C) | Several cm | ~μM | € |
| OI | 5 mm | mm–cm | pM–nM | € - €€€ |
| PAI | ≤0.1 mm | Several cm | pM | € |
| SPECT | 0.5–2 mm (PC)<br>5–12 mm (C) | No limit | <pM | €€ |
| PET | 1–2 mm (PC)<br>3–6 mm (C) | No limit | fM | €€€ |

*Source:* Adapted from Ref. [1].
MRI: Magnetic Resonance Imaging; CT: Computed Tomography; US: UltraSound;
OI: Optical Imaging; PAI: Photoacoustic Imaging; SPECT: Single Photon Emission Computed Tomography; PET: Positron Emission Tomography.
PC=preclinical scanner; C=clinical scanner.

## 2 CURRENT STATUS OF MONOMODAL IMAGING TECHNIQUES

Currently, the most common monomodal imaging techniques used in medical imaging are X-ray, CT, MRI, PET, and ultrasound. Each of these techniques has properties suited for specific purposes. While X-ray and CT are great for the determination of hard tissue injuries, MRI provides a remarkable soft tissue resolution. On the other hand, PET has limited resolution but offers unprecedented functional information at the molecular level. The main advantages and disadvantages are summarized in Table 2.

In terms of economic costs, sonography and X-rays are the most accessible and hence the most widespread procedures; in contrast, the price of an MRI scanner can be as high as 3 million USD, making it hardly accessible.

The field of imaging is developing rapidly, as can be seen in current trends. Most notably, artificial intelligence (AI) using deep learning for the processing and analysis of imaging data is being employed to provide accurate data analysis and forecast the treatment outcome [2–4].

## 3 MULTIMODAL IMAGING TECHNIQUES

The dynamic field of molecular imaging is revolutionizing our understanding of molecular processes [5–8]. With its precision in targeting receptors and cellular

**TABLE 2**
**Most Commonly Used Monomodal Imaging Techniques in Medicine and Their Characteristics**

| Imaging Technique | Source of Signal | Advantages | Disadvantages |
|---|---|---|---|
| Ultrasound | Sound waves with very high frequencies (3–10 MHz) | Inexpensive, availability, no exposure to ionizing radiation, real-time imaging | Low resolution, limited tissue penetration |
| X-Ray | High energy electromagnetic radiation (20 keV–150 keV) | Quick and inexpensive, availability | Radiation burden, limited soft tissue contrast |
| CT | High energy electromagnetic radiation (20 keV–150 keV) | Quick, high resolution images | Radiation burden, limited soft tissue contrast |
| PET | Photons produced by annihilation event of positron | Functional imaging of even molecular processes, whole body imaging | Radiation burden, poor anatomical detail |
| MRI | Magnetic field (most commonly 1.5 or 3 T) and radiofrequency pulse | Excellent soft tissue resolution, variety of MRI techniques, deep tissue penetration, no radiation burden | Price, long scan times, contraindications (implants) |

mechanisms, molecular imaging is indispensable in precision medicine. It offers real-time, quantitative insights into the molecular intricacies of diseases, from cardiovascular issues to inflammation and cancer. Consequently, there has been a push toward developing integrated systems that combine molecular, functional, and anatomical information in precision imaging. The use of AI in medical imaging is directly connected to quantitative imaging for obtaining quantitative, comprehensive data that can be used for diagnosis, monitoring, and treatment planning by translating images into quantitative data [9,10]. This approach is pivotal for a comprehensive grasp of complex biological phenomena [11].

## 3.1 PET/CT

The PET/CT system consists of two separate consecutively installed devices with an integrated bed that allows the patient to be moved during the examination. The standard PET/CT protocol initiates with a comprehensive "scout" whole-body scan to select the scanning area precisely. Following this, a tomography is performed, accompanied by the injection of a PET radioactive tracer. The final phase of the examination entails a dedicated PET scan (Figure 1) [12]. Following data collection, the user corrects and reconstructs the CT data to generate an image.

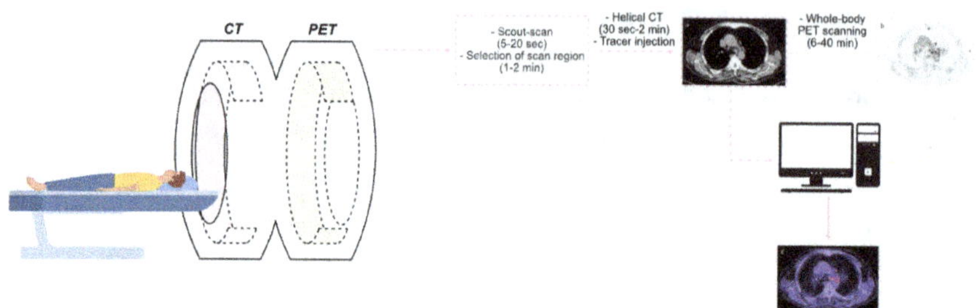


**FIGURE 1** Illustration of PET/CT workflow. (Adapted from Triviño-Ibáñez et al. [14].)

Attenuation correction is then executed utilizing the CT image, which is downscaled to PET resolution and transformed (scaled using a bilinear or hybrid method) to 511 keV, representing the rest of the energy of an electron or positron. The converted image is projected into the PET data format and used to correct the original PET emission data. The attenuation-corrected PET data is reconstructed on a PET processor. Finally, the images are overlaid and displayed on a common console [13].

Arguably, the main limitation on PET/CT is the low soft tissue contrast of CT. This has led to the development of hybrid PET/MRI systems where the high spatial resolution of MRI overcomes the low resolution of CT.

## 3.2 PET/MRI

Although the concept of hybrid PET/MRI was developed earlier than PET/CT, its implementation was delayed due to technical problems associated with the integration of PET within an intense magnetic field [15,16] and the high cost associated.

Currently, commercially available systems include consistent models like the ingenuity TF PET/MR system and fully integrated systems. These systems vary in the materials utilized for the photomultiplier and scintillation detectors within the PET component and in the order of their interconnection and the device's electronics (Figure 2).

Recently, there has been a consensus recommendation for the use of PET/MRI in oncology [17]. However, unified standards or guidelines for PET/MRI are scarce. As a result, the protocols for conducting examinations vary, and there are differences in the scan times and sequence of steps performed during the scan. This variability is dependent on the types of tissues and diseases under investigation [18]. For example, neurology MRI protocols included diffusion-weighted imaging [19,20], fluid attenuation inversion recovery, and $T_1$ and $T_2$ acquisition [21]. For the investigation of tumor imaging, diffusion-weighted imaging frequently included Dixon or volumetric interpolated breath-hold examination sequences [22,23].

### 3.2.1 Current Implementation of Clinical PET/MRI Scanners

From a technical point of view, the design of a PET insert for preclinical imaging requires surmounting multiple challenges due to the high static magnetic field, the

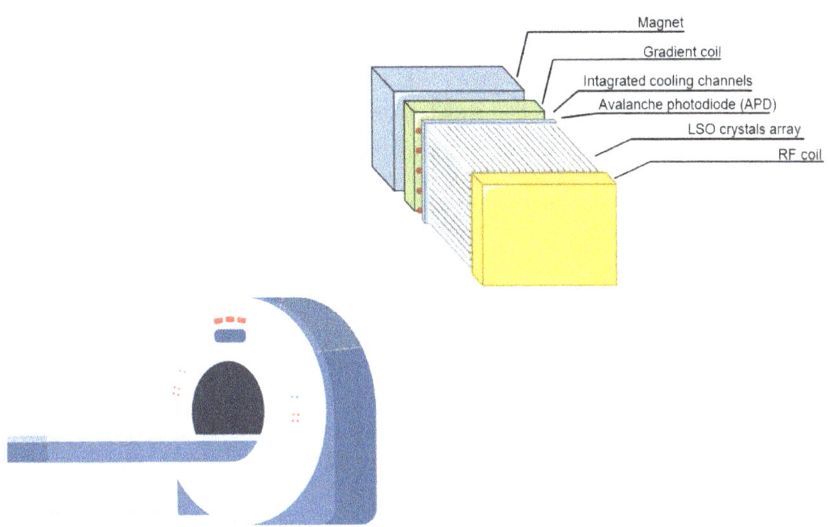


**FIGURE 2** Example of the integrated commercially available system "Biograph mMR".

fast switching of the gradient coils, and the interaction of the radiofrequency field with the PET electronics. One of the significant hurdles stands in the development and implementation of PET/MRI scanners in the clinic, and ways to solve them are discussed in the following points:

1. *Design and implementation of MR-based attenuation correction.* Attenuation correction, an essential part of PET images for PET/CT, is performed using CT data based on X-ray absorption. It cannot be used for PET/MRI, as MRI image intensity is not directly related to tissue attenuation. Instead, the methods rely on tissue segmentation, i.e., obtaining different classes of tissue types, which are then assigned a specific attenuation factor, or Atlas-based approaches [24–26].
2. *The influence of MRI on PET.* The combination of MRI technology and traditional PET technology, which was based on coupling bismuth oxide-based scintillation crystal detectors with photomultiplier tubes (PMTs), is not possible due to their sensitive to the strong magnetic field generated by the superconducting primary electromagnetic coil [27]. The previous generation of photomultiplier tubes used avalanche photodiodes (APDs), but they proved imperfect as they were thermally sensitive and required additional electronic signal amplification. This problem has been solved in a new generation of PET detectors—silicon photomultipliers (SiPM), which is essentially a silicon APD with a slightly modified structure, a matrix of integrated independent single-photon avalanche diodes (SPADs) with individual quenching resistors and

readout electronics, providing high resolution, high sensitivity, and lower magnetic susceptibility [28,29].

3. *The influence of PET on MRI.* PET detectors caused interference in MR magnetic field homogeneity, radiofrequency, and gradient systems. At the same time, magnetic field strengths above 5 Tesla (T) can further improve the quality of PET images by reducing the average free path of positrons between annihilation and emission [30,31]. To date, studies have also confirmed the effect of the magnetic field on positron emitters and show that the improvement, in contrast with increasing field strength, increases exponentially rather than linearly. Thus, images obtained with 9.4 T MRI have significantly better spatial resolution [32]. What provides even more opportunities for multidimensional visualization [33] and is separately used in research neuroimaging and cognitive neuroscience is the blood oxygenation level-dependent functional MR method (BOLD-fMRI-PET) [34]. Currently, the possible implementation of ultra-high field MRI technology [35,36] in clinical trials is only expected. Commercially available for human scanning are devices with a magnetic field strength of 7 T, but today, they retain their experimental status, and with 9.4 T, there are only small devices for animals up to 3 kg (MRS*PET/MR 9.4 T, BioSpec 94/20 PET/MR systems), while the clinic devices with a maximum magnetic field strength of 4 T are approved for use [37]. Only recently, the International Electrotechnical Commission approved (IEC 60601-2-33) an increase in the use of static magnetic fields in MRI up to 8 Tesla (first-level controlled mode).

Finally, there exist technologies that are complemented by modern commercially available devices. "Time of flight" (TOF-camera) technology, which is usually installed in PET machines, was integrated for the first time in the 3 T PET/MR scanner (Signa PET/MR system), which allows for non-contact and marker-free calculation of the multidimensional respiratory signal in different locations (e.g., chest and abdomen). TOF cameras provide better detection of lesions [38], shorter scan times or reduced dose [39], and more accurate measurement of uptake by the affected area. This is especially important for obtaining radiation data for upper body tumors and reducing artifacts.

## 4 PET/MRI CONTRAST AGENTS

### 4.1 Basic Concepts

It is important to clarify some terminology and review the current doses administered for PET, MRI, and PET/MRI contrast. The source of the signal in PET is the PET radiotracer. The PET radiotracer contains a radionuclide that decays, emitting positrons. A positron subsequently produces two orthogonal photons in interaction with an electron in a process known as annihilation, which are detected by the PET scanner to create the image. On the other hand, MRI often uses contrast agents to enhance the signal. A contrast agent is, by definition, *a substance used to increase the contrast of structures of fluids within the body in medical imaging* [40]. Most of the clinically used MRI contrast agents fall into the category of GBCAs (gadolinium-based contrast agents), whose contrast enhancement stems from the paramagnetic metal gadolinium, which affects the relaxation rate of surrounding water molecules.

While the dose approved for the clinical use of GBCAs is typically between 0.1 and 0.3 mmol $kg^{-1}$ body weight [41], the dose recommended for [$^{18}$F]FDG ([$^{18}$F]fluorodeoxyglucose), the most used PET radiotracer for PET imaging, varies between 185 and 370 MBq in adults [42]. A lower dose of 96.2 MBq is used for pediatric patients. In PET/MRI examinations, the PET tracer has always been present to yield the image from the PET modality, while contrast enhancement with an MRI contrast agent is optional. However, PET/MRI examinations are routinely done in the presence of an MRI contrast agent, as can be seen in the case of PET/MRI examinations in brain oncology, where more than 50% of examinations in the period from 2018 to 2021 were performed with an MRI contrast agent [43]. The dose of GBCAs provided to patients generally ranges from 0.05 mmol $kg^{-1}$ (gadoterate meglumine) to 0.1 mmol $kg^{-1}$ (gadopentetate dimeglumine) [44,45]. For the [$^{18}$F]FDG, the administered doses range from 3 to 5 MBq $kg^{-1}$ and 153 to 503 MBq, respectively [46–49]. However, recently, new approaches to lower the radiation burden for pediatric patients during PET/MRI examinations have been investigated, as current dosing guidelines recommend 3.5–5.3 MBq $kg^{-1}$ of [$^{18}$F]FDG [50].

## 4.2 Rationale and Challenges for Bimodal PET/MRI Contrast Agents

### 4.2.1 Small Molecules

The availability of bimodal PET/MRI contrast agents with low molecular weight is quite limited, primarily due to the challenges involved in their preparation. The combination of PET and MRI modalities presents a specific challenge: creating a contrast agent where the MRI component needs to be present in a significantly larger excess ($10^6$–$10^7$ times) compared to the PET component. To address this concentration scaling challenge for low molecular PET/MRI bimodal contrast agents, the solution is to prepare a mixture that combines a radioactive (hot) tracer with a substantial excess of its non-radioactive (cold) counterpart, as shown in Figure 3.

Additional challenges must be addressed in preparing bimodal PET/MRI contrast agents, not limited to the differing levels of the PET and MRI components. As previously stated, the PET co-agent and MRI contrast agent need to be ideally the same molecule, with the only difference being the isotope triggering the PET or MR response. For example, Frullano et al. reported a bimodal contrast agent $^{18}$F/Gd PET co-agent and MRI $^{19}$F/Gd with $^{18-19}$F triggering the response in PET and MRI, respectively [51]. This approach is advantageous because it

**FIGURE 3** A simplified representation shows a mixture of the radioactive (hot) and non-radioactive (cold) counterparts of the PET/MRI bimodal tracer, highlighting the significant excess of the MRI component over the PET component. From a chemical viewpoint, these two components exhibit identical physiochemical behavior, differing in their isotope.

allows for the straightforward preparation of both "cold" and "hot" counterparts. This convenience also lies in the widespread availability and well-documented chemistry of $^{18}F$ [52]. Another interesting approach is using two isotopes of the same element. This case has been approached using $^{52}Mn$ as the PET component and $^{55}Mn$ as a paramagnetic metal [53,54]. This strategy ensures that both molecules in the PET/MRI bimodal "cocktail" possess similar physiochemical properties, as their coordination environment and kinetic inertness should be nearly identical. However, bimodal PET/MRI agents are often obtained by combining two different elements. The most common combinations are $^{68}Ga$/Gd and $^{64}Cu$/Gd, respectively [55,56]. These combinations are well-suited for PET/MRI, but they result in a mixture of two non-identical entities or molecules. 1,4,7,10-Tetraazacyclododecane-1,4,7,10-tetrayl)tetraacetic acid (DOTA) molecule is the frequent choice as a chelating agent for low molecular bimodal PET/MRI contrast agents. Gd-DOTA, specifically DOTAREM, is the most stable MRI contrast agent used clinically [57]. DOTA provides a thermodynamic and kinetically stable complex of coordination 9 with $Gd^{3+}$ (Figure 4). However, the situation is quite different in the coordination with PET radionuclides as $^{68}Ga$ or $^{64}Cu$. The Ga(III) preferred coordination number is 6. Hence, it is not unexpected that the complexation with DOTA, tailored explicitly for lanthanides, will result in a structure where two pendant arms are exposed and not coordinated (Figure 4). This phenomenon yields a pair of non-equivalent molecules where one ($^{68}Ga$-DOTA complex) has different physiochemical properties due to the two uncoordinated acetate pendant arms.

The use of DOTA for $^{68}Ga$ chelation presents another critical challenge, namely the conditions for labeling. Typically, labeling procedures occur at high temperatures and a pH of 4.6. This can pose significant problems when using targeted probes containing biomolecules, as they are likely to undergo decomposition during the process. To address this issue, 1,4,7-triazacyclononane-1,4,7-triacetic acid (NOTA) macrocycles can be employed, allowing for the rapid chelation of $^{68}Ga$ at room temperature [58].

Another important issue is the difference in stability of the DOTA complexes depending on the (radio)metals that can lead to different pharmacokinetics *in vivo*.

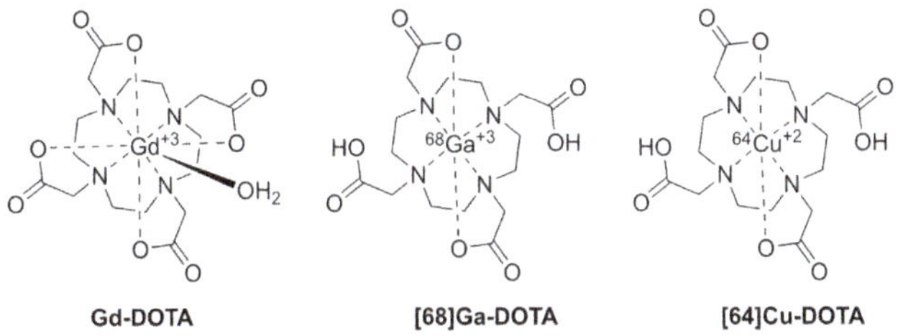

Gd-DOTA [68]Ga-DOTA [64]Cu-DOTA

**FIGURE 4** Structures of Gd-DOTA, [$^{68}$]Ga-DOTA, and [$^{64}$]Cu-DOTA.

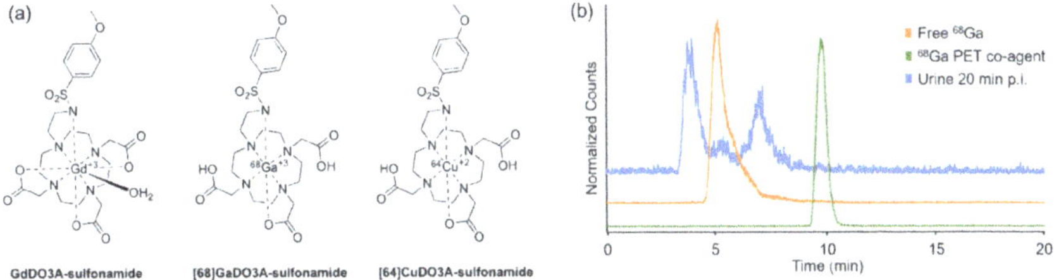


**FIGURE 5** (a) PET/MRI pH-responsive contrast agents. Structures of pH-responsive MRI contrast agents and PET-co agents were used in the study. (b) HPLC radioactivity chromatograms of PET co-agents and free $^{68}$Ga. (Figure reproduced with permission from [55].)

Pollard et al. observed adequate stability of the Gd-DOTA complex but a significant degradation of the $^{68}$Ga-DOTA PET co-agent *in vivo* (Figure 5a and b) [55]. The authors prepared the $^{64}$Cu version of the same co-agent to address this issue. However, poor *in vivo* stability was also observed with a complete de-chelation of $^{64}$Cu from the PET co-agent confirmed by the urine analysis.

The same phenomenon was observed in a study conducted by Uppal et al., where a bimodal PET/MRI contrast agent for thrombus targeting was created from the fibrin-targeting peptide EP-2104R. This was achieved by partially removing gadolinium from Gd-DOTA units by DTPA, followed by the complexation of the free units with $^{64}$Cu to produce the bimodal PET/MRI contrast agent. *Ex vivo* biodistribution analysis revealed the highest levels of gadolinium in the kidneys, while the highest levels of $^{64}$Cu were also observed in the kidneys, with significant accumulation in the liver. Interestingly, the percentage of the injected dose per gram of tissue for $^{64}$Cu exceeded that of gadolinium by a factor of 30. Notably, $^{64}$Cu also exhibited significantly higher accumulation (two- to fivefold) in the spleen, intestine, and stomach [56]. This suggests that $^{64}$Cu may have undergone metabolism by liver enzymes, possibly binding to superoxide dismutase [59,60]. It can also be speculated how the non-selective decomplexation of Gd-DOTA units affected the overall pharmacokinetics, as the mixture likely consisted of molecules with non-identical numbers of $^{64}$Cu atoms in various positions.

Notably, the challenges related to kinetic inertness and chelate stability are not only limited to systems involving different metals. Brandt et al. developed a bimodal PET/MRI contrast agent using the $^{52g}$Mn/$^{55}$Mn pair and the trans-CDTA (*trans*-1,2-diaminocyclohexane-*N,N,N′,N′*-tetraacetic acid) chelator [53]. Considering that both MRI and PET components of the pair are chemically identical, one would expect them to exhibit the same *in vivo* pharmacokinetics. However, limitations in the stability of the [Mn(CDTA)] complexes were found. In previous literature, the same authors showed good radiochemical stability after the incubation of [[$^{52g}$Mn]Mn(CDTA)]$^{2-}$ in human blood serum. However, subsequent radiocomplex analysis revealed a significant degree of decomplexation, highlighting the relevance of producing kinetically stable complexes [54].

Generally, the preferred excretion pathway for contrast agents is renal excretion through the kidneys into the bladder unless the contrast agent is intentionally designed for hepato-biliary excretion. This is particularly crucial for potential bimodal PET/MRI contrast agents, as prolonged retention leads to an undesirable increase in radiation exposure for the patient, resulting in an elevated effective dose. The same principle applies to the accumulation in any organ.

### 4.2.2 Nanoparticles

Different considerations must be considered when designing nanoparticles for use as PET/MRI contrast agents. One crucial yet often underestimated aspect is the difference in biodistribution between the nanoparticle and the free/unchelated radionuclide. Some PET radionuclides exhibit similar biodistribution patterns in their unchelated form as nanoparticles, such as [$^{64}$Cu]$CuCl_2$, [$^{68}$Ga]$GaCl_3$, and [$^{52}$Mn]$MnCl_2$, leading to potential misinterpretation of imaging signals when the free radionuclide signal is attributed to the nanoparticles [61–63]. The choice of radionuclide is frequently based on its availability within a specific facility rather than its suitability for the imaging application. However, understanding the radionuclide biodistribution can aid in selecting the most appropriate radionuclide for the specific application, minimizing signal overlap between the free radionuclide and the radiolabeled nanomaterial. Different biodistribution also introduces another critical concept: the radiochemical stability (RCS). RCS measures the strength of the bond between the nanoparticle and radionuclide after the radiolabeling reaction. It indicates whether the radionuclide will detach from the nanoparticle when injected *in vivo*. Therefore, evaluating this parameter is essential in every case, with a high RCS (indicative of a stable radionuclide–nanoparticle conjugate) being a prerequisite for further use [64].

When considering the concentration of material required to produce efficient signals in both PET and MRI, combining nanoparticles with a radionuclide allows for appropriate doses to achieve optimal *in vivo* responses. A PET emitter is incorporated into nanoparticles in trace amounts, involving only a few radioactive atoms integrated into the nanoparticles. This minimal incorporation, while preserving the integrity of the nanoparticles, proves sufficient to generate a suitable signal due to the high sensitivity of PET. Furthermore, maintaining the integrity of the nanoparticle after radiolabeling is crucial to preserve the pharmacokinetics and contrast capabilities in MRI. The decay of specific metal radionuclides into stable isotopes can generate new metals within the nanoparticles. Hence, these new metals have the potential to impact the relaxivity of the nanoparticles, which might affect their effectiveness as contrast agents in MRI. On the other hand, PET/MRI nanoparticles are, in most cases, developed from paramagnetic and superparamagnetic nanoparticles such as $Fe_3O_4$, $Fe_2O_3$, $Gd_2O_3$, $Mn_3O_4$, $HoF_3$, or $Dy_2O_3$ that generate a suitable contrast enhancement in MRI [65]. The presence of the magnetic atoms in the nanoparticles is undoubtedly significant as they form the crystal lattice of the nanoparticles. Therefore, the large proportion of magnetic atoms per nanoparticle allows moderate doses to enhance contrast in MRI.

### 4.2.3 Quantitative Imaging

Quantification plays a fundamental role in medical imaging as it is crucial for accurate diagnosis, effective treatment planning, research advancements in precision imaging, and overall improvement of patient care. However, both PET and MRI face particular challenges regarding quantification and precision [66–68]. The quantification challenges in PET include partial volume effects due to limited spatial resolution [69], attenuation correction due to variations in tissue density that can lead to errors in SUV [24], and motion artifacts influenced by respiratory or cardiac motion [70]. Meanwhile, challenges in quantitative detection for MRI include susceptibility artifacts in tissues due to magnetic susceptibility and motion effects [71], which can lead to artifacts in MRI, affecting the accuracy of anatomical delineation and potentially impacting quantification, and B0 and B1 inhomogeneities that can lead to non-uniform images [72].

These challenges are particularly relevant in precision and molecular imaging when using contrast agents to detect relevant biomarkers of a disease accurately. Hybrid PET/MRI systems offer a synergistic solution for quantifying the contrast media *in vivo*. For example, attenuation correction can be applied to PET using MRI Dixon maps [73]. MRI contrast agents can be tracked and quantified by PET, improving the quantification of physiological and metabolic biomarkers and reducing false positives and negatives [66,74]. Furthermore, the hybrid PET/MRI detection of contrast media allows for multiplex and multiparametric data analysis, including perfusion, metabolism, and structural information, enhancing diagnostic accuracy and providing a platform for precision imaging [75–77].

## 4.3 Small Molecules *vs* Nanoparticles

### 4.3.1 Small Molecules

Most compounds used as PET tracers or MRI contrast agents are small molecules. The only clinically used nanoparticle MRI diagnostic is off-label use of Ferumoxytol [78], which the FDA approved for anemia treatment. Conversely, most published examples of PET/MRI bimodal contrast agents are nanoparticle-based. This reflects that the preparation, purification, and formulation of nanoparticle-based bimodal PET/MRI contrast agents are more rapid and straightforward. However, there are undeniable disadvantages to the small-molecule-based bimodal PET/MRI contrast agents. The main disadvantage is the biodistribution and pharmacokinetics of nanoparticles, where their large size results in liver and spleen accumulation and hepatobiliary excretion, prolonging the radiation burden on the examined organism. Another disadvantage is the resulting MRI contrast provided by these nanoparticles, as most examples are based on SPIOs (superparamagnetic iron oxides), which yield $T_2$ contrast, darkening the image.

On the other hand, their easy preparation and modification make them great scaffolds for parallel drug delivery and theranostic use. Moreover, unless in the case of the small-molecule PET/MRI bimodal contrast agents, their labeling can be performed without a chelator. Another disadvantage that balances the ease of preparation and purification is the reproducibility of preparation between batches. One must take extra caution to keep all the parameters the same, especially the nanoparticle's size and degree of labeling. These precautions are not important in the case of small-molecule-based agents because they possess a well-defined and easy-to-control chemical structure. Also, the small molecules have well-known pharmacokinetics with preferred renal excretion through the kidneys to the bladder, where this way of excretion minimizes the radiation burden on the examined organism. However, unless in the case of nanoparticles, the PET/MRI agent is not present as a single entity (single particle) but as a "cocktail" of MRI and PET co-agents because of the vast difference in needed concentrations for the PET and MRI modality. This results in a more demanding design and synthesis of low-molecular PET/MRI contrast agents, where one must always design and synthesize both pairs of the co-agents.

### 4.3.2 Nanoparticles

In the past decade, there has been a significant rise in using nanoparticles as contrast agents for molecular imaging [79]. This surge in popularity can be attributed to the size-dependent properties of these materials. One crucial characteristic of nanoparticles is their large surface-to-volume ratio, allowing for a specific surface area that facilitates the incorporation of multiple bioactive species, imaging tags, and therapeutic drugs within a single material. This unique property grants nanoparticle multifunctional capabilities and enables precise control over the pharmacokinetics of the tracer, a level of control challenging to achieve with small-molecule contrast agents. Generally, nanoparticles larger than 5–6 nm are quickly engulfed by macrophages (Kupffer cells) in the liver and spleen, a process that timely depends on the adequate particle size upon particle interaction with blood proteins (resulting in protein corona formation) [80]. Hence, particle size and surface composition significantly influence nanoparticles' pharmacokinetics. Particle size can be easily manipulated during particle preparation, while surface composition can be adjusted by conjugating specific vectors and fouling/anti-fouling compounds onto the particle surface. The simpler and quicker preparation protocols are another potential advantage of nanoparticles over small molecules. Due to their multifunctionality, PET/MRI nanoparticles can be synthesized with high yields in 1–2 steps. Moreover, their nanoscale dimensions facilitate rapid and reliable size-exclusion purification techniques, eliminating the need for the more complicated and time-consuming traditional methods employed with small molecules.

The use of nanoparticles as PET/MRI tracers also presents potential drawbacks. First, there are inherent concerns about the toxicity of nanoparticles when employed in human subjects. This concern is significantly heightened in the case of inorganic magnetic nanoparticles such as $Fe_3O_4$, $Mn_3O_4$, $MnO_2$, $Dy_2O_3$, and $Gd_2O_3$, which have limited examples in clinical practice thus far [81]. Additionally, fully characterizing a nanoparticle-based tracer, quantifying active biomolecules on the particle, and determining the appropriate dosage for administration remain challenging. Furthermore, the administration of nanoparticles often leads to significant accumulation in the liver and spleen for prolonged periods regardless of the composition and pharmacokinetics of the particles. The next section will discuss synthetic approaches to synthesize bimodal PET/MRI agents.

## 4.4 Preparation of Bimodal PET/MRI Contrast Agents

### 4.4.1 Low Molecular Bimodal PET/MRI Contrast Agents

Due to the limited availability of low-molecular bimodal PET/MRI contrast agents in the literature, the preparation of each one will be further described in chronological order.

**[$^{18}$F]GdDOTA-4AMP-F:** The first example of a pH-responsive PET/MRI contrast agent was presented by Frullano et al. by attaching a pH-responsive MRI contrast agent Gd-DOTA-4AmP with radiolabeled [$^{18}$F] Fluoroethylazide using CuAAC (copper(I)-catalyzed alkyne-azide cycloaddition (Figure 6). Due to the different levels of concentration demanded by both PET and MRI, a non-radioactive variant with the stable $^{19}$F isotope was also synthesized. Phantom measurements were performed, and images from both modalities were obtained. Total radiosynthesis time, from radiolabeling of azido-fragment to click reaction, took 4 h, and radiochemical yield was 0.6% (product activity of 10.2 kBq starting from 1.80 GBq) [51]. The total radiosynthesis time and radiochemical yield

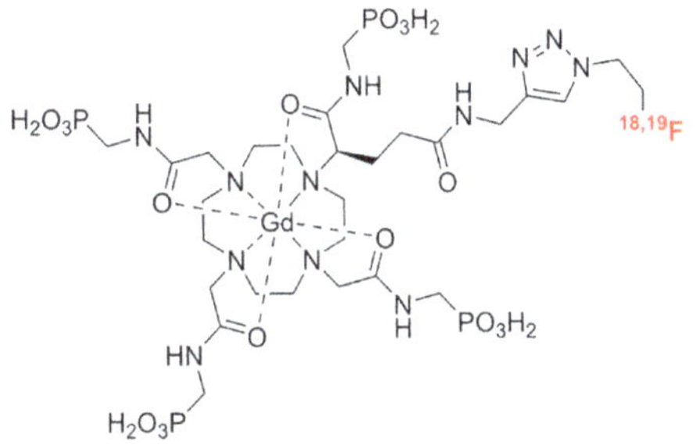

**FIGURE 6** The structure of the bimodal PET/MRI contrast agent published in the pioneering work of Frullano et al. [51].

demonstrate how hard and demanding the preparation of a bimodal PET/MRI low molecular contrast agent is.

**[$^{64}$Cu]GdEP-2104R:** Another approach to achieving the bimodal PET/MRI contrast agent is combining the radioactive and paramagnetic metals as a PET signal source and for MRI contrast. The first published case from Uppal et al. was realized by exchanging a fraction of gadolinium ions for $^{64}$Cu in the fibrin-targeted probe EP-2104R (Figure 7) for further application in thrombi detection in rats. These studies showed that detecting the bimodal probe in the thrombus was feasible [56]. This approach, although with positive results, has its limitations. The structural motif of the DOTA chelator is not ideal for the chelation of the $^{64}$Cu due to the low kinetic inertness of the Cu-DOTA chelate. This inadequacy can lead to *in vivo* $^{64}$Cu leaching [60]. This happened in this work, where the authors observed the deposition of unchelated $^{64}$Cu in the liver, spleen, intestine, and stomach. In addition, this synthetic approach relies on uncontrolled radiolabeling that may lead to products with a different number of ligands chelated, which may result in multiple products with different pharmacokinetics.

**[$^{68}$Ga]TRAP(HMDA–GdDOTA)3:** Notni et al. reported a bimodal PET/MRI probe using a gallium selective TRAP chelator (3,3′,3″-((1,4,7-triazonane-1,4,7-triyl)tris(methylene))tris(hydroxyphosphoryl))-tripropanoic acid) (Figure 8). This TRAP chelator acted as a core for binding three DOTA units. First, the chelator was fully complexed with gadolinium ions and then treated with DTPA (2-[bis({2-[bis(carboxymethyl)amino]ethyl})amino]acetic acid) to remove the gadolinium from the central TRAP chelator. The partially chelated product was subsequently complexed with the $^{68}$Ga and combined with the non-radioactive variant (containing natural non-radioactive Ga) in a $2.5 \times 10^7$ excess over the hot compound. This blend was later injected into mice, and the signals from MRI and PET were simultaneously acquired [82].

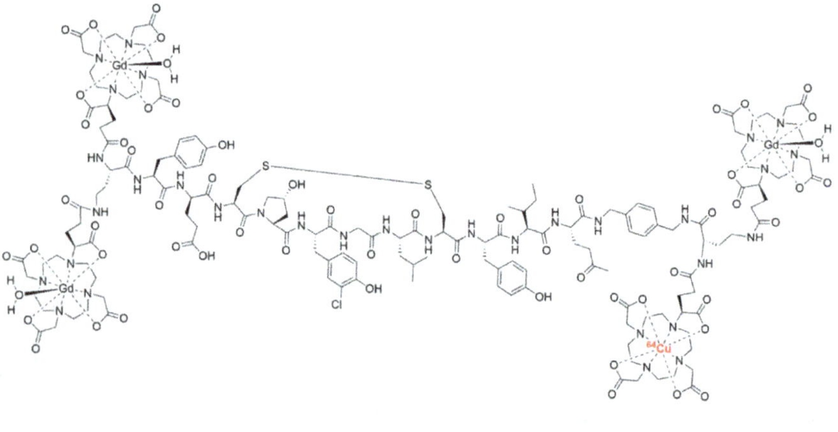

**FIGURE 7** Chemical structure of fibrin targeted bimodal PET/MRI probe prepared from EP2104R by fractional exchange of Gd for $^{64}$Cu.

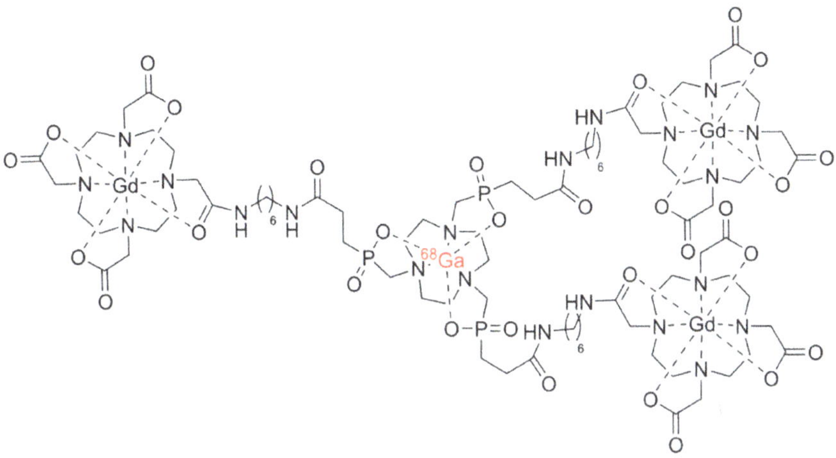

**FIGURE 8** Structure of the bimodal PET/MRI contrast agent prepared by Notni et al. The contrast agent consists of a selective Ga(III) TRAP moiety with three gadolinium chelates connected by pendant arms. (Reproduced with permission from Ref. [82] Copyright 2013, Wiley-VCH.)

**[$^{68}$Ga]Gd6LGa:** Kumar et al. synthesized a dendrimeric bimodal PET/MRI probe consisting of six Gd-DOTA moieties and one central NOTA $^{67/68}$Ga-chelating moiety (Figure 9) [83]. Initially, the probe was fully chelated with gadolinium ions. Then, an excess of DTPA was used to de-chelate the central NOTA moiety. Initially, the central NOTA moiety was complexed with $^{67}$Ga for SPECT imaging to determine the pharmacokinetics and biodistribution profile of the probe. The probe was primarily excreted through the kidneys, with prolonged uptake in the kidneys (43.0±5.1%ID/g at 24h post-injection), likely due to the free amino groups on the agent. In the subsequent experiment, the c(RGDyK), an $\alpha_v\beta_3$ integrin targeting peptide, was conjugated to the chelator and labeled with $^{68}$Ga for successful PET/MRI imaging of mice bearing integrin $\alpha_v\beta_3$ positive U87MG tumors.

**[[$^{52g}$Mn]Mn(CDTA)]$^{2-}$:** Due to concerns regarding gadolinium deposition and toxicity, research in the field of manganese-based MRI contrast agents has gathered increased attention. Vanasschen et al. have contributed to this field by developing a bimodal PET/MRI contrast agent through a rapid and convenient method. They employed an isotopic mixture of $^{52g}$Mn/$^{55}$Mn$^{+2}$ along with CDTA (*trans*-1,2-diaminocyclohexane-*N,N,N′,N′*-tetraacetic acid) chelator to formulate this bimodal PET/MRI contrast agent (Figure 10). This approach resulted in a chemically identical MRI contrast agent and PET co-agent, with the only distinction being the isotope of manganese. However, this concept has limitations in the stability of the [[$^{52g}$Mn]Mn(CDTA)]$^{2-}$ chelate. The authors incubated [[$^{52g}$Mn]Mn(CDTA)]$^{2-}$ in human blood serum at 37°C and performed radio-HPLC after 1, 3, 5, 18, and 24h to determine the proportion of the intact complex. The percentage of intact complex was as follows: ≥95% after 1h, 87% after 3h, 78% after 5h, 33% after 18h, and 28% after 24h [84].

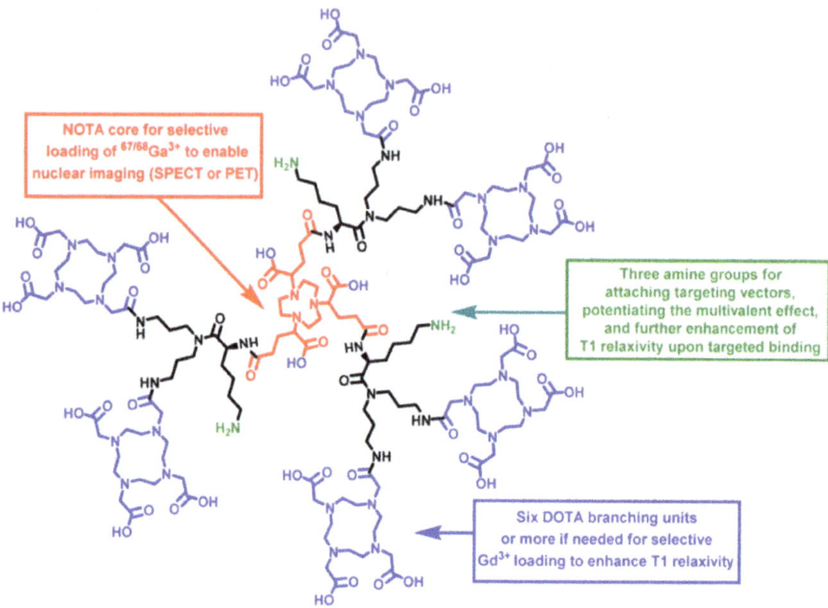


**FIGURE 9** Chemical structure of chelator for preparation of PET/MRI dendrimeric bimodal contrast agent and description of its parts function, prepared by Kumar et al. (Figure reproduced with permission from [83].)

**FIGURE 10** Chemical structure of PET/MRI bimodal probe $[[^{52g}Mn]Mn(CDTA)]^{2-}$ synthesized by Vanasschen et al. [84].

**$[^{52g/55}Mn][Mn(7a)]$:** Another example of a bimodal PET/MRI bimodal probe that uses the $^{52g}Mn/^{55}Mn$ pair employs a structure with multiple manganese chelation sites built around a benzene core (Figure 11) [53]. The benzene core in this structure was functionalized by three *trans*-CDTA ligands (*trans*-1,2-Diamino cyclohexane-*N,N,N′,N′*-tetraacetic acid) by the CuAAC reaction. These ligands were used to introduce more sites for manganese chelation, aiming to improve the relaxivity. Finally, the deprotected ligand with three *trans*-CDTA units was isotopically labeled by paramagnetic Mn spiked with the positron-emitting radionuclide $^{52g}Mn$.

**Radiometal-Based PET/MRI Contrast Agents for Sensing Tumor Extracellular pH:** Lowe et al. developed a bimodal PET/MRI contrast agent based on

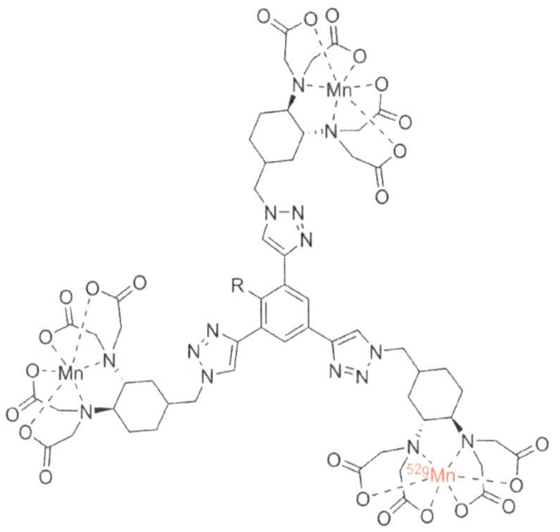

**FIGURE 11** Structure of labeled $^{52g}Mn/^{55}Mn$ PET/MRI bimodal probe synthesized by Brandt et al. [85].

a previously reported pH-responsive MRI contrast agent, GdDO3A-sulfonamide [86]. The responsiveness of this agent stems from the presence of an arylsulfonamide pendant arm that, at lower pH levels, undergoes decoordination, allowing water molecules to coordinate with gadolinium and thereby increasing relaxivity. Conversely, the pendant arm becomes coordinated when the pH increases, blocking access to water molecules to the gadolinium core. The authors used $^{68}Ga$ or $^{64}Cu$ complexes of the pH-responsive MRI probe as PET co-agents (Figure 12) and were able to determine the pH of the solution with their PET/MRI agents. However, significant de-chelation was observed for the $^{64}Cu$ complex and some associated degradation with the $^{68}Ga$ PET co-agent.

### 4.4.2 Nanoparticles

A popular strategy for synthesizing PET/MRI nanoparticles involves using magnetic nanoparticles capable of providing $T_1$, $T_2$, or $T_1/T_2$ contrast in MRI, which are then radiolabeled with a positron emitter for PET imaging [87]. These magnetic nanoparticles can be radiolabeled using chelator-based or non-chelator approaches, depending on the particle composition and synthesis protocol (Figure 13). In chelator-based methods, a bifunctional ligand is conjugated to the particle's surface before or after coordination with the radionuclide (Figure 13a) [87]. This approach is widely used for PET/MRI contrast agents due to its versatility, although it requires careful surface functionalization to preserve the physicochemical properties of the particle. Additionally, suitable ligands must be chosen to ensure stable coordination complexes, preventing potential radionuclide detachment under *in vivo* conditions. Different chelators have been successfully employed for the radiolabeling of magnetic nanoparticles. One of the first examples was reported by Lee et al. using 1,4,7,10-tetraazacyclododecane-1,4,7,10-tetraacetic

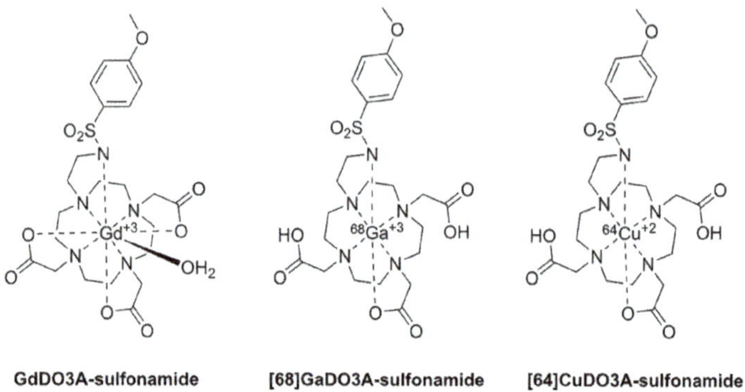


**FIGURE 12** Chemical structures of compounds used to prepare PET/MRI pH-responsive contrast agent. The GdDO3A-sulfonamide was used as a pH-responsive MRI agent, and [$^{68}$Ga]GaDO3A-sulfonamide or [$^{64}$Cu]CuDO3A-sulfonamide were used as PET co-agents.

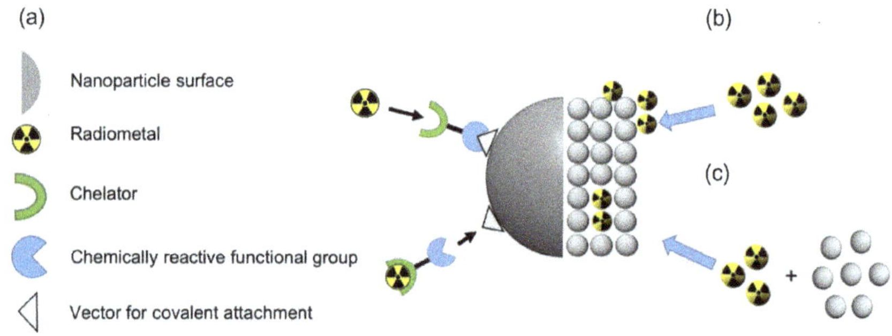


**FIGURE 13** Schematic of the three main methods applied for radiolabeling magnetic nanoparticles with positron emitters: (a) Chelator-based strategy, (b) chemical adsorption, and (c) hot+cold precursors.

acid (DOTA) as a chelating agent for subsequent radiolabeling with $^{64}$Cu allowing PET/MRI detection of angiogenesis in tumors [88]. Zhan et al. utilized a similar approach for radiolabeling $Mn_3O_4$ nanoparticles with $^{64}$Cu by coordinating with 1,4,7-triazacyclononane-1,4,7-triacetic acid (NOTA) chelator [89]. In a more recent example, Thomas et al. radiolabeled iron oxide nanoparticles with $^{64}$Cu using 3,4-dihydroxy-l-phenylalanine (LDOPA), obtaining good responses in both PET and MRI [90].

Non-chelator approaches, a more recent development, leverage the physicochemical properties of the nanoparticles to integrate the radionuclide within the core or surface without extra modifications [91]. An extended method employing this approach for PET/MRI nanoparticles is the radio-mineralization of iron oxide nanoparticles with metallic radionuclides under high temperatures (Figure 13b).

In this process, radiometals such as $^{89}Zr$, $^{68}Ge$, and $^{68}Ga$ are deposited onto the nanoparticles' surface by simply mixing them with aqueous solutions of the radiometal at high temperatures (~90°C), yielding high radiolabeling efficiency and radiochemical stability [92]. An alternative approach in non-chelator methods involves integrating the radiometal directly into the crystal structure of the particles (Figure 13c). This strategy, known as the hot+cold precursors reaction, entails adding the radiometal and particle precursors to obtain the radiolabeled particle in a single step. This method has been employed for synthesizing and radiolabeling different nanoparticles, including gold, up-converting, quantum dots, cerium oxide, and silver nanoparticles [1]. For PET/MRI nanoparticles, this technique was initially applied by Wong et al., who incorporated $^{64}Cu$ into the core of dextran-coated iron oxide nanoparticles using a microwave-assisted protocol to create a $T_2$-MRI/PET contrast agent [93]. A few years later, Pellico et al. reported a similar approach for synthesizing $^{68}Ga$ core-doped iron oxide nanoparticles for $T_1$-MRI/PET [94].

Although less commonly used, another approach involves utilizing nanoparticles as a scaffold for integrating magnetic metals and radionuclides, enabling responses in PET/MRI. In this vein, Truillet et al. applied a polysiloxane matrix to coordinate $Gd^{3+}$ using 1,4,7,10-tetraazacyclododecane-1-glutaric anhydride-4,7,10-triacetic acid (DOTAGA) and 2,2′-(7-(1-carboxy-4-((2,5-dioxopyrrolidin-1-yl)oxy)-4-oxobutyl)-1,4,7-triazonane-1,4-diyl)diacetic acid) (NODAGA) for coordination with $^{68}Ga^{3+}$, achieving high radiolabeling yields and robust responses in both PET and MRI [95]. In another example, Senders et al. incorporated a perfluoro-crown ether into a phospholipid-based nanoparticle for $^{19}F$-MRI and conjugated the particle with deferoxamine (DFO) for subsequent radiolabeling with $^{89}Zr$ [96].

# 5 CURRENT PRECLINICAL APPLICATIONS OF PET/MRI CONTRAST AGENTS

## 5.1 Cardiovascular Applications

In recent years, the proliferation of clinical PET/MRI scanners has underscored their effectiveness in diagnosing conditions such as inflammation, ischemia, reperfusion, and calcification, all of which are linked to the development of cardiovascular diseases (CVD) like coronary artery disease, atherosclerosis, and cardiac sarcoidosis [97–99]. Despite their potential, most of these clinical applications rely on common PET tracers such as [$^{18}F$]FDG, [$^{18}F$]FNa, [$^{13}N$]$NH_3$, [$^{68}Ga$]FAPI, and [$^{15}O$]Water in combination with non-contrasted or Gd-based contrasted MRI for detailed anatomical reference [100,101]. The lack of suitable contrast agents capable of providing adequate responses in both imaging techniques simultaneously remains challenging. This scarcity can be attributed to the relative novelty of PET/MRI systems, resulting in a

**TABLE 3**
**PET/MRI Nanoparticles for Cardiovascular Diseases**

| Formulation | PET | MRI | Application | Reference |
|---|---|---|---|---|
| $^{89}Zr$-$^{19}F$-HDL | $C_{34}$-DFO-$^{89}Zr$ | Perfluoro-15-crown-5-ether | Ischemic heart disease | [96] |
| $^{68}Ga$-IONPs-Alendronate | $^{68}Ga$ core-doped $Fe_2O_3$ NPs | Citric acid-coated IONPs | Vascular calcifications in atherosclerosis | [103] |
| MDIO-$^{64}Cu$-DOTA | $^{64}Cu$-DOTA | Dextran-coated IONPs | Targeting macrophages in atherosclerosis | [104] |
| $^{64}Cu$-TNP | $^{64}Cu$-DTPA | Dextran-coated IONPs | Targeting macrophages in atherosclerosis | [102] |
| $^{64}Cu$-NOTA-IONP@MMP2c-PEG2K | $^{64}Cu$-NOTA | Carboxylate-coated IONPs | Matrix Metalloproteinase-2 in atherosclerosis | [105] |
| $^{18}F$-Macroflor | $^{18}F$-P3-C#C | MPO-Gd | Ischemic heart disease | [106] |
| $^{68}Ga$-IONPs-Tz | $^{68}Ga$ core-doped $Fe_2O_3$ NPs | Citric acid-coated IONPs | Targeting oxidized Phospholipids in atherosclerosis | [107] |

limited number of optimal preclinical PET/MRI contrast agents compared to monomodal PET and MRI contrast agents.

The only example of a PET/MRI contrast agent based on a small molecule for CVD in the literature is the fibrin target EP-2104R probe described in previous sections. Using this probe, the authors successfully detect arterial thrombus in the carotid of rats by both PET and MRI.

Despite the limited number of examples (as shown in Table 3), the use of nanoparticles with responses in both PET and MRI for diagnosing CVDs is more prevalent than molecular probes. In 2008, Nahrendorf et al. reported the first PET/MRI nanoparticle-based probe designed to target macrophages in atherosclerosis [102]. In this research, the authors harnessed the ability of the particles to generate negative contrast in MRI and radiolabeled them with $^{64}Cu$ using a DTPA chelator for subsequent PET studies. Biodistribution studies conducted *in vivo* demonstrated a substantial uptake of these nanoparticles in phagocytic cells within regions exhibiting significant atherosclerosis, such as the aortic root and arch. The remarkable uptake can be attributed to the extended circulation time of the nanoparticles, coupled with the strong affinity of phagocytic cells for dextran-coated nanoparticles. More recently, Senders

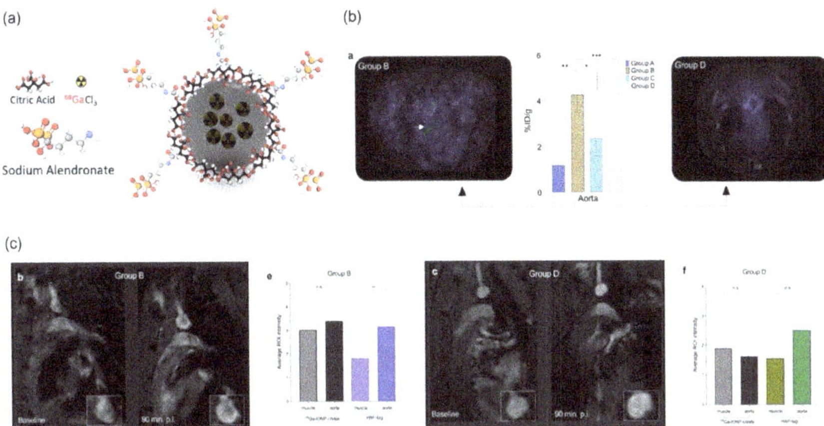


**FIGURE 14** (a). Chemical representation of $^{68}$Ga-IONP-Alendronate (HAP-Multitag Probe) (b). Aorta uptake of HAP-multitag and PET/CT images of group B (16 weeks old and 8 weeks high-fat diet, HFD, ApoE $^{-/-}$ mice) and group D (26 weeks old and 18 weeks HFD, ApoE $^{-/-}$ mice), (c). $T_1$-weighted MRI before (baseline) and 90 min after i.v. injection of HAP-multitag for groups B and D, and average ROI intensity ($n$ = 10 per mouse) in the muscle and aorta 90 min after intravenous injection of $^{68}$Ga-IONP-citrate or HAP-multitag. $^{**}P < 0.01$, one-way ANOVA; error bars indicate s.d.

et al. introduced a lipoprotein-based nanotracer for simultaneous $^{19}$F-MRI and $^{89}$Zr-PET tracking of myeloid cells in mouse atherosclerosis and myocardial infarction models [96]. In this study, a perfluoro-crown ether was combined with phospholipids to provide a robust response in $^{19}$F-MRI. Additionally, the nanoparticles were radiolabeled with $^{89}$Zr via chelation with desferoxamine B (DFO) for short-term pharmacokinetic studies. The significant uptake of the nanotracer by myeloid cells enabled the *in vivo* evaluation, shedding light on their dynamics in ischemic heart disease. The latest reported example in PET/MRI nanoparticle applications for cardiovascular diseases involves the use of $^{68}$Ga radiolabeled iron oxide nanoparticles to detect vascular microcalcification in atherosclerosis [103]. In this work, researchers developed a one-step microwave synthetic protocol for the nanoparticles, incorporating $^{68}$Ga into the crystal lattice and functionalization with a calcium-binding bisphosphonate (Figure 14a). The exceptional sensitivity of PET enabled precise quantification of calcium targeting by the nanoparticles, correlating with the atherosclerosis progression (Figure 14b). Simultaneously, the small core of the iron oxide nanoparticles with a paramagnetic-like response allowed for positive contrast in MRI, providing highly detailed anatomical references at various stages of the disease progression (as shown in Figure 14c).

## 5.2 Imaging Infection and Inflammation

There is significant clinical evidence that combined PET/MRI is beneficial in the diagnosis and evaluation of infection and inflammatory diseases [108]. The use of bimodal agents, however, has been scarce and mostly confined to the pre-clinical area. A notable example from Patel et al. reports on a bimodal agent based in iodine-125 labeled VCAM-1 targeting superparamagnetic microparticle conjugates [109]. This report, although not strictly PET/MR but SPECT/MRI, demonstrated the value of combining the two techniques into a single probe for cross-validation and synergistic information. Following previous reports that showed vascular cell adhesion molecule-1 (VCAM-1) targeted micron-sized particles of iron oxide (MPIOs) allowed detecting and locating brain inflammation [110], the authors included radiolabeling of the VCAM-1-MPIO system with iodine-125 (a gamma-emitter). This provided excellent whole-body biodistribution imaging and quantification properties to the VCAM-1-MPIOs. Using a rat model of cerebral inflammation based on intracerebral injection of tumor necrosis factor-alpha and a rat model of status epilepticus, the authors also noted differences in the quantification of the two signals/modalities, highlighting the utility of the bimodal approach for cross-validation and maximizing the synergistic information obtained.

## 5.3 Oncology

In oncology, PET/MRI bimodal probe applications are scarce, often requiring sequential contrast media for disease detection and localization. Noteworthy applications include characterizing the tumor microenvironment by monitoring hypoxia, angiogenesis, and cellular density [74,111,112]. Additionally, other PET/MRI probes prove valuable in assessing treatment response, capturing changes in both metabolic activity and morphological characteristics through PET and MRI [47]. An interesting simultaneous detection approach involves employing hyperpolarized 1-$^{13}$C-pyruvate alongside [$^{18}$F]FDG in dogs with sarcomas and carcinomas. While the individual imaging techniques failed to differentiate between cancer phenotypes, the combined analysis of PET/MRI revealed a significant statistical difference [113–115]. Another exciting application was reported recently by Shalaby et al., which employs gene-encoded cell engineering techniques. This approach utilized the human organic anion transporter polypeptide 1B3 (OATP1B3), which is involved in internalizing the clinical MRI contrast agent gadolinium ethoxybenzyl-diethylenetriaminepentaacetic acid (Gd-EOB-DTPA), and the human sodium iodide symporter (NIS), enabling the uptake of the PET tracer [$^{18}$F] tetrafluoroborate ([$^{18}$F] TFB). The authors demonstrated the ability to non-invasively track tumor cells engineered with this dual MRI/PET reporter system *in vivo* [116].

Regarding the application of small molecule bimodal PET/MRI probes in oncology, the only report found in the literature is the cocktail application of Gd/$^{64}$Cu-DO3A-sulfonamide or Gd/$^{68}$Ga-DO3A-sulfonamide (vide 2.5.1) for monitoring pH *in vivo* using a subcutaneously implanted MIA PaCa-2 pancreatic cancer mouse model [55].

The advancement of PET/MRI nanoparticle development in oncology also showed markable progress. Among the key innovations are iron oxide nanoparticles labeled with various isotopes, such as $^{64}$Cu [91], $^{68}$Ga [117], and $^{124}$I [118]. These isotopes contribute to the versatility and functionality of the nanoparticles, enabling a spectrum of imaging options for different cancer applications. In particular, the incorporation of specific targeting moieties enhances the precision of these nanoparticles. The use of RGD (Arg–Gly–Asp) and prostate-specific membrane antigen (PSMA) demonstrates a strategic approach to achieve targeted imaging. RGD, known for its affinity to integrins overexpressed in tumor vasculature, ensures a targeted approach for certain cancer types. On the other hand, PSMA targeting is particularly relevant in the context of prostate cancer imaging [89,119]. This dual-pronged strategy, combining isotopic diversity with selective targeting, represents a cutting-edge paradigm in PET/MRI nanoparticle development for oncological applications.

### 5.4 Neuroimaging

In the realm of neurology, the applications of PET/MRI are diverse and impactful; one can find relevant studies in neurotransmitter imaging [120], functional connectivity [97,121], amyloid and alpha-synuclein PET/MRI correlations with cellularity and perfusion [122], and brain inflammation [108]. However, the inability to cross the blood–brain barrier hinders the development of successful probes that can be used *in vivo* [123]. A rare successful example of intact brain imaging function by simultaneous PET/MRI detection is reported by Miloushev and Qin et al. with the use of hyperpolarized MRI $^{13}$C-metabolites and [$^{18}$F]FDG PET [124,125].

## 6 CONCLUSIONS

Despite years of research, using bimodal PET/MRI agents is still challenging for chemists and molecular imaging scientists. For the former, designing and synthesizing molecules or agents that can achieve contrast and provide information in both imaging techniques is not trivial due to several inherent limitations that impose significant synthetic barriers. For the latter, finding accurate and relevant preclinical and clinical applications for these agents is difficult. Quantification of imaging agents is an essential aspect of molecular imaging. Therefore, a straightforward application of the bimodal approach is cross-validating quantification data between different modalities. Additionally, PET-guided MRI optimizes imaging sessions and utilizes the high sensitivity and whole-body detectability of PET and the high spatial resolution of MRI.

It is worth noting that despite considerable initial interest, this area of research has arguably not progressed as much as other areas of molecular imaging in the last ca. 10 years. This could be attributed to the challenges described above, which are inherent to the bimodal concept, but also to the relatively low number of preclinical and clinical PET/MR scanners available compared to PET/CT systems and preventing a larger number of researchers exploring innovative applications of bimodal agents. It is hoped that as these systems are more widely available, the field will be able to explore some of the promising applications identified in this literature review in more detail.

## ACKNOWLEDGMENTS

We gratefully acknowledge the Swiss Werner Siemens Foundation and the Deutsche Forschungsgemeinschaft (DFG, German Research Foundation) under Germany's Excellence Strategy—EXC 2180–390900677 and by the Alexander von Humboldt Foundation through the Sofja Kovalevskaja award.

## REFERENCES

1. J. Pellico, P. J. Gawne, R. T. M. de Rosales, *Chem. Soc. Rev.* **2021**, *50*, 3355–3423. https://doi.org/10.1039/D0CS00384K.
2. G. Currie, K. E. Hawk, E. Rohren, A. Vial, R. Klein, *J. Med. Imaging Radiat. Sci.* **2019**, *50*, 477–487. https://doi.org/10.1016/j.jmir.2019.09.005.
3. N. H. Shah, A. Milstein, S. C. Bagley, *JAMA* **2019**, *322*, 1351. https://doi.org/10.1001/jama.2019.10306.
4. I. Castiglioni, L. Rundo, M. Codari, G. Di Leo, C. Salvatore, M. Interlenghi, F. Gallivanone, A. Cozzi, N. C. D'Amico, F. Sardanelli, *Phys. Med.* **2021**, *83*, 9–24. https://doi.org/10.1016/j.ejmp.**2021**.02.006.
5. P. J. Cassidy, G. K. Radda, *J. R. Soc. Interface* **2005**, *2*, 133–144. https://doi.org/10.1098/rsif.2005.0040.
6. M. A. Pysz, S. S. Gambhir, J. K. Willmann, *Clin. Radiol.* **2010**, *65*, 500–516. https://doi.org/10.1016/j.crad.2010.03.011.
7. J.-W. Bai, S.-Q. Qiu, G.-J. Zhang, *Signal Transduct. Target. Ther.* **2023**, *8*, 89. https://doi.org/10.1038/s41392-023-01366-y.
8. K. He, S. Zeng, L. Qian, *J. Pharm. Anal.* **2020**, *10*, 397–413. https://doi.org/10.1016/j.jpha.2020.07.006.
9. R. A. Gatenby, O. Grove, R. J. Gillies, *Radiology* **2013**, *269*, 8–14. https://doi.org/10.1148/radiol.13122697.
10. W. Rogers, S. Thulasi Seetha, T. A. G. Refaee, R. I. Y. Lieverse, R. W. Y. Granzier, A. Ibrahim, S. A. Keek, S. Sanduleanu, S. P. Primakov, M. P. L. Beuque, D. Marcus, A. M. A. van der Wiel, F. Zerka, C. J. G. Oberije, J. E. van Timmeren, H. Woodruff, P. Lambin, *Br. J. Radiol.* **2020**, *93* (1108), 20190948. https://doi.org/10.1259/bjr.20190948.
11. Z. Han, M. Ke, X. Liu, J. Wang, Z. Guan, L. Qiao, Z. Wu, Y. Sun, X. Sun, *Mol. Imaging Biol.* **2022**, *24*, 8–22. https://doi.org/10.1007/s11307-021-01631-y.
12. B. J. Pichler, M. S. Judenhofer, C. Pfannenberg, in *Molecular Imaging I*, Eds.W. Semmler, M. Schwaiger, *Handbook of Experimental Pharmacology*; Springer, Berlin, Heidelberg, 2008; pp 109–132. https://doi.org/10.1007/978-3-540-72718-7_6.
13. A. M. Alessio, P. E. Kinahan, P. M. Cheng, H. Vesselle, J. S. Karp, *Radiol. Clin. North Am.* **2004**, *42*, 1017–1032. https://doi.org/10.1016/j.rcl.2004.08.001.
14. E. M. Triviño-Ibáñez, B. M. Jiménez-Rodríguez, T. Rudolphi-Solero, E. Y. García-Rivero, A. Rodríguez-Fernández, J. M. Llamas-Elvira, M. Gómez-Río, C. Morales-García, *Diagnostics* **2022**, *12*, 835. https://doi.org/10.3390/diagnostics12040835.

15. H. Herzog, *Z. Für Med. Phys.* **2012**, *22*, 281–298. https://doi.org/10.1016/j.zemedi.2012.07.003.
16. S. Vandenberghe, P. K. Marsden, *Phys. Med. Biol.* **2015**, *60*, R115. https://doi.org/10.1088/0031-9155/60/4/R115.
17. P. Veit-Haibach, H. Ahlström, R. Boellaard, R. C. Delgado Bolton, S. Hesse, T. Hope, M. W. Huellner, A. Iagaru, G. B. Johnson, A. Kjaer, I. Law, U. Metser, H. H. Quick, B. Sattler, L. Umutlu, G. Zaharchuk, K. Herrmann, *Eur. J. Nucl. Med. Mol. Imaging* **2023**, *50*, 3513–3537. https://doi.org/10.1007/s00259-023-06406-x.
18. W. P. Fendler, J. Czernin, K. Herrmann, T. Beyer, *J. Nucl. Med.* **2016**, *57*, 2016–2021. https://doi.org/10.2967/jnumed.116.174169.
19. S. Gaddamanugu, O. Shafaat, H. Sotoudeh, A. H. Sarrami, A. Rezaei, Z. Saadatpour, A. Singhal, *Neuroradiology* **2022**, *64*, 15–30. https://doi.org/10.1007/s00234-021-02819-3.
20. M. Drake-Pérez, J. Boto, A. Fitsiori, K. Lovblad, M. I. Vargas, *Insights Imaging* **2018**, *9*, 535–547. https://doi.org/10.1007/s13244-018-0624-3.
21. G. S. Young, M. D. Geschwind, N. J. Fischbein, J. L. Martindale, R. G. Henry, S. Liu, Y. Lu, S. Wong, H. Liu, B. L. Miller, W. P. Dillon, *AJNR Am. J. Neuroradiol.* **2005**, *26*, 1551–1562.
22. S. C. Peter, E. Wenkel, E. Weiland, M. Dietzel, R. Janka, A. Hartmann, J. Emons, M. Uder, S. Ellmann, *Eur. Radiol.* **2020**, *30*, 2761–2772. https://doi.org/10.1007/s00330-019-06608-8.
23. M. Kataoka, H. Ueda, T. Koyama, S. Umeoka, K. Togashi, R. Asato, S. Tanaka, J. Ito, *Am. J. Roentgenol.* **2005**, *184*, 313–319. https://doi.org/10.2214/ajr.184.1.01840313.
24. G. Krokos, J. MacKewn, J. Dunn, P. Marsden, *EJNMMI Phys.* **2023**, *10*, 52. https://doi.org/10.1186/s40658-023-00569-0.
25. E. Schreibmann, J. A. Nye, D. M. Schuster, D. R. Martin, J. Votaw, T. Fox, *Med. Phys.* **2010**, *37*, 2101–2109. https://doi.org/10.1118/1.3377774.
26. M. Hofmann, I. Bezrukov, F. Mantlik, P. Aschoff, F. Steinke, T. Beyer, B. J. Pichler, B. Schölkopf, *J. Nucl. Med.* **2011**, *52*, 1392–1399. https://doi.org/10.2967/jnumed.110.078949.
27. B. J. Pichler, H. F. Wehrl, A. Kolb, M. S. Judenhofer, *Semin. Nucl. Med.* **2008**, *38*, 199–208. https://doi.org/10.1053/j.semnuclmed.2008.02.001.
28. S. Gundacker, A. Heering, *Phys. Med. Biol.* **2020**, *65*, 17TR01. https://doi.org/10.1088/1361-6560/ab7b2d.
29. P. Lecoq, S. Gundacker, *Eur. Phys. J. Plus* **2021**, *136*, 292. https://doi.org/10.1140/epjp/s13360-021-01183-8.
30. B. E. Hammer, N. L. Christensen, B. G. Heil, *Med. Phys.* **1994**, *21*, 1917–1920. https://doi.org/10.1118/1.597178.
31. H. Herzog, H. Iida, C. Weirich, L. Tellmann, J. Kaffanke, S. Spellerberg, L. Caldeira, E. R. Kops, N. J. Shah, "Influence from High and Ultra-High Magnetic Field on Positron Range Measured with a 9.4TMR-BrainPET," IEEE Nuclear Science Symposuim & Medical Imaging Conference, **2010**, pp. 3410–3413. https://doi.org/10.1109/NSSMIC.2010.5874439.
32. N. J. Shah, H. Herzog, C. Weirich, L. Tellmann, J. Kaffanke, L. Caldeira, E. R. Kops, S. M. Qaim, H. H. Coenen, H. Iida, *PLoS One* **2014**, *9*, e95250. https://doi.org/10.1371/journal.pone.0095250.
33. C.-H. Choi, C. Stegmayr, A. Shymanskaya, W. A. Worthoff, N. A. da Silva, J. Felder, K.-J. Langen, N. J. Shah, *EJNMMI Phys.* **2020**, *7*, 50. https://doi.org/10.1186/s40658-020-00319-6.
34. S. H. Maramraju, S. D. Smith, S. S. Junnarkar, D. Schulz, S. Stoll, B. Ravindranath, M. L. Purschke, S. Rescia, S. Southekal, J.-F. Pratte, P. Vaska, C. L. Woody, D. J. Schlyer, *Phys. Med. Biol.* **2011**, *56*, 2459. https://doi.org/10.1088/0031-9155/56/8/009.

35. M. J. P. van Osch, A. G. Webb, *Curr. Radiol. Rep.* **2014**, *2*, 61. https://doi.org/10.1007/s40134-014-0061-0.
36. M. Mühlenweg, G. Schaefers, S. Trattnig, *Radiology* **2015**, *55*, 691–696. https://doi.org/10.1007/s00117-015-2859-z.
37. Health, C. for D. and R. Criteria for Significant Risk Investigations of Magnetic Resonance Diagnostic Devices - Guidance for Industry and Food and Drug Administration Staff. https://www.fda.gov/regulatory-information/search-fda-guidance-documents/criteria-significant-risk-investigations-magnetic-resonance-diagnostic-devices-guidance-industry-and (accessed 2023-11-12).
38. R. Minamimoto, C. Levin, M. Jamali, D. Holley, A. Barkhodari, G. Zaharchuk, A. Iagaru, *Mol. Imaging Biol.* **2016**, *18*, 776–781. https://doi.org/10.1007/s11307-016-0939-8.
39. M. A. Queiroz, G. Delso, S. Wollenweber, T. Deller, K. Zeimpekis, M. Huellner, F. de G. Barbosa, G. von Schulthess, P. Veit-Haibach, *PLoS One* **2015**, *10*, e0128842. https://doi.org/10.1371/journal.pone.0128842.
40. No. *Dorlands Illustrated Med. Dict.*
41. K. Leung, *Rep. Med. Imaging* **2012**, *15*. https://doi.org/10.2147/RMI.S20665.
42. R. D. Niederkohr, S. P. Hayden, J. J. Hamill, J. P. Jones, J. D. Schaefferkoetter, E. Chiu, *Am. J. Nucl. Med. Mol. Imaging* **2021**, *11*, 428–442.
43. A. Smeraldo, A. M. Ponsiglione, A. Soricelli, P. A. Netti, E. Torino, *Int. J. Nanomed.* **2022**, *17*, 3343–3359. https://doi.org/10.2147/IJN.S362192.
44. O. Martin, B. M. Schaarschmidt, J. Kirchner, S. Suntharalingam, J. Grueneisen, A. Demircioglu, P. Heusch, H. H. Quick, M. Forsting, G. Antoch, K. Herrmann, L. Umutlu, *J. Nucl. Med.* **2020**, *61*, 1131–1136. https://doi.org/10.2967/jnumed.119.233940.
45. O. A. Catalano, E. Nicolai, B. R. Rosen, A. Luongo, M. Catalano, C. Iannace, A. Guimaraes, M. G. Vangel, U. Mahmood, A. Soricelli, M. Salvatore, *Br. J. Cancer* **2015**, *112*, 1452–1460. https://doi.org/10.1038/bjc.2015.112.
46. A. J. Theruvath, F. Siedek, K. Yerneni, A. M. Muehe, S. L. Spunt, A. Pribnow, M. Moseley, Y. Lu, Q. Zhao, P. Gulaka, A. Chaudhari, H. E. Daldrup-Link, *Radiol. Artif. Intell.* **2021**, *3*. https://doi.org/10.1148/ryai.2021200232.
47. V. Mistry, J. R. Scott, T.-Y. Wang, P. Mollee, K. A. Miles, W. P. Law, G. Hapgood, *Cancer Imaging* **2023**, *23*, 11. https://doi.org/10.1186/s40644-023-00520-7.
48. C. A. Marschner, F. Aloufi, M. Aitken, E. Cheung, P. Thavendiranathan, R. M. Iwanochko, M. Balter, Y. Moayedi, J. Duero Posada, K. Hanneman, *Radiol. Cardiothorac. Imaging* **2023**, *5*. https://doi.org/10.1148/ryct.220292.
49. K. Westphal, M. Eiber, M. Henninger, K. Scheidhauer, A. J. Beer, W. Thaiss, C. Rischpler, *Medicine (Baltimore)* **2023**, *102*, e33533. https://doi.org/10.1097/MD.0000000000033533.
50. J. P. Schmall, S. Surti, H. J. Otero, S. Servaes, J. S. Karp, L. J. States, *J. Nucl. Med.* **2021**, *62*, 123–130. https://doi.org/10.2967/jnumed.119.240127.
51. L. Frullano, C. Catana, T. Benner, A. D. Sherry, P. Caravan, *Angew. Chem. Int. Ed.* **2010**, *49*, 2382–2384. https://doi.org/10.1002/anie.201000075.
52. O. Jacobson, D. O. Kiesewetter, X. Chen, *Bioconjug. Chem.* **2015**, *26*, 1–18. https://doi.org/10.1021/bc500475e.
53. M. R. Brandt, C. Vanasschen, J. Ermert, H. H. Coenen, B. Neumaier, *Dalton Trans.* **2019**, *48*, 3003–3008. https://doi.org/10.1039/C8DT04996C.
54. C. Vanasschen, M. Brandt, J. Ermert, H. H. Coenen, *Dalton Trans.* **2016**, *45*, 1315–1321. https://doi.org/10.1039/C5DT04270D.
55. A. C. Pollard, J. de la Cerda, F. W. Schuler, T. R. Pollard, A. Kotrotsou, F. Pisaneschi, M. D. Pagel, *Biosensors* **2022**, *12*, 134. https://doi.org/10.3390/bios12020134.
56. R. Uppal, C. Catana, I. Ay, T. Benner, A. G. Sorensen, P. Caravan, *Radiology* **2011**, *258*, 812–820. https://doi.org/10.1148/radiol.10100881.

57. T. Frenzel, P. Lengsfeld, H. Schirmer, J. Hütter, H.-J. Weinmann, *Invest. Radiol.* **2008**, *43*, 817–828. https://doi.org/10.1097/RLI.0b013e3181852171.
58. D. J. Berry, Y. Ma, J. R. Ballinger, R. Tavaré, A. Koers, K. Sunassee, T. Zhou, S. Nawaz, G. E. D. Mullen, R. C. Hider, P. J. Blower, *Chem. Commun.* **2011**, *47*, 7068. https://doi.org/10.1039/c1cc12123e.
59. L. A. Bass, M. Wang, M. J. Welch, C. J. Anderson, *Bioconjug. Chem.* **2000**, *11*, 527–532. https://doi.org/10.1021/bc990167l.
60. C. A. Boswell, X. Sun, W. Niu, G. R. Weisman, E. H. Wong, A. L. Rheingold, C. J. Anderson, *J. Med. Chem.* **2004**, *47*, 1465–1474. https://doi.org/10.1021/jm030383m.
61. J. D. Steinberg, A. Raju, P. Chandrasekharan, C.-T. Yang, K. Khoo, J.-P. Abastado, E. G. Robins, D. W. Townsend, *EJNMMI Res.* **2014**, *4*, 15. https://doi.org/10.1186/2191-219X-4-15.
62. R. Chakravarty, S. Chakraborty, A. Dash, *Mol. Pharm.* **2016**, *13*, 3601–3612. https://doi.org/10.1021/acs.molpharmaceut.6b00582.
63. S. A. Graves, R. Hernandez, J. Fonslet, C. G. England, H. F. Valdovinos, P. A. Ellison, T. E. Barnhart, D. R. Elema, C. P. Theuer, W. Cai, R. J. Nickles, G. W. Severin, *Bioconjug. Chem.* **2015**, *26*, 2118–2124. https://doi.org/10.1021/acs.bioconjchem.5b00414.
64. X. Sun, W. Cai, X. Chen, *Acc. Chem. Res.* **2015**, *48*, 286–294. https://doi.org/10.1021/ar500362y.
65. J. Pellico, C. M. Ellis, J. J. Davis, *Contrast Media Mol. Imaging* **2019**, *2019*, 1–13. https://doi.org/10.1155/2019/1845637.
66. M. S. Judenhofer, H. F. Wehrl, D. F. Newport, C. Catana, S. B. Siegel, M. Becker, A. Thielscher, M. Kneilling, M. P. Lichy, M. Eichner, K. Klingel, G. Reischl, S. Widmaier, M. Röcken, R. E. Nutt, H.-J. Machulla, K. Uludag, S. R. Cherry, C. D. Claussen, B. J. Pichler, *Nat. Med.* **2008**, *14*, 459–465. https://doi.org/10.1038/nm1700.
67. J. H. Jung, Y. Choi, K. C. Im, *Nucl. Med. Mol. Imaging* **2016**, *50*, 3–12. https://doi.org/10.1007/s13139-016-0393-1.
68. A. Korde, R. Mikolajczak, P. Kolenc, P. Bouziotis, H. Westin, M. Lauritzen, M. Koole, M. M. Herth, M. Bardiès, A. F. Martins, A. Paulo, S. K. Lyashchenko, S. Todde, S. Nag, E. Lamprou, A. Abrunhosa, F. Giammarile, C. Decristoforo, *EJNMMI Radiopharm. Chem.* **2022**, *7*, 18. https://doi.org/10.1186/s41181-022-00168-x.
69. V. Bettinardi, I. Castiglioni, E. De Bernardi, M. C. Gilardi, *Clin. Transl. Imaging* **2014**, *2*, 199–218. https://doi.org/10.1007/s40336-014-0066-y.
70. A. Gillman, J. Smith, P. Thomas, S. Rose, N. Dowson, *Med. Phys.* **2017**, *44*, e430–e445. https://doi.org/10.1002/mp.12577.
71. M. Zaitsev, J. Maclaren, M. Herbst, *J. Magn. Reson. Imaging* **2015**, *42*, 887–901. https://doi.org/10.1002/jmri.24850.
72. B. Belaroussi, J. Milles, S. Carme, Y. M. Zhu, H. Benoit-Cattin, *Med. Image Anal.* **2006**, *10*, 234–246. https://doi.org/10.1016/j.media.2005.09.004.
73. Y. Berker, J. Franke, A. Salomon, M. Palmowski, H. C. W. Donker, Y. Temur, F. M. Mottaghy, C. Kuhl, D. Izquierdo-Garcia, Z. A. Fayad, F. Kiessling, V. Schulz, *J. Nucl. Med.* **2012**, *53*, 796–804. https://doi.org/10.2967/jnumed.111.092577.
74. F. C. Michelotti, G. Bowden, A. Küppers, L. Joosten, J. Maczewsky, V. Nischwitz, G. Drews, A. Maurer, M. Gotthardt, A. M. Schmid, B. J. Pichler, *Theranostics* **2020**, *10*, 398–410. https://doi.org/10.7150/thno.33410.
75. J. Schmitz, J. Schwab, J. Schwenck, Q. Chen, L. Quintanilla-Martinez, M. Hahn, B. Wietek, N. Schwenzer, A. Staebler, U. Kohlhofer, O. H. Aina, N. E. Hubbard, G. Reischl, A. D. Borowsky, S. Brucker, K. Nikolaou, C. la Fougère, R. D. Cardiff, B. J. Pichler, A. M. Schmid, *Cancer Res.* **2016**, *76*, 5512–5522. https://doi.org/10.1158/0008-5472.CAN-15-0642.

76. J. A. Disselhorst, M. A. Krueger, S. M. M. Ud-Dean, I. Bezrukov, M. A. Jarboui, C. Trautwein, A. Traube, C. Spindler, J. M. Cotton, D. Leibfritz, B. J. Pichler, *Proc. Natl. Acad. Sci.* **2018**, *115*, E2980–E2987. https://doi.org/10.1073/pnas.1718304115.
77. P. Katiyar, J. Schwenck, L. Frauenfeld, M. R. Divine, V. Agrawal, U. Kohlhofer, S. Gatidis, R. Kontermann, A. Königsrainer, L. Quintanilla-Martinez, C. la Fougère, B. Schölkopf, B. J. Pichler, J. A. Disselhorst, *Nat. Biomed. Eng.* **2023**, *7* , 1014–1027. https://doi.org/10.1038/s41551-023-01047-9.
78. Y. Huang, J. C. Hsu, H. Koo, D. P. Cormode, *Theranostics* **2022**, *12*, 796–816. https://doi.org/10.7150/thno.67375.
79. M. Woźniak, A. Płoska, A. Siekierzycka, L. W. Dobrucki, L. Kalinowski, I. T. Dobrucki, *Int. J. Mol. Sci.* **2022**, *23*, 2658. https://doi.org/10.3390/ijms23052658.
80. W. Poon, Y.-N. Zhang, B. Ouyang, B. R. Kingston, J. L. Y. Wu, S. Wilhelm, W. C. W. Chan, *ACS Nano* **2019**, *13*, 5785–5798. https://doi.org/10.1021/acsnano.9b01383.
81. Yu. A. Koksharov, S. P. Gubin, I. V. Taranov, G. B. Khomutov, Yu. V. Gulyaev, *J. Commun. Technol. Electron.* **2022**, *67*, 101–116. https://doi.org/10.1134/S1064226922020073.
82. J. Notni, P. Hermann, I. Dregely, H. Wester, *Chem. Eur. J.* **2013**, *19*, 12602–12606. https://doi.org/10.1002/chem.201302751.
83. A. Kumar, S. Zhang, G. Hao, G. Hassan, S. Ramezani, K. Sagiyama, S.-T. Lo, M. Takahashi, A. D. Sherry, O. K. Öz, Z. Kovacs, X. Sun, *Bioconjug. Chem.* **2015**, *26*, 549–558. https://doi.org/10.1021/acs.bioconjchem.5b00028.
84. C. Vanasschen, M. Brandt, J. Ermert, H. H. Coenen, *Dalton Trans.* **2016**, *45*, 1315–1321. https://doi.org/10.1039/C5DT04270D.
85. M. R. Brandt, C. Vanasschen, J. Ermert, H. H. Coenen, B. Neumaier, *Dalton Trans.* **2019**, *48*, 3003–3008. https://doi.org/10.1039/C8DT04996C.
86. M. P. Lowe, D. Parker, O. Reany, S. Aime, M. Botta, G. Castellano, E. Gianolio, R. Pagliarin, *J. Am. Chem. Soc.* **2001**, *123*, 7601–7609. https://doi.org/10.1021/ja0103647.
87. J. Lamb, J. P. Holland, *J. Nucl. Med.* **2018**, *59*, 382–389. https://doi.org/10.2967/jnumed.116.187419.
88. H.-Y. Lee, Z. Li, K. Chen, A. R. Hsu, C. Xu, J. Xie, S. Sun, X. Chen, *J. Nucl. Med.* **2008**, *49*, 1371–1379. https://doi.org/10.2967/jnumed.108.051243.
89. Y. Zhan, S. Shi, E. B. Ehlerding, S. A. Graves, Shreya Goel, J. W. Engle, J. Liang, J. Tian, W. Cai, *ACS Appl. Mater. Interfaces* **2017**, *9*, 38304–38312. https://doi.org/10.1021/acsami.7b12216
90. G. Thomas, J. Boudon, L. Maurizi, M. Moreau, P. Walker, I. Severin, A. Oudot, C. Goze, S. Poty, J.-M. Vrigneaud, F. Demoisson, F. Denat, F. Brunotte, N. Millot, *ACS Omega* **2019**, *4*, 2637–2648. https://doi.org/10.1021/acsomega.8b03283.
91. F. Chen, P. A. Ellison, C. M. Lewis, H. Hong, Y. Zhang, S. Shi, R. Hernandez, M. E. Meyerand, T. E. Barnhart, W. Cai, *Angew. Chem. Int. Ed.* **2013**, *52*, 13319–13323. https://doi.org/10.1002/anie.201306306.
92. P. S. Patrick, L. K. Bogart, T. J. Macdonald, P. Southern, M. J. Powell, M. Zaw-Thin, N. H. Voelcker, I. P. Parkin, Q. A. Pankhurst, M. F. Lythgoe, T. L. Kalber, J. C. Bear, *Chem. Sci.* **2019**, *10*, 2592–2597. https://doi.org/10.1039/C8SC04895A.
93. R. M. Wong, D. A. Gilbert, K. Liu, A. Y. Louie, *ACS Nano* **2012**, *6*, 3461–3467. https://doi.org/10.1021/nn300494k.
94. J. Pellico, J. Ruiz-Cabello, M. Saiz-Alía, G. del Rosario, S. Caja, M. Montoya, L. Fernández de Manuel, M. P. Morales, L. Gutiérrez, B. Galiana, J. A. Enríquez, F. Herranz, *Contrast Media Mol. Imaging* **2016**, *11*, 203–210. https://doi.org/10.1002/cmmi.1681.
95. C. Truillet, P. Bouziotis, C. Tsoukalas, J. Brugière, M. Martini, L. Sancey, T. Brichart, F. Denat, F. Boschetti, U. Darbost, I. Bonnamour, D. Stellas, C. D. Anagnostopoulos, V. Koutoulidis, L. A. Moulopoulos, P. Perriat, F. Lux, O. Tillement, *Contrast Media Mol. Imaging* **2015**, *10*, 309–319. https://doi.org/10.1002/cmmi.1633.

96. M. L. Senders, A. E. Meerwaldt, M. M. T. van Leent, B. L. Sanchez-Gaytan, J. C. van de Voort, Y. C. Toner, A. Maier, E. D. Klein, N. A. T. Sullivan, A. M. Sofias, H. Groenen, C. Faries, R. S. Oosterwijk, E. M. van Leeuwen, F. Fay, E. Chepurko, T. Reiner, R. Duivenvoorden, L. Zangi, R. M. Dijkhuizen, S. Hak, F. K. Swirski, M. Nahrendorf, C. Pérez-Medina, A. J. P. Teunissen, Z. A. Fayad, C. Calcagno, G. J. Strijkers, W. J. M. Mulder, *Nat. Nanotechnol.* **2020**, *15*, 398–405. https://doi.org/10.1038/s41565-020-0642-4.
97. G. Argalia, M. Fogante, N. Schicchi, F. M. Fringuelli, P. Esposto Pirani, C. Cottignoli, C. Romagnolo, A. Palucci, G. Biscontini, L. Balardi, G. Argalia, L. Burroni, *Clin. Transl. Imaging* **2023**. https://doi.org/10.1007/s40336-023-00586-0.
98. M. L. Senders, C. Calcagno, A. Tawakol, M. Nahrendorf, W. J. M. Mulder, Z. A. Fayad, *Nat. Biomed. Eng.* **2022**, *7*, 202–220. https://doi.org/10.1038/s41551-022-00970-7.
99. H. S. A. Tingen, G. D. van Praagh, P. H. Nienhuis, A. Tubben, N. D. van Rijsewijk, D. ten Hove, N. A. Mushari, T. S. Martinez-Lucio, O. I. Mendoza-Ibañez, J. van Sluis, C. Tsoumpas, A. W. J. M. Glaudemans, R. H. J. A. Slart, *Br. J. Radiol.* **2023**. https://doi.org/10.1259/bjr.20230704.
199. B. Whittington, M. R. Dweck, E. J. R. van Beek, D. Newby, M. C. Williams, *J. Magn. Reson. Imaging* **2023**, *57*, 1301–1311. https://doi.org/10.1002/jmri.28554.
101. R. Cardoso, T. M. Leucker, *PET Clin.* **2020**, *15*, 509–520. https://doi.org/10.1016/j.cpet.2020.06.007.
102. M. Nahrendorf, H. Zhang, S. Hembrador, P. Panizzi, D. E. Sosnovik, E. Aikawa, P. Libby, F. K. Swirski, R. Weissleder, *Circulation* **2008**, *117*, 379–387. https://doi.org/10.1161/CIRCULATIONAHA.107.741181.
103. J. Pellico, I. Fernández-Barahona, J. Ruiz-Cabello, L. Gutiérrez, M. Muñoz-Hernando, M. J. Sánchez-Guisado, I. Aiestaran-Zelaia, L. Martínez-Parra, I. Rodríguez, J. Bentzon, F. Herranz, *ACS Appl. Mater. Interfaces* **2021**, *13*, 45279–45290. https://doi.org/10.1021/acsami.1c13417.
104. C. Tu, T. S. C. Ng, R. E. Jacobs, A. Y. Louie, *JBIC J. Biol. Inorg. Chem.* **2014**, *19*, 247–258. https://doi.org/10.1007/s00775-013-1054-9.
104. Y. Tu, X. Ma, H. Chen, Y. Fan, L. Jiang, R. Zhang, Z. Cheng, *Int. J. Nanomed.* **2022**, *17*, 6773–6789. https://doi.org/10.2147/IJN.S385679.
106. E. J. Keliher, Y.-X. Ye, G. R. Wojtkiewicz, A. D. Aguirre, B. Tricot, M. L. Senders, H. Groenen, F. Fay, C. Perez-Medina, C. Calcagno, G. Carlucci, T. Reiner, Y. Sun, G. Courties, Y. Iwamoto, H.-Y. Kim, C. Wang, J. W. Chen, F. K. Swirski, H.-Y. Wey, J. Hooker, Z. A. Fayad, W. J. M. Mulder, R. Weissleder, M. Nahrendorf, *Nat. Commun.* **2017**, *8*, 14064. https://doi.org/10.1038/ncomms14064.
107. J. Pellico, I. Fernández-Barahona, M. Benito, Á. Gaitán-Simón, L. Gutiérrez, J. Ruiz-Cabello, F. Herranz, *Nanomed. Nanotechnol. Biol. Med.* **2019**, *17*, 26–35. https://doi.org/10.1016/j.nano.2018.12.015.
108. M. Kirienko, P. A. Erba, A. Chiti, M. Sollini, *Semin. Nucl. Med.* **2023**, *53*, 107–124. https://doi.org/10.1053/j.semnuclmed.2022.10.005.
109. N. Patel, B. A. Duffy, A. Badar, M. F. Lythgoe, E. Årstad, *Bioconjug. Chem.* **2015**, *26*, 1542–1549. https://doi.org/10.1021/acs.bioconjchem.5b00380.
110. M. A. McAteer, N. R. Sibson, C. von zur Muhlen, J. E. Schneider, A. S. Lowe, N. Warrick, K. M. Channon, D. C. Anthony, R. P. Choudhury, *Nat. Med.* **2007**, *13*, 1253–1258. https://doi.org/10.1038/nm1631.
111. H. Schmidt, C. Brendle, C. Schraml, P. Martirosian, I. Bezrukov, J. Hetzel, M. Müller, A. Sauter, C. D. Claussen, C. Pfannenberg, N. F. Schwenzer, *Invest. Radiol.* **2013**, *48*, 247–255. https://doi.org/10.1097/RLI.0b013e31828d56a1.

112. J. Griessinger, J. Schwab, Q. Chen, A. Kühn, J. Cotton, G. Bowden, H. Preibsch, G. Reischl, L. Quintanilla-Martinez, H. Mori, A. N. Dang, U. Kohlhofer, O. H. Aina, A. D. Borowsky, B. J. Pichler, R. D. Cardiff, A. M. Schmid, *NPJ Breast Cancer* **2022**, *8*, 1–13. https://doi.org/10.1038/s41523-022-00398-x.
113. H. Gutte, A. E. Hansen, S. T. Henriksen, H. H. Johannesen, J. Ardenkjaer-Larsen, A. Vignaud, A. E. Hansen, B. Børresen, T. L. Klausen, A.-M. N. Wittekind, N. Gillings, A. T. Kristensen, A. Clemmensen, L. Højgaard, A. Kjær, *Am. J. Nucl. Med. Mol. Imaging* **2014**, *5*, 38–45.
114. H. Gutte, A. E. Hansen, M. M. E. Larsen, S. Rahbek, S. T. Henriksen, H. H. Johannesen, J. Ardenkjaer-Larsen, A. T. Kristensen, L. Højgaard, A. Kjær, *J. Nucl. Med.* **2015**, *56*, 1786–1792. https://doi.org/10.2967/jnumed.115.156364.
115. A. E. Hansen, H. Gutte, P. Holst, H. H. Johannesen, S. Rahbek, A. E. Clemmensen, M. M. E. Larsen, C. Schøier, J. Ardenkjaer-Larsen, T. L. Klausen, A. T. Kristensen, A. Kjaer, *Eur. J. Radiol.* 2018, *103*, 6–12. https://doi.org/10.1016/j.ejrad.**2018**.02.028.
116. N. Shalaby, J. Kelly, F. Martinez, M. Fox, Q. Qi, J. Thiessen, J. Hicks, T. J. Scholl, J. A. Ronald, *Mol. Imag. Biol.*, **2022**, *24*, 341 351. https://doi.org//10.1007/s11307-021-01697-8
117. T. Almasi, N. Gholipour, M. Akhlaghi, A. Mokhtari Kheirabadi, S. M. Mazidi, S. H. Hosseini, P. Geramifar, D. Beiki, N. Rostampour, D. Shahbazi Gahrouei, *Int. J. Polym. Mater. Polym. Biomater.* **2021**, *70*, 1077–1089. https://doi.org/10.1080/00914037.2020.1785451.
118. J. Choi, J. C. Park, H. Nah, S. Woo, J. Oh, K. M. Kim, G. J. Cheon, Y. Chang, J. Yoo, J. Cheon, *Angew. Chem. Int. Ed.* **2008**, *47*, 6259–6262. https://doi.org/10.1002/anie.200801369.
119. C. Liolios, T. S. Koutsikou, E.-A. Salvanou, F. Kapiris, E. Machairas, M. Stampolaki, A. Kolocouris, E. K. Efthimiadou, P. Bouziotis, *Int. J. Pharm.* **2022**, *624*, 122008. https://doi.org/10.1016/j.ijpharm.2022.122008.
120. J. Ceccarini, H. Liu, K. Van Laere, E. D. Morris, C. Y. Sander, *Front. Physiol.* **2020**, *11*,792
121. T. M. Ionescu, M. Amend, R. Hafiz, B. B. Biswal, H. F. Wehrl, K. Herfert, B. J. Pichler, *NeuroImage* **2021**, *236*, 118045. https://doi.org/10.1016/j.neuroimage.2021.118045.
122. F. C. Maier, H. F. Wehrl, A. M. Schmid, J. G. Mannheim, S. Wiehr, C. Lerdkrai, C. Calaminus, A. Stahlschmidt, L. Ye, M. Burnet, D. Stiller, O. Sabri, G. Reischl, M. Staufenbiel, O. Garaschuk, M. Jucker, B. J. Pichler, *Nat. Med.* **2014**, *20*, 1485–1492. https://doi.org/10.1038/nm.3734.
123. J. Debatisse, O. F. Eker, O. Wateau, T.-H. Cho, M. Wiart, D. Ramonet, N. Costes, I. Mérida, C. Léon, M. Dia, M. Paillard, J. Confais, F. Rossetti, J.-B. Langlois, T. Troalen, T. Iecker, D. Le Bars, S. Lancelot, B. Bouchier, A.-C. Lukasziewicz, A. Oudotte, N. Nighoghossian, M. Ovize, H. Contamin, F. Lux, O. Tillement, E. Canet-Soulas, *Brain Commun.* **2020**, *2*, fcaa193. https://doi.org/10.1093/braincomms/fcaa193.
124. V. Z. Miloushev, K. L. Granlund, R. Boltyanskiy, S. K. Lyashchenko, L. M. DeAngelis, I. K. Mellinghoff, C. W. Brennan, V. Tabar, T. J. Yang, A. I. Holodny, R. E. Sosa, Y. W. Guo, A. P. Chen, J. Tropp, F. Robb, K. R. Keshari, *Cancer Res.* **2018**, *78*, 3755–3760. https://doi.org/10.1158/0008-5472.CAN-18-0221.
125. H. Qin, V. N. Carroll, R. Sriram, J. E. Villanueva-Meyer, C. von Morze, Z. J. Wang, C. A. Mutch, K. R. Keshari, R. R. Flavell, J. Kurhanewicz, D. M. Wilson, *Sci. Rep.* **2018**, *8*, 7928. https://doi.org/10.1038/s41598-018-26296-6.

# Index

Note: **Bold** page numbers refer to tables and *italic* page numbers refer to figures.

For Product Safety Concerns and Information please contact our EU representative GPSR@taylorandfrancis.com
Taylor & Francis Verlag GmbH, Kaufingerstraße 24, 80331 München, Germany

www.ingramcontent.com/pod-product-compliance
Lightning Source LLC
LaVergne TN
LVHW020558110826
845149LV00002B/307

* 9 7 8 1 0 3 2 4 4 9 4 5 6 *